VECTOR-BORNE PATHOGENS: INTERNATIONAL TRADE AND TROPICAL ANIMAL DISEASES

Caribbean Heartwater (by Nicolas Barré).

ANNALS OF THE NEW YORK ACADEMY OF SCIENCES
Volume 791

VECTOR-BORNE PATHOGENS: INTERNATIONAL TRADE AND TROPICAL ANIMAL DISEASES

Edited by Emmanuel Camus, James A. House, and Gerrit Uilenberg

The New York Academy of Sciences
New York, New York
1996

Library of Congress Cataloging-in-Publication Data

Vector-borne pathogens : international trade and tropical animal
    diseases / edited by Emmanuel Camus, James A. House, and Gerrit
    Uilenberg.
        p.   cm. — (Annals of the New York Academy of Sciences ; v.
    791)
    Proceedings of a symposia entitled Vector-Borne Pathogens:
    Challenges for the 21st Century and International Trade and Animal
    Diseases, held on May 8-12, 1996 in San José, Costa Rica.
    Includes bibliographical references (p.   ) and index.
    ISBN 0-89766-955-X (cloth : alk. paper). — ISBN 0-89766-956-8
    (paper : alk. paper)
    1. Ticks as carriers of disease—Tropics—Congresses.   2. Tick-
borne diseases in animals—Tropics—Congresses.   3. Insects as
carriers of disease—Tropics—Congresses.   4. Livestock—Tropics—
Diseases—Congresses.   5. Ticks—Control—Tropics—Congresses.
6. Vector control—Tropics—Congresses.   7. Veterinary tropical
medicine—Congresses.   8. International trade—Congresses.
I. Camus, Emmanuel.   II. House, James A.   III. Uilenberg, G.
(Gerrit)   IV. Vector-Borne Pathogens: Challenges for the 21st
Century and International Trade and Animal Diseases (1996 : San
José, Costa Rica)   V. Series.
Q11.N5 vol. 791
[SF810.T5]
636.2′0894433—dc20   96-31285
                                                                  CIP

                              PCP
          Printed in the United States of America
             ISBN 0-89766-955-X (cloth)
             ISBN 0-89766-956-8 (paper)
                  ISSN 0077-8923

ANNALS OF THE NEW YORK ACADEMY OF SCIENCES

Volume 791
July 23, 1996

VECTOR-BORNE PATHOGENS: INTERNATIONAL TRADE AND TROPICAL ANIMAL DISEASES[a]

Editors
EMMANUEL CAMUS, JAMES A. HOUSE, and GERRIT UILENBERG

Conference Organizers
JAMES A. HOUSE, EMMANUEL CAMUS, DAN SHEESLEY, HUGO VEIT, LUIS RODRIGUES, DAVE WILSON, KATHY M. KOCAN, NELL AHL, TOM WALTON, JIM C. WILLIAMS, FRANS JONGEJAN, EDMOUR BLOUIN, and CHRISTOPHER M. GROOCOCK

CONTENTS

[a] This volume represents the proceedings of the third biennial meeting of the Society for Tropical Veterinary Medicine comprising the symposia entitled Vector-Borne Pathogens: Challenges for the 21st Century and International Trade and Animal Diseases and contributed papers from the general sessions; this meeting was held on May 8–12, 1995 in San José, Costa Rica.

Symposium: International Trade and Animal Diseases

Financial support for this meeting was provided by:

- BAYER AG
- FOOD AND AGRICULTURE ORGANIZATION OF THE UNITED NATIONS
- INTER-AMERICAN INSTITUTE FOR COOPERATION ON AGRICULTURE
- MALLINCKRODT VETERINARY INC.
- PFIZER INC ANIMAL HEALTH
- RHONE MERIEUX
- UNITED STATES DEPARTMENT OF AGRICULTURE–
  ANIMAL AND PLANT HEALTH INSPECTION SERVICE
    INTERNATIONAL SERVICES
    VETERINARY SERVICES
  NATIONAL VETERINARY SERVICES LABORATORIES

Preface

The Society for Tropical Veterinary Medicine (STVM) held its third biennial meeting at the Inter-American Institute for Cooperation on Agriculture (IICA) in Cornado, Costa Rica on May 8-12, 1995. The meeting reflected the growing horizons of the STVM as it encompassed three major areas: Vector-borne pathogens: challenges for the twenty-first century; international trade and animal diseases; and general sessions.

The meeting was dedicated to two prominent scientists whose careers had a significant impact on tick-borne disease research, Dr. Alan Young and Dr. R. A. I. Norval.

Since the last meeting of the STVM in Guadeloupe (F.W.I.) two years ago, significant progress has been made concerning heartwater, a rickettsial disease introduced from Africa into the Caribbean. The specificity of serological diagnosis has been greatly improved and now a tool is available for epidemiological surveys. Research on heartwater, like almost every other disease, has gone into the molecular era: The polymerase chain reaction can now be used to detect low levels of infection in the tick vector. The protection against heartwater by vaccination with inactivated *Cowdria ruminantium,* first realized in goats in Guadeloupe, has been confirmed in sheep and in cattle. After years of discussion and meetings, the eradication of the tropical Bont tick vector of heartwater eventually began in the Lesser Antilles. The growing importance of acaricide resistance in another tick, *Boophilus microplus,* in South America, will require particular attention in the Caribbean. Concerning the diagnosis and control of other tick-borne diseases, the most significant progress is the successful *in vitro* cultivation of *Anaplasma marginale,* a major breakthrough in anaplasmosis research. New strategies to control anaplasmosis are focused on the tick vector by trying to develop vaccines against haemoparasites in ticks. There is a continuing effort to study the immunological aspects of *Babesia bovis,* looking for immunogenic proteins such as RAP 1 and apical complex-specific proteins and trying to better understand the immune response. In the absence of good vaccination against bovine dermatophilosis, the selection of resistant cattle based on genetic markers could be a solution. The screwworm eradication program in Central America by dispersion of sterile flies is progressing. It is very encouraging for the tick eradication campaign to have the successful example of a regional eradication program, even if the difficulties and the methods are different. The title of this symposium alluded to challenges in the next century. Several challenges will possibly be met by the end of *this* century: the Bont tick eradication program and the screwworm eradication from Central America. Other challenges will persist into the twenty-first century, like the control of heartwater by vaccination in Africa, the development of an efficient recombinant vaccine against babesiosis and anaplasmosis, and the control of trypanosomiasis in Africa.

Beside the symposium, there were a general session and trade session at the meeting.

The **International Trade and Animal Diseases** symposium presented an opportunity to discuss the risk analysis and see examples of its growing relevance with the implementation of NAFTA and the GATT.

The **general session** was devoted to the subject of tropical veterinary medicine. Papers were presented on diagnostic tests developed for various diseases, with particular emphasis on PCR and on blocking-ELISA: bovine neosporosis, pasteurellosis, Ehrlichiosis, African horsesickness, and foot and mouth disease were among the diseases addressed.

The STVM is very grateful to the support of several international organizations for their contributions toward the success of our meeting. We take this opportunity to sincerely thank Bayer AG; the Centre de Cooperation Internationale en Recherche Agronomique pour le Development; the Food and Agriculture Organization of the United Nations (FAO); the Inter-American Institute for Cooperation on Agriculture; Mallinckrodt Inc.; Pfizer Inc Animal Health; Rhone Merieux; the United States Department of Agriculture (USDA), Animal and Plant Health Inspection Service (APHIS), International Services; and USDA APHIS Veterinary Services, National Veterinary Services Laboratories.

One of us (J.A.H.), as President Elect of the STVM, had the responsibility for organizing the meeting. He takes this opportunity to gratefully thank Dr. Emmanuel Camus for his effective organization of the scientific sessions and for serving as the Scientific Chair of the meeting. No meeting can be a success without the dedication and persistence of a local organizing committee for the successful execution of the critical details necessary to make the meeting flow smoothly; we gratefully thank Dr. David Wilson, Local Organization Chair, who was ably assisted by Ms. Pilar Fernandez, Ms. Laura Cartin, and Ms. Roxana Montero all of IICA. The STVM especially thanks the IICA and its director, Dr. Carlos Aquino, for generously and graciously providing such an excellent site for our STVM '95 meeting. Sincere thanks go to Dr. Emmanuel Camus, who chaired the symposium entitled "Vector-borne Pathogens: Challenges for the Twenty-first Century," Dr. Dan Sheesley, who ably chaired the symposium entitled "International Trade and Animal Diseases," and Dr. Hugo Veit, who chaired the general sessions. The council of Dr. Kathy Kocan, as President, was invaluable to the success of organizing the meeting and is sincerely appreciated. The local scientific contacts and input of Dr. Luis Rodrigues is greatly appreciated. The moral support and input of Dr. Nell Ahl (Newsletter Editor), Dr. Thomas Walton (Treasurer), Dr. Jim Williams (Past President), Drs. Edmour Blouin, Christopher Groocock, Frans Jongejan (Councilors) is greatly appreciated. The STVM also greatly appreciates the persistent, effective help of Claudia Pruitt, who assisted with numerous details of mailings, preparation of documents, and manuscript organization for publication.

Finally, we are very grateful to the editorial staff of the New York Academy of Sciences, and particularly to the subeditor of the volume, the late Mrs. Sheila Treitler, for seeing this book through the press in such a thoroughly professional manner.

—EMMANUEL CAMUS
—JAMES A. HOUSE
—GERRIT UILENBERG

Progress and Priorities in Research on Heartwater

G. UILENBERG [a]

CIRAD-EMVT[b]
10, rue Pierre-Curie
94704 Maisons-Alfort, France

INTRODUCTION

I was asked to give a plenary address on the history of heartwater, past and current research, and projections of the future, and I set out trying to do so. The first title I tried was "Heartwater" Yesterday, Today, Tomorrow," which not only resembles that of the first international workshop on heartwater, held in South Africa in 1986,[1] but is also the English name for a tropical flowering bush, *Brunfelsia calycina.* In any case, I quickly found that it would be difficult to give an even slightly original complete coverage, when there are so many recent reviews of heartwater, some more extensive than others.[1-8] Moreover, a complete review would certainly not be possible in half an hour.

I therefore decided to change the title and be selective, concentrating on areas in which real progress has been made in the last decade or so, highlighting in particular new aspects that would seem to be of importance from a practical (control) point of view, and to point out some priority areas for research. The areas I chose, some possibly in a somewhat arbitrary manner, are etiology, distribution, economic importance, epidemiology, immunology, diagnosis, and immunization. For example, clinical signs, lesions, pathogenesis, treatment, and vector control will not or hardly be mentioned, for various reasons. Coverage of the literature is also selective. I am sure that other participants will complete my introduction and present the latest data.

It is perhaps superfluous to recall once more that most of the progress in the last decade would not have been possible without the success in culturing *Cowdria ruminantium* in the laboratory, obtained at last 10 years ago in South Africa,[9] and subsequent progress in this field.[10]

I am grateful to those scientists who have allowed me access to unpublished research results.

ETIOLOGY

We can only guess when and where the group of organisms that now constitute the Ehrlichiaceae originated and how and when *Cowdria,* with its particular site of

---

[a] Centre de coopération internationale en recherche agronomique pour le développement—Département d'élevage et de médecine vétérinaire.

[b] Present affiliation: "A Surgente," Route du Port, 20130 Cargèse (Corse), France.

1

multiplication in vacuoles in the cytoplasm of endothelial cells, evolved in this group. The relationship of *C. ruminantium* to certain insect symbionts[11,12] may mean that this group originated from such microorganisms. Members of the genus *Ehrlichia* have a worldwide distribution, but as far as is known, *Cowdria* is limited to subsaharan Africa, and to the Antilles where it was transported by human intervention.

Until a few years ago, *Cowdria* was *Cowdria,* and *Ehrlichia* was *Ehrlichia.* But then three new discoveries were made: (1) cross-reactions in serological tests were shown to exist between *C. ruminantium* and certain species of *Ehrlichia;*[13–16] (2) agents that started out as apparent *Ehrlichia* could change into organisms indistinguishable from *C. ruminantium;*[17–19] (3) the close relationship between *C. ruminantium* and species of *Ehrlichia* was confirmed by comparing 16S ribosomal DNA sequences.[20] An undeniable conclusion is that *C. ruminantium* is very closely related to some species of *Ehrlichia.* But should we go as far as scientists[21] who suggest that *Cowdria ruminantium* should be called *Ehrlichia ruminantium?* As long as *Ehrlichia* was defined as a parasite of white blood cells, and *Cowdria* as inhabiting endothelial cells (although it also parasitizes white blood cells), there could be no confusion, but things are different if the one can indeed change into the other. However, as Du Plessis himself stated,[19] he was starting out with *putative* ehrlichial agents, not with straightforward *Ehrlichia,* and further studies on these startling results are essential before we can accept that the heartwater agent can be transmitted by *Hyalomma* and *Rhipicephalus* ticks and that the tick genus has a radical influence on its behavior. Future research should compare the results of PCR on these agents with primers used for *C. ruminantium* before and after they are passaged by *Amblyomma* ticks, and if possible compare their nucleotide sequences with those of *C. ruminantium* and various *Ehrlichia* species. A start on such comparisons appears to have been made.[22] Another logical approach is of course to try and change these agents back into *Ehrlichia.* This is undoubtedly a priority area for research, both from a scientific point of view and a practical one.

DISTRIBUTION

Heartwater appears to be of African origin. In subsaharan Africa, it seems to occur wherever the *Amblyomma* vectors exist. Although the disease has apparently not yet been reported in Guinea, Sierra Leone, and Togo, this does not necessarily mean that it is absent there. [Positive results in the indirect fluorescent antibody test have been obtained with ruminant sera in Guinea (A. M. Barry, unpublished data, 1994), but as we shall see, the available serological tests for heartwater are insufficiently specific for definite conclusions.] It has been found on most of the islands around subsaharan Africa, Madagascar, La Réunion, Mauritius, the Comoro islands, Zanzibar, and São Tomé. African vectors are established in parts of the Arabian peninsula too, and in Yemen, positive results have been obtained with an ELISA (K. J. Sumption, M. Al-Gadafi and A. Williams, unpublished data, 1994), but as stated, this fact does not constitute conclusive proof.

Outside the African region *Cowdria ruminantium* is only known to exist with certainty on three Caribbean islands, Guadeloupe, Marie-Galante, and Antigua. Positive serological results in ruminants on other islands, in the absence of clinical evidence, are inconclusive. Reports of the disease in southeast Europe (one as recent as 1991),[23] Turkey, Iran, all regions without *Amblyomma* ticks, and southeast Asia, should be disregarded as long as no serious evidence is put forward.

The suspected role of the cattle egret in the spread of *Amblyomma variegatum* (the tropical bont tick) in the Caribbean[24] is of great importance and provides an added stimulus for the eradication program of this introduced African vector from the western hemisphere.

ECONOMIC IMPORTANCE

The disease has been regarded as having an important economic impact. But concepts have been influenced to a considerable extent by the fact that most of the initial research was carried out in South Africa, which had early on a large and highly susceptible ruminant population of European origin. In fact, the infection generally appears to be of low importance to local populations of ruminants, in particular cattle, in endemic areas, presumably due to a long natural selection over centuries of coexistence (see below, under epidemiology).

The real economic impact on well-supervised, fully susceptible livestock is relatively easy to determine (loss of production and mortality, cost of control), but we still have very few reliable figures on the real significance of heartwater for local livestock populations in various situations and circumstances. This would seem to be one of the priority areas for research, as the admissible cost of control surely depends on the economic importance of the disease.

However, here is another aspect to the economic importance of cowdriosis (and of certain other diseases as well): the productivity of resistant local livestock is often low, but the disease prevents the introduction of more productive but also more susceptible exotic breeds. An important loss of potential production exists therefore in endemic areas, and in situations where this can be overcome by intensive tick control and good veterinary care, the high cost of prevention takes away a good part of the benefits of increased production.[25]

EPIDEMIOLOGY

The disease was first described in South Africa in the 19th century, and may have been observed there already in 1838, not surprisingly as it is above all a clinical disease of animals imported into endemic areas, and European settlers introduced susceptible European livestock in South Africa at an early date. There is no time to go into all the details of the early history of the disease, and I would like to refer to a number of reviews dealing with this point.[1,2,7,8,26–28]

Edington[29] and Dixon[30] first showed in 1898 that the disease could be transmitted by inoculating blood from sick animals, and Lounsbury[31] was the first to confirm that *Amblyomma* ticks could transmit the disease.

Hosts

Heartwater was classically considered as an infection of ruminants only, until it was shown that (certain) carnivores, rodents, reptiles, birds, and lagomorphs could be infected too,[3,32–34] the infection being mostly subclinical in such hosts, with notable exceptions occurring in rodents, as first reported by Du Plessis and Kümm.[35] Infectivity and pathogenicity for various animal species and groups are strain-linked. Surprising differences in susceptibility to various stocks of *C. ruminantium* exist between species of one group, e.g., rodents,[36] and even ruminants (for instance the Kümm stock is highly pathogenic to goats, but causes no disease in cattle).[7]

The fact that *C. ruminantium* can be grown in human endothelial cells[37] raises the question whether some strains might be infective for human beings, although so far there is no evidence for this.[38]

Transmission

Ten African *Amblyomma* species are now known to be capable of transmitting *C. ruminantium,* although for various reasons they are not all of equal importance as natural vectors:

A. hebraeum[31]	Only southern Africa.
A. variegatum[39]	Most of tropical Africa, Caribbean.
A. pomposum[40]	Only southern-central Africa.[c]
A. gemma[42]	Dry areas of eastern Africa.
A. lepidum[43]	Dry areas of eastern Africa.
A. tholloni[44]	Adults almost solely on elephants.
A. sparsum[45]	Seldom found on domestic animals
A. astrion[46]	Limited distribution.
A. cohaerens[47]	Limited distribution.
A. marmoreum[48]	Adults almost solely on tortoises.

The combination of the three facts that a species of which the adults occur primarily on tortoises, *A. marmoreum,* has been shown to be a vector, that an African tortoise species can be infected with the heartwater agent and remain a latent carrier, infective to ticks for at least three months, and that immatures of *A. marmoreum* are found frequently on ruminants, and those of the main vector in southern Africa, *A. hebraeum,* occur on tortoises,[49] makes a strong case for a potential role of reptiles in the epidemiology of heartwater (as well as a very long sentence!). (Nevertheless, there is considerable evidence that the disease maintains itself quite well in the absence of wild animals.)

Although not a priority, a few more African *Amblyomma* species should be assessed for their vector potential, not only species primarily occurring on ruminants such as *A. eburneum,* and *A. splendidum,* but also those adapted to reptiles in view of their potential role in epidemiology. One experiment with an African reptile tick, *A. nuttalli,* has given a negative result,[50] but so did several attempts using *A. marmoreum,*[50,51] until it was at last shown that the latter could transmit the infection after all.[3]

---

[c] This is presumably the tick considered by Walker and Olwage[41] to be *A. pomposum.*

Three American species have also been shown to be experimental vectors:

A. maculatum[52] Good potential vector.
A. cajennense[47] *Poor potential vector.*
A. dissimile[53] Potential vector, not normally found on domestic animals.

The infection is not yet known to exist in any area where these ticks occur. *A. cajennense,* the most widely spread, fortunately appears to be a poor vector, and it is unlikely that the amphibian and reptile tick *A. dissimile* would be able to play a significant role; but *A.maculatum* has shown itself to be quite an efficient vector.[47,53] All American *Amblyomma* species, in particular those that occur on ruminants, should be tested for their potential vector role; the negative results obtained with *A. americanum, A. neumanni* and *A. imitator*[54] are not conclusive. However, should the present campaign for the eradication of the tropical bont tick in the Caribbean succeed, exploring the possible vector role of American *Amblyomma* species would be less of a priority. Asian *Amblyomma* ticks have not yet been examined for their vector capability.

Studies on the role of pheromones attracting and stimulating the attachment of adult *Amblyomma* vectors are of interest in the epidemiology of the disease and might prove to be of value as part of an integrated approach to control.[55,56]

Innate Resistance and Endemic Stability

The low susceptibility of local livestock is particularly striking in cattle; local populations of small ruminants sometimes show varying degrees of susceptibility, for instance, the Red Sokoto goat in Nigeria,[57] the Creole goat on Guadeloupe,[58] and local sheep and goats in the Sudan.[59]

Wherever there is no adequate veterinary infrastructure, the most logical approach to the control of heartwater would appear to be the use of resistant livestock, even if the productivity of such animals is limited. Genetic markers could be useful for the transfer of resistance to more productive breeds, either by classical cross-breeding or by genetic engineering, although this could presumably only be envisaged if resistance were determined by one or a very limited number of genes. Because of various problems, preliminary studies on the heritability of resistance in Creole goats on Guadeloupe[60] have so far not yielded conclusive results, as Emmanuel Camus will tell you; also, the criteria used for distinguishing resistant animals from susceptible ones are questionable.

Endemic stability without significant losses to heartwater may occur in cattle exposed to large numbers of infected *Amblyomma* ticks. It has been known for a long time[61] that young calves possess a natural resistance, or rather tolerance, to the infection. A high degree of tolerance may last for only a few weeks in highly susceptible exotic breeds, but probably much longer in local cattle possessing innate resistance, as is evident for instance from the work of Soldan *et al.*[62] If all calves are infected before this tolerance wanes, the entire cattle population acquires immunity without a significant disease problem. In situations where the infection rate of the tick population is low, such as Madagascar[63] and Guadeloupe,[58] this may of course mean that the numbers of ticks necessary to achieve endemic stability are so high that unacceptable direct damage by the ticks occurs, such as secondary abscesses leading to loss of teats, and lameness, screwworm infestation, and severe cases of dermatophilosis in susceptible breeds in certain regions.[64–66]

Endemic stability in small ruminants is less well documented, but it is undeniable that local populations survive in circumstances where exotic stock cannot be kept without preventive measures.

Furthermore, very recent research results indicate that vertical transmission of the infection from dams to calves may be common (in 9 of 18 calves), and also that colostrum may after all play a role in calfhood immunity (S. L. Deem, as yet unpublished data from a Ph.D. thesis, 1994).[d] On the other hand, as colostrum appeared to be possibly involved in vertical transmission, while it is known that even calves of uninfected dams possess the inverse age resistance to infection, further research is important to untangle the various factors that may be involved in these important findings.

Antigenic Diversity

While it was believed for a long time that there were no significant antigenic differences between strains of *C. ruminantium*,[67] apart from a few considered to be abnormal as they were pathogenic to mice,[47] concepts have changed since then. It is now known that antigenic diversity is extensive, irrespective of pathogenicity for mice.[68–71] No pattern in antigenic diversity has been detected,[70,71] and as recently reviewed,[6] cross-protection patterns of various stocks differ in different hosts. Antigenic diversity may be responsible for otherwise unexplained breakdowns in immunity in the field, especially in highly susceptible breeds of goats where the partial cross-immunity differences may be most apparent.[69]

So far there is no other way of typing strains than by cross-protection tests *in vivo*, highly inaccurate unless large numbers of experimental ruminants can be used per test and becoming quickly unworkable with the increasing numbers of isolates to be tested. Methods to determine the antigenic profile of new isolates based on molecular biology, such as restriction fragment length polymorphisms (RFLP) and random amplification of polymorphic DNA (RAPD) are being studied (D. Martinez and A. Bensaid, unpublished data).

The Carrier State

Among several important recent findings that have changed our knowledge of the epidemiology of heartwater is certainly also the discovery that recovered ruminants may remain long-term carriers of rickettsia,[72,73] intermittently infective to ticks. In fact, this is not really surprising, considering what happens in infections with related

---

[d] Papers based on a 1994 PhD thesis of the University of Florida, Gainesville, are in press in *Vet. Parasitol.*

organisms, such as *Ehrlichia* and *Anaplasma.* Retrospectively, it is quite astonishing that the discovery has taken so long, but then of course attempts to transmit the infection from recovered animals have been few in number.

IMMUNOLOGY

It has been known for over 60 years[50] that circulating antibodies by themselves have no effect on the disease, but further knowledge on the nature of the immunology in heartwater has remained for a long time very superficial. Neutralizing effect of antibodies could not be demonstrated in several experiments.[47,74–76] Although colostrum-derived antibodies are found in calves and lambs born to immunized mothers, passive immunity has not been demonstrated either, and the age-linked resistance of young animals is not related to the immune status of the dam and occurs in offspring of noninfected dams as well. (Research results by Sharon Deem, mentioned above, could lead to a change in concepts.)

However, du Plessis[77] could inhibit the infectivity for mice of the Kümm stock by incubation with hyperimmune serum, provided complement was present, and Byrom *et al.*[76] showed a neutralizing role of antibodies, which diminished the infectivity of *C. ruminantium* for endothelial cell cultures. In spite of these recent findings, the immunity in heartwater appears to be largely cell mediated. Athymic nude mice are unable to mount an immune response to infection, and there is some evidence of adoptive transfer of immunity with CD8[+] cells of immune mice.[75] The role of interferons and other cytokines in the resistance to cowdriosis is being investigated.[6,78–80]

Further knowledge of the nature of the immunity is a priority, as it should constitute a rational basis for the development of vaccines based on genetic engineering.

DIAGNOSIS

Diagnosis of the active disease in live animals is still problematic. Clinical symptoms are never conclusive, and microscopic confirmation is so far only possible on brain biopsy smears,[81] a method that livestock owners are not likely to favor very much. One of the recommendations of a recent FAO Expert Consultation is: "The development of a rapid diagnostic test for heart-water, based on *Cowdria* antigen or DNA detection, should be attempted."[82] The word "rapid" is especially important, as the course of heartwater is commonly so rapid that no one in his right mind would await the result of a lengthy laboratory test before applying treatment to an animal suspected on clinical symptoms. As Bezuidenhout, discussing DNA probes, put it: "One should guard against unrealistic expectations regarding the utilization of these probes as diagnostic tools, especially in certain countries of Africa. Under these circumstances, farmers and stockmen get to know the disease so well that they seldom bother to have the diagnosis confirmed. The chance that they will pay for such a diagnosis is remote. Also the absence of a proper infrastructure makes the collection and testing of material, no matter what the origin may be, very difficult."[5] Therefore, unless a really rapid crush-side test can be developed, research on diagnosis of the disease in live animals is not in fact a priority. Diagnostic tests for heartwater remain mainly tools for research and surveillance.

The first workable serological test was a big step forward.[83] However, although this and subsequent tests (indirect immunofluorescent tests, indirect and competitive ELISA, and western blotting[84–89] have been used in epidemiological surveys and serosurveillance, it has become apparent that their specificity leaves much to be desired, and that cross-reactions with certain species of *Ehrlichia* occur.[13–16] This is true even for tests based on the use of a monoclonal antibody directed against the major antigenic protein (MAP1), because this MAP (or at least a protein with similar epitopes) is conserved in these *Ehrlichiae*.[15] There are also false-positive reactions in all these tests using *C. ruminantium* antigens in areas where it has not (yet?) been proved that they are due to *Ehrlichia*.[89–91] Recent studies have led to the development of a more specific (but possibly less sensitive) ELISA, using only a fragment of the MAP1, produced by recombinant technology;[92] this ELISA is in the process of being evaluated.

DNA probes and PCR techniques have also been developed[93–95] and so far appear to be specific, although further evaluation is needed. Such techniques are not yet universally applicable; radioactive material is needed for present DNA probes, and extraneous contamination remains a danger in the PCR; both are still relatively expensive. But we all know how fast new techniques, about which most scientists are at first sceptical, may develop into universally accepted procedures. PCR is the more sensitive[95] and depending on the primers used could be the most promising approach to a sensitive and specific research tool for demonstrating the presence of *C. ruminantium* in recovered carrier animals, and in ticks.

IMMUNIZATION

The infection and treatment method, using intravenous inoculation of virulent material followed by treatment, or even without treatment in young animals with reverse age-linked resistance, was developed over half a century ago.[61] It is still the only way of immunization in practical use, in spite of several well-documented drawbacks. This method has been reviewed by Bezuidenhout.[96]

Improvements in the infection and treatment method have been proposed recently, such as:

- a slow-release implant of doxycycline, administered at the time of infection, which should make further monitoring unnecessary.[97] Further evaluation of the efficacy of this method is needed, as experiments in Malawi tend to indicate that such implants may inhibit immunity in some (zebu-cross) cattle (J. A. Lawrence, K. Tjornehoj, A. P. Whiteland and P. T. Kafuwa, unpublished data, 1994);

- freeze-drying of infected blood allows the material to be kept in normal refrigerators for significant periods of time,[98] whereas previously all infective material had to be kept at −70°C or less.

- use of highly infective endothelial cell culture, instead of blood or tick-derived material,[99] might be cheaper, would be easier to standardize, carry less risk of contamination by other pathogens, and reduce the number of experimental animals to be used.

So far, these approaches do not appear to have been followed up, even in the country where they originated, apparently because of sheer human inertia (C. E. Yunker, personal communication).

Two new approaches are presently being explored, the use of attenuated organisms and of inactivated ones. These have only become available after *in vivo* culture made it possible to obtain large amounts of infective elementary bodies.

Jongejan[100] discovered that a Senegalese stock had become attenuated in culture; intravenous inoculation gave a high level of homologous protection in sheep and goats. This has been confirmed, using the same stock, by D. Martinez in Guadeloupe (unpublished). A field trial in Senegal with the attenuated material was promising.[101] However, attenuation of other stocks has not yet been reliably attained, and use of the attenuated material does not solve the problems of antigenic diversity.[102] Further studies are needed.

Martinez *et al.*[103] in Guadeloupe and Mahan *et al.*[104] in Zimbabwe have shown that the injection of inactivated elementary bodies obtained by *in vitro* culture, with an adjuvant, induces homologous immunity, and Martinez' team has found that a commercial mineral-oil-based adjuvant can be used instead of Freund's adjuvants and that the immunity persists for at least a year (unpublished data). These preliminary results give hope that safe vaccines can be developed, although the problem of antigenic diversity is also unlikely to be solved by this method, and a mixture of stocks may have to be used.

It is hoped that further studies of the immune response and identification of the protective antigens and the encoding genes involved may in the end lead to a modern molecular vaccine, although the extensive antigenic diversity in *C. ruminantium* will probably constitute a serious complication. Relevant work on the molecular biology of *C. ruminantium* has been started.[105–107]

CONCLUSIONS

Some of the important research results of the last decade or so are:

1. *In vitro* culture of *C. ruminantium*.
2. The close relationship of *Cowdria* to *Ehrlichia*.
3. The suspected role of the cattle egret in the spread of *A. variegatum* in the Caribbean and the potential role of American *Amblyomma* ticks.
4. Vertebrates other than ungulates may be infected and possibly involved in the epidemiology (tortoises!).
5. The extent of antigenic diversity.
6. The carrier state after recovery from cowdriosis.
7. Vertical transmission from dam to calf.
8. First results from immunology studies.

9. Development of diagnostic tests.
10. First results from molecular biology studies.
11. The role of attraction-attachment pheromones in the biology of *Amblyomma* vectors.
12. Immunization with attenuated and inactivated material.

As far as the future is concerned, some priority areas for research on heartwater are:

1. Can certain *Ehrlichia* species change into *Cowdria* (and vice versa)?
2. What is the economic impact of cowdriosis on local livestock populations in various circumstances?
3. Development of *in vitro* methods for typing antigenic profiles of isolates.
4. Role of vertical transmission in calfhood immunity.
5. Development and validation of specific and sensitive tests for epidemiological studies and surveillance.
6. Development of effective, safe, and affordable vaccines.
7. Eradication of the tropical bont tick from the western hemisphere.

ACKNOWLEDGMENTS

I wish to thank Drs. James House and Katherine Kocan, President-Elect and President of the Society for Tropical Veterinary Medicine, for inviting and sponsoring me to contribute to this meeting. I am also grateful to Drs. Nicolas Barré and Emmanuel Camus of the Guadeloupe team of CIRAD-EMVT, my former institute, for their sponsorship.

REFERENCES

1. BIGALKE, R. D., C. M. CAMERON, F. M. C. GILCHRIST, A. J. MORREN, T. W. NAUDÉ, R. K. REINECKE, R. C. TUSTIN, H. VAN ARK, A. J. M. VERSTER, D. W. VERWOERD & J. B. WALKER, Eds. 1987. Heartwater: past, present and future. Proceedings of a Workshop held at Berg en Dal, Kruger National Park, Sept. 8-16, 1986. Onderstepoort J. Vet. Res. **54:** 161-546.
2. CAMUS, E. & N. BARRÉ. 1988. La cowdriose (heartwater). Revue générale des connaissances. Etudes et Synthèses de l'IEMVT no. 4. 2nd edition. IEMVT. Maisons-Alfort, 152 pp.
3. BEZUIDENHOUT, J. D. 1990. Recent research findings on cowdriosis. *In* Ehrlichiosis. A Vector-Borne Disease of Animals and Humans. J. C. Williams & I. Kakoma, Eds.: 125-135. Kluwer Academic Publishers. Dordrecht, the Netherlands.
4. NORVAL, R. A. I., H. R. ANDREW, C. E. YUNKER & M. J. BURRIDGE. 1992. Biological processes in the epidemiology of heartwater. *In* Tick Vector Biology. Medical and Veterinary Aspects. B. H. Fivaz, T. N. Petney & I. G. Horak, Eds.: 71-86. Springer-Verlag. Berlin, Germany.
5. BEZUIDENHOUT, J. D. 1993. The significance of recent highlights in heartwater research. Rev. Elev. Méd. Vét. Pays Trop. **46:** 101-108.
6. UILENBERG, G. & E. CAMUS. 1993. Heartwater (cowdriosis). *In* Rickettsial and Chlamydial Diseases of Domestic Animals. Z. Woldehiwet & M. Ristic, Eds.: 293-332. Pergamon Press. Oxford, New York, Seoul & Tokyo.

7. BEZUIDENHOUT, J. D., L. PROZESKY, J. L. DU PLESSIS & S. R. VAN AMSTEL. 1994. Heartwater. *In* Infectious Diseases of Livestock. With Special Reference to Southern Africa. J. A. W. Coetzer, G. R. Thomson, R C. Tustin & N. P. Kriek, Eds.: 351–370. Oxford University Press. Oxford, England.

8. CAMUS, E., N. BARRÉ, D. MARTINEZ & G. UILENBERG. 1996. Heartwater (Cowdriosis). A Review. 2nd edit. Office International des Epizooties. Paris, France.

9. BEZUIDENHOUT, J. D., C. L. PATERSON & B. J. H. BARNARD 1985. *In vitro* cultivation of *Cowdria ruminantium*. Onderstepoort J. Vet. Res. **52:** 113–120.

10. YUNKER, C. E. 1995. Current status of in vitro cultivation of *Cowdria ruminantium*. Vet. Parasitol. **57:** 205–211.

11. STOUTHAMER, R., J. A. J. BREEUWERT, R. F. LUCK & J. H. WERREN. 1993. Molecular identification of microorganisms associated with parthenogenesis. *Nature* **361:** 66 68.

12. WERREN, J. H., G. D. D. HURST, W. ZHANG, J. A. J. BREEUWER, R. STOUTHAMER, & M. E. N. MAJERUS. 1994. Rickettsial relative associated with male killing in the ladybird beetle (*Adalia bipunctata*). J. Bacteriol. **176:** 388–394.

13. LOGAN, L. L., C. J. HOLLAND, C. A. MEBUS & M. RISTIC. 1986. Serological relationship between *Cowdria ruminantium* and certain *Ehrlichia* species. Vet. Rec. **119:** 458–459.

14. DU PLESSIS, J. L., E. CAMUS, P. T. OBEREM & L. MALAN. 1987. Heartwater serology: some problems with the interpretation of results. Onderstepoort J. Vet. Res. **54:** 327–329.

15. JONGEJAN, F., N. DE VRIES, J. NIEUWENHUIJS, A. H. M. VAN VLIET & L. A. WASSINK. The immunodominant 32-kilodalton protein of *Cowdria ruminantium* is conserved within the genus *Ehrlichia*. Rev. Elev. Méd. Vét. Pays Trop. **46:** 145–152.

16. KELLY, P. J., L. A. MATHEWMAN, S. M. MAHAN, S. SEMU, T. PETER, P. R. MASON, P. BROUQUI & D. RAOULT. 1994. Serological evidence for antigenic relationships between *Ehrlichia canis* and *Cowdria ruminantium*. Res. Vet. Sci. **56:** 170–174.

17. DU PLESSIS, J. L. 1990. Increased pathogenicity of an *Ehrlichia*-like agent after passage through *Amblyomma hebraeum:* a preliminary report. Onderstepoort J. Vet. Res. **57:** 233–237.

18. DU PLESSIS, J. L. 1992. Increased pathogenicity of *Ehrlichia*-like agents after passage through *Amblyomma hebraeum*. J. S. Afr. Vet. Assoc. **63:** 88–89.

19. DU PLESSIS, J. L. 1993. The relationship between *Cowdria* and *Ehrlichia:* change in the behaviour of ehrlichial agents passaged through *Amblyomma hebraeum*. Rev. Elev. Méd. Vét. Pays Trop. **46:** 131–143.

20. VAN VLIET, A. H. M., F. JONGEJAN & B. A. M. VAN DER ZEIJST. 1992. Phylogenetic position of *Cowdria ruminatium* (Rickettsiales) determined by analysis of amplified 16S ribosomal DNA sequences. Int. J. Syst. Bacteriol. **42:** 494–498.

21. BROUQUI, P., M. L. BIRG & D. RAOULT. 1994. Cytopathic effect, plaque formation, and lysis of *Ehrlichia chaffeensis* grown on continuous cell lines. Infect. Immun. **62:** 406–411.

22. YUNKER, C. E. & B. A. ALLSOPP. 1994. Modern biotechnological methods for diagnosis of cowdriosis. Present and future trends. Use of Applicable Biotechnological Methods for Diagnosing Haemoparasites. *In* G. Uilenberg, A. Permin & J. W. Hansen. Proceedings of an Expert Consultation, Mérida, Mexico, Oct. 4–6, 1993. FAO, Rome: 143–155.

23. BERBEROVIC, M., B. PRAJNINGER, M. GLADAN & G. ALTABARI. 1991. The nephroso-nephritis syndrome in the clinical picture of bovine cowdriosis. Veterinarski Glasnik **45:** 249–254. (In Serbo-Croatian.)

24. CORN, J. L., N. BARRE, B. THIEBOT, T. E. CREEKMORE, G. I. GARRIS & V. F. NETTLES. 1993. Potential role of cattle egrets, *Bubulcus ibis* (Ciconiformes: Ardeidae), in the dissemination of *Amblyomma variegatum* (Acari: Ixodidae) in the eastern Caribbean. J. Med. Entomol. **30:** 1029–1037.

25. UILENBERG, G. 1995. International collaborative research: significance of tick-borne hemoparasitic diseases to world animal health. Vet. Parasitol. **57:** 19–41.

26. CURASSON G. 1943. Traité de protozoologie vétérinaire et comparée, **3:** 359-378. Vigot Frères. Paris, France.

27. HENNING, M. W. 1956. Animal Diseases in South Africa. 3rd Edition: 1155-1178. Central News Agency Ltd. Pretoria, South Africa.

28. NEITZ, W. O. 1968. Heartwater. Bull. Off. Int. Epiz. **70:** 329-336.

29. EDINGTON, A. 1898. Heartwater. Agric. J. Cape of Good Hope **12:** 748-754.

30. DIXON, R. W. 1898. Heartwater experiments. Agric. J. Cape of Good Hope **12:** 754-760.

31. LOUNSBURY, C. P. 1900. Tick-heartwater experiment. Agric. J. Cape of Good Hope **16:** 682-687.

32. MASON, J. H. & R. A. ALEXANDER. 1940. The susceptibility of the ferret to heartwater. J. S. Afr. Vet. Med. Assoc. **11**(3): 98-107.

33. HUDSON, J. R. & R. M. HENDERSON. 1941. Some preliminary experiments on the survival of heartwater "virus" in rats. J. S. Afr. Vet. Med. Assoc. **12**(2): 39-49.

34. OBEREM, P. T. & J. D. BEZUIDENHOUT. 1987. Heartwater in hosts other than domestic ruminants. Onderstepoort J. Vet. Res. **54:** 271-275.

35. DU PLESSIS, J. L. & N. A. L. KÜMM. 1971. The passage of *Cowdria ruminantium* in mice. J. S. Afr. Vet. Med. Assoc. **42:** 217-221.

36. STEWART, C. G. 1989. *Cowdria ruminantium* stocks: how significant are they? Zimbabwe Vet. J. **20:** 149-154.

37. TOTTE, P., D. BLANKAERT, T. MARIQUE, C. KIRKPATRICK, J. P. VAN VOOREN & J. WERENNE. 1993. Bovine and human endothelial cell growth on collage microspheres and their infection with the rickettsia *Cowdria ruminantium:* prospects for cells and vaccine production. Rev. Elev. Méd. Vet. Pays Trop. **46:** 153-156.

38. KELLY, P. J., C. E. YUNKER, P. R. MASON, M. TAGWIRA & L. A. MATTHEWMAN. 1992. Absence of antibody to *Cowdria ruminantium* in sera from humans exposed to vector ticks. S. Afr. Med. J. **81:** 578.

39. DAUBNEY, R. 1930. Natural transmission of heart-water of sheep by *Amblyomma variegatum* (Fabricius 1794). Parasitology **22:** 260-267.

40. NEITZ, W. O. 1947. The transmission of heartwater by *Amblyomma pomposum* Dönitz, 1909. South Afr. Sci. **1:** 83. (In Afrikaans.)

41. WALKER, J. B. & A. OLWAGE. 1987. The tick vectors of *Cowdria ruminantium* (Ixodoidea, Ixodidae, genus *Amblyomma*) and their distribution. Onderstepoort J. Vet. Res. **54:** 353-379.

42. LEWIS, E. A. 1949. Annual Report for 1947: 51. Department of Veterinary Services. Kenya.

43. KARRAR, G. 1966. Further studies on the epizootiology of heartwater in the Sudan. Sudan J. Vet. Sci. Anim. Husb. **6:** 83-85.

44. MACKENZIE, P. K. I. & R. A. I. NORVAL. 1980. The transmission of *Cowdria ruminantium* by *Amblyomma tholloni*. Vet. Parasitol. **7:** 265-268.

45. NORVAL, R. A. I. & P. K. I. MACKENZIE. 1981. The transmission of *Cowdria ruminantium* by *Amblyomma sparsum*. Vet. Parasitol. **8:** 189-191.

46. UILENBERG, G. & T. A. NIEWOLD. 1981. *Amblyomma astrion* Dönitz, 1909 nouveau vecteur expérimental de la cowdriose. Rev. Elev. Méd. Vét. Pays Trop. **34:** 267-270.

47. UILENBERG, G. 1983. Heartwater (*Cowdria ruminantium* infection): current status. Adv. Vet. Sci. Comp. Med. **27:** 427-480.

48. BEZUIDENHOUT, J. D. 1987. Natural transmission of heartwater. Onderstepoort J. Vet. Res. **54:** 349-351.

49. PETNEY, T. N. & I. G. HORAK. 1988. Comparative host usage by *Amblyomma hebraeum* and *Amblyomma marmoreum* (Acari, Ixodidae), the South African vectors of the disease heartwater. J. Appl. Entomol. **105:** 490-495.

50. ALEXANDER, R. A. 1931. Heartwater. The Present State of Our Knowledge of the Disease: 89-150. 17th Report of the Director of Veterinary Services and Animal Industry. South Africa.

51. NORVAL, R. A. I. 1975. Studies on the ecology of *Amblyomma marmoreum* Koch 1844 (Acarina: Ixodidae). J. Parasitol. **61:** 737-742.

52. UILENBERG, G. 1982. Experimental transmission of *Cowdria ruminantium* by the Gulf Coast tick *Amblyomma maculatum:* danger of introducing heartwater and benign African theileriasis onto the American mainland. Am. J. Vet. Res. **43:** 1279-1282.

53. JONGEJAN, F. 1992. Studies on the transmission of *Cowdria ruminantium* by African and American *Amblyomma* ticks. *In* Procedures and Abstracts of the 1st International Conference on Tick-Borne Pathogens at the Host-Vector Interface, Sept. 15-18, 1992, Saint Paul, Minnesota: 143.

54. UILENBERG, G., E. CAMUS & N. BARRÉ. 1985. Quelques observations sur une souche de *Cowdria ruminantium* isolée en Guadeloupe (Antilles françaises). Rev. Elev. Méd. Vét. Pays Trop. **38:** 34-42.

55. NORVAL, R. A. I., H. R. ANDREW & C. E. YUNKER. 1989. Pheromone-mediation of host-selection in bont ticks (*Amblyomma hebraeum* Koch). Science **243:** 364-365.

56. PAVIS, C. & N. BARRE. 1993. Kinetics of male pheromone production by *Amblyomma variegatum* (Acari: Ixodidae). J. Med. Entomol. **30:** 961-965.

57. ILEMOBADE, A. A. 1977. Heartwater in Nigeria. I. The susceptibility of different local breeds and species of domestic ruminants to heartwater. Trop. Anim. Health Prod. **9:** 177-180.

58. CAMUS, E. 1989. Etude épidémiologique de la cowdriose à *Cowdria ruminantium* en Guadeloupe. Etudes et synthèses de l'IEMVT no. 33. IEMVT. Maisons-Alfort, France.

59. KARRAR, G. 1960. Rickettsial infection (heartwater) in sheep and goats in the Sudan. Br. Vet. J. **116:** 105-114.

60. MATHERON, G., N. BARRE, E. CAMUS & J. GOGUE. 1987. Genetic resistance of Guadeloupe native goats to heartwater. Onderstepoort J. Vet. Res. **54:** 337-340.

61. NEITZ, W. O. & R. A. ALEXANDER. 1941. The immunisation of calves against heartwater. J. S. Afr. Vet. Med. Assoc. **12**(4): 103-111.

62. SOLDAN, A. W., T. L. NORMAN, S. MASAKA, E. A. PAXTON, R. M. EDELSTEN & K. J. SUMPTON. 1993. Seroconversion to *Cowdria ruminantium* of Malawi zebu calves, reared under different tick control strategies. Rev. Elev. Méd. Vét. Pays Trop. **46:** 171-177.

63. UILENBERG, G. 1971. Etudes sur la cowdriose à Madagascar. Première partie. Rev. Elev. Méd. Vét. Pays Trop. **24:** 239-249.

64. ASSELBERGS, M. J. & C. M. LOPES PEREIRA. 1990. Damage by *Amblyomma hebraeum* in local (*Bos indicus*) cattle in Mozambique. *In* Livestock Production in the Tropics. H. Kuil, R. W. Paling & J. E. Huhn, Eds.: 358-360. Faculty of Veterinary Medicine. Utrecht, the Netherlands.

65. NORVAL, R. A. I., R. W. SUTHERST, O. G. JORGENSEN, J. D. GIBSON & J. D. KERR. 1989. The effect of the bont tick (*Amblyomma hebraeum*) on the weight gain of Africander steers. Vet. Parasitol. **33:** 329-341.

66. CAMUS, E. & N. BARRE. 1990. *Amblyomma variegatum* and associated diseases in the Caribbean: strategies for control and eradication in Guadeloupe. Parassitologia **32:** 185-193.

67. NEITZ, W. O. 1939. The immunity in heartwater. Onderstepoort J. Vet. Sci. Anim. Husb. **13:** 245-283.

68. JONGEJAN, F., G. UILENBERG, F. F. J. FRANSSEN, A. GUEYE & J. NIEUWENHUIJS. 1988. Antigenic differences between stocks of *Cowdria ruminantium*. Res. Vet. Sci. **44:** 186-189.

69. JONGEJAN, F., M. J. C. THIELEMANS, C. BRIERE & G. UILENBERG. 1991. Antigenic diversity of *Cowdria ruminantium* isolates determined by cross-immunity. Res. Vet. Sci. **51:** 24-28.

70. DU PLESSIS, J. L., L. VAN GAS, J. A. OLIVIER & J. D. BEZUIDENHOUT. 1989. The heterogenicity of *Cowdria ruminantium* stocks: cross-immunity and serology in sheep and pathogenicity to mice. Onderstepoort J. Vet. Res. **56:** 195-201.

71. Du Plessis, J. L., F. T. Potgieter & L. Van Gas. 1990. An attempt to improve the immunization of sheep against heartwater by using different combinations of 3 stocks of *Cowdria ruminantium*. Onderstepoort J. Vet. Res. **57:** 205–208.

72. Andrew, H. R. & R. A. I. Norval. 1989. The carrier status of sheep, cattle and African buffalo recovered from heartwater. Vet. Parasitol. **34:** 261–266.

73. Camus, E. 1992. Le portage asymptomatique de bovins et chèvres Créole guéris de la cowdriose en Guadeloupe. Rev. Elev. Méd. Vét. Pays Trop. **45:** 133–135.

74. Yunker, C. E. 1988. Current research on heartwater in Zimbabwe. Zimbabwe Vet. J. **19:** 25–31.

75. Du Plessis, J. L., P. Berche & L. Van Gas. 1991. T cell–mediated immunity to *Cowdria ruminantium* in mice: the protective role of Lyt-2$^+$ T cells. Onderstepoort. J. Vet. Res. **58:** 171–179.

76. Byrom, B., S. M. Mahan & A. F. Barbet. 1993. The development of antibody to *Cowdria ruminantium* in mice and its role in heartwater disease. Rev. Elev. Méd. Vét. Pays Trop. **46:** 197–201.

77. Du Plessis, J. L. 1993. An *in vitro* test to demonstrate the inhibitory effect of homologous immune serum on the infectivity of *Cowdria ruminantium*. Onderstepoort J. Vet. Res. **60:** 69–73.

78. Totte, P., D. Blankaert, P. Zilimwabagabo & J. Werenne. 1993. Inhibition of *Cowdria ruminantium* infectious yield by interferons alpha and gamma in endothelial cells. Rev. Elev. Méd. Vét. Pays Trop. **46:** 189–194.

79. Totte, P., F. Jongejan, A. L. W. De Gee & J. Werenne. 1994. Production of alpha interferon in *Cowdria ruminantium*-infected cattle and its effect on infected endothelial cell cultures. Infect. Immun. **62:** 2600–2604.

80. Mahan, S. M., G. E. Smith & B. Byrom. 1994. Conconavalin A–stimulated bovine T-cell supernatants inhibit growth of *Cowdria ruminantium* in bovine endothelial cells in vitro. Infect. Immun. **62:** 747–750.

81. Synge, B. A. 1978. Brain biopsy for the diagnosis of heartwater. Trop. Anim. Health Prod. **10:** 45–48.

82. Uilenberg, G., A. Permin & J. W. Hansen, Eds. 1994. Use of applicable biotechnological methods for diagnosing haemoparasites. *In* Proceedings of an Expert Consultation, Mérida, Mexico, Oct. 4–6, 1993. FAO, Rome: 6.

83. Du Plessis, J. L. 1981. The application of the indirect fluorescent antibody test to the serology of heartwater. *In* Tick Biology and Control. Proceedings of the International Congress on Tick Biology and Control, Jan. 27–29, 1981, Grahamstown, South Africa: 47–52.

84. Logan, L. L., T. C. Whyard, J. C. Quintero & C. A. Mebus. 1987. The development of *Cowdria ruminantium* in neutrophils. Onderstepoort J. Vet. Res. **54:** 197–204.

85. Martinez, D., J. Swinkels, E. Camus & F. Jongejan. 1990. Comparaison de trois antigènes pour le sérodiagnostic de la cowdriose par immunofluorescence indirect. Rev. Elev. Méd. Vét. Pays Trop. **43:** 159–166.

86. Jongejan, F., M. J. C. Thielemans, M. De Groot, P. J. S. Van Kooten & B. A. M. Van Der Zeijst. 1991. Competitive enzyme-linked immunosorbent assay for heartwater using monoclonal antibodies to a *Cowdria ruminantium*-specific 32-kilodalton protein. Vet. Microbiol. **28:** 199–211.

87. Martinez, D., S. Coisne, C. Sheikboudou, M. Moutoussamy, B. Mari & T. Vidalenc. 1993. The development of ELISA tests for the serodiagnosis of dermatophilosis and cowdriosis and their use in epidemiological studies. *In* Resistance or Tolerance of Animals to Disease and Veterinary Epidemiology and Diagnostic Methods. Proceedings EEC Contract. Workshops, November 2–6, 1992, Rethymno, Crete, Greece. CIRAD-EMVT, Maisons-Alfort, France: 123–131.

88. Martinez, D., S. Coisne, C. Sheikboudou & F. Jongejan. 1993. Detection of antibodies to *Cowdria ruminantium* in the serum of domestic ruminants by indirect ELISA. Rev. Elev. Méd. Vét. Pays Trop. **46:** 115–120.

89. MULLER KOBOLD, A., D. MARTINEZ, E. CAMUS & F. JONGEJAN. 1992. Distribution of heartwater in the Caribbean determined on the basis of detection of antibodies to the conserved 32-kilodalton protein of *Cowdria ruminantium*. J. Clin. Microbiol. **30:** 1870-1873.

90. DU PLESSIS, J. L., J. D. BEZUIDENHOUT, M. S. BRETT, E. CAMUS, F. JONGEJAN, S. M. MAHAN & D. MARTINEZ. 1993. The sero-diagnosis of heartwater: a comparison of five tests. Rev. Elev. Méd. Vét. Pays Trop. **46:** 123-129.

91. MAHAN, S. M., N. TEBELE, D. MUKWEDEYA, S. SEMU, C. B. NYATHI, L. A. WASSINK, P. J. KELLY, T. PETER & A. F. BARBET. 1993. An immunoblotting diagnostic assay for heartwater based on the immunodominant 32-kilodalton protein of *Cowdria ruminantium* detects false positives in field sera. J. Clin. Microbiol. **31:** 2729-2737.

92. VAN VLIET, A. H. M., S. M. MAHAN, D. MARTINEZ, E. CAMUS, B. A. M. VAN DER ZEIJST & F. JONGEJAN. 1994. Development of a *Cowdria ruminantium* specific ELISA based on recombinant antigens. *In* Application of Biotechnology: 29-32. Faculty of Veterinary Medicine. Utrecht, the Netherlands.

93. WAGHELA, S. D., F. R. RURANGIRWA, S. M. MAHAN, C. E. YUNKER, T. B. CRAWFORD, A. F. BARBET, M. J. BURRIDGE & T. C. McGUIRE. 1991. A cloned DNA probe identifies *Cowdria ruminantium* in *Amblyomma variegatum* ticks. J. Clin. Microbiol. **29:** 2571-2577.

94. MAHAN, S. M., S. D. WAGHELA, T. C. McGUIRE, F. R. RURANGIRWA, L. A. WASSINK & A. F. BARBET. 1992. A cloned DNA probe for *Cowdria ruminantium* hybridizes with eight heartwater strains and detects infected sheep. J. Clin. Microbiol. **30:** 981-986.

95. PETER, T. F., S. L. DEEM, A. F. BARBET, R. A. I. NORVAL, B. H. SIMBI, P. J. KELLY & S. M. MAHAN. 1995. Development and evaluation of PCR assay for detection of low levels of *Cowdria ruminantium* infection in *Amblyomma* ticks not detected by DNA probe. J. Clin. Microbiol. **33:** 166-172.

96. BEZUIDENHOUT, J. D. 1989. *Cowdria* vaccines. *In:* Veterinary Protozoan and Hemoparasite Vaccines. I. G. Wright, Ed.: 31-42. CRC Press. Boca Raton, FL.

97. OLIVIER, J. A., A. P. LÖTTER, W. G. E. SCHWULST & O. MATTHEE. 1988. Chemoprophylaxis of heartwater by using a subcutaneous pill or paste. Newsletter of the Onderstepoort Veterinary Institute **6**(1): 7-9. (In Afrikaans.)

98. DU PLESSIS, J. L., L. VAN GAS, F. J. LABUSCHAGNE & S. WIJMA. 1990. The freeze-drying of *Cowdria ruminantium*. Onderstepoort J. Vet. Res. **57:** 145-149.

99. BRETT, S. 1993. Heartwater tissue culture vaccine. In Biennial Report of the Veterinary Research Institute Onderstepoort, 1990/1991: 19. Department of Agricultural Development. Pretoria, South Africa.

100. JONGEJAN, F. 1991. Protective immunity to heartwater (*Cowdria ruminantium* infection) is acquired after vaccination with in vitro-attenuated rickettsiae. Infect. Immun. **59:** 729-731.

101. GUEYE, A., F. JONGEJAN, M. MBENGUE, A. DIOUF & G. UILENBERG. 1994. Essai sur le terrain d'un vaccin atténué contre la cowdriose. Rev. Elev. Méd. Vét. Pays Trop. **47:** 401-404.

102. JONGEJAN, F., S. W. VOGEL, A. GUEYE & G. UILENBERG. 1993. Vaccination against heartwater using *in vitro* attenuated *Cowdria ruminantium* organisms. Rev. Elev. Méd. Vét. Pays Trop. **46:** 223-227.

103. MARTINEZ, D., J. C. MAILLARD, S. COISNE, C. SHEIKBOUDOU & A. BENSAID. 1994. Protection of goats against heartwater acquired by immunization with inactivated elementary bodies of *Cowdria ruminantium*. Vet. Immunol. Immunopathol. **41:** 153-163.

104. MAHAN, S. M., H. R. ANDREW, N. TEBELE, M. J. BURRIDGE & A. F. BARBET. 1995. Immunisation of sheep against heartwater with inactivated *Cowdria ruminantium*. Res. Vet. Sci. **58:** 46-49.

105. VAN VLIET, A. H. M., F. JONGEJAN, M. VAN KLEEF & B. A. M. VAN DER ZEIJST. 1994. Molecular cloning, sequence analysis, and expression of the gene encoding the

immunodominant 32-kilodalton protein of *Cowdria ruminantium.* Infect. Immun. **62:** 1451–1456.

106. MAHAN, S. M., T. C. McGUIRE, S. M. SEMU, M. V. BOWIE, F. JONGEJAN, F. R. RURANG-IRWA & A. F. BARBET. 1994. Molecular cloning of a gene encoding the immunogenic 21 kDa protein of *Cowdria ruminantium.* Microbiology **140:** 2135–2142.

107. MAHAN, S. M. 1995. Review of the molecular biology of *Cowdria ruminantium.* Vet. Parasitol. **57:** 51–56.

Uncharacterized *Ehrlichia* Spp. May Contribute to Clinical Heartwater

B. A. ALLSOPP, M. T. ALLSOPP, J. H. DU PLESSIS,
AND E. S. VISSER

Department of Molecular Biology
Onderstepoort Veterinary Institute
Onderstepoort 0110, South Africa

RELATIONSHIPS AMONG THE *RICKETTSIALES*

Heartwater is a disease of domestic ruminants normally caused by the rickettsial pathogen *Cowdria ruminantium*. Traditional taxonomy[1] places the order *Rickettsiales* in the α-subdivision of the phylum originally called the purple bacteria, now known as the *Proteobacteria*. The *Rickettsiales* are further subdivided into three families, the *Rickettsiaceae*, the *Bartonellaceae*, and the *Anaplasmataceae*. The family *Rickettsiaceae* is subdivided into three tribes, the *Ehrlichieae*, the *Rickettsieae*, and the *Wolbachieae*. The *Ehrlichieae* contains three genera, *Ehrlichia, Cowdria,* and *Neorickettsia. Cowdria ruminantium* is classified as the sole species of its genus (FIG. 1).

During the past decade, there has been a fundamental change in classification brought about by molecular phylogenetics. The molecule most extensively used is the small-subunit ribosomal RNA (srRNA) gene, and there is a large body of data based on the sequence of this gene from a wide variety of organisms.[2] This is known as the ribosomal database project (RDP); it has transformed phylogenetic studies, and the greatest effect has probably been upon bacterial phylogenetics. We extracted from the RDP database the srRNA data for 21 *Rickettsiales* isolates; these included 11 *Ehrlichia* species, 4 *Cowdria* stocks, 3 *Rickettsia* species, and *Anaplasma marginale*. Additional information on the *Ehrlichia* organisms is presented in FIGURE 2. We used the sequence alignment as provided by the RDP and inferred phylogenetic trees using the Phylip 3.5 phylogeny package[3,4] including *Escherichia coli* as an outgroup species to root the trees. Three different algorithms were employed to infer Fitch-Margoliash, maximum likelihood, and maximum parsimony trees.

The resulting maximum likelihood tree is shown in FIGURE 3. The Fitch-Margoliash and maximum parsimony trees had identical topologies (data not shown). The trees suggest that all *Ehrlichia* species had a common ancestor which was not shared with the genus *Rickettsia*. The ancestral *Ehrlichia* has given rise to three different taxa. The most deeply branching group, containing the human and equine pathogens *E. sennetsu* and *E. risticii,* we have called group I *Ehrlichia*. The second *Ehrlichia*-containing taxon includes *Anaplasma marginale;* we have called this group II *Ehrlichia*. The third *Ehrlichia*-containing group, which includes *Cowdria ruminantium,* shared a common ancestor with group II *Ehrlichia*. We have referred to this third *Ehrlichia*-containing taxon as group III *Ehrlichia*.

This picture is very different from the traditional classification of the *Rickettsiales,* which includes all *Ehrlichia* species in one genus, places *Anaplasma marginale* in

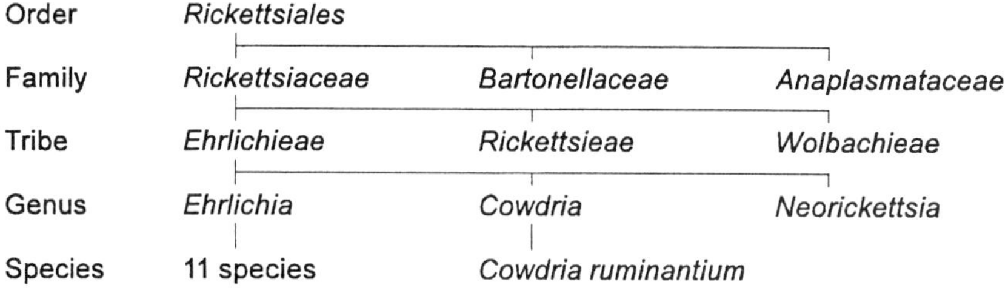

FIGURE 1. Traditional classification of the *Rickettsiales*.

a completely different family, and creates a separate genus for *Cowdria*. It would appear to be necessary to create two new genera for the *Anaplasma*-related (group II) and *Cowdria*-related (group III) *Ehrlichia* respectively. There are no obvious similarities of life cycle that distinguish the three groups of *Ehrlichia* species; they infect a range of hosts and several different cell types within these hosts, but the differences do not correlate with the phylogenetic groupings (FIG. 2.).

We should note that there is no consensus on what constitutes a species solely at the srRNA level. The two *E. canis* isolates share 99.9% sequence identity and are considered to be of the same species. Compare this with the three species *E. equi*,

Species	Stock Identifier	Taxon	Accession Number	Host species	Host cell type	Ref.
E. canis	Florida	Group III	M73226	Canine	Monocytes	20
	Oklahoma	*Cowdria* related	M73221			
E. chaffeensis	Arkansas		M73222	Human	Monocytes	20
E. ewingi			M73227	Canine	Granulocytes	21
E. muris			U15527	Murine		22
E. equi		Group II	M73223	Equine	Granulocytes	20
E. bovis	Lydenburg	*Anaplasma* related	U03775	Bovine	Monocytes	9
E. phagocytophila	OS		M73220	Bovine Ovine Caprine Cervine	Granulocytes	20
	FG		M73224			
E. platys			M82801	Canine	Platelets	21
Ehrlichia sp.			U02521	Human	Granulocytes	24
E. risticii	Illinois	Group I *Ehrlichia*	M21290	Equine	Epithelial cells, mast cells, monocytes	23
E. sennetsu	11908		M73225	Human	Monocytes	20
	Miyayama		M73219			

FIGURE 2. Biological data from eleven *Ehrlichia* species having srRNA sequence entries in GenBank™.

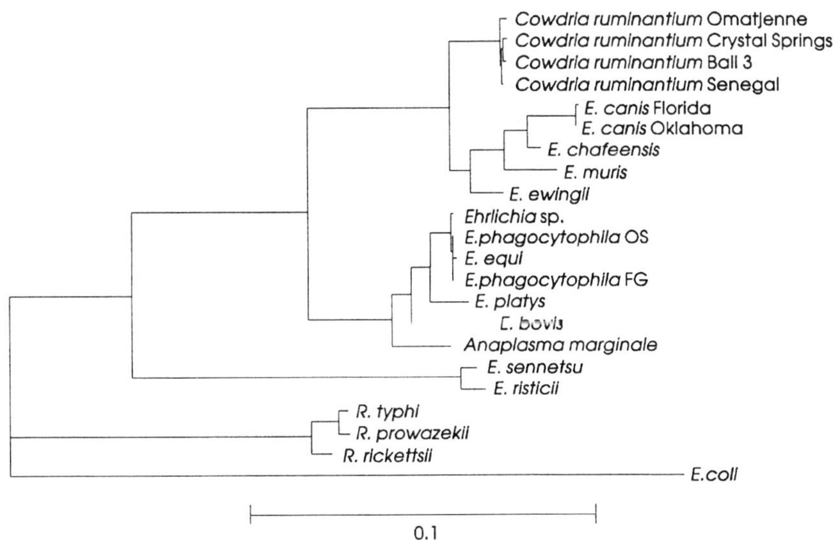

FIGURE 3. Maximum likelihood phylogenetic tree of some *Rickettsiales.* The scale bar represents 10 nucleotide substitutions/100 bases.

E. phagocytophila, and *E.* sp., where each shares 99.9% identity with each of the others. In this second example there are host and antigenic differences that have led to the identification of the three as different species. The four different *Cowdria* genotypes share identities in the range 98.8%-99.2% so it seems pointless at this stage to discuss whether or not they should be considered to constitute a single species. It does appear, however, that there is justification for assigning *Cowdria ruminantium* to the same genus as the other group III *Ehrlichia* species.

A VARIETY OF BACTERIA ARE ASSOCIATED WITH HEARTWATER

The influence of molecular phylogenetics has led to a number of studies in which bacteriologists have searched environmental niches for bacterial srRNA sequences using the polymerase chain reaction (PCR). The data generated indicate that there is a multiplicity of previously unknown, and as yet uncultured, bacterial organisms in many different environments.[5-7] In one instance an uncultured disease-causing bacterium was identified in human tissue by the same means.[8] We have used this strategy to investigate a range of heartwater-infective material.[9]

From the srRNA gene sequence of *C. ruminantium* (Crystal Springs)[10] we designed two oligonucleotides that would specifically amplify a portion of the rickettsial gene in the presence of an excess of mammalian DNA. The chosen fragment, 323 bp in length, spans the hypervariable V1 loop. We conducted PCR with these primers using DNA from a range of samples taken from clinical heartwater cases, from apparently healthy sheep and goats from both heartwater-free and heartwater-endemic areas, tissue culture material infected with *Cowdria*, and tick material from heartwater-free and heartwater-endemic areas.[9] We obtained 278 sequences that were homologous to known bacterial sequences; they were grouped into *Rickettsiales* (193 clones), *Pseudomonadaceae* (62 clones), and other miscellaneous sequences (23 clones). Among the 193 *Rickettsiales* sequences, 141 closely related clones were *Cowdria* genotypes, 10 clones were related to *E. platys*, 2 to *E. equi*, 6 to *E. canis*, and 6 to *A. marginale*. We decided to examine in more detail the *E. platys* and *E. canis* homologues.

EHRLICHIA SP. (GERMISHUYS)

In 1984, a sheep died showing symptoms of heartwater on a farm in the northeastern Transvaal, South Africa. The animal was autopsied, and brain smears were prepared and Giemsa stained. The postmortem signs were typical of heartwater, and the brain smear showed colonies that were considered to be *C. ruminantium*. The diagnosis was heartwater, and a lymph node suspension was prepared and frozen and referred to as the Germishuys isolate of *C. ruminantium*.[11] Subsequently, sheep were inoculated with this material and infected blood was passaged into other sheep. Sheep may react severely to infection with Germishuys blood stabilate; on occasion they may self-cure, but on other occasions the infection is lethal. In one cross-protection trial, 3 out of 3 sheep died when challenged with this isolate although the animals had been immunized against the Kümm stock of *C. ruminantium*.[11] Infected sheep develop a high antibody titer in the indirect fluorescent antibody (IFA) test for *C. ruminantium*.[12]

During the PCR amplification experiments described in the previous paragraph, we determined the sequence of the V1 loop fragment amplified from Germishuys-infected material. The closest known homologue was found to be *E. canis*, but we were unable to detect any sequences homologous to *Cowdria*. It has also not been possible to infect bovine endothelial cells with the isolate under the normal *C. ruminantium* tissue culture conditions.[13] We therefore decided to amplify the whole srRNA gene of this unknown organism, using PCR primers fD1 (CCG AAT TCG TCG ACA ACA GAG TTT GAT CCT GGC TCA G)[14] and B (CCC GGG ATC CAA GCT TGA TCC TTC TGC AGG TTC ACC TAC)[15] which hybridize to the 5′ and 3′ ends of the gene respectively. The procedures are described elsewhere.[16] The PCR product was cloned into M13 and sequenced, and the full sequence was found to be 99.9% identical to the sequences of *E. canis* (Oklahoma isolate) and *E. canis* (Florida isolate).

It appears, therefore, that the Germishuys agent is a previously unknown group III *Ehrlichia* species, closely related to *E. canis*. This organism is infectious to sheep and goats, the infection may be fatal, the symptoms are not easily distinguished from "normal" heartwater caused by *C. ruminantium*, and cured or self-cured animals may show a high antibody titer in the IFA test for *C. ruminantium*. This organism is obviously a source of epidemiological confusion. For the time being, we are referring to it as *Ehrlichia* sp. (Germishuys).

EHRLICHIA SP. (OMATJENNE)

In 1987, 43 cattle sera were collected on the farm "Omatjenne" in the Otjiwarongo district of Namibia.[17] This area is free from *Amblyomma* ticks, although *Rhipicephalus* and *Hyalomma* species are common. The sera were examined by the IFA test for *C. ruminantium*,[17] and 81% of the samples were found to be positive. A number of ticks were collected from the cattle, individually homogenized, and each injected into a mouse. Four weeks later, sera from the mice were also examined by the IFA test and one sample, from a mouse inoculated with a *Hyalomma truncatum* tick, was positive. Spleen homogenate taken from this mouse was passaged through 3 cycles of infection, *Amblyomma hebraeum* to sheep, and on the third passage a sheep died with symptoms of heartwater. Geimsa-stained peritoneal macrophages from the mice and endothelial cells from the sheep showed the presence of organisms indistinguishable from *C. ruminantium*. Because this "Omatjenne agent" had originated from a *Hyalomma* tick, it was assumed to be an *Ehrlichia* species.

During the PCR amplification experiments described above, we obtained V1 loop sequences amplified from the original mouse infected with the Omatjenne agent and the sheep that had died at passage three. In the mouse we found only a single sequence, closely related to *C. ruminantium* (Crystal Springs), and we refer to this as *C. ruminantium* (Omatjenne). In the sheep we found both *C. ruminantium* (Omatjenne) and a sequence closely related to *E. platys*. We decided to obtain the full srRNA sequence of this *Ehrlichia* species, and we inoculated the last of the stored infective sheep material into a fresh sheep in order to obtain a further stock of the organism. The inoculated sheep developed an elevated temperature and reacted severely, blood samples were taken, and the animal was treated to prevent expected death. After amplification, we found only *C. ruminantium* (Omatjenne) sequences in the new material, the *Ehrlichia* organism had apparently failed to establish itself on passage. Unfortunately we now have no material known to contain the original Omatjenne *Ehrlichia* organism.

In an effort to recover further examples of the organism, we collected more *H. truncatum* ticks from the Otjiwarongo district and searched by PCR amplification for *Ehrlichia* species. In this case we found only sequences identical to the tick symbiont *Wolbachia persica*. After widening the search, we eventually recovered the Omatjenne *Ehrlichia* V1 loop sequence from two sheep blood samples from Klerksdorp, Western Transvaal. These sheep, although apparently perfectly healthy, were IFA positive for heartwater and also yielded *C. ruminantium* (Omatjenne) sequences. Almost all the srRNA sequence variation between closely related *Ehrlichia* species is to be found in the V1 loop, with an occasional 1 or 2 differences in the remainder of the molecule. This made it highly probable that the Klerksdorp sheep samples contained the unknown Omatjenne *Ehrlichia,* and we therefore amplified, cloned, and sequenced the whole srRNA gene as already described for *Ehrlichia* sp. (Germishuys). The sequence was found to be 99.5% identical to the sequence of *E. platys*.

It seems that this organism, either alone or in combination with *C. ruminantium* (Omatjenne), can cause death with heartwater-type symptoms, although sheep may also carry this combination of organisms and remain apparently healthy. Infected animals show a high antibody titer in the IFA test for *C. ruminantium,* although we do not know whether this is caused by the *Cowdria* or the *Ehrlichia* component of the infection. For the moment we refer to this organism as *Ehrlichia* sp. (Omatjenne).

THE DIAGNOSIS OF HEARTWATER

The pathological changes associated with heartwater are considered insufficiently specific for definitive diagnosis,[18] and confirmation has traditionally relied upon microscopic visualization of rickettsial bodies in brain smears. However it is difficult, if not impossible, to distinguish between all *Cowdria* and *Ehrlichia* organisms by light microscopy. All presently used serological tests for *Cowdria* have been found to cross-react with one or more species of *Ehrlichia* or have given positive results with samples from areas where heartwater has never been observed.[19] We have now discovered two new *Ehrlichia* spp. that sometimes give rise to heartwater-type symptoms but on other occasions are to be found in asymptomatic animals. At least one of these organisms can, on its own, induce a high antibody titer in the IFA test for *C. ruminantium*. Until more specific serological tests become available, it appears that oligonucleotide probes based upon srRNA sequences can provide the most reliable diagnostic information. We have designed four probes that are specific for each of the *Cowdria* genotypes shown in FIGURE 3, and two probes that detect all group II and all group III *Ehrlichia* species respectively. The use of these probes for epidemiological surveys will enable us to determine the relative importance of the different *Cowdria* and *Ehrlichia* species that we have found to be associated with heartwater.

REFERENCES

1. RIKIHISA, Y. 1991. The tribe *Ehrlichieae* and ehrlichial diseases. Clin. Microbiol. Rev. **4:** 286-308.
2. LARSEN, N., G. J. OLSEN, B. L. MAIDAK, M. J. McCAUGHEY, R. OVERBEEK, T. J. MACKE, T. L. MARSH, & C. R. WOESE. 1993. The ribosomal database project. Nucleic Acids Res. **21**(Supplement): 3021-3023.
3. FELSENSTEIN, J. 1989. PHYLIP: Phylogeny Inference Package (version 3.2). Cladistics **5:**164-166.
4. FELSENSTEIN, J. 1993. PHYLIP Version 3.5.1. University of Washington, Seattle, WA.
5. WARD, D. M., R. WELLER & M. M. BATESON. 1990. 16S rRNA sequences reveal numerous uncultured microorganisms in a natural community. Nature **345:** 63-65.
6. FUHRMAN, J. A., K. McCALLUM & A. A. DAVIS. 1993. Phylogenetic diversity of subsurface marine microbial communities from the Atlantic and Pacific Oceans. Appl. Environ. Microbiol. **59:** 1294-1302.
7. BARNS, S. M., R. E. FUNDYGA, M. W. JEFFRIES & N. R. PACE. 1994. Remarkable archaeal diversity detected in a Yellowstone National Park hot spring environment. Proc. Natl. Acad. Sci. USA. **91:** 1609-1613.
8. RELMAN, D. A. 1993. The identification of uncultured microbial pathogens. J. Infect. Dis. **168:** 1-8.
9. ALLSOPP, M. T. E. P., E. S. VISSER, J. L. DU PLESSIS, S. W. VOGEL & B. A. ALLSOPP. Molecular characterisation of heartwater-associated organisms and the development of probes for epidemiological studies. (Unpublished).
10. DAME, J. B., S. M. MAHAN & C. A. YOWELL. 1992. Phylogenetic relationship of *Cowdria ruminantium,* agent of heartwater, to *Anaplasma marginale* and other members of the

order *Rickettsiales* determined on the basis of 16S rRNA sequence. Int. J. Systematic Bacteriol. **42:** 270–274.

11. DU PLESSIS, J. L., L. VAN GAS, J. A. OLIVIER & J. D. BEZUIDENHOUT. 1989. The heterogenicity of *Cowdria ruminantium* stocks: cross immunity and serology in sheep and pathogenicity to mice. Onderstepoort J. Vet. Res. **56:** 195–201.

12. DU PLESSIS, J. L. & L. MALAN. 1987. The application of the indirect fluorescent antibody test in research on heartwater. Onderstepoort J. Vet. Res. **54:** 319–325.

13. YUNKER, C. E. Personal communication.

14. WEISBURG, W. G., S. M. BARNS, D. A. PELLETIER & D. J. LANE. 1991. 16S ribosomal DNA amplification for phylogenetic study. J. Bacteriol. **173;** 697–703.

15. SOGIN, M. L. 1990. Amplification of ribosomal RNA genes for molecular evolution studies. *In* PCR Protocols. M. A. Innis, D. H. Gelfand, J. J. Sninsky & T. J. White, Eds.: 307–314. Academic Press. San Diego, CA.

16. ALLSOPP, M. T. E. P., T. CAVALIER-SMITH, D. T. DE WAAL & B. A. ALLSOPP. 1994. Phylogeny and evolution of the piroplasms. Parasitology **108:** 147–152.

17. DU PLESSIS, J. L. 1990. Increased pathogenicity of an *Ehrlichia*-like agent after passage through *Amblyomma hebraeum;* a preliminary report. Onderstepoort J. Vet. Res. **57:** 233–237.

18. PROZESKY, L. 1987. The pathology of heartwater. III. A review. Onderstepoort J. Vet. Res. **54:** 281–286.

19. YUNKER, C. E. & B. A. ALLSOPP. 1994. Modern biotechnological methods for diagnosis of cowdriosis—present and future trends. *In* Proceedings of the FAO Expert Consultation, Merida, Mexico, 1993. Use of Applicable Biotechnological Methods for Diagnosing Haemoparasites. FAO. Rome, Italy.

20. ANDERSON, B. E., J. E. DAWSON, D. C. JONES & K. H. WILSON. 1991. *Ehrlichia chaffeensis,* a new species associated with human ehrlichiosis. J. Clin. Microbiol. **29:** 2838–2842.

21. GREENE, G. E., D. C. JONES & J. E. DAWSON. 1992. *Ehrlichia ewingii* sp. nov., the etiologic agent of canine granulocytic ehrlichiosis. Int. J. Syst. Bacteriol. **42:** 299–302.

22. WEN, B., Y. RIKIHISA, J. MOTT, P. A. FUERST, M. KAWAHARA & C. SUTO. *Ehrlichia muris* sp. nov., a new species of *Ehrlichia,* identified on the basis of 16S rRNA base sequences, serological, morphological, and biological characteristics. Int. J. Syst. Bacteriol. (In press.)

23. WEISBURG, W. G., M. E. DOBSON, J. E. SAMUEL & G. A. DASCH. 1989. Phylogenetic diversity of the Rickettsiae. J. Bacteriol. **171:** 4202–4206.

24. CHEN, S. M., J. S. DUMLER, J. S. BAKKEN & D. H. WALKER. 1994. Identification of a granulocytotropic *Ehrlichia* species as the etiologic agent of human disease. J. Clin. Microbiol. **32:** 589–595.

Amino Acid and Protein Depletion in Medium of Cell Cultures Infected with *Cowdria ruminantium*

A. W. H. NEITZ [a] AND C. E. YUNKER [b,c]

[a]Department of Biochemistry
University of Pretoria
Pretoria 0002, Republic of South Africa

[b]Protozoology Section
Onderstepoort Veterinary Institute
Private Bag X5
Onderstepoort 0110, Republic of South Africa

Heartwater (*Cowdria ruminantium* infection) is an important, often fatal rickettsial disease of ruminants that is transmitted in nature by ixodid ticks of the genus *Amblyomma* parasitic on a variety of vertebrates. The distribution of the disease includes the sub-Saharan region of the African continent, some islands off its eastern and western coasts, and certain ones in the Caribbean Sea to where it was transported by man.[1] In the last-named focus, heartwater represents a distinct threat to the livestock industries of the New World by virtue of the presence of proven vector ticks capable of being introduced by migrating birds.[2]

Until recently, progress in research on diagnosis, epidemiology, and control was impeded by lack of a means for production of large amounts of cell-free organisms for use in diagnostic tests and as immunogens. In 1985, the first successful continuous culture of *Cowdria ruminantium* was reported by workers at the Veterinary Research Institute, Onderstepoort, Republic of South Africa.[3] This long-awaited discovery enabled researches to produce and isolate large quantities of organisms, antigens, proteins, and other products, including monoclonal antibodies and DNA libraries, which resulted in the development of various serodiagnostic tests[4–8] and gene probes.[9–11] Less than a decade after development of the culture system for *C. ruminantium*, the majority of current heartwater research reports are based on culture-produced organisms or their products.[12]

Despite the promise that the *in vitro* system offers, *C. ruminantium* remains a difficult organism to cultivate, with significant discrepancies existing, both within and among various laboratories, in the ability to successfully and repeatedly produce cultured organisms.[12] Most likely, these discrepancies are largely due to the inadequacy of the culture medium, inasmuch as the nutritional requirements and metabolic processes of the heartwater rickettsia are poorly known. Consequently, media currently in use for cultivation of the organism have been chosen arbitrarily or empirically.

[c]Present address: 230 Pioneer Drive, Port Ludlow, WA 98365.

In this study, we attempted to determine utilization of amino acids (AAs) and proteins by *C. ruminantium* grown in the simplest medium in use, Glasgow minimum essential medium (GMEM), in the hope of identifying substrate deficiencies and gaining an insight into the metabolic processes of the organism.

MATERIALS AND METHODS

Cells and Culture Medium

Two continuous lines of vascular endothelial cells derived from ruminants were used: sheep brain endothelial (SBE-189)[13] and bovine saphenous vein endothelial (BSV-793)[14] cells. Cultures were grown to confluency in 800 cm^2 roller bottles in the presence of GMEM containing 20% heat-inactivated bovine serum and the antibiotics sodium benzyl penicillin (200 U/ml), streptomycin sulfate (200 μg/L), and Fungizone (5.0 μg/L). The pH of the complete medium was approximately 6.7. Uninfected SBE-189 and both infected and uninfected BSV-189 cells were screened for the presence of mycoplasmal contaminants by staining with bis-benzamide (Hoechst 33258) and examination of slides with fluorescence microscopy. BSV-793 cells were also tested by cultivation in PPLO broth and agar plates. No evidence of mycoplasmal contamination was seen in samples from either cell line.

Infection and Sampling of Cultures

Duplicate roller bottle cultures were infected with cell-free supernatants collected from late-stage (5 days postinfection) cultures of the Welgevonden stock of *C. ruminantium.* Immediately before infection and at 24-h intervals thereafter for 5 days, 10 ml samples were taken from each bottle, pooled according to treatment, and frozen at −20 °C for analyses. Other roller bottle cultures not infected with the rickettsia were used as negative culture controls. Medium controls consisted of samples of the medium taken before application to the cultures.

Amino Acid Analyses

Samples were thawed and passed through Millipore low amino-acid-binding cellulose membranes with 10,000 nominal molecular weight limit in order to remove proteins. Amino acids analyses were carried out with a Waters Automated Analyzer (Millipore Corp.), using Pico-Tag phenyl isothiocyanate (PTC) prederivitization methodology.[15] The Pico-Tag method involves two steps: (1) precolumn derivatization of the sample and (2) analysis by reverse phase, high-performance liquid chromatography (HPLC). In the first step, AAs are derivatized with phenyl isothiocyanate to produce PTC-AAs. In the second, these derivatives are then analyzed by HPLC in amounts as low as 1 picomole.

Protein Analyses

High-performance capillary electrophoresis separation of proteins was carried out on a Beckman P/ACE System 2100, utilizing System Gold software for control and data collection.[16] A polyimide-coated, fused silica capillary with 75 μ id and 57 cm length from inlet to detector was used. Borate buffer of high pH (Beckman Instruments) was used for the separations, which were performed at 30 kV constant voltage. Capillary temperature was maintained at 25 °C, and detection was by absorbance at 214 nm.

RESULTS AND DISCUSSION

Amino Acids

Decline in AA concentrations over the 5-day course of infection occurred only with arginine and glutamine. In both lines of cells, as well as in uninfected and *Cowdria*-infected cultures, an almost complete depletion of arginine in culture medium was seen as early as the first (BSV-793) or second (SBE-189) day following application to the cells (TABLES 1 and 2). By the third day, less than 5% of the original amount was detectable. Glutamine also decreased markedly in all cases over the 5-day course, with 87% of the original concentration being lost in infected and 71% lost in uninfected SBE-189 cells, while 38% of glutamine disappeared from infected and 63.5% from uninfected BSV-793 cells. These results reveal that, for purposes of culturing *C. ruminantium*, GMEM may be deficient in arginine and only marginally adequate in glutamine. Both of these are essential AAs in the sense that omission of either from the medium results in death of most or all cultured animal cells thus far studied.[17] In addition, at least one rickettsia has been shown to readily utilize arginine.[18] Obviously, if initial concentrations of either AA are low, there is a possibility that parasite and host will compete for them. Glasgow MEM,[19] originally used for growing *C. ruminantium*,[20] contains only twice the minimal concentration of arginine required by cultured cells. Glutamine is labile and will lose up to 5% of its original concentration at mammalian cell incubation temperatures.[21] Thus care should be taken to insure that *Cowdria* cultures are given an adequate initial supply of this rapidly depleted AA. Supplementation of cultures with glutamine at intervals might also serve to improve the growth of the heartwater organism, especially in long-term cultures.

In general, increases in some AAs were seen by the first day of exposure of cells to medium, regardless of cell line or infection status. This may be due to the release of cell-surface-bound AAs upon exposure. Certain other AAs (e.g., glutamic acid, alanine) increased over the 5-day period in both infected and uninfected cells of either line, implying *de novo* synthesis or proteolysis on the part of the cells and/or the rickettsia. One means by which glutamic acid could increase is conversion from glutamine. Whereas asparagine, which is closely related to glutamine, increased markedly in both infected and uninfected SBE-189 cell medium, its increase in BSV-793 cell medium was insignificant. Asparagine, though not an essential AA for cultured cells, is involved in nitrogen metabolism of prokaryotes.

TABLE 1. Concentration of Free Amino Acids (FAA) in Culture Medium of *Cowdria ruminantium*–Infected (P) and Uninfected (N) Sheep Brain Endothelial (SBE-189) Cells

		Concentration (nanomoles/100 µl) on Day (D) after Infection				
FAA[a]	Medium[b]	D1 (P)	D2 (P)	D3 (P)	D4 (P)	D5 (P)
D	9.1	9.7	10.5	11.7	12.7	16.2
E	20.5	27.6	31.9	40.9	49.4	67.5
S	11.8	18.1	22.1	25.1	29.7	37.1
G	–	–	–	21.4	24.9	33.2
H	13.1	18.6	22.1	20.7	23.0	25.8
R	34.3	23.9	1.9	1.2	1.6	1.5
T	46.5	44.7	46.0	46.9	47.8	54.5
A	21.6	32.2	37.4	40.6	47.7	54.9
P	6.9	24.3	31.5	36.9	40.0	50.9
Y	20.7	23.7	25.4	26.0	27.6	30.2
V	81.9	89.2	95.4	97.0	103.2	110.7
I	50.0	53.2	58.0	60.1	62.3	69.5
L	87.5	91.3	94.3	95.1	96.8	103.9
F	39.2	42.0	44.8	45.9	47.1	50.4
K	66.7	71.8	76.6	77.2	81.7	86.3
M	18.4	24.0	25.3	25.0	25.9	27.9
▲	1.2	2.6	2.2	3.4	3.0	1.5
N	3.8	7.5	9.9	10.6	11.5	13.3
Q	77.1	63.3	55.8	38.2	18.8	10.9

		Concentration (nanomoles/µl) on Day (D) after Infection				
FAA[a]	Medium[b]	1 (N)	2 (N)	3 (N)	4 (N)	5 (N)
D	9.1	9.8	11.0	11.0	10.2	10.4
E	20.5	40.9	50.1	50.8	48.4	51.2
S	11.8	26.9	36.4	35.7	32.6	34.6
G	–	–	27.1	27.4	29.5	33.5
H	13.1	16.8	21.1	20.6	23.1	24.1
R	34.3	21.2	0.8	0.8	1.1	1.4
T	46.5	47.2	51.0	50.0	45.0	44.8
A	21.6	45.6	54.0	53.9	51.5	55.9
P	6.9	39.4	52.9	53.7	51.2	59.9
Y	20.7	24.5	27.8	27.4	27.2	27.8
V	81.9	94.3	103.5	102.0	98.5	101.7
I	50.0	57.2	62.3	60.6	58.2	59.7
L	87.5	93.3	98.2	97.2	93.2	95.1
F	39.2	41.7	46.1	45.5	42.4	44.0
K	66.7	75.0	81.8	79.9	76.0	80.2
M	18.4	19.7	22.3	23.7	22.3	23.7
▲	1.2	0.2	0.2	0.2	0.2	0.2
N	3.8	7.8	9.7	9.8	10.5	10.2
Q	77.1	50.4	23.8	22.5	28.3	20.2

[a] Amino acids: D = aspartic acid, E = glutamic acid, S = serine, G = glycine, H = histidine, R = arginine, T = threonine, A = alanine, P = proline, Y = tyrosine, V = valine, I = isoleucine, L = leucine, F = phenylalanine, K = lysine, M = methionine, W = tryptophan, N = asparagine, Q = glutamine, ▲ = hydroxyproline.

[b] Medium control, sampled immediately prior to infection.

TABLE 2. Concentration of Free Amino Acids (FAA) in Culture Medium of *Cowdria ruminantium*-infected (P) and Uninfected (N) Bovine Saphenous Vein (BSC-793) Cells

		Concentration (nanomoles/100 μl) on Day (D) after Infection				
FAA[a]	Medium[b]	D1 (P)	D2 (P)	D3 (P)	D4 (P)	D5 (P)
D	8.9	3.4	4.4	5.4	5.3	5.8
E	23.7	21.0	26.8	40.4	44.3	56.8
S	15.7	33.9	35.8	40.8	38.3	42.6
G	–	59.0	63.7	73.2	68.5	77.2
H	17.6	20.8	21.6	24.2	22.0	23.3
R	39.3	1.6	1.4	1.8	1.5	1.5
T	46.4	62.3	65.4	79.6	70.9	79.4
A	29.1	21.1	28.4	39.9	42.4	53.3
P	26.1	14.1	16.1	19.9	21.0	24.6
Y	38.9	54.7	54.3	57.5	53.0	55.4
V	98.2	95.7	95.4	105.4	94.2	100.7
I	53.3	82.1	81.3	88.8	79.0	85.0
L	98.2	90.6	89.4	98.3	86.3	92.8
F	44.0	44.7	44.5	49.4	45.1	48.8
K	57.4	82.0	82.4	92.0	57.1	88.5
M	21.8	23.4	23.2	25.4	20.2	21.7
W	9.9	31.8	33.4	38.5	39.1	43.4
N	8.8	4.0	4.7	5.8	5.9	6.9
Q	352.0	318.0	286.9	283.2	231.5	218.8

		Concentration (nanomoles/100 μl) on Day (D) after Infection				
FAA[a]	Medium[b]	D1 (N)	D2 (N)	D3 (N)	D4 (N)	D5 (N)
D	8.9	11.8	14.4	11.7	11.9	13.2
E	23.7	39.1	56.0	57.4	63.9	70.5
S	15.7	25.2	31.6	26.8	29.0	37.8
G	23.7	39.1	56.0	57.4	63.9	70.5
H	17.6	20.7	25.8	21.4	21.7	20.9
R	39.3	9.2	3.8	3.6	3.5	3.5
T	46.4	51.6	59.5	53.1	52.9	52.0
A	29.1	42.3	61.3	61.2	67.0	74.3
P	26.1	49.4	69.7	68.5	72.7	76.1
Y	38.9	37.4	45.2	40.1	39.9	38.9
V	98.2	108.2	137.2	121.0	122.2	122.2
I	53.3	58.5	74.1	64.2	65.4	64.9
L	98.2	107.8	134.7	117.0	116.7	115.1
F	44.0	48.5	61.1	53.7	54.6	55.3
K	57.4	67.0	85.0	75.8	76.3	75.6
M	21.8	27.9	37.1	31.1	33.4	30.0
W	9.9	51.7	51.9	75.6	67.5	76.2
N	8.8	11.8	15.9	14.5	15.5	15.9
Q	352.0	227.7	245.0	175.8	153.1	128.4

[a] For amino acid names, see TABLE 1.
[b] Medium control, sampled immediately before infection.

TABLE 3. Concentrations of Major Protein Fractions (1–5) in Culture Medium of *Cowdria ruminantium*–Infected (P) and Uninfected (N) SBE-189 Cells[a]

Days PI	Peak (absorbance 214 nm)				
	1	2	3	4	5
0 (C)	0.018	0.035	0.008	0.010	0.009
1 (P)	0.014	0.034	0.006	0.010	0.010
2 (P)	0.014	0.036	0.006	0.010	0.010
3 (P)	0.014	0.040	0.006	0.010	0.013
4 (P)	0.014	0.039	0.006	0.008	0.010
5 (P)	0.014	0.040	0.006	0.007	0.010
1 (N)	0.023	0.033	0.012	0.016	0.018
2 (N)	0.024	0.043	0.010	0.014	0.015
3 (N)	0.026	0.046	0.010	0.013	0.015
4 (N)	0.027	0.044	0.011	0.014	0.020
5 (N)	0.030	0.047	0.011	0.015	0.024

[a] SBE = sheep brain endothelium; PI = postinfection; C = medium control sampled immediately before infection.

Some differences in concentrations of AAs between infected and uninfected cultures were seen over time (FIG. 1). Serine, threonine, and isoleucine were at higher concentrations in infected cells than in controls in both cell lines, whereas alanine and proline were at lower concentrations in infected cells than in uninfected ones. It is possible that the former AAs are synthesized by *C. ruminantium*, while the latter are utilized by the organism.

TABLE 4. Concentrations of Major Protein Fractions (A–D) in Culture Medium of *Cowdria ruminantium*–Infected (P) and Uninfected (N) BSV-793 Cells[a]

Days PI	Peak (absorbance 214 nm)			
	A	B	C	D
0 (MC)	0.023	0.027	0.053	0.043
1 (P)	0.044	0.045	0.000	0.024
2 (P)	0.050	0.055	0.000	0.020
3 (P)	0.045	0.047	0.000	0.026
4 (P)	0.055	0.053	0.000	0.030
5 (P)	0.051	0.052	0.000	0.025
1 (N)	0.043	0.043	0.050	0.035
2 (N)	0.042	0.042	0.053	0.038
3 (N)	0.048	0.047	0.050	0.035
4 (N)	0.055	0.054	0.054	0.030
5 (N)	0.042	0.053	0.070	0.038

[a] BSV = bovine saphenous vein; PI = postinfection; MC = medium control sampled immediately before infection.

nmole/100µl

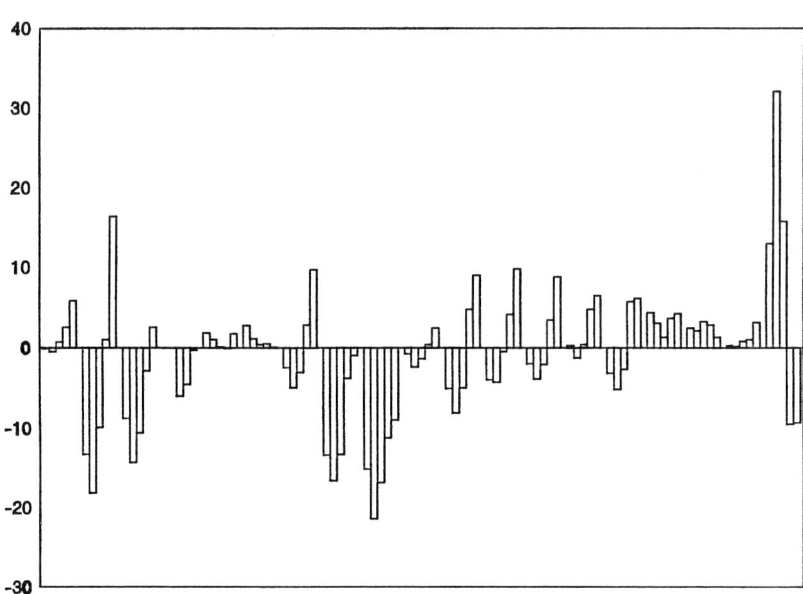

FIGURE 1. Concentrations of free amino acids in culture medium of *Cowdria ruminantium*-infected and uninfected SBE-189 cells. (Expressed as nmoles/100µl in infected cultures minus nmoles/100µl in uninfected cultures. For amino acid abbreviations, see TABLE 1.)

Proteins

In contrast to the SBE-189 cells infected with *C. ruminatium,* for which a decrease in concentration of all protein fractions was observed, for infected BSV-793 cells only one protein fraction showed significant reduction in concentration. In the latter cell line, two protein fractions present in both infected and uninfected cells increased from day 1 postinfection to day 2, while another fraction decreased. After this, they remained fairly constant. The observed decreases in protein concentration could be due to proteolysis or adsorption onto the cell membranes. The increases may be due to release of proteins by the cells.

These results, which reveal a constant increase in certain AAs, either through protein degradation or *de novo* synthesis or both, as well as decreases in major proteins of *Cowdria*-infected cultures, warrant further study. Such studies may reveal the biochemical strategies of this obligate parasite of eukaryotic cells, possibly leading to an understanding of the pathogenesis of heartwater infection. Our subsequent studies will deal with the augmentation of culture medium with specific metabolites with the aim of designing a medium specific for *C. ruminantium.*

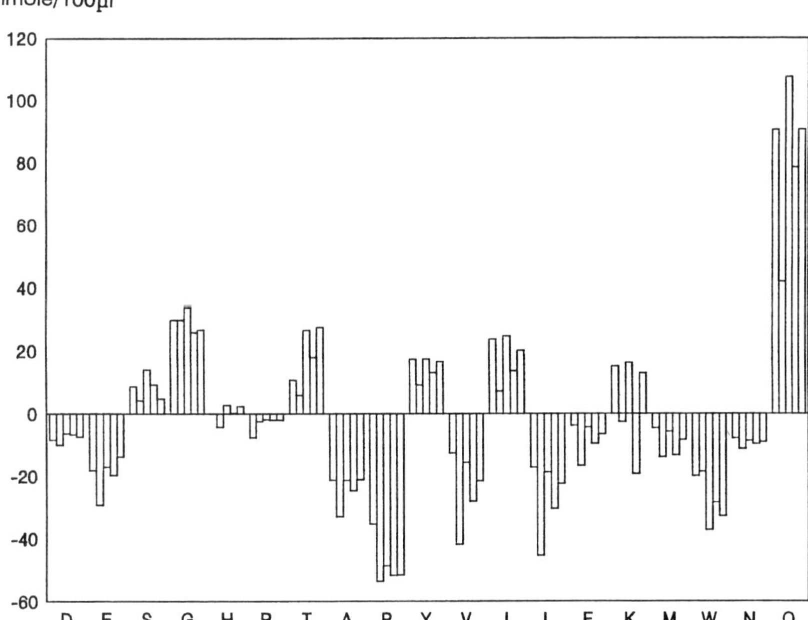

FIGURE 2. Concentrations of free amino acids in culture medium of *Cowdria ruminantium*-infected and uninfected BSV-793 cells. (See explanation for FIG. 1.)

SUMMARY

Cowdria ruminantium (Rickettsiales) causes heartwater in ruminants of Africa, and some islands off Africa and in the Caribbean Sea. The *in vitro* culture method for the organism devised in 1985, which provided for the first time a means for production of adequate quantities of live organisms and their products, is erratic and requires improvement. We studied depletion of amino acids (AAs) and major proteins in culture medium taken daily from infected and uninfected ovine and bovine vascular endothelial cell cultures. AAs of these samples were analyzed by Pico Tag reversed phase HPLC precolumn derivatization, and major proteins determined by capillary electrophoresis using a 57 cm × 75 µ fused silica tube at high pH. In both ovine and bovine cell cultures, significant depletion of arginine and glutamine occurred over a 5-day observation period regardless of whether they were infected or uninfected. This indicates that supplementation of nutrient media with these AAs might improve conditions for growth of the organism. Both AAs are essential for survival of cultured cells, and probably for the rickettsia (although the metabolism of *C. ruminantium* is poorly understood). Concentrations of several AAs increased in infected cultures, implying *de novo* synthesis and/or proteolysis on the part of the organism. In fact, several protein fractions did decrease in culture medium throughout the course of infection, while increasing or remaining unchanged in uninfected control cultures.

Absorbance (214 nm)

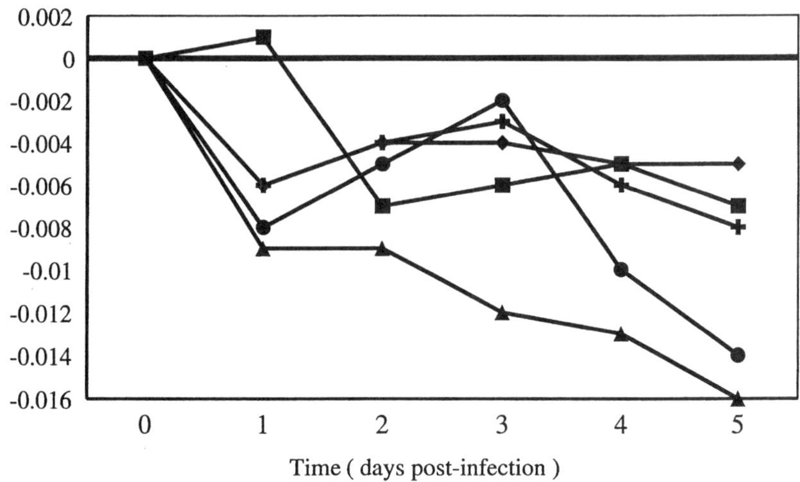

FIGURE 3. Concentrations of major protein fractions in culture medium of *Cowdria ruminantium*-infected and uninfected SBE-189 cells from day 1 to day 5 postinfection (infected minus uninfected; triangles = protein 1, squares = protein 2, diamonds = protein 3, plus signs = protein 4, circles = protein 5).

Absorbance (214 nm)

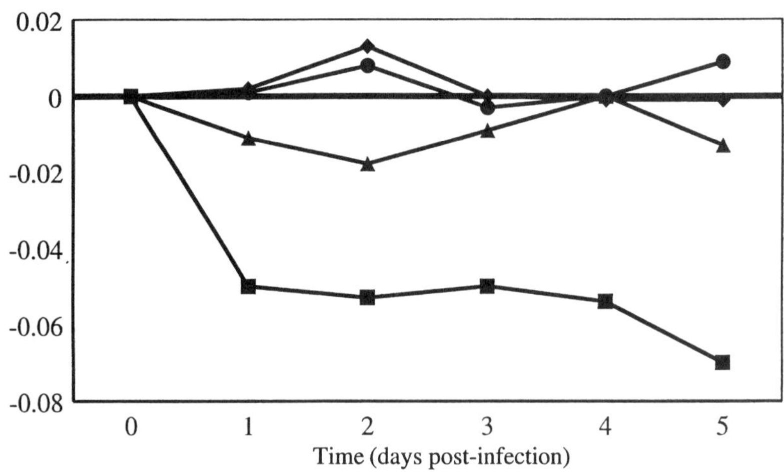

FIGURE 4. Concentrations of major protein fractions in culture medium of *Cowdria ruminantium*-infected and uninfected BSV-793 cells from day 1 to day 5 postinfection (infected minus uninfected; circles = protein A, diamonds = protein B, squares = protein C, triangles = protein D).

Proteolytic activity by *C. ruminantium* may be essential for nitrogen metabolism by the organism. It is suggested that studies such as these will facilitate the development of a specific medium for optimal *in vitro* growth of the heartwater organism, and may also lead to an understanding of the metabolic stratagem of *C. ruminantium*. This knowledge, in turn, could reveal the mechanism for pathogenesis of heartwater, with implications for control.

ACKNOWLEDGMENTS

We thank Mesdames E. Horn and S. Brett, Protozoology Division, Onderstepoort Veterinary Institute, for producing the cultures and Mr. N. J. Taljaard, Department of Biochemistry, University of Pretoria, for help with analyses of amino acids. Dr. J. C. Williams, Scientist-Director, US Public Health Service (Ret.) kindly provided helpful information concerning rickettsial metabolism.

REFERENCES

1. CAMUS, E. & N. BARRÉ. 1988. Heartwater. A Review. Publ. Off. Int. Epizooties. Paris, France.

2. BARRÉ N., G. UILENBERG, P. C. MOREL & E. CAMUS. 1987. Danger of introducing heartwater onto the American mainland: potential role of indigenous and exotic *Amblyomma* ticks. Onderstepoort J. Vet. Res. **54:** 405-417.

3. BEZUIDENHOUT, J. D., C. L. PATTERSON & B. H. J. BARNARD. 1986. *In vitro* cultivation of *Cowdria ruminantium,*. Onderstepoort J. Vet. Res. **54:** 113-120.

4. NEITZ, A. W. H., G. J. VILJOEN, J. D. BEZUIDENHOUT, P. T. OBEREM, W. VAN WYNGAARDT & N. M. J. VERMEULEN. 1986. The detection of antibodies to *Cowdria ruminantium* in serum and *C. ruminantium* antigen in *Amblyomma hebraeum* by an enzyme-linked immunosorbent assay. Onderstepoort J. Vet. Res. **53:** 39-41.

5. YUNKER, C. E., B. BYROM & S. SEMU. 1988. Cultivation of *Cowdria ruminantium* in bovine vascular endothelial cells. Kenya Veterinarian **12:** 12-16.

6. MARTINEZ, D., J. T. SWINKELS, E. CAMUS & F. JONGEJAN. 1990. Comparaison de trois antigènes pour les sèrodiagnostique de la cowdriose. Rev. Élev. Méd. Vét. Pays Trop. **43:** 159-166.

7. JONGEJAN, F. 1992. Serodiagnosis of *Cowdria ruminantium:* current status. *In* Recent Developments in the Control of Anaplasmosis, Babesiosis and Cowdriosis: 67-77. Publ. ILRAD. Nairobi, Kenya.

8. SEMU, S., S. M. MAHAN, C. E. YUNKER & M. J. BURRIDGE. 1992. Development and persistence of *Cowdria ruminantium* antibodies following experimental infection of cattle, as detected by the indirect fluorescent antibody test. Vet. Immunol. Immunopathol. **33:** 339-352.

9. WAGHELA, S. D., F. R. RURANGIRWA, S. M. MAHAN, C. E. YUNKER, T. G. CRAWFORD, A. F. BARBET, M. J. BURRIDGE & T. C. McGUIRE. 1991. A cloned DNA probe identifies *Cowdria ruminantium* in *Amblyomma variegatum* ticks. J. Clin. Microbiol. **29:** 2571-2577.

10. MAHAN, S. M., S. D. WAGHELA, T. C. McGUIRE, F. R. RURANGIRWA, L. A. WASSINK & A. F. BARBET. 1992. A cloned DNA probe for *Cowdria ruminantium* hybridizes with eight heartwater strains and detects infected sheep. J. Clin. Microbiol. **39:** 981-986.

11. YUNKER, C. E., S. M. MAHAN, S. D. WAGHELA, T. C. McGUIRE, F. R. RURANGIRWA, A. W. BARBET & L. A. WASSINK. 1993. Detection of *Cowdria ruminantium* by means of a DNA probe, pCS20 in infected bont ticks, *Amblyomma hebraeum*, the major vector of heartwater in Southern Africa. Epidemiol. Infect. **110:** 95-104.

12. YUNKER, C. E. Current status of *in vitro* cultivation of *Cowdria ruminantium.* Vet. Parasitol. (In press.)

13. BEZUIDENHOUT, J. D. & S. BRETT. 1992. Progress with the cultivation of *Cowdria ruminantium* in endothelial cells. *In* Recent Developments in the control of Anaplasmosis, Babesiosis and Cowdriosis: 141–147. Publ. ILRAD. Nairobi, Kenya.

14. YUNKER, C. E., S. M. VOGEL, J. L. DU PLESSIS, E. HORN & A. JOSEMANS. 1995. An *in vitro* infectivity assay for *Cowdria ruminantium:* some parameters of infection of cultured endothelial cells. Abstr. 2nd International Conference on Tick-Borne Pathogens at the Host-Vector Interface, Kruger Nat. Park, South Africa, 28 Aug.–1 Sept.

15. BIDLINGMEYER, B. A., S. A. COHEN & . T. L. TARVIN. 1984. Rapid analysis of amino acids using pre-column derivatization. J. Chromatogr. **336:** 93–104.

16. SCHWARTZ, H. E., R. H. PALMIERI, J. A. NOLAN & R. BROWN. 1982. Introduction to capillary electrophoresis of proteins and peptides. Publication of Beckman Instruments.

17. EAGLE, H. 1959. Amino acid metabolism in cell cultures. Science **130:** 432–437.

18. WILLIAMS, J. C. 1975. Unpublished information. As cited in WILLIAMS, J. C., E. M. PETERSON & J. C. COOLBAUGH. *In* Rickettsiae and Rickettsial Diseases. J Kazar, R. A. Ormsbee & I. N. Tarasevich, Eds.: 106. Publishing House of the Slovak Academy of Sciences. Bratislava, Slovakia.

19. MACPHERSON, I. & M. STOKER. 1961. Polyoma transformation of hamster clones—an investigation of genetic factors affecting cell competence. Virology **16:** 147–151.

20. BEZUIDENHOUT, J. D. 1987. The present state of *Cowdria ruminantium* cultivation in cell lines. Onderstepoort J. Vet. Res. **54:** 205–210.

21. 1993. Sigma Cell Culture Catalogue: 185–187. Publication of The Sigma Chemical Company, St. Louis, MO.

Recombinant Expression and Use in Serology of a Specific Fragment from the *Cowdria ruminantium* MAP1 Protein[a]

ARNOUD H. M. van VLIET,

BERNARD A. M. van der ZEIJST,

EMMANUEL CAMUS,[b] SUMAN M. MAHAN,[c]

DOMINIQUE MARTINEZ,[b] AND FRANS JONGEJAN[d]

Institute of Infectious Diseases and Immunology
Department of Bacteriology
Faculty of Veterinary Medicine
Utrecht University
Post Office Box 80.165
3508 TD Utrecht, the Netherlands

[b]IEMVT-CIRAD
BP 1232
97185 Pointe-à-Pitre Cedex
Guadeloupe, French West Indies

[c]University of Florida/USAID/SADC Heartwater Research Project
Post Office Box CY551 Causeway
Harare, Zimbabwe

INTRODUCTION

Cowdria ruminantium is the causative agent of cowdriosis (heartwater), a rickettsial disease of domestic and wild ruminants. The disease is transmitted by ticks of the genus *Amblyomma*.[1] Cowdriosis is endemic in sub-Saharan Africa, and there it is a major obstacle for upgrading local breeds of livestock with more productive susceptible exotic breeds.[1] The disease is also present on at least three Caribbean islands,[2–4] and poses a threat to livestock production on the North and South American mainland.[5]

Several serological tests have been developed for detection of antibodies to *C. ruminantium*. These include immunofluorescence assays,[6–8] an indirect ELISA based on *in vitro* cultured *C. ruminantium* organisms,[9] and two assays based on the immunodominant major antigenic protein of *C. ruminantium* (MAP1,[10] formerly called Cr32[11]).

[a]This work was supported by the Commission of the European Union, Directorate General XII, STD-3 program under contract no. TS3*-CT91-0007.

[d]Author to whom correspondence should be addressed.

These assays were either a competitive ELISA[12] or an immunoblotting assay.[13,14] However, cross-reactivity between *C. ruminantium* antigens and antibodies to several *Ehrlichia* species has been observed in all these tests.[13-17] This cross-reactivity is problematic, as shown by discrepancies between the distribution of *Amblyomma variegatum* ticks and the presence of antibodies to *C. ruminantium*.[13,18-20] As a consequence, a serological test with improved specificity was urgently required for a better understanding of the epidemiology of heartwater.

The recent cloning and expression of the MAP1 gene of *C. ruminantium*[21] allowed identification of immunogenic regions on the MAP1 protein, and evaluation of their use in the construction of a specific serological assay for cowdriosis. This study describes the development and validation of such an ELISA based on recombinant MAP1 antigens.

MATERIALS AND METHODS

Rickettsial and Bacterial Strains

C. ruminantium isolates used in this study were from Senegal (Senegal),[22] South Africa (Welgevonden),[23] and Guadeloupe (Gardel).[24] *C. ruminantium* was cultivated in bovine umbilical endothelial cells as described previously.[25] The *E. ovina* isolate has been described previously.[14] *Escherichia coli* strain M15[pREP4] was used as host strain for plasmid pQE9 (Qiagen Inc., Chatsworh, CA) and grown in LB-broth supplemented with 50 μg/ml kanamycin and 100 μg/ml ampicillin.

Molecular Techniques

All DNA manipulations, cloning experiments, and SDS-polyacrylamide gel electrophoresis were carried out according to standard methods.[26]

Antisera

Antibodies to *C. ruminantium* or *E. ovina* were raised in experimentally infected sheep by an infection and treatment method.[27] Infection with *C. ruminantium* or *E. ovina* was done by intravenous injection of virulent blood. Field serum panels used were obtained from Zimbabwe and several Caribbean islands. The Zimbabwean panel consisted of 55 immunoblot-negative and 111 immunoblot-positive sheep sera from heartwater-free areas on the highveld of Zimbabwe.[13] This panel was supplemented with 10 serum samples from sheep experimentally infected with *C. ruminantium* (Welgevonden isolate). The panel from the Caribbean islands consisted of 81 *C. ruminantium* indirect ELISA positive sheep sera and 7 indirect ELISA negative sheep sera.[9,20] This panel was supplemented with one serum sample from a sheep experimentally infected with *C. ruminantium* (Gardel isolate).

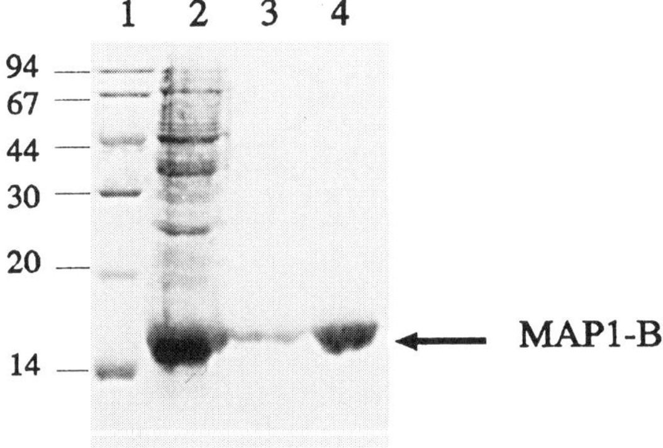

FIGURE 1. Purification of recombinant MAP1-B, demonstrated by using SDS-polyacrylamide gel electrophoresis. Lane 1, molecular weight markers; lane 2, *E. coli* cells expressing MAP1-B; lane 3, MAP1-B eluted at pH 8.0; lane 4, MAP1-B eluted at pH 6.3.

Serology

Identification and expression of MAP1-A and MAP1-B immunogenic regions are described elsewhere.[28] Briefly, PCR fragments encoding either amino acids 1-152 or 47-152 of the mature MAP1 protein[21] fused with 6 consecutive histidine residues were expressed in *E. coli* using the QiaExpress vectors (Qiagen Inc., Chatsworth, CA). Recombinant MAP1-A and MAP1-B fusion proteins were purified using Ni^{2+}-NTA agarose under denaturing conditions as described by the manufacturer (Qiagen Inc., Chatsworth, CA). Immunoblotting[14] and MAP1-B ELISAs[28] were performed as described elsewhere. Reactivity of antisera in the MAP1-B ELISA was expressed as a percentage of a high-titered positive control serum (% C++, see FIGS. 3, 4, and 5) as recommended elsewhere.[29]

RESULTS

Specificity of Immunogenic Regions on the C. ruminantium *MAP1 Protein*

Immunogenic regions on the MAP1 protein were identified after recombinant expression of overlapping fragments of the MAP1 protein in *E. coli* and subsequent immunoblotting with antisera to *C. ruminantium* and *E. ovina*.[28] Two immunogenic regions were identified, designated MAP1-A and MAP1-B.[28] They were expressed with 6 consecutive histidine residues (6*His) added at the N-terminus to obtain them in a suitable form for a serological assay. Using this system, recombinant MAP1 proteins could be easily purified from contaminating *E. coli* proteins due to binding of the 6*His tag to nickel. FIGURE 1 shows a typical purification of MAP1-B, with no contaminating proteins visible after elution from Ni^{2+}-NTA agarose (FIG. 1). FIGURE

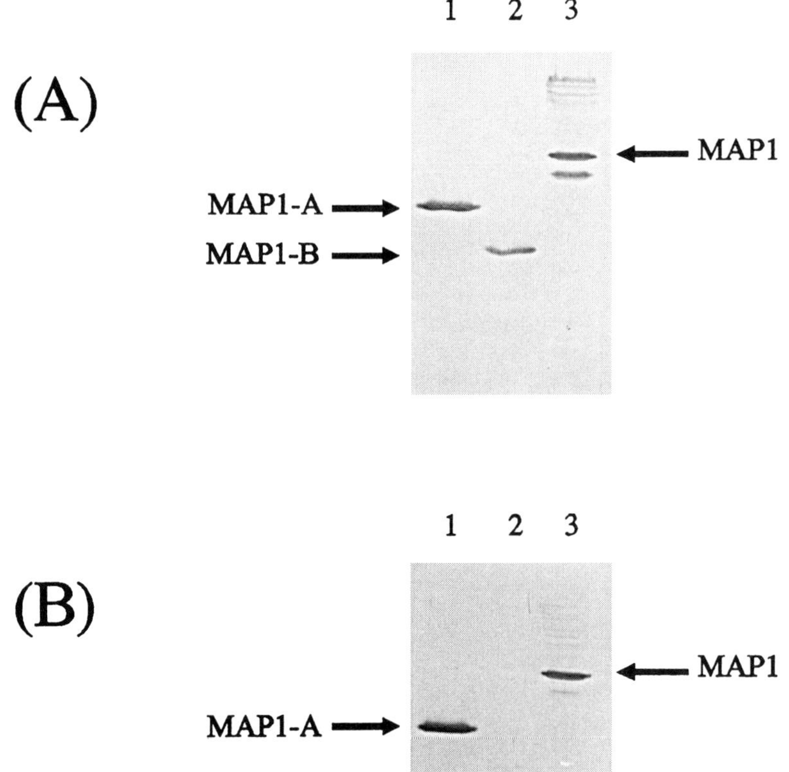

FIGURE 2. Reactivity of recombinant MAP1-A and MAP1-B on immunoblots with antisera to (A) *Cowdria ruminantium* (Senegal isolate) and (B) *Ehrlichia ovina*. Lane 1, MAP1-A; lane 2, MAP1-B; lane 3, *C. ruminantium* (Senegal isolate) antigens from endothelial cell cultures.

2 shows immunoblots of recombinant MAP1-A and MAP1-B reacted with antibodies to *C. ruminantium* and *E. ovina*. The MAP1-A region (which includes MAP1-B) contains one or more conserved epitopes which are recognized by antisera to *C. ruminantium, E. ovina, E. bovis, E. canis, E. chaffeensis,* and three MAP1-specific monoclonal antibodies (data not shown).

The MAP1-B region contains one or more epitopes with a high specificity for *C. ruminantium* antibodies. Only antibodies to *C. ruminantium, E. canis,* and *E. chaffeensis* reacted with MAP1-B. However, antisera to all *Ehrlichia* species that infect ruminants (*E. ovina, E. bovis,* and *E. phagocytophila*) did not react with MAP1-B.[28] The usefulness of a MAP1-B-based ELISA was evaluated with sera from sheep goats, and cattle experimentally infected with 9 different isolates of *C. ruminantium.* Reactivity of antisera with MAP1-B was expressed as a percentage of the reactivity of a serum sample of a sheep infected with *C. ruminantium* (C++). The cutoff value was fixed at twice the mean percentage reactivity of serum samples from noninfected control animals.[29,30] The mean percentage reactivity of negative sheep sera was 14.5

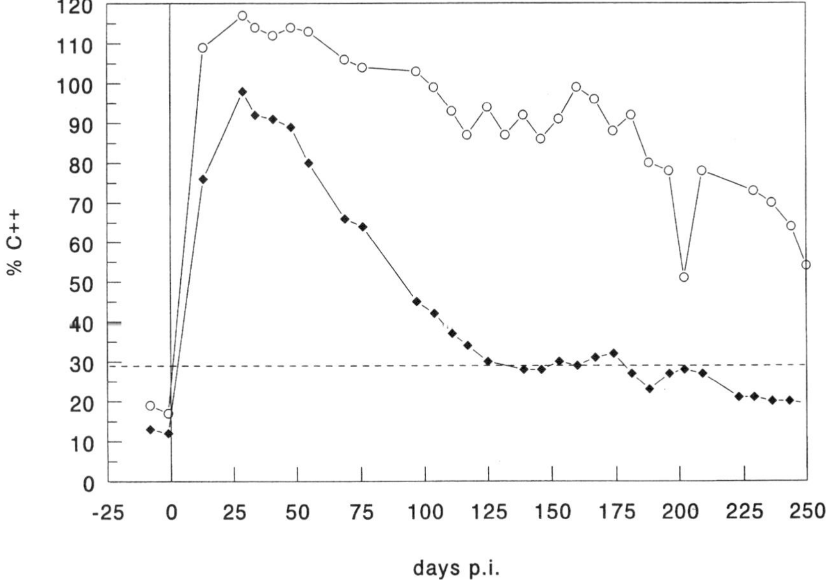

FIGURE 3. Reactivity of sequential serum samples obtained from two sheep experimentally infected with *C. ruminantium* (Senegal isolate). Reactivity with MAP1-B is expressed as percentage reactivity when compared to a positive control (C++). The dotted line represents the cutoff value. Circles: samples from sheep 110; diamonds: samples from sheep 144.

± 4.5 (*n* = 6). The cutoff value was therefore fixed at 29% for sheep sera. FIGURE 3 shows two representative examples of the reactivity of sheep antisera with MAP1-B, as a function of time postinfection (pi). To assess the sensitivity of the MAP1-B ELISA, serum samples of 64 experimental ruminants were taken 4 to 8 weeks after infection with *C. ruminantium,* and their reactivity with MAP1-B was determined. All serum samples were positive (data not shown).

Reactivity of Zimbabwean Sheep Sera with MAP1-B

It has been previously shown that sheep sera from areas in Zimbabwe that were considered to be free from *Amblyomma* ticks reacted with the MAP1 protein on immunoblots.[13] These reactions were considered to be false positives: 111 of these false-positive sheep sera, together with 55 negative sera and 10 sera from sheep experimentally infected with *C. ruminantium,* were retested using the MAP1-B antigen. The results obtained with these sheep sera are shown in FIGURE 4. False positivity was dramatically reduced, as only 9 out of 111 immunoblot-positive sera were detected by MAP1-B ELISA. Reactivity of the remaining false-positive sera ranged from 29%–43%. None of the 55 negative control sera exceeded the cutoff value, whereas all 10 positive controls reacted.

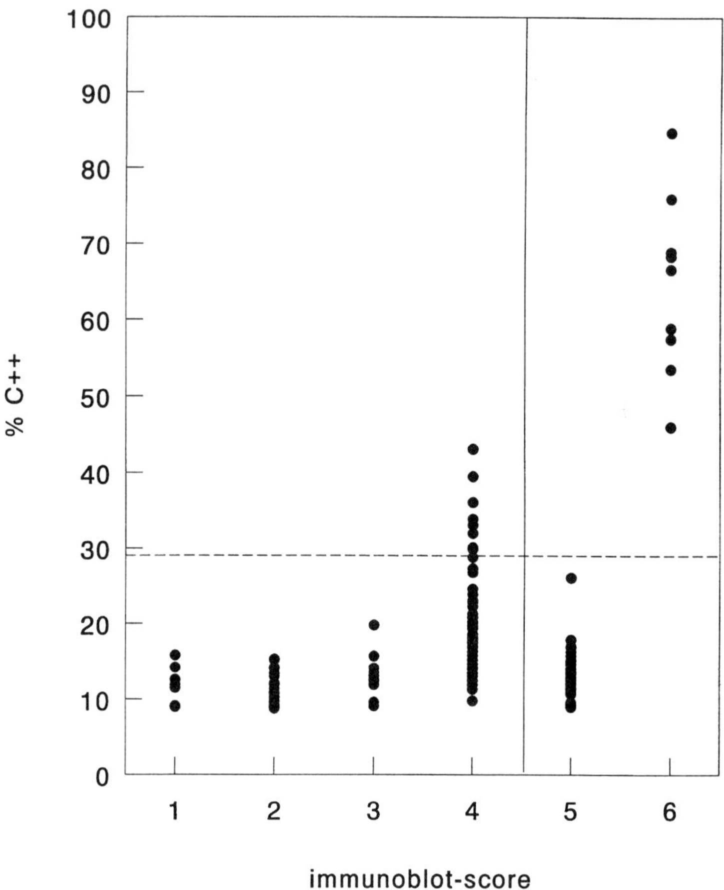

FIGURE 4. Reactivity of Zimbabwean sheep sera with MAP1-B in ELISA. Sera are grouped according to their immunoblot score.[13] Negative and positive controls are included in lanes 5 and 6, respectively. The dotted line represents the cutoff value. Lane 1, very weak positive sera ($n = 8$); lane 2, weak positive sera ($n= 26$); lane 3, positive sera ($n = 15$); lane 4, strongly positive sera ($n = 62$); lane 5, negative sera ($n = 55$); lane 6, positive controls ($n = 10$): sera from sheep experimentally infected with *C. ruminantium* (Welgevonden isolate).

Reactivity of Caribbean Sheep Sera with MAP1-B

Using competitive ELISA[18] and indirect ELISA,[20] sera from islands with no record of *C. ruminantium* infection and with low *A. variegatum* infestation[18,20] were positive. A panel of sheep sera previously tested with an indirect ELISA[20] were reacted with MAP1-B. FIGURE 5 shows the reactivity of these sera with MAP1-B. Sera are grouped per island. A high number of positive sera was found only with sheep from Antigua, where *C. ruminantium* has been isolated previously.[2] Using sera from islands where

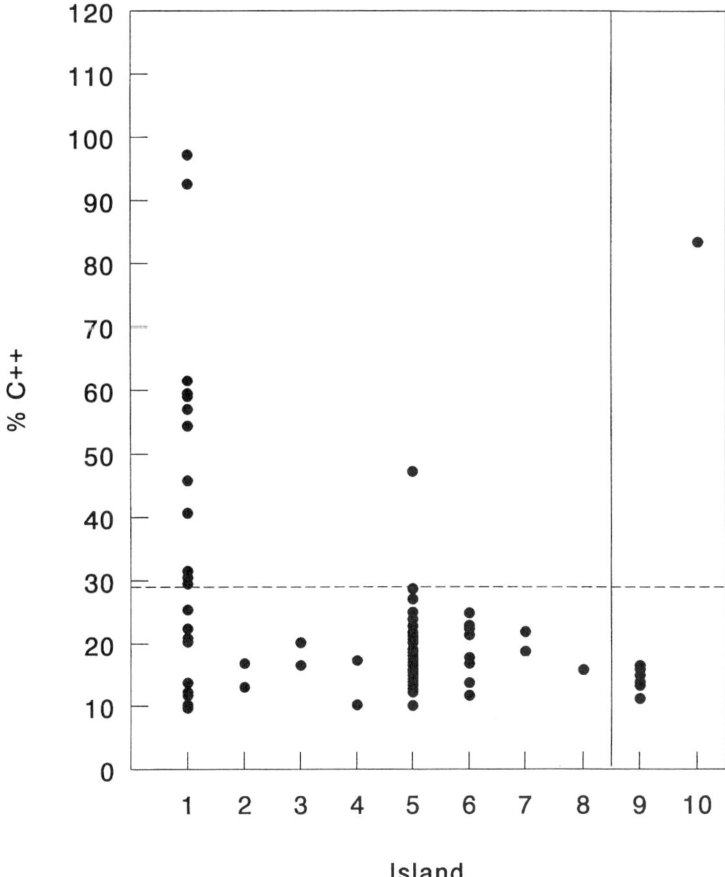

FIGURE 5. Reactivity of sheep sera from the Caribbean islands with MAP1-B in ELISA. Sera are grouped per island. Controls are included in lanes 9 and 10. The broken line represents the cutoff value. Lane 1, Antigua (n = 23); lane 2, Barbados (n = 3); lane 3, Dominica (n = 2); lane 4, Grenada (n = 2); lane 5, Martinique (n = 39); lane 6, Montserrat (n = 9); lane 7, St. Kitts (n = 2); lane 8, St. Vincent (n = 1); lane 9, negative control sera (n = 7); lane 10, positive control serum from a sheep experimentally infected with *C. ruminantium* (Gardel isolate).

C. ruminantium has not been isolated,[20] only one serum sample obtained from a sheep on the island of Martinique was positive.

DISCUSSION

An accurate assessment of the current distribution of *C. ruminantium* in Africa and in the Caribbean region is hampered by false-positive reactions in serological tests, probably due to cross-reactive antibodies to *Ehrlichia* species. Positive sera were found in areas or on islands with no record of *Amblyomma* tick infestation. In order to obtain a serological assay with an improved specificity, the MAP1 protein of *C. ruminantium* was analyzed for the presence of immunogenic regions. Two of these regions were identified, and designated MAP1-A and MAP1-B. MAP1-A contained one or more epitopes responsible for cross-reactivity between MAP1 and antisera to *Ehrlichia* species, whereas MAP1-B contained one or more *C. ruminantium*-specific epitopes. Antibodies cross-reacting with MAP1-B were found in antisera to *E. chaffeensis* and *E. canis*. No cross-reactive antibodies were found in antisera to other *Ehrlichia* species. The observed pattern of cross-reactivity correlates with relationships based on rRNA sequence level, indicating that *C. ruminantium*, *E. canis,* and *E. chaffeensis* form a cluster separate from all other *Ehrlichia* species.[31] As *E. canis* or *E. chaffeensis* have never been isolated from ruminants, the cross-reactivity between these two *Ehrlichia* species and *C. ruminantium* MAP1-B is not expected to hamper heartwater serology.

Recombinant MAP1-B antigen is recognized by antibodies to 9 isolates of *C. ruminantium* originating from different geographic regions.[28] In addition, in a MAP1-B-based ELISA, as in other tests, antibodies to *C. ruminantium* were detected in sera from experimentally infected cattle, sheep, and goats at 12-24 days pi, and these sera remained positive for 50-250 days pi.[28] The MAP1-B-based ELISA had an excellent sensitivity, as sera from 64 ruminants experimentally infected with *C. ruminantium* all reacted with MAP1-B. The improved specificity of a MAP1-B based serological test was demonstrated using sera from areas or islands considered to be heartwater free, which were positive in other serological tests. False positivity of Zimbabwean sera was reduced to 9 out of 111 sera, and false positivity of Caribbean sera was reduced to 1 out of 58. The reason for these remaining false positives is not known. However, the reactivity with MAP1-B of most of these sera only slightly exceeded the cutoff value, indicating the possibility that a more accurate determination of the cutoff value after testing of more serum samples may lead to a further elimination of false-positive reactions. Finally, isolation and characterization of the organism that causes these putative false-positive reactions will be required.

SUMMARY

The major antigenic protein (MAP1) of *Cowdria ruminantium* was screened for immunogenic regions by expression of overlapping recombinant DNA clones of the gene encoding the MAP1 protein. Two regions, designated MAP1-A and MAP1-B, were recognized by all antisera to 9 different isolates of *C. ruminantium*. MAP1-A contained one or more epitopes responsible for false-positive reactions with *Ehrlichia* antisera in several serological tests for cowdriosis. Cross-reactivity with MAP1-B was limited to antisera to *Ehrlichia chaffeensis* and *Ehrlichia canis*. Antisera to *Ehrlichia* species that infect ruminants (*E. bovis, E. ovina,* and *E. phagocytophila*) did not recognize MAP1-B. The sensitivity of an indirect ELISA based on MAP1-B was found to be excellent, since all sera from animals experimentally infected with *C. ruminantium* (64 out of 64) reacted with MAP1-B. Validation of this ELISA was carried out with field sera obtained from sheep raised in heartwater-free areas in Zimbabwe and from several Caribbean islands. Only 9 out of 111 samples from Zimbabwe, and 1 out of 58 samples from the Caribbean islands, which were considered to be false positives by immunoblot or indirect ELISA, reacted with MAP1-B. Thus, the ELISA based on MAP1-B is at present the most specific and sensitive serological test for cowdriosis.

ACKNOWLEDGMENTS

We thank Hans Nieuwenhuijs, Lilian Wassink, and Ron Dwinger for technical assistance.

REFERENCES

1. UILENBERG, G. & E. CAMUS. 1993. Heartwater (Cowdriosis). *In* Rickettsial and Chlamydial Diseases of Domestic Animals. Z. Woldehiwet & M. Ristic, Eds.: 293–332. Pergamon Press, Oxford, England.
2. BIRNIE, E. F., M. J. BURRIDGE, E. CAMUS & N. BARRÉ. 1985. Heartwater in the Caribbean: isolation of *Cowdria ruminantium* from Antigua. Vet. Rec. **116:** 121–123.
3. UILENBERG, G., N. BARRÉ, E. CAMUS, M. J. BURRIDGE & G. I. GARRIS. 1984. Heartwater in the Caribbean. Prev. Vet. Med. **2:** 255–267.
4. PERREAU, P., P. C. MOREL, N. BARRÉ & P. DURAND. 1980. Existence de la cowdriose à *Cowdria ruminantium*, chez les petits ruminants des Antilles Francaises (La Guadeloupe) et des Mascareignes (La Reunion et Ile Maurice). Rev. Élev. Méd. Vét. Pays Trop. **33:** 21–22.
5. BARRÉ, N., G. UILENBERG, P. C. MOREL & E. CAMUS. 1987. Danger of introducing heartwater onto the American mainland: potential role of indigenous and exotic *Amblyomma* ticks, Onderstepoort. J. Vet. Res. **54:** 405–417.
6. DU PLESSIS, J. L. & L. MALAN. 1987. The application of the indirect fluorescent antibody test in research on heartwater. Onderstepoort. J. Vet. Res. **54:** 319–325.
7. JONGEJAN, F., L. A. WASSINK, M. J. C. THIELEMANS, N. M. PERIÉ & G. UILENBERG. 1989. Serotypes in *Cowdria ruminantium* and their relationship with *Ehrlichia phagocytophila* determined by immunofluorescence. Vet. Microbiol. **21:** 31–40.
8. SEMU, S. M., S. M. MAHAN, C. E. YUNKER & M. J. BURRIDGE. 1992. Development and persistence of *Cowdria ruminantium* specific antibodies following experimental infection of cattle, as detected by the indirect fluorescent antibody test. Vet. Immunol. Immunopathol. **33:** 339–352.
9. MARTINEZ, D., S. COISNE, C. SHEIKBOUDOU & F. JONGEJAN. 1993. Detection of antibodies to *Cowdria ruminantium* in the serum of domestic ruminants by indirect ELISA. Rev. Élev. Méd. Vét. Pays Trop. **46:** 115–120.
10. BARBET, A. F., S. M. SEMU, N. CHIGAGURE, P. J. KELLY, F. JONGEJAN & S. M. MAHAN. 1994. Size variation of the major immunodominant protein of *Cowdria ruminantium*. Clin. Diagn. Lab. Immunol. **1:** 744–746.
11. JONGEJAN, F. & M. J. C. THIELEMANS. 1989. Identification of an immunodominant antigenically conserved 32-kilodalton protein from *Cowdria ruminantium*. Infect. Immun. **57:** 3243–3246.
12. JONGEJAN F., M. J. C. THIELEMANS, M. DE GROOT, P. J. VAN KOOTEN & B. A. M. VAN DER ZEIJST. 1991. Competitive enzyme-linked immunosorbent assay for heartwater using monoclonal antibodies to a *Cowdria ruminantium*-specific 32-kilodalton protein. Vet. Microbiol. **28:** 199–211.
13. MAHAN, S. M., N. TEBELE, D. MUKWEDEYA, S. SEMU, C. B. NYATHI, L. A. WASSINK, P. J. KELLY, T. PETER & A. F. BARBET. 1993. An immunoblotting diagnostic assay for

heartwater based on the immunodominant 32-kilodalton protein of *Cowdria ruminantium* detects false positives in field sera. J. Clin. Microbiol. **31:** 2729-2737.

14. JONGEJAN, F., N. DE VRIES, J. NIEUWENHUIJS, A. H. M. VAN VLIET & L. A. WASSINK. 1993. The immunodominant 32-kilodalton protein of *Cowdria ruminantium* is conserved within the genus *Ehrlichia*. Rev. Élev. Méd. Vét. Pays Trop. **46:** 145-152.

15. DU PLESSIS, J. L., J. D. BEZUIDENHOUT, M. S. BRETT, E. CAMUS, F. JONGEJAN, S. M. MAHAN & D. MARTINEZ. 1993. The sero-diagnosis of heartwater: a comparison of five tests. Rev. Élev. Méd. Vét. Pays Trop. **46:** 123-129.

16. LOGAN, L. L., C. J. HOLLAND, C. A. MEBUS & M. RISTIC. 1986. Serological relationship between *Cowdria ruminantium* and certain *Ehrlichia*. Vet. Rec.**119:** 458-459.

17. DU PLESSIS, J. L., E. CAMUS, P. T. OBEREM & L. MALAN. 1987. Heartwater serology: some problems with the interpretation of results. Onderstepoort J. Vet. Res. **54:** 327-329.

18. MULLER KOBOLD, A., D. MARTINEZ, E. CAMUS & F. JONGEJAN. 1992. Distribution of heartwater in the Caribbean determined on the basis of detection of antibodies to the conserved 32-kilodalton protein of *Cowdria ruminantium*. J. Clin. Microbiol. **30:** 1870-1873.

19. DE VRIES, N., S. M. MAHAN, U. USHEWOKUNZE-OBATOLU, R. A. I. NORVAL & F. JONGEJAN. 1993. Correlation between antibodies to *Cowdria ruminantium* (*Rickettsiales*) in cattle and the distribution of *Amblyomma* vector ticks in Zimbabwe. Exp. Appl. Acarol. **17:** 799-810.

20. CAMUS, E., D. MARTINEZ, L. BEAUPERTHUY, A. BENDERDOUCHE, S. COISNE, C. CORBETTE, N. DENORMANDIE, G. GARRIS, E. HARRIS, T. KING, L. LANNOY, A. LOUVET, E. LUNEL, M. MONTROSE, B. NISBETT, B. NYACK, J. ROBINSON, P. ROUCHOSSE, G. B. SWANSTON, B. THIÉBOT & G. D. THYE. 1983. Heartwater in Guadeloupe and in the Caribbean. Rev. Élev. Méd. Vét. Pays Trop. **46:** 109-114.

21. VAN VLIET, A. H. M., F. JONGEJAN, M. VAN KLEEF & B. A. M. VAN DER ZEIJST. 1994. Molecular cloning, sequence analysis, and expression of the gene encoding the immunodominant 32-kilodalton protein of *Cowdria ruminantium*. Infect. Immun. **62:** 1451-1456.

22. JONGEJAN, F., G. UILENBERG, F. F. J. FRANSSEN, A. GUEYE & J. NIEUWENHUIJS. 1988. Antigenic differences between stocks of *Cowdria ruminantium*. Res. Vet. Sci. **44:** 186-189.

23. DU PLESSIS, J. L. 1985. A method for determining the *Cowdria ruminantium* infection rate of *Amblyomma hebraeum:* effects in mice injected with tick homogenates. Onderstepoort. J. Vet. Res. **52:** 55-61.

24. UILENBERG, G., E. CAMUS & N. BARRÉ. 1985. Quelques observations sur une souche de *Cowdria ruminantium* isolée en Guadeloupe (Antilles francaises). Rev. Élev. Méd. Vét. Pays Trop. **38:** 34-42.

25. JONGEJAN, F. 1991. Protective immunity to heartwater (*Cowdria ruminantium* infection) is acquired after vaccination with in vitro-attenuated rickettsiae. Infect. Immun. **59:** 729-731.

26. SAMBROOK, J., E. F. FRITSCH & T. MANIATIS, Eds. 1989. Molecular Cloning, a Laboratory Manual. 2nd Edit. Cold Spring Harbor Laboratory. Cold Spring Harbor, NY.

27. JONGEJAN, F., M. J. C. THIELEMANS, C. BRIÈRE & G. UILENBERG. 1991. Antigenic diversity of *Cowdria ruminantium* isolates determined by cross-immunity. Res. Ves. Sci. **51:** 24-28.

28. VAN VLIET, A. H. M., B. A. M. VAN DER ZEIJST, E. CAMUS, S. M. MAHAN, D. MARTINEZ & F. JONGEJAN. 1995. Use of a specific immunogenic region on the *Cowdria ruminantium* MAP1 protein in a serological assay. J. Clin. Microbiol. **33:** 2405-2410.

29. WRIGHT, P. F., E. NILSSON, E. M. A. VAN ROOIJ, M. LELENTA & M. H. JEGGO. 1993. Standardisation and validation of enzyme-linked immunosorbent assay techniques for the detection of antibody in infectious disease diagnosis. Rev. Sci. Tech. Off. Int. Epiz. **12:** 435-450.

30. MARTINEZ, D., B. MARI, G. AUMONT & T. VIDALENC. 1993. Development of a single dilution ELISA to detect antibody to *Dermatophilus congolensis* in goat and cattle sera. Vet. Microbiol. **34:** 47–62.
31. VAN VLIET, A. H. M., F. JONGEJAN & B. A. M. VAN DER ZEIJST. 1992. Phylogenetic position of *Cowdria ruminantium* (*Rickettsiales*) determined by analysis of amplified 16S ribosomal DNA sequences. Int. J. Syst. Bacteriol. **42:** 494–498.

Genetic Resistance of Creole Goats to Cowdriosis in Guadeloupe

Status in 1995

E. CAMUS,[a] J. C. MAILLARD,[a] G. RUFF,[b] L. PEPIN,[c]
M. NAVES,[d] AND G. MATHERON [e]

[a]*CIRAD-EMVT*
BP 515
97165 PAP Cedex
Guadeloupe, French West Indies

[b]*University of Berne*
Berne, Switzerland

[c]*National Institutes of Health*
Bethesda, Maryland

[d]*INRA*
Guadeloupe, French West Indies

[e]*CIRAD*
Montpellier, France

INTRODUCTION

A genetic predisposition to resistance (R) and to susceptibility (S) for cowdriosis has been demonstrated in certain Creole goat lines from guadeloupe.[1,2] After experimental infection, the observed resistance rate (in terms of survival) of goats removed decades ago from an endemic area was 0–25%, while it was 54% in a population isolated from the disease for 10 years and reached 78% in flocks raised in an endemic area. In an experiment conducted on 198 goats between 2 and 8 months old from the same flock, the resistance varied greatly (22 to 88%) depending on the sire, with a heritability of 0.26 for paternal half-sibs and 0.55 for full-sibs.

In order to follow the transmission of the R or S traits in the offspring and to identify genetic markers of R/S, a complete crossbreeding scheme between R and S males and females was applied.

MATERIAL AND METHODS

Goats

All goats used in the experiment were Creole goats.[3] Goats from the island of "Les Saintes" close to Guadeloupe, have been evolving in an environment free of cowdriosis and *Amblyomma variegatum* for the last 100 years. Previous experimental infection with ticks infected as larvae and fed on goats at the nymphal stage gave a 100% mortality in these goats from Les Saintes.[2] They represent the "presumed susceptible" group (PS) in the crossbreeding.

Creole goats from the Duclos farm that survived after an experimental infection with ticks or with blood were considered resistant (R) to cowdriosis.

Experimental Design

1	R	male	n. 84647	× 15	R	females
1	R	male	n. 84619	× 15	PS	females
1	PS	male	n. 88961	× 15	R	females
1	PS	male	n. 88956	× 15	PS	females

The goats were bred three times in January 1991, 1992, and 1993. The offspring were weighed monthly.

The 6-month-old kids (age of maturity for the immune system) were experimentally infected to determine their resistance or susceptibility to cowdriosis. Later, all the PS dams and sires were also experimentally infected.

Experimental Infection

Amblyomma variegatum larvae were fed on goats experimentally infected 8 days before with the local Gardel stock of *Cowdria ruminantium*. The tick blood meal lasted for an average of 8 days, covering the rickettsaemia[4] (at least from day 12 to 16 or more).

Engorged larvae were then maintained at 25 °C and 90% humidity, and moulted into nymphs. Batches of 20 nymphs were used as infecting material by feeding on the experimental goats. Daily rectal temperature was recorded. In case of death brain smears were carefully examined under microscope after Giemsa staining.

In case of survival, antibodies to *Cowdria* were looked for with an indirect ELISA test.[5] Clinical reactions, incubation period, and rectal temperature were registered on the offspring. The results were analyzed together with those of previous experiments.[2]

Genetic Markers

Polymorphisms in several immunogenetic systems were investigated and compared to experimental infection issue.

- 3 blood proteins (hemoglobin, transferrin, and X protein) were analyzed by electrophoresis techniques.

- 6 blood groups (A, B, C, R, E, and F) were serotyped using 28 monospecific reagents.

- 1 milk protein (the alpha S1 casein), which is highly polymorphic, was studied by microsatellites.

- the Caprine leucocyte antigens system (CLA) in the MHC, was typed using purified reagents detecting 27 class I and 7 class II specificities.

- 9 polymorphic microsatellites in genomic DNA (INRA 003, 005, 006, 011, 023, 036, 040, 063, ILSTS 08) were compared using molecular techniques.[6,7]

Sperm Collection

In order to store semen from 9 PS and R sires, before the experimental infection their sperm was collected for storage in deep-frozen form.

Sires were reared in individual pens and separately presented to a female. During a training period, an assessment of their behavior and semen quality was realized. The preparation of the semen for storage was decided after 5 days of successful training. Semen preparation was performed using the French technique.[8]

Controls of the semen after collection included the measurement of the ejaculate volume and the concentration of the ejaculate; the total number of spermatozoa was calculated from these two measures.

Visual assessment of the sperm motility and of the percentage of living cells was done before and after storage.

RESULTS

Parents

After the last crossbreeding in 1993, the PS parents were infected:

- Of 28 PS dams, 7 resisted the experimental infection (R) and 21 died from cowdriosis (S), meaning a resistance rate of 25% among the dams.

- Of 2 PS sires, 1 resisted the experimental infection (R) and the other one died before challenge and could not be classified as R or S; its descendants could therefore not be included in the results.

Offspring

From 160 kids, only 41 could be used for analysis because:

- a sire could not be classified as R or S.

- the mortality rate due to hemonchosis was high and difficult to control (anthelmintic resistance), particularly because the goats were bred outside in very humid conditions (annual rain fall of 3000 mm).

TABLE 1. Clinical Reaction and Other Parameters According to R/S Goats

	R	S
n	99	83
Incubation period (days)	15.10 ± 5.76	15.50 ± 5.44
Rectal temp. (°C)	41.57 ± 0.52	41.53 ± 0.88
Birth mass (kg)	1.70	1.62
Daily mass gain (g)	58.8	59.7
Resistance rate		
Sex ⎰ Males	53%	47%
⎱ Females	58	42%
Litter size 1	58%	42%
2–3	54%	46%

- several offspring showed incompatible genetic markers (blood groups and CLA) compared to their parents.

Because of the absence of S sire, neither SS nor SR group could be investigated so far. Only two groups of offspring could be evaluated:

20 RS (resistant sire, susceptible dams)
21 RR (resistant sire, resistant dams)

Of 20 RS kids, 11 died after the experimental infection, 9 survived. The resistance rate was 45%.

Of 21 RR kids, 5 died and 16 survived. The resistance rate was 76%.

The difference between resistance rates in both groups was significant (chi^2 = 4.2; $\propto$ = 4%).

Genetic Transmission of R/S Character

Twenty-thee dams had 2 successive offsprings:

of 15 dams giving one R kid, 14 gave a second R kid.
of 8 dams giving an S kid, only 2 gave a similar one.

Clinical Reaction and Other Parameters

The results are presented in TABLE 1.
There were no statistically significant differences between R and S for these parameters.

TABLE 2. CLA Markers in R/S goats

	Specificities	Frequencies		Probabilities		
		R $n = 114$	S $n = 74$	χ^2	p (1 l.d.)	Signif.
R	CLY panleuco	0.25	0.04	11.47	<0.001	+++
	Be 1	0.42	0.22	8.38	<0.001	+++
	Be 9	0.07	0.22	8.58	<0.001	+++
S	Be 23	0.04	0.16	7.63	<0.001	+++
	Be 22	0.08	0.17	4.06	<0.05	++

Genetic Markers

Among the 4 families observed, 188 animals including parents and offspring were challenged with *Cowdria ruminantium*, and classified as resistant (114) or susceptible (74).

The only significant differences between R and S were observed on the CLA system (TABLE 2).

The most significant difference occurred with the CLY specificity which is a panleucocyte antigen. It appears highly correlated with the R character.

Concerning the CLA specificities, the Be 1 specificity correlated with the R character and 3 specificities Be 9, Be 22, and Be 23 correlated with the S one.

Concerning the other genetic systems studied, no differences were observed between R and S goats.

Sperm Collection

The sperm was collected from 9 sires, 5 R and 4 S. The mean semen characteristics (volume: 0.82 ml; concentration: $4.81-10^9$ spermatozoa/ml) were similar for both groups. A total of 43 ejaculates were successfully deep frozen. Postthawing motility and percentage of living cells were correct for all the bucks (3.2 and 76% respectively).

Between 72 and 140 straws per buck were stored in liquid nitrogen, representing a total of 490 straws from R sires and 387 from S sires.

DISCUSSION

There is a clear difference between the mortality rate of RR kids compared to RS ones. Unfortunately we lack SS and SR groups, because one of our "presumed susceptible" sires was resistant, the other could not be tested.

The only way to get SS and SR kids is to inseminate goats with sperm from PS and R sires and, after birth, to check the resistance/susceptibility of the dams by experimental infection. Goats from "Les Saintes" are more likely susceptible, but even with experimental tick infection, they are not 100% susceptible, contrary to what was observed previously[2] (mortality rate of 22/22).

The other possibility will be embryo splitting. Embryos from PS dams inseminated by sperm from S sires, could be split. One half of the embryo could then be used to test for susceptibility, the other half to procreate. This strategy could ensure the status of progenitors for further breeding experiments.

Among the 114 animals classified as R, 5 males and 17 females showing simultaneously the markers of resistance CLY and Be 1, without any marker of susceptibility, were selected to constitute the basis of a possible program of selection of animals genetically resistant to cowdriosis.

The mode of infection with infected ticks, used because of its similarity with the natural infection and because it was supposed to induce a 100% mortality on Creole goats from "Les Saintes," is tedious and unstandardizable. A better approach would be to use a known number of *Cowdria* elementary bodies from cell culture, inducing an incubation period of approximately 12 days comparable to the natural disease (2–3 days for the nymph to attach and to inoculate the organisms and then 12 days before the first rise of temperature). This aspect of standardization of the dose could well be important because there could be a link between the dose inoculated and the intensity of the reaction.

There is no difference between R and S goats in terms of clinical reaction after experimental infection: the reaction is equally severe for both groups and the only difference lies in death/survival. It means that the incubation period and the level of rectal temperature cannot be used to treat goats before death. Among the biological parameters, alpha interferon observed in resistant cattle[9] has not been detected in goats. It would probably be interesting to look for biological parameters associated with nervous symptoms, which are always followed by death.

Compared to previous results,[10] and with more data now, CLY panleuco and Be 1 are confirmed markers of R and Be 9 and Be 22 markers of S. An additional marker of S is Be 23. The use of microsatellites as genetic markers in goats is just beginning, with microsatellites found in cattle. Many other bovine microsatellites remain to be tested, first to assess their presence in goats, and then to look for a possible correlation with resistance/susceptibility to cowdriosis. These markers are promising because of their high polymorphism and because of the number of microsatellites all along the genomic DNA.

Collection and deep freezing of the semen were correctly achieved for all the bucks, but better results were obtained with younger bucks, between 10 and 17 months old.

Few references exist on deep frozen storage of Caprine semen from a native breed in a tropical country, and most are from India.[11] The technique described in the FAO training manual proved to be useful in the small laboratory installed for this purpose. Some problems of adaptation occurred, such as the availability of pure water, resolved by the use of mineral water.

CONCLUSION

The resistance of creole goats to cowdriosis can be improved by breeding resistant sires to resistant dams.

The R character seems to be better genetically transmitted by dams than the S one.

Sperm has been collected from 4 S sires and 3 R sires, and stored in liquid nitrogen in order to create additional SR and SS progeny and to complete our experimental design.

Several CLA seem to be genetic markers of R (Be 1, CLY) and of S (Be 9, Be 22, Be 23).

Further powerful tools like microsatellites and probes defining other interesting loci possibly involved in disease pathogenesis are now at hand and will be included in future experiments.

When Creole goats resistant to cowdriosis will be selected, it will be necessary to check their resistance to hemonchosis which is the most pathogenic disease for goats in the Lesser Antilles.

SUMMARY

A genetic predisposition to resistance (R)/susceptibility (S) has been demonstrated for cowdriosis in certain goat lines. In order to identify genetic markers of R/S and to follow their transmission to the offspring, 4 groups of sires and dams were crossbred in 1991, 1992, and 1993: Rr, RS, SR, and SS. The offspring were challenged at the age of six months with subsequent challenge of the S parent. From 28 presumed S dams, 7 turned out R and from 2 presumed S sires, 1 was R and the other undefinable.

The resistance rate was 76% for 21 RR kids and 45% for 20 RS kids. The R character seems to be better transmitted by dams than the S one. No difference was observed between R and S goats in terms of incubation period and body temperature level after challenge. There was no statistical effect found of sex, litter size, or birth mass on the R/S character; only a slight effect on daily mass gain was observed. Sperm has been collected from 4 S sires (died after challenge) and 5 R sires, and stored in liquid nitrogen in order to conceive additional SR and SS progeny. Several Caprine leucocyte antigens seem to be genetic markers of R (Be 1, CLY) and of S (Be 9, Be 22, Be 23). Further powerful tools like microsatellites and probes defining other interesting loci possibly involved in disease pathogenesis are now at hand and will be included in future experiments.

REFERENCES

1. MATHERON, G., N. BARRE, E. CAMUS & J. GOGUE. 1987. Genetic resistance of Guadeloupe native goats to heartwater. Onderstepoort J. Vet. Res. **54:** 337-340.
2. MATHERON, G., E. CAMUS, N. BARRE & J. GOGUE. 1991. Résistance à la cowdriose de chèvres créoles en Guadeloupe. Bilan en 1988. Rev. Elev. Méd. Vét. Pays Trop., (n. spécial): 145-153.
3. CHEMINEAU, P., Y. COGNIE, A. XANDE, F. PEROUX, G. ALEXANDRE, F. LEVY, E. SHITALOU, J. M. BECHE, D. SERGENT, E. CAMUS, N. BARRE & J. THIMONIER. 1984. Le "cabrit" créole de Guadeloupe et ses caractéristiques zootechniques: monographie. Rev. Elev. Méd. Vét. Pays Trop. **37:** 225-238.
4. CAMUS, E. 1987. Etude épidémiologique de la cowdriose (heartwater) en Guadeloupe. Thèse de doctorat ès Sciences. Paris. Sud.

5. MARTINEZ, D., S. COISNE, C. SHEIKBOUDOU & F. JONGEJAN. 1993. Detection of antibodies to *Cowdria ruminantium* in the serum of domestic ruminants by indirect ELISA. Rev. Elev. Méd. Vét. Pays Trop. **46:** 115-120.
6. PEPIN, L. 1994. Recherche des polymorphismes génétiques chez les caprins: application à l'étude de la diversité des populations, au contrôle de filiation et à la resistance á la cowdriose. Thése de doctorat. Paris.
7. PEPIN, L., Y. AMIGUES, A. LEPINGLE, J. L. BERTHIER, A. BENSAID & D. VAIMAN. 1995. Sequence conservation of microsatellites between *Bos taurus* (cattle), *Capra hircus* (goat) and related species. Examples of use in percentage testing and phylogency analysis. Heredity **74:** 53-61.
8. FAO. 1991. Training Manual on Artificial Insemination in Sheep and Goats. Animal Production and Health Paper n. 83. FAO. Rome, Italy.
9. TOTTE, P., F. JONGEJAN, A. L. W. DE GEE & J. WERENNE. 1994. Production of alpha interferon in *Cowdria ruminantium* infected cattle and its effect on infected endothelial cell culture. Infect. Immun. **62:** 2600-2604.
10. RUFF, G., J. C. MAILLARD, E. CAMUS, G. MATHERON & E. DEPRES. 1993. Occurrence of Caprine leucocyte antigen (CLA) in Creole goats susceptible/resistant to heartwater. Rev. Elev. Méd. Vét. Pays Trop. **46:** 205-207.
11. SAHNI, K. L. 1987. Practical aspects of artificial insemination of goats in India. Proceedings of the IV International Conference on Goats. Brasilia, March 8-13, I: 549-569.

Appropriate Strategies for the Control or Eradication of Ticks and Tick-Borne Diseases

A. WILSON

*Scientific Services
Cooper (Zimbabwe) Ltd.
Post Office Box 2699
Harare, Zimbabwe*

INTRODUCTION

The Effects of Ticks and Tick-Borne Diseases

Over 80% of the world's cattle population of 1281 million are at risk from ticks and tick-borne diseases (TBD).[1] Over a decade ago, McCosker estimated global costs of control and productivity losses as US $7000 million annually (US $7/head/annum).[2] In spite of vast sums of money spent on tick control, the problem is as large as ever.

Ticks can cause physical damage which affects productivity and they transmit the TBD anaplasmosis, babesiosis, heartwater (HW), and theileriosis including east coast fever (ECF).

Endemic Stability

Endemic stability to TBD develops when there is sufficient tick transmission to infect all calves while they are protected from clinical disease by passive calfhood immunity. Endemically unstable situations arise when there are insufficient ticks to ensure that all calves receive challenge.[3] In the presence of endemic stability, tick control can be minimal and strategic to control the physical effects of the ticks.

If susceptible cattle are introduced into an endemically stable situation, they must be protected by vaccination against the prevailing TBD or total tick control must be practiced.[1]

Eradication of TBD

The eradication of a TBD from a region is technically possible, and it has been demonstrated in the cases of ECF and HW in Zimbabwe[5] and possibly in the case of HW from some islands in the Caribbean. HW was introduced into Guadaloupe about 1830 with cattle from Senegal infested with *Amblyomma variegatum* ticks.[6]

Eradication of these ticks from the Caribbean would enable an increase in animal productivity and freedom from HW and dermatophilosis, of particular benefit to exotic highly productive breeds. The threat of the spread of HW to the mainland countries of North, Central, and South America would be removed. In this context, it is of interest to look at the experience of Zimbabwe while remembering that technology in Africa may not be applicable ten thousand miles away.

The Eradication of ECF from Zimbabwe

An example of the spread of a TBD in a susceptible cattle population has been seen in Zimbabwe (formerly Rhodesia) where the ecology is suitable for tick life and single and multihost ticks are numerous. Prior to the arrival of the first white settlors in 1890, some TBD were present, but they were not a problem because the cattle were endemically stable for these diseases. In 1901, east coast fever caused by *Theileria parva* was introduced with a consignment of cattle from Tanganyika. The local cattle were fully susceptible, the vector *Rhipicephalus appendiculatus* was present, and the disease spread to almost every part of the country carried by the trek oxen which were the sole means of transport at that time. The result was devastating, and up to 95% of some herds died in the first wave of infection.[5] Stringent quarantine measures were imposed, and all movement of cattle was prohibited except with an official written permit. In Zimbabwe, this permit system is still in force for general disease control.

Acaracide Use in ECF Control

It was found that ECF was transmitted by *Rhipicephalus appendiculatus,* and by this time, arsenical acaracides for the treatment of cattle in plunge dip tanks were available. In 1914, a compulsory dipping ordinance was introduced which required all cattle in designated areas to be dipped in 0.16 to 0.20% arsenious oxide solutions at intervals of 7 days or less throughout the year.

The quarantine, movement of livestock, and dipping regulations were introduced by government at the request of farmers.

After the first onslaught, the situation stabilized. Compulsory dipping at weekly intervals during the summer tick season and two weekly in the winter was extended to most of the country. Zimbabwe probably had the most intensive legally enforced and best supervised dipping system in the world. It was an offense to own tick-infested cattle and a more serious offense if they were in a public place such as a sale yard. All cattle in Zimbabwe are still officially inspected every 4 to 16 weeks depending on the disease situation. By diligent application and regular inspection, ECF was eradicated from most of the country by 1935, but the disease persisted in the Melsetter/Chipinga District on the eastern border of the country. This remaining focus remained a threat from which spread could occur.

Very Intensive Tick Control

A policy of very intensive supervision and tick control was introduced on these farms and ranches. Inspectors attended every dipping to count and record the cattle.

If even one animal was missing, all cattle had to be held until it was found. All sickness or death had to be reported immediately so that a veterinary official could take blood, lymph node, and spleen smears for diagnosis or to detect ECF. If an animal was missing or died and became putrid before effective specimens could be taken, the farm was quarantined for two years.[8] These draconian policies were highly unpopular, but the Veterinary Department kept up the pressure and in 1954, 50 years after the outbreak began, the last case of ECF was diagnosed by H. S. de Bruijn the Superintending Inspector, confirmed by the writer, Veterinary Officer in an adjoining district, and then by Dudley Lawrence, an authority on the theilerial diseases, at that time Director of Veterinary Services.

Eradication of the Disease but Not the Vector

The disease was declared eradicated in 1960, but according to de Bruijn and Williamson (who had been a District Veterinary Officer in the area),[8] the vector was not eradicated except in one or two areas where the ecology was unsuitable for its survival.

A reservoir of vector ticks persisted in a small, rarely seen, population of antelope.

It also appeared that for the time being, *Boophilus decoloratus* was eradicated and most cattle were susceptible to babesiosis and anaplasmosis.[7]

This situation has been described in detail to show that in this case, very intensive long-term control eradicated the disease but not the vector.

The Benefits from Tick Control

Conscientious tick control has been practiced by most cattle owners in Zimbabwe for the past 70 years. The national herd has grown from 20,000 in 1905 to 6 million and the number of dip tanks and mechanical spray races has increased from zero to 6000 (FIG. 1).[5]

The national herd was built around the dip tank and the legislation that enforced the control of ticks and TBD and controlled the movement of cattle from place to place. However, for political and economic reasons, this strict legal tick control is no longer enforced, but it is still illegal to move tick-infested cattle from place to place.

HEARTWATER IN ZIMBABWE

Similar success was achieved in controlling HW, but the control was lost due to changes in circumstances. This disease was first confirmed in the southern part of Zimbabwe in 1927.[5] It spread widely until 1935 but then, by regular dipping and quarantine, the affected areas were reduced and by 1975 they were relatively small as shown in FIGURE 2.[5]

It was considered impossible to eradicate from the remaining areas because of numerous alternate wildlife hosts both large and small.

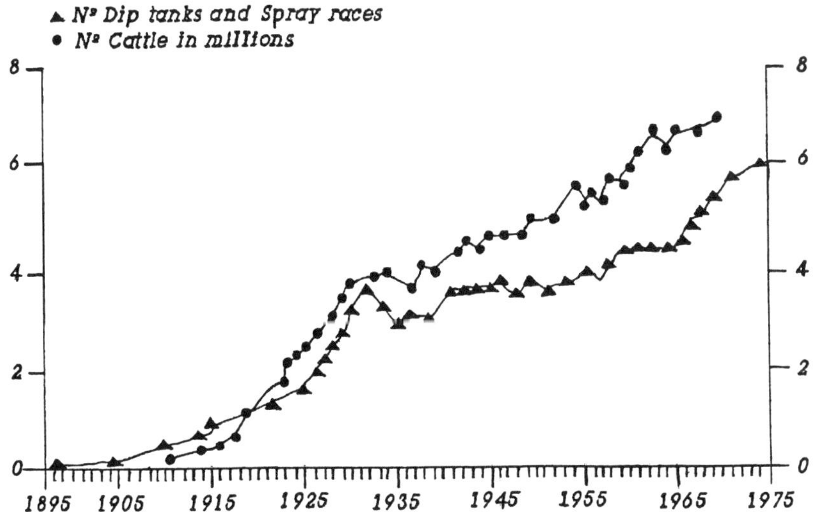

FIGURE 1. Graph showing the growth of the national cattle herd and the numbers of dip tanks and spray races in use between 1895 and 1975.[5] Triangles = number of dip tanks and spray races; circles = number of cattle in millions.

Most people (the writer included) believed that *Amblyomma* could not survive in the high rainfall areas, and in fact over the years a few isolated incursions had been easily eradicated by dipping in acaracides.

However, from 1975 onward, the bont tick and heartwater have spread to numerous areas of Zimbabwe where they are now endemic and causing economic loss. This has been due to a number of factors including:

a. During the disturbances that occurred from 1975 to 1980, prior to independence, dipping ceased or was reduced particularly in the communal dip tanks in tribal areas.

b. There was frequent unauthorized movement of *Amblyomma*-infested cattle to areas in the north which established foci of disease and ticks which often spread to adjoining farms.

c. Very laudably it became customary for most commercial farmers to conserve wildlife. Many such animals were bought and transported from the heartwater areas and, in spite of regulations, may have brought infected ticks. The wildlife animals have multiplied and are available as alternate hosts not subjected to tick control.

d. At the same time many scientists were advocating less intensive tick control in commercial and tribal sectors.[1,9] As a result some, but not all, farmers resorted

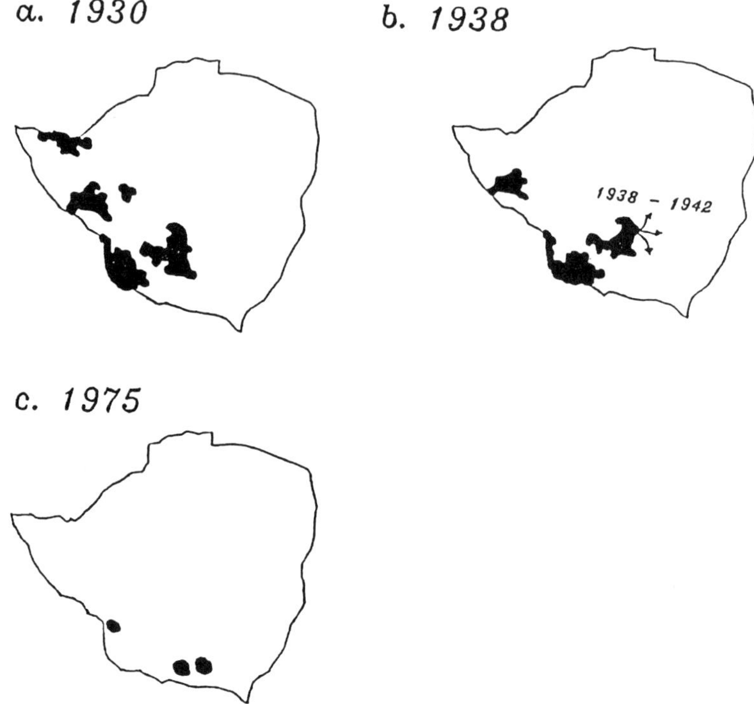

FIGURE 2. Maps showing the distribution of heartwater in Rhodesia in 1930 (upper left), 1938 (upper right), and 1975 (bottom).[5]

to strategic dipping, long interval use of pour-ons or the use of homemade pour-ons and hand sprays which were not fully effective.

A complex situation has emerged where heartwater exists on many farms, some endemically stable but many endemically unstable. Few if any farms have been affected where conventional regular dipping is practiced.

This example of the spread of HW is given to show that HW can be controlled but, with reduced control, it can spread rapidly.

Eradication Trials

The writer and colleagues are involved in eradication field trials in which different herds on adjacent farms totaling about 3000 head of cattle, well managed and kept in paddocks but badly affected by heartwater, will be dipped weekly in amitraz or synthetic pyrethroid-based acaracides for 24 months. There are small or moderate numbers of wild ungulates present. The results may have some indications for the proposed Caribbean program.

THE PROBLEM IN THE CARIBBEAN

This has been well documented.[6,7,10] The disease and/or the vector could spread and adversely affect animal health and productivity on all the islands particularly with an increased incidence of dermatophilosis, and more seriously HW could spread to the mainland where there is a vast susceptible ruminant population.[7]

PROGRAM FOR THE ERADICATION OF *AMBLYOMMA VARIEGATUM* FROM THE CARIBBEAN

The Project Document

The excellent project document describes the introduction into Guadaloupe of *Amblyomma variegatum* with cattle from Senegal about 1830, the spread to other islands, the effects of HW and dermatophilosis, and the serious threat of spread to the mainland countries. The extent of infestation is evaluated and the objectives and strategies for eradication are set out.[10]

The following ideas and comments are submitted:

To be successful, the target must be the participation of every animal owner at every treatment. Legislation to enforce this and to only allow the movement of tick-free animals from place to place should be available. However, this is not the complete answer, and it would be preferable if all or the majority of the population were in favor and enthusiastic. It is suggested that the priority should be public relations and the promotion of the benefits, using television, radio, the press, road shows, videos, and films for the public and for schools. Evident political support is desirable.

It is suggested that worthwhile prizes or incentives should be put up for performance, particularly for groups. Group enthusiasm can be used to overcome unwilling individuals. Project funds could be provided and company sponsorship sought.

Identification of Animals and Recording

It is considered that ease and accuracy of identification are essential to ensure that every owner produces animals on the appointed day. Metal tags are often difficult to read. Plastic ear tags are better for large animals. However, in spite of the cost, implanted electronic microchip identification should be considered for ease, accuracy, and freedom from tampering. The technique would also make it easier to identify animals illegally moved from place to place or from island to island.

The Choice of Topical Acaracides

The organochlorines and toxaphene are unsuitable because of widespread tick resistance, and they are environmentally unacceptable because they leave residues in meat and milk.

The organophosphate chlorfenvinphos is effective against *Amblyomma,* but it has a very short residual effect of approximately 3 days, and it is becoming environmentally suspect for animal treatment.

Amitraz is a very effective acaracide and has a residual effect of 7 days or more when used as a dip or as a spray.[11,12]

Some synthetic pyrethroids are effective against *Amblyomma* and have a long residual effect. Flumethrin and deltamethrin are available as dips or sprays and as pour-ons which have an effective residual effect of 7 days as a wash or 14 days when applied as a pour-on at 1 ml per 10 kg body weight.

The Application of Acaracides

Dipping is the most effective and least costly and, where dip tanks are available, should be used. Methods for maintaining wash concentrations can be demonstrated. Hand spraying is effective for small groups of animals, but it is wasteful as the surplus wash runs off and is not recycled.

Pour-ons are convenient but comparatively very costly, perhaps three times the cost of dipping or spraying, but some are highly effective when applied at the full dose rate every two weeks or less.[13] The most effective acaracides with long residual effect are some synthetic pyrethroids (SP) and amitraz. Resistance of ticks to SPs could develop, but this is less likely in the case of amitraz.

Animals Treated and Intervals of Treatment

Cattle, goats, and sheep are essential subjects as they carry 100% of adult ticks and the majority of immatures.[15] Weekly application of wash or fortnightly application of pour-ons is recommended. Pigs and donkeys should also be treated fortnightly with pour-on or spray wash. Dogs appear frequently to carry immatures or even adults, and in some areas are often heavily infested,[14] and it is suggested that small dog dips be set up at treatment points. Old domestic bathtubs are effective.

Dogs, Domestic and Feral

Domestic dogs should be identified with collars and possibly microchips. Feral dogs should be destroyed as they can act as alternative untreated hosts for ticks.

Antelopes and Wild Goats

They too could act as reservoir hosts and should be destroyed, but keeping a nucleus for reestabishment where appropriate.

Alternative Methods of Tick Control

Pheromone-Baited Acaracide-Treated Tags

It wa shown that ethylvinyl neck bands and tail tags containing 10% amitraz but without pheromone effectively reduced populations of lone star ticks.[15] Pheromone acaracide mixtures were tried in South Africa in 1978,[16] and now tags evolved by the University of Florida baited with the male *Amblyomma variegatum* aggregation attachment pheromone and impregnated with a suitable acaracide effectively attract and kill female *Amblyomma*. The writer has no personal experience of them but believes that the method is promising for tick control or as an aid to control.

Aerial Spraying

It is estimated that 250 g of synthetic pyrethroid per hectare applied sequentially would be required to kill ticks. It would be technically difficult, very costly and environmentally disastrous.

Integrated Management Strategies (IMP)

Economically feasible IMP strategies do not seem suitable for an eradication program as workers have shown they only reduce *Amblyomma* populations by 89% after 5 years.[17]

Pasture Spelling

The removal of all animals from areas for at least two years would be required, and for socioeconomic reasons this is impossible.

Injections and Oral Applications

The ivermectins on a sustained release basis have been suggested, but the results shown in the literature[18,19] are not encouraging. In addition, resistance and environmental problems might result.

The Duncan Applicator

This was invented in Zimbabwe. Acaracide pour-on slowly flows from a reservoir down a column against which animals are encouraged to rub by the provision of feed cubes in the container at the base.[20,21] In game ranching, the device is very helpful in reducing infestation but not considered fully effective for an eradication program, but it could be employed as an aid where some animals wild or domestic are not under control.

Pilot Schemes

It is suggested that whichever procedures are adopted, pilot schemes should be set up on one or two islands. These could be given full attention, potential snags ironed out, and perhaps improved methods found.

When Is Success Achieved?

When regular treatment of all animals has been in progress for a few months, it is probable that few if any *Amblyomma* ticks will be seen even on diligent inspection. Naturally it would be premature to claim success at this stage, but if there is apparent success after 12 months it is suggested that sentinel animals should be clearly earmarked and left untreated for 12 weeks or more and periodic detailed inspections carried out with the animals fully restrained in recumbency.

Where no *Amblyomma* ticks are found on monitoring, it is suggested that the intensive application of acaracide should continue for a further 12 months and the monitoring process be repeated then and from time to time.

Serology could also be carried out on selected young stock to ascertain if they are acquiring HW antibodies.

Regular vigilance should be maintained for the foreseeable future.

CONCLUSIONS

The difficulties of achieving eradication are enormous, but so is the reward for success in that it would remove the risk of spread of HW to other islands and the mainland and improve animal health and productivity in the islands, through the eradication of ticks, tick-borne protozoal diseases, and dermatophilosis. It is considered that the technical and scientific aspects are the least difficult. The main problems are social and organizational, requiring sustained and devoted application by the project staff at all levels and the willing cooperation and assistance of the public officials, and politicians.

ACKNOWLEDGMENTS

The write gratefully acknowledges the valuable editorial and scientific advice willingly given by Dr. J. A. Lawrence.

REFERENCES

1. PEGRAM, R. G., R. J. TATCHELL, J. J. DE CASTRO, H. G. B. CHIZYUKA, M. J. CREEK, P. J. McCOSKER, M. C. MORAN & G. NIGARURA. 1993. Tick control: new concepts. World Animal Rev. **74-75:** 2-14.
2. McCOSKER, P. J. 1979. Global aspects of the management and control of ticks of veterinary importance. Recent Advances in Acarology **II:** 45-53. Academic Press. New York, NY.
3. NORVAL, R. A. I. 1981. A re-assessment of the role of dipping in the control of tick borne disease in Zimbabwe. Proceedings of the International Conference on Tick Biology and Control: 87-90. Grahamstown, South Africa.
4. FAO. 1984. Ticks and Tick Borne Disease Control. A Practical Field Manual. Vol II. Tick Borne Disease Control. Food and Agriculture Organization of the United Nations. Rome, Italy.
5. LAWRENCE, J. A. & R. A. I. NORVAL. 1979. A history of ticks and tick borne diseases of cattle in Rhodesia. Rhod. Vet. J. **10:** 28-40.

6. CAMUS, E. & N. BARRÉ. 1987. Epidemiology of heartwater in Guadaloupe and in the Caribbean. Onderstepoort J. Vet. Res. **54:** 419-426.

7. BARRÉ, N., G. UILENBERG, P. C. MOREL & E. CAMUS. 1987. Danger of introducing heartwater onto the American mainland: potential role of indigenous and exotic *Amblyomma* ticks. Onderstepoort J. Vet. Res. **54:** 405-417.

8. DE BRUIJN, H. S. & J. M. WILLIAMSON. 1995. Personal communication.

9. NORVAL, R. A. I. 1983. Arguments against intensive dipping. Zimbabwe Vet. J. **14**(1/4): 19-25.

10. FAO. 1994. Programme for the Eradication of *Amblyomma variegatum* from the Caribbean. Phase II.

11. BOKMA, B. H. & J. L. SHAW. 1993. Eradication of a new focus of *Amblyomma variegatum* in Puerto Rico. Rev. Elev. Méd Vét Pays Trop. **46:** 355-358.

12. DAKER, J. A. F., R. J. TAYLOR & G. D. STANFORD. 1973. A new triazapentadiene compound active against major cattle ticks in South Africa. In Proceedings 7th British Insecticide and Fungicide Conference: 291-300.

13. HAMEL, H. D., 1987. The use of flumethrin 1% pour on for the control of *Amblyomma SPP* in various southern african countries. Onderstepoort J. Vet. Res. **54:** 521-524.

14. CAMUS, E. & N. BARRÉ. 1992. The role of *Amblyomma variegatum* in the transmission of heartwater with special reference to Guadaloupe. Ann. N.Y. Acad. Sci.: 33-41.

15. GEORGE, J. E. & J. A. MILLER. 1992. The efficacy of a combination neck band and tail tag containing amitraz against lone star ticks on pastured cattle. Proceedings of 36th Animal Livestock Insect Workers Conference.

16. RECHAV, Y. & G. B. WHITEHEAD. 1978. Field trials with pheromone—acaracide mixtures for the control of *Amblyomma hebraeum*. J. Econ. Entomol.: 149.

17. BARNARD, D. R., G. A. MOUNT, D. G. HAILE & E. DANIELS. 1994. Integrated management strategies for *Amblyomma americanum* (Acari:Ixodidae) on pastured beef cattle. J. Med. Entomol. **31**(4): 571-585.

18. SOLL, M. D., I. H. CARMICHAEL, G. E. SWAN & S. J. GROSS. 1989. Control of induced infestations of three African multihost tick species with sustained release ivermectin, Exp. Appl. Acarol. **1:** 121-130.

19. NOLAN, J., H. J. SCHNITZERLNG & P. BIRD. 1985. The use of ivermectin to cleanse tick infested cattle. Australian Vet. J. **62**(II): 386-388.

20. DUNCAN, I. M. & N. MONKS. 1992. Tick control on eland (*Taurotragus oryx*) and buffalo (*Syncerus caffer*) with flumethrin 1% pour on through a Duncan applicator. J. S. Afr. Vet. Assoc. **63**(1): 7-10.

21. DUNCAN, I. M. 1992. Tick control on cattle with flumethrin pour on through a Duncan applicator. J. S. Afr. Vet. Assoc. **63**(3): 125-127.

Tropical Bont Tick Eradication Campaign in the French Antilles

Current Status

N. BARRÉ,[a] E. CAMUS,[a] J. FIFI,[b] P. FOURGEAUD,[c]
G. NUMA,[d] F. ROSE-ROSETTE,[c] AND H. BOREL [e]

[a]*CIRAD-EMVT*
BP 515
97165 Petit Bourg
Guadeloupe, French West Indies

[b]*DSV*
Jardin d'Essai
97100 Basse Terre

[c]*DSV*
BP 671
97262 Fort de France

[d]*Conseil Général*
97110 Abymes

[e]*INRA*
BP 515
97165 Petit Bourg
Guadeloupe, French West Indies

INTRODUCTION

The tropical Bont tick, *Amblyomma variegatum,* is an African tick species. It was introduced in the central islands of the Eastern Caribbean during the 19th century, with infested cattle imported from the Guinea Coast, presumably from Senegal. In all the islands where it is established, the tick is associated with dermatophilosis, a skin bacterial disease responsible for heavy losses in susceptible cattle breeds. In three islands, the tick is the vector of heartwater, a nervous rickettsial disease of small ruminants and European cattle or crossbreds. Heartwater has been discovered relatively recently (1980) in Guadeloupe,[1] despite important losses induced in goats and the fact that it was probably introduced with the tick nearly two centuries ago.

For at least 25 years, the tick had been controlled in the French islands of the Caribbean by the application of acaricides on livestock on a voluntary basis. This method is considered satisfactory by most of the French Antilles livestock owners. The recent discovery of heartwater, the confirmation of the close relationship between the tick and severe dermatophilosis,[2,3] and the increasing spread of the tick in the Caribbean for the last 30 years[4,5] worry agricultural authorities of the region. This situation stimulated reflections and decisions of the veterinary services of the Carib-

TABLE 1. Islands of the Caribbean Infested by *Amblyomma variegatum:* Year of Introduction of the Tick and Class of Risk (see text for class definition)

	Year of First Infestation	Risk Class
Guadeloupe (F)	1828 ?	IV
Marie Galante (F)	1835 ?	IV
Antigua	1895	IV
Martinque (F)	1948	III
St Croix	1967	II (eradicated ?)
St Lucia	1970	III
Puerto Rico	1974	III (eradicated ?)
Nevis	1977	IV
St Kitts	1978	IV
St Martin (F)/St Marteen	1979	III
Vieques	1981	II
Anguilla	1982	III
Monserrat	1982	III
La Désirade (F)	1982	III
Dominica	1983	III
Barbados	1984	III
Culebra	1987	II
St Vincent	1988	II

bean and the mainland, agricultural international organizations, and funding institutions to solve definitively the problem.[6]

In this paper, the authors will recall the history of the problem and the current situation in the French Antilles.

CURRENT *AMBLYOMMA VARIEGATUM* DISTRIBUTION IN THE CARIBBEAN

At present, 18 islands of the Caribbean are or have been infested.[6,7] Three islands were infested during the 19th century (TABLE 1), presumably in 1828 for Guadeloupe,[8] but possibly much earlier.[9] Martinique was contaminated in the middle of the current century. This situation, with 4 islands infested, remained apparently unchanged for the next 20 years until the tick was discovered on St Croix.[10] This was the signal of an accelerating increase of the spread of the tick which invaded one new island every two years for the next 25 years.[4] The role of the cattle egret—a migrating bird recently established in the Caribbean—in the spread of the tick has been suspected.[7,11]

The estimated risk of infestation and the present infestation level of each island by the tropical bont tick have been classified in 4 classes.[6]

Class I: islands and mainland countries free of ticks and at lower risk of infestation.
Class II: islands uninfested, but at higher risk (IIa), or where ticks have been reported but are not considered established (IIb).

TABLE 2. Annual Cost of *Amblyomma variegatum* in Guadeloupe[a]

	Annual Cost
Current tick control (salaries, equipments, supplies, maintenance)	$820,000
Losses in cattle	$1,120,000
Losses in small ruminants	$220,000
Total	$2,160,000

[a] From Camus and Stachurski.[13,14]

Class III: tick established but limited in distribution.
Class IV: tick widespread.

Considering the 7 French islands: Les Saintes and St Barthelemy are in class IIa; Martinique, Desirade, and St Martin in class III; Guadeloupe and Marie Galante in class IV.

JUSTIFICATIONS FOR AN ERADICATION PROGRAM

The tick is a vector or is associated with two major diseases of ruminants, dermatophilosis and heartwater. In the Caribbean, various sources[2,3] give evidence of the extremely high economic incidence of dermatophilosis on naive local cattle breeds in the years following the introduction of the tick, or on exotic cattle breeds introduced in an infested area. Heartwater also induces losses in local small ruminants (and exotic breeds), reaching 10 to 20% of the flocks each year. Moreover, the current tick-control campaign is costly and will continue permanently if nothing more is done.

An example of the current cost of the tick in Guadeloupe is given in TABLE 2. Creole cattle, a breed highly resistant to tick-borne diseases, are predominant in this island; exotic or crossbreds representing only 7% of livestock. One can postulate that tick eradication will allow an increase of exotic cattle (estimated 15%) and then an increase in annual production of $580,000 which is not taken into account in the global cost reported in TABLE 2. The estimated cost per head of cattle is relatively low in Guadeloupe, compared with other Caribbean islands, as most livestock is of Creole breed, highly resistant to tick-borne diseases. It is much higher in islands like Martinique and St Martin where Brahman or European breeds highly susceptible to dermatophilosis and tick-borne diseases respectively are dominant. In such islands, all the animals of many herds died or were slaughtered *in extremis* because of acute dermatophilosis. In Nevis, dermatophilosis reduced the size of livestock to a magnitude of 75%.[12]

Another argument to justify the urgent setting up of an eradication program is the recent accelerating spread of the tick: 14 islands were infested in the last 25 years, compared with 4 in the previous two centuries, and there is no reason to believe that this propagation will stop by itself. The tick is transported by illegal movements of livestock and probably also by migrating birds, especially the cattle egret, a bird widespread and numerous in the Caribbean and in North, Central, and South America. The biological requirements of the tick allow its establishment in the Greater Antilles and on the mainland from the south of Mexico to the south of Brazil.[15,16] Once introduced on the mainland, it will be impossible to prevent its spread all over the large areas of livestock production in America, climatically suitable for the development of the tick. Losses induced by dermatophilosis on the highly susceptible naive livestock will be dramatic.

Finally, the biology of the tick in the Caribbean—hosts, survival period, duration of blood meal—is well known,[17,18] and its susceptibility to acaricides has been extensively studied. There are no major technical obstacles to succeed in the eradication. As a proof, the tick has been eliminated from Puerto Rico and Vieques[19] and from St Croix.[10,20] It is also easily eradicated from experimental herds in Guadeloupe,[21] at least until reinfestation from neighboring untreated animals occurs. The difficulties are of another nature: success requires an appropriate budget, a strict organization, and a consistent animal-owners' information effort. The critical point will be to convince all animal owners to support the constraints of repeated treatments of their livestock for a 3-year period of time.

HISTORY OF THE PROGRAM

The most salient points for the initiation of the eradication program are listed below:

1980: discovery of heartwater in the Western Hemisphere.[1]

1982: *Amblyomma maculatum* is an experimental vector of heartwater.[22]

1984: role of *A. variegatum* in the appearance of clinical cases of dermatophilosis suspected[2,3,23] and proved.[24]

1985: potential geographical distribution of *A. variegatum* established.[15]

1985; recommendation of chief veterinary officers of CARICOM to eradicate the tick (Kingston, Jamaica, 24–25 May, 1985).

1986: technical meeting in Washington: feasibility, organization, cost, and economic justification of a regional eradication program.[6]

1989: biology of *A. variegatum* intensively studied.[17,18,25]

1989: *A. variegatum* apparently successfully eradicated from Puerto Rico.[19]

1991: eradication program considered a high priority by CARICOM ministries of agriculture. Funds required presented to donors meetings organized by FAO in St Lucia and Rome.

1992: request presented to EEC by French overseas departments.

1993: demonstration of the increasing spread of the tick, and suspected role of the cattle egret in the dissemination: islands are not isolated.[11]

Trials on acaricide susceptibility.[21,26]

1994: EEC decides to support part of the program in Guadeloupe and Martinique on the POSEIDOM budget. Campaign starts officially 1st April, 1995.

CURRENT SITUATION IN GUADELOUPE AND MARTINIQUE

Background

The main characteristics of Guadeloupe and Martinique in regard to tick eradication efforts are reported in TABLE 3. Guadeloupe has 5 dependencies of various sizes,

TABLE 3. General Characteristics of the Eradication Program in Guadeloupe and Martinique ($1 = 5 FF)[a]

	Martinique	Guadeloupe
Number of islands	1	6
Surface	1080 km^2	1750 km^2
Agricultural area infested	30%	95%
Livestock situation		
Cattle	28,200	65,000
Small ruminants	52,500	28,500
Others[a]	3,000	3,800
Number cattle owners	5,430	10,130
Spraying personnel		
Current	9	14
Recruited 1994	6	14
Total	15	28
Staff		
Total	3	5
Spraying vehicles		
Current	8	9
Purchased 1995	6	19
Total	14	28
Total 5 years budget expected	$7,200,000	$9,900,000
	24% still uncertain	10% still uncertain
Cost per cattle equivalent	$170	$135

[a] Pigs tethered in the field, equines.

Marie-Galante with about 10,000 cattle is the largest of them. Livestock is composed of 93,200 cattle and 81,000 small ruminants usually managed in small herds with an average of 6 cattle per owner. French islands have the largest ruminant population of the Lesser Antilles with 76% of cattle and 55% of small ruminants. A voluntary tick control program has been established in both islands for more than 25 years. About 30-40% of animal owners adhere to the current control campaign and, in Guadeloupe, pay a share of about $4 per animal for 26 treatments a year. However, other animal owners apply acaricides by themselves, but do it at their own convenience, more or less regularly.

Phasing

As for other islands concerned, programs in the French islands will be carried out in three phases:

- Preparatory phase (1 year, April 1994–April 1995): recruitment and training of staff and personnel, information, legislation, surveys, and complementary experiments, animal census and identification, purchase of equipment and supplies.

- Eradication phase (3 years, April 1995–April 1998): mandatory acaricide treatment of all livestock and dogs, evaluation of progress.

- Surveillance phase (1 year): monitoring of eradication efficiency, search for foci of persistent infestation, specific treatment of foci.

Strategy

The official objective of the campaign in the French islands defined by veterinary experts of EEC is the eradication of tick-borne diseases: heartwater and babesiosis (Journal Officiel des Communautés Européennes, n. L 65/S, 9/3/94). The strategy proposed to reach this objective and for which budget, organization procedures, legislation, and research were established is the eradication of the tick vectors.

Organization

Both in Guadeloupe and Martinique, a steering committee including agricultural and veterinary services, animal-owner representatives, local sanitary organization involved in tick control campaign, scientists involved in research on ticks, and donors has been established. The committee meets every month. It recruits the personnel, orders equipment and supplies, organizes training and information, and makes various decisions.

During the first year of the preparatory phase, 14 and 6 animal health personnel for Guadeloupe and Martinique programs, respectively, have been recruited. They will work in addition to a total of 23 persons already employed in the voluntary campaign, under the supervision of 1 project leader in each island and 1 and 4 assistants and technicians for Martinique and Guadeloupe, respectively. One to two secretaries for each island are employed by the project. If possible, unemployed workers will reinforce the teams to have at least two people per team. The high cost of salaries in French islands does not allow more than two persons per team.

In the initial project, it was planned to establish 41 teams in Guadeloupe and 25 in Martinique. Due to budget restrictions, the number of teams had to be reduced to 28 and 14 teams for these two islands respectively. It means an average of 2700 to 3000 cattle and cattle equivalent (4 small ruminants equal 1 bovine) to be treated by each team every 2 weeks, compared with 1700–1800 in the initial project.

Budget of the Campaign

The budget is not completely provided for (TABLE 4). Financial support of local assemblies is expected from Martinique and, in both islands, participation of animal owners has to be confirmed. It should be mentioned that EEC support is revised every year considering the results obtained, and that the French Ministry of Agriculture has to advance EEC funds to the project at the beginning of each year. Delay in

TABLE 4. Financial Participation of Donors to the Eradication Campaign

	Martinique	Guadeloupe
Total budget	$7,200,000	$9,900,000
EEC	33%	26%
Ministry of Agriculture	11%	17%
General council	32%	30%
Regional council	15%	8%
Animal owners	9%	11%
Others		8%

payment induces difficulties in the management of the project, particularly for payment of salaries. Local assemblies (general and regional councils) are the most important contributors with 38-45% of the budget, followed by EEC (26-33%), French Ministry of Agriculture (11-17%) and animal owners (9-11%). Thirty-three percent of the 1994 budget covers salaries, 25% the purchase of acaricides, 22% is to be used for equipment, and 11% is operating funds. Research and surveillance activities concern 1% of the budget.

Activities Conducted during the Preparatory Phase

Recruitment of Personnel, Purchase of Material and Equipment

Twenty-six people were recruited at the end of 1994, and a preliminary short training organized. Twenty-five cars were purchased at the beginning of 1995 and spraying material ordered. A logo has been designed and glued on cars. Due to late recruitments, mapping and census of livestock and registration of animal owners, as well as identification of cattle, started late in 1994 and have only been partially implemented during this first year of preparation. They should be completed during the second year of the project. In Martinique, current tick distribution is under examination. A workshop was organized in Martinique which joined together 200 representatives of agricultural organization, animal owners, administration, research, to explain the objectives and the workplan of the project.

Tenders for acaricides have been published, and, in addition, complementary trials on their efficacy on *A. variegatum* were performed.

Surveys were conducted in Guadeloupe and Martinique to estimate the will of owners to collaborate and the obstacles to the success, as well as the most appropriate media channels to use for promoting communication.

In the meantime, contacts were established with international agricultural institutions (FAO, IICA) to examine the possibilities of coordination of the French project with the other regional project. Some particular comments on the choice of acaricides and on surveys amongst animal owners are given below.

TABLE 5. Acaricides Registered in France that Were Proposed by Manufacturers Who Tendered for Acaricide Contract in Martinique. Technical Aspects.

Commercial Name (Application)	Concentration of Active Ingredient	Used in the FW1	Efficacy on *A. variegatum* (Guadeloupe)	Withholding Period (Milk: m; Meat: M)	Precautions
Asuntol 50 (spray)	500 g/kg of Cou-maphos	+++ 20 years	ND	m: 0 M: 15 days	
Sebacil (spray)	500 g/l of Phoxim	No	90%	m: forbidden M: 28 days	Forbidden on lactating cows
Taktik 12.5% (spray)	125 g/l of Amitraz	+++ 10 years	100%	m: 1 day M: 14 days	Toxic for horses Stability
Butox 50 (spray)	50 g/l of Delta-methrin	+++ 10 years	85%	m: 0 M: 0	
Butox Pour-on (pour & spot)	10 g/l of Delta-methrin	+ 5 years	100% (spot)	m: 0 M: 0	
Bayticol Pour-on (pour & spot)	10 g/l of Flu-methrin	+ 5 years	99% (pour-on) 100% (spot)	m: 0 M: 0	

Acaricide Candidates for the Eradication Campaign in Guadeloupe and Martinique

Considering the budget available for acaricides ($180,000 for a first delivery in Martinique), a tender had to be published in an official journal (BOAMP: Official bulletin of public markets announcements).

The tender is formulated separately for Guadeloupe and Martinique, and its objective is to provide the amount of acaricides necessary for one year of treatment. The tender (and then the choice of acaricides) should be repeated each year, for the 3 years of acaricide application. It is specified in the tender that acaricides have to be registered in France. Due to limited budget and necessity to modify progressively the animal owners' habits, it was decided to favor spraying formulations, a method already in use for about 25 years in the French West Indies (FWI), but which has various inconveniences. In Martinique, only 5% of cattle would be treated with pour-on formulations in 1995, the remaining animals being sprayed.

Technical characteristics of acaricides are reported in TABLE 5. The efficacy of some acaricides has been tested in Guadeloupe, on naturally infested cattle, following a standardized experimental protocol. Deltamethrin (Butox) and Flumethrin (Bayticol) pour-on formulations used as pour-on or spot-on at the dose recommended, indicated in TABLES 5 and 6, are highly efficient. In this trial, Amitraz (Taktik) applied by spray killed 100% of the adult ticks attached on the animals. This acaricide appears more efficient than Butox 50, which leaves 15% of adult ticks attached, even in the days following the treatment. A slightly higher concentration for this compound

TABLE 6. Dilution, Concentration, and Dose of Acaricides Recommended by Manufacturers and Comparative Prices of Treatments (Delivery Price Martinique, $1US = 5 FF).

Commercial Name (Application)	Dilution and Concentr. Recommended	Price/Liter or kg	Number Treatments/ Liter of Commercial Compound	Cost/Ttmt	Cost for 78 Ttmts of 115,000 Cattle Equivalents[a]
Asuntol 50 (spray)	1 kg/1000 l 500 mg/l	$66	333 (31/cattle)	0.20	$1,794,000
Sebacil (spray)	1 l/2000 l 250 mg/l	$60.6	666 (31/cattle)	0.09	$807,300
Taktik 12.5% (spray)	1 l/500 l 250 mg/l	$27.4	166 (31/cattle)	0.16	$1,480,600
Butox 50 (spray)	1 l/1330 l 37.5 mg/l	$100	443 (31/cattle)	0.22	$2,024,800
Butox pour-on (pour & spot)		$52	33.3 (30 ml/cattle 250 kg)	1.56	$14,007,200
Bayticol pour-on (pour & spot)		$44.4	40 (25 ml/cattle 250 kg)	1.11	$9,956,700

[a] Cattle of an average 250 kg or 4 small ruminants.

should be tested (50 mg/l). Pyrethroid spray formulations present important advantages over organophosphates and amidines (toxicity, withholding period) and would have their place in the range of acaricides.

Withholding period represents a problem, especially for Phoxim (Sebacil). Animals should not be treated 1 month before slaughtering, a period long enough to allow female ticks to feed and detach before the animal can be killed and the skin destroyed. Also, this compound should not be used on lactating cows. These animals are few in the FWI (1 to 2% of all cattle) and they will have to be treated with pyrethroids. Amitraz also presents some drawbacks: it is unsafe to use on horses (but they are less than a thousand in the FWI), it induces some sleepiness in susceptible humans and animals, and shows a slow unstability when stored for a long period of time.

Commercial aspects are reported in TABLE 6. Unit prices for the treatment of a bovine weighing 250 kg indicate a large variation between compounds: Butox 50 is at a comparable price with other spray acaricides, while pour-on formulations are at least 5 times more expensive. Supposing a whole eradication campaign is carried out with the same compound, comprising 3 years of treatments every 14 days (78 treatments) of the 115,000 cattle and cattle equivalent of Guadeloupe and Martinique, the total cost varies from $800,000 to $14 M depending on the acaricide chosen. This last price goes far beyond the total budget of the campaign and justifies the recourse to the cheapest compounds or the association of different compounds.

TABLE 7. Answers from 301 Animal Owners of Guadeloupe Concerning Their Interest in the Eradication Campaign and Pointing Out Different Obstacles and Constraints Limiting the Chances of Success.

Animal retention	Tethered: 95%	Nothing: 3%
	Corral/crush: 2%	
Accessibility to animals	Close to road: 86%	100–300 m: 11%
		300–3000 m: 3%
Availability of owners	Present all time: 28%	
	Present morning h: 81%	Absent morning: 19%
	Present evening h: 42%	Absent evening: 58%
Accept treatment when absent	Yes: 86%	No: 13%
Accept other methods than spray	Yes: 91%	No: 4%
Accept to participate financially ($4/cattle/ year)	Yes: 99%	No: 1%
Consider eradication desirable	Yes: 96%	No: 1%
Consider eradication feasible	Yes: 67%	No: 16%
Read news papers	Regularly: 20%	Irregularly: 32%
		Never: 48%
Watch TV	Regularly: 53%	Irregularly: 37%
		Never: 10%
Listen to radio	Regularly: 63%	Irregularly: 26%
		Never: 11%

Survey among Animal Owners

A survey on a sample of 301 animal owners proportional to their numbers in each commune of Guadeloupe, chosen randomly, was conducted in December 1994. The same questionnaire is being used in Martinique where it has been answered already by 110 owners. The objectives were to determine the reactions of owners to the project, their will to collaborate, the constraints induced by their activities or mentalities, and the most appropriate channels to communicate and spread information and publicity.

Questionnaires only for Guadeloupe have been analyzed. Ninety-eight percent of livestock owners have cattle (7.6 heads per owner), 48.5% have goats (8.9 heads), 46.5% have pigs (3.8) and dogs (1.8). Our sample reflects the known situation of livestock: small units, unspecialized, owners occupied in various activities.

The other information contained in the survey is listed in TABLE 7. They indicate the feeling of animal owners before any structured campaign of sensitization. Their interest in the campaign is high, and they agree massively to support the program financially, even if only a small majority of owners (67%) believes that the eradication is feasible. Their adaptability to new technologies to combat ticks is encouraging.

Some animals are far from roads, and some other owners will refuse their animals to be treated if they are absent. In these cases, at least, it will be justified to leave acaricides at a definite place for the owner to use by himself.

Considering communication, it appears clearly that radio and television would be the most appropriate supports to establish publicity for the campaign and programs of information. Seventy percent of owners watch television in the evening at local news hours.

DISCUSSION–CONCLUSION

The preparatory phase of the tick eradication campaign was started in Guadeloupe and Martinique in April 1994. During this first year of preparatory phase, a steering committee was established and a project leader designated. Personnel has been recruited, vehicles purchased, and tenders have been broadcast to potential suppliers. Efforts were made to complete the budget and to interest complementary donors to the program. Budget is limited and does not allow the use of the most efficient, safe, easy to use acaricides and the use of the most appropriate number of spraying teams. It is absolutely necessary for the success of the campaign that funds be allocated for the whole program and not be put up for discussion each year. Some personnel training and animal-owner information were organized and must increase during the next year. Thought was given to defining the most efficient methods to computerize field data and to establish control of the development and progress of the program. Livestock owners' registration and animal census and identification were initiated and must be intensified. Legislation was prepared and published, particularly to transform the current voluntary control program to a mandatory campaign, and to reduce livestock movements. Complementary trials were made to compare the efficiency and ease of application of various acaricides. Their choice will depend on their quality and price. Surveys were conducted among owners to appreciate their will to participate in the program, the difficulties to execute it, and the media channels used by owners through which messages and information should be spread.

In parallel with this program in the French islands, it is expected that a similar campaign will start rapidly and simultaneously in neighboring infested islands of the Caribbean to have a chance to solve definitively the problem in the Western Hemisphere, and to prevent the spread to the Mainland and the Greater Antilles.

SUMMARY

The *Amblyomma variegatum* eradication campaign was officially started in April 1994 in Martinique and in Guadeloupe, including its dependencies of Marie Galante, Desirade and St Martin. A budget of $10.5 and $5.9 million for Guadeloupe and Martinique, respectively, was initially (1991) calculated and considered necessary to achieve the program.

However, EEC, the most important donor, estimated that 75% only of this proposal was acceptable, on which it agrees to support a maximum of 50%. The balance had to be provided by French government, local political institutions, and animal owners.

The current budget actualized in 1995 by veterinary authorities of Martinique and Guadeloupe is $7,200,000 and $9,900,000 respectively. The program will take 5 years: preparation (1 year), acaricide application (3 years), surveillance (1 year). During this first year, a pilot committee was established, and a project leader was designated for each island. Efforts were essentially oriented to organize the program in the field and to obtain funds from French and local counterparts. Funds allowed the recruitment and training of 19 new agents in Guadeloupe and 9 in Martinique, in addition to personnel already involved in the tick control program. Census of animal owners and identification of cattle started or were developed. Tenders were invited to provide vehicles and acaricides. For the latter, choices were made considering the prices and efficacy as well as funds available. Due to a limited budget and cost of pour-on formulations, pour-on acaricides will be used on a maximum of 30% of animals only. In order to investigate animal owners' reticences and the most appropriate media channels to establish a communication and a training program, a survey was conducted in Guadeloupe on a sample of 301 animal owners. The results are presented. In order to strengthen the chances of success of the campaign in the French islands, it is expected that a similar program will take place soon in infested neighboring islands.

REFERENCES

1. PERREAU, P., P. C. MOREL, N. BARRE & P. DURAND. 1980. Existence de la cowdriose (heartwater) à *Cowdria ruminantium* chez les petits ruminants des Antilles françaises (la Guadeloupe) et des Mascareignes (la Réunion et île Maurice). Rev. Elev. Méd. Vét. Pays Trop. **33:** 21–22.
2. BURRIDGE, M. J., N. BARRE, E. F. BIRNIE, E. CAMUS & G. UILENBERG. 1984. Epidemiological studies on heartwater in the Caribbean, with observations on tick associated dermatophilosis. Proceedings of the XIIIth World Congress on Diseases of Cattle, Durban, South Africa, **1:** 542–546.
3. UILENBERG, G., N. BARRE, E. CAMUS, M. J. BURRIDGE & G. I. GARRIS. 1984. Heartwater in the Caribbean. Prev. Vet. Med. **2:** 255–267.
4. ALDERINK, F.J. & E. H. McCAULEY. 1988. The probability of the spread of *Amblyomma variegatum* in the Caribbean. Prev. Vet. Med. **6:** 285–294.
5. UILENBERG, G. 1990. Extension de la tique *Amblyomma variegatum* dans les Antilles: comment expliquer cette grave menace et que faire? Rev. Elev. Méd. Vét. Pays Trop. **43:** 297–299.
6. FAO, USDA, IICA. 1987. Management of the Tropical Bont Tick (*Amblyomma variegatum*) and associated diseases in the Caribbean. A Feasibility Proposal. Report.
7. BARRE, N., G. GARRIS & E. CAMUS. 1995. Propagation of the tick *Amblyomma variegatum* in the Caribbean: Rev. sci. tech. off. int. Epiz. **14:** 841–855.
8. CURASSON, G. 1943. *Trypanosoma vivax* et variétés. Traité de Protozoologie humaine et comparée. Tome I. Paris Vigot Edit.: 267–269.
9. MAILLARD, N. & J. C. MAILLARD. Historique du peuplement bovin dans les îles françaises des Antilles. Synthèse bibliographique. Historia Medicinae Veterinariae. (In press.)
10. HOURRIGAN, J. L., O. L. KELSEY, C. C. CRAGO & D. J. GILHOOLY. 1969. Eradication efforts against tropical bont tick *Amblyomma variegatum,* in the Virgin islands. JAVMA **154**(3): 540–545.
11. CORN, J., N. BARRE, B. THIEBOT, T. E. CREEKMORE, G. I. GARRIS & V. F. NETTLES. 1993. The potential role of cattle egrets. *Bubulcus ibis,* (Ciconiformes: Ardeiae) in the dissemination of *Amblyomma variegatum* (Acari: Ixodidae) in the Eastern Caribbean. J. Med. Entomol. **6:** 1029–1037.

12. HADRILL, D. J., R. BOID, T. W. JONES & L. BELL-SAKYI. 1990. Bovine babesiosis on Nevis. Implications for tick control. Vet. Rec. **126:** 403-404.
13. CAMUS, E. 1987. Contribution à l'étude épidémiologique de la cowdriose (*Cowdria ruminantium*) en Guadeloupe. Thèse Doc. Etat. Paris-Sud. Orsay, France.
14. STACHURSKI, F. 1988. Impact économique actuel et intérêt de l'éradication de la tique *Amblyomma variegatum* pour l'élevage bovin guadeloupéen. Thèse Doc. vét. Créteil, France.
15. SUTHERST, R. W. & G. F. MAYWALD. 1985. A computerised system for matching climates in ecology. Agriculture, Ecosystems and Environment **13:** 281-299.
16. BARRE, N., G. UILENBERG, P. C. MOREL & E. CAMUS. 1987. Danger of introducing heartwater onto the American mainland: potential role of indigenous and exotic *Amblyomma* ticks. Onderstepoort J. Vet. Res. **54:** 405-417.
17. GARRIS, G. I. 1984. Colonization and life cycle of *Amblyomma variegatum* (Acari: Ixodidae) in the laboratory in Puerto Rico. J. Med. Entomol. **1:** 86-90.
18. BARRE, N. 1989. Biologie et écologie de la tique *Amblyomma variegatum* (Acarina: Ixodina) en Guadeloupe (Antilles françaises). Thése Doc. Etat. Paris-Sud. Orsay, France.
19. GARRIS, G., B. H. BOKMA, R. K. STRICKLAND & G. P. COMBS. 1989. Evaluation of the eradication program for *Amblyomma variegatum* (Acari: Ixodidae) on Puerto Rico. Exp. Appl. Acarol. **6:** 67-76.
20. BOKMA, B.H. & J. L. SHAW. 1993. Detection and eradication of a new focus of *Amblyomma variegatum* in Puerto Rico. Rev. Elev. Méd. Vét. Pays Trop. **46:** 355-358.
21. BARRE, N., G. I. GARRIS & R. APRELON. 1993. Acardicides for eradication of the tick *Amblyomma variegatum* in the Caribbean. Rev. Elev. Méd. Vét. Pays Trop. **46:** 349-354.
22. UILENBERG, G. 1982. Experimental transmission of *Cowdria ruminantium* by the Gulf Coast tick, *Amblyomma maculatum:* danger of introducing heartwater and benign African theileriasis onto the American mainland. Am. J. Vet. Res. **43:** 1279-1282.
23. MATHERON, G., N. BARRE, F. ROGER, B. ROGEZ, D. MARTINEZ & C. SHEIKBOUDOU. 1989. La dermatophilose des bovins à *Dermatophilus congolensis* dans les Antilles françaises. III: comparaison entre élevages infectés et indemnes. Rev. Elev. Méd. Vét. Pays Trop. **42:** 331-347.
24. MARTINEZ, D., N. BARRE, B. MARI & T. VIDALENC. 1992. Studies on the role of *Amblyomma variegatum* in the transmission of *Dermatophilus congolensis*. *In* Tick Vector Biology, Medical and Veterinary Aspects. B. Fivaz, T. Petney & I. Horak, Eds: Springer-Verlag. Berlin, Germany. 97-99.
25. BARRE, N. & G. GARRIS. 1991. Biology, ecology and management of the tick *Amblyomma variegatum* in the Caribbean. *In* CTA Seminar Proceeding. Cowdriosis and Dermatophilosis of livestock in the Caribbean region. CARDI, CTA. Antigua. 63-84.
26. GARRIS, G. & N. BARRE. 1990. Acaricide susceptibility of *Amblyomma variegatum* (Acari: Ixodidae) from Puerto Rico and Guadeloupe. Exp. Appl. Acarol. **12:** 171-179.

Potential Impact of Wildlife on the Tropical Bont Tick Eradication Program in the Caribbean[a]

JOSEPH L. CORN,[b] NICOLAS BARRÉ,[c]
GLEN I. GARRIS,[d] AND VICTOR F. NETTLES [b]

[b]Southeastern Cooperative Wildlife Disease Study
College of Veterinary Medicine
The University of Georgia
Athens, Georgia 30602

[c]Centre de Coopération Internationale en Recherche
Agronomique pour le Développement
Département d'élevage et de Médecine Vétérinaire
BP 1232, 91785
Pointe-a-Pitre Cedex
Guadeloupe, French West Indies

[d]United States Department of Agriculture
Animal and Plant Health Inspection Service
Veterinary Services, National Animal Health Programs
4700 River Road, Unit 43
Riverdale, Maryland 20737

The tropical bont tick (*Amblyomma variegatum*) was introduced into the Caribbean on cattle brought from West Africa to Guadeloupe in 1830, and at present is found on numerous islands in the region. In Africa, domestic and wild hosts for the tick are abundant; however, the diversity of potential wildlife hosts in the Caribbean is limited. Because the survival of populations of this tick in the Caribbean apparently is dependent on the presence of livestock, eradication can be approached through the treatment of livestock. However, wildlife infestations may be of importance relative to the length of local treatment programs, and to the timing of treatment programs throughout the region.

WILDLIFE AS HOSTS FOR *A. VARIEGATUM*

Amblyomma variegatum is a 3-host tick with a wide host range. In Africa, adults of *A. variegatum* feed on a number of species of large mammals, especially buffalo

[a] Support was received through Cooperative Agreement 94-9-6-13-0032-CA, Veterinary Services, Animal and Plant Health Inspection Service, US Department of Agriculture, and the Inter-American Institute for Cooperation on Agriculture.

(*Syncerus caffer*) and several species of antelope (Bovidae). Other hosts include giraffes (Giraffidae), zebras (Equidae), elephants (*Loxodonta africana*), and bush pigs (*Potamochoerus porus*). Birds such as the gray hornbill (*Lophocerus nasutus*) also are parasitized by the adult tick. Nymphs feed on a variety of birds and medium to large-sized mammals. Mammalian hosts for nymphs include several species of antelope, insectivores, carnivores, lagomorphs, and rodents. Avian hosts for nymphs include hornbills (Bucerotidae), bustards (Otididae), vultures (Accipitridae), and storks (Ciconiidae). Larvae parasitize mostly birds and small mammals. Avian hosts for larvae are as listed for nymphs; mammalian hosts include hares (Leporidae), reedbucks (Bovidae), jackals (Canidae), and mongooses (Viverridae).[1]

In comparison with the numbers of host species present in Africa, potential wildlife species that may serve as hosts for the three parasitic stages of *A. variegatum* in the West Indies are limited. Terrestrial mammals present on islands where the tick is present in the Caribbean include the manicou opossum (*Didelphis marsupialis*), African green monkey (*Cercopithecus sabaeus*), black rat (*Rattus rattus*), Norway rat (*Rattus norvegicus*), house mouse (*Mus musculus*), agouti (*Dasyprocta leporina*), raccoon (*Procyon minor*), small Indian mongoose (*Herpestes auropunctatus*), fallow deer (*Dama dama*), and white-tailed deer (*Odocoileus virginianus*).[2-7] However, most of these wildlife species are present only on one or a small number of islands in the region. The avifauna of the West Indies is much more diverse; over 400 species of birds are listed as occurring regularly in the region.[8]

Data on the status of wildlife as hosts for *A. variegatum* in the West Indies are limited. In Guadeloupe, where the infestation of cattle averaged about 43 adult ticks per cow, 65 of 124 mongooses, 21 of 80 cattle egrets (*Bubulcus ibis*), 29 of 62 Carib grackles (*Quiscalus lugubris*), 1 of 14 common ground doves (*Columbina passerina*), and 1 of 20 black-faced grassquits (*Tiaris bicolor*) were found with varied degrees of infestation. Twenty black rats, 48 house mice, 24 zenaida doves (*Zenaida aurita*), and 256 birds in the order Charadriiformes were not infested.[9] In Antigua, larvae of *A. variegatum* were recovered from 40 of 540 mongooses, 5 of 212 feathered or live cattle egrets, and 3 of 60 Carib grackles; nymphs were recovered from 1 cattle egret, 4 of 540 mongooses, and from a single feral cat. Specimens were not recovered from 5 Norway rats, 27 house mice, and 707 additional birds representing 16 species. Cattle infestations averaged from 3 to 16 adult ticks per cow.[10,11] In Puerto Rico, larvae of *A. variegatum* were found on 1 of 64 mongooses, 2 of 534 house mice, and 1 of 183 black rats; not infested were 50 Norway rats, 73 cattle egrets, 6 greater Antillean grackles (*Quiscalus niger*), 4 mourning doves (*Zenaida macroura*), and 2 common ground doves. The infestation of cattle on Puerto Rico averaged about 20 adult ticks per cow.[12] In the Virgin Islands, surveys conducted after the detection of *A. variegatum* on St. Croix indicated that mongooses, but not white-tailed deer, were infested.[13]

WILDLIFE INFESTATIONS DURING AN
ERADICATION PROGRAM

In Guadeloupe, all parasitic life stages of *A. variegatum* are dependent on domestic animals as hosts.[9] Cattle and goats were hosts for 95% of the larvae, 97% of the nymphs, and all of the adults of *A. variegatum* found on hosts in Guadeloupe.[9] Nevertheless, the infestation of mongooses by *A. variegatum* may be important as a short-term alternative host system for larvae and nymphs of this tick during an eradication program. The effect of acaricide treatments of cattle on the infestation of mongooses by *A. variegatum* was evaluated during a field evaluation of selected acaricides in Guadeloupe.[14] Periodic examinations of mongooses during the treatment program indicated that the mean prevalence and mean relative density of infestation of mongooses by larvae decreased, but the tick was not eliminated. Data on nymphal infestations were inconclusive, and mongooses were not infested by adults. The continued infestation of mongooses probably was a result of the short duration of the treatment periods. The initial treatment period lasted for one year, while the second lasted for four months.[14]

Larvae from female ticks that become engorged and drop from cattle could be present in the environment for up to 11 months after the initial detachment of an adult female from a host.[15] Nymphs can survive for 15 months; thus larvae and nymphs that feed on mongooses could produce adults of *A. variegatum* 26 months after the original female had left its host. The use of mongooses or other wildlife as hosts by larvae or nymphs could potentially result in the presence of small numbers of adults of *A. variegatum* for about 46 months.[15] Whether mongooses could help maintain *A. variegatum* during an eradication program is unknown. Of note, the eradication program against *A. variegatum* in St. Croix apparently resulted in the elimination of the tick in eight months. Mongooses were present on St. Croix; however, the infestation was of limited extent and the eradication program included both premise treatments and dipping of livestock.[13,16] In Puerto Rico, ticks sporadically appeared on livestock in several areas during the eradication program for over three years after treatment of livestock began.[17]

Many of the mammals and ground-dwelling birds present in the Caribbean could serve as hosts for larvae and nymphs of *A. variegatum*, and thus temporarily may serve as an alternative host system for larvae and nymphs during an eradication program. Surveillance of wildlife could be used to determine if larvae and nymphs are present on these alternative hosts at any stage of an eradication program. Mongooses are present on many of the islands in the region, are abundant where present, and have been demonstrated to be hosts in all previous surveys for *A. variegatum* in the region. Where mongooses are present, they would be an appropriate animal for general surveillance.

Where large wildlife species are present in tick-infested areas, they may be of importance as hosts for adults of *A. variegatum*. In Africa, where both untreated wildlife and treated cattle are present on infested pastures, unfed adults and nymphs attach in larger numbers on the untreated wildlife than on the treated cattle.[18] The potential for such wildlife involvement in the Caribbean apparently is limited, and may only involve the white-tailed deer on St. Kitts. White-tailed deer on Culebra were not infested when examined in 1989 (Southeastern Cooperative Wildlife Disease Study, unpublished data), and the eradication on this island apparently was successful in the presence of the deer.[17] White-tailed deer also were not a factor in the previous eradication of *A. variegatum* from St. Croix.[13] The fallow deer on Antigua are present only on the offshore islet of Guiana Island and are not in contact with tick-infested areas. However, the presence and infestation of feral livestock, including burros and swine, and also feral dogs, have received little attention, and may be of significance. Surveillance of potential wildlife or feral hosts can be used to determine the abundance of animals such as the white-tailed deer on St. Kitts, their potential for contact with infested pastures, and their status as hosts. If animals other than domestic animals are found to serve as a temporary alternative host system for adults during an eradication program, it may be necessary to extend the length of the treatment program at that location. If these animals independently are supporting a population of *A. variegatum,* a treatment program directed towards these hosts or their environment may be needed.

CATTLE EGRETS AS POTENTIAL DISSEMINATORS OF *A. VARIEGATUM*

Birds that are hosts for ticks also could serve to disperse ticks geographically. Collections of ticks from outside of their range include ticks on birds migrating between Africa, Asia, and Europe;[19–22] birds of eastern Mediterranean and sub-Saharan origin examined in Cypress[23] and spring migrants arriving in Sweden[24,25] and northern Italy.[26] Movement of wildlife infested by immature and adult stages of *A. variegatum* among infested and uninfested islands in the Caribbean could result in the dissemination of this tick to previously uninfested areas, or in the reinfestation of areas from which the tick had been eradicated.[27]

Next to domestic animals, cattle egrets may provide the most efficient means of dissemination of *A. variegatum* among islands in the Caribbean and to areas outside the region.[27] Cattle egrets are commonly associated with pastures and cattle,[28–30] and this association puts them in contact with *A. variegatum.* The cattle egret is native to Africa,[31] and it became established in the Western Hemisphere during this century. Early sightings of cattle egrets in the Western Hemisphere were from British Guiana and Surinam in 1877, 1882, 1911, and 1912.[32,33] The first specimen was collected in British Guiana in 1937.[34] Cattle egrets probably were first observed in the United States in Florida in 1941.[35] Cattle egrets became established in the eastern Caribbean beginning in the 1950s. First reports of cattle egrets on selected islands in the Caribbean occurred as follows (the approximate date of the introduction of *A. variegatum* is given in parenthesis): Jamaica, 1948 (tick absent); Puerto Rico, 1948 (1974, 1993); Antigua, 1950 (1880s); Trinidad, 1951 (tick absent); Cuba, 1954 (tick absent); St. Croix, 1955 (1967, 1987); Barbados, 1956 (1983); Grenada, 1956 (tick absent); Hispaniola, 1956 (tick absent); Martinique, 1958 (1948); Guadeloupe, 1959 (1830); St. Vincent, 1962 (tick absent); St. Lucia, 1963 (1970); St. Martin, 1964 (1978); Montserrat, 1966 (1983); La Desirade, 1969 (1982); St. Kitts, 1973 (1978); and Barbuda, 1976 (tick absent).[15,30,36] It is of note that the recent expansion in the range of *A. variegatum* followed the establishment of cattle egrets in the region.

The feeding periods for larvae and nymphs of *A. variegatum* are 6-13 days and 5-10 days, respectively.[37] Larvae and nymphs of *A. variegatum* that have fed on cattle egrets are viable[38] thus these ticks could be expected to survive for several days during the dispersal or migration of cattle egrets. The host status and movement patterns of cattle egrets in the Caribbean were evaluated in a study of birds from Antigua and Guadeloupe.[10] Prevalences of infestation by larvae ranged from 0 to

27% during the three-year study, and cattle egrets marked in Antigua and Guadeloupe subsequently were observed on 14 Caribbean Islands and in the Florida Keys. The proportion of birds observed on islands other than the island on which the birds were marked ranged from 1.2 to 12.9% per observation period (observation periods were from 7 to 11 months in length). Estimates of the numbers of infested cattle egrets emigrating from Antigua and Guadeloupe ranged from 0 to 0.24% of the current populations, or from 0 to 26 infested birds per observation period.[10]

At present, the source of infested cattle egrets probably is limited to those islands with abundant ticks, large cattle egret populations, and high rates of cattle egret emigration. This combination of factors exists in Antigua and Guadeloupe, and may exist on other islands including St. Kitts and Nevis. However, if the number of infested islands increases and the tick populations become more abundant and widespread, the potential number of islands that can serve as sources of infested cattle egrets will increase. About half of the emigration of cattle egrets detected in the study mentioned above was to islands within 100 km of the islands on which the birds were marked.[10] Thus, the potential for dissemination of the tick by cattle egrets is greatest within the eastern Caribbean. Islands within this region would risk reinfestation if the tick was eradicated from some, but not all, of the source islands in the region. However, as this source region expands, the potential for infestation of islands at the limit of the ticks range and of more distant areas will increase.[10]

The probability of the small numbers of larvae or nymphs of *A. variegatum* carried by cattle egrets being introduced to an uninfested island, surviving, and generating a viable population is low.[10] A variety of factors would be involved in successful colonization, i.e., environmental conditions, host availability, and the ability to find a mate and to reproduce. Also, since nymphs have rarely been found on cattle egrets, it seems unlikely that the heartwater agent, *Cowdria ruminantium,* could be introduced in this way. Other potential mechanisms of dissemination exist and probably are of greater importance. Legal and illegal movement of cattle and other domestic livestock, pets, and animal hides all may be of significance, in part because adult ticks would be involved.[27]

With a potential for the dissemination of *A. variegatum* by cattle egrets demonstrated, and other mechanisms also possible, the eradication of *A. variegatum* from only one or a few islands, while other islands with heavy infestations remained, would result in a relatively high probability of reinfestation and do little to stop the spread of the tick. Simultaneous eradication of *A. variegatum,* especially on the islands with widespread and abundant tick populations, would greatly reduce or eliminate the potential for both the reinfestation of islands on which eradication programs had been completed, and the infestation of new islands.

SUMMARY

Wildlife are hosts for *A. variegatum* throughout its range in Africa,[1] and have been demonstrated to serve as hosts for larvae and nymphs on several islands in the Caribbean.[9-14] Studies conducted in the Caribbean have indicated that most larvae and nymphs and all adults of the tick feed on livestock; therefore, eradication can be approached through the treatment of livestock.[9,15] However, since small numbers of animals such as the mongoose are infested with larvae and nymphs, these animals may represent a short-term alternative host system.[14] Such infestations potentially could result in a presence of small numbers of ticks for up to 46 months after the onset of a treatment program and could serve as a source to reinfest livestock during the eradication process.[15] Wildlife surveys have been conducted in St. Croix, Puerto Rico, Guadeloupe, and Antigua;[9-14] but the host status of species present on other islands such as the white-tailed deer on St. Kitts has not been evaluated. As part of the eradication program, the species of wildlife present on each island and the association of these species with infested pastures should be determined. Surveys of wildlife can be used during the eradication to determine if wildlife are serving as an alternative host system.

Cattle egrets may serve to disseminate the tick among islands in the region, and to areas outside the region. Cattle egrets have been shown to serve as hosts for the tick,[9,10] and to move throughout the eastern Caribbean and to North America.[10] The potential for interisland dissemination of ticks by cattle egrets, as well as by other means, substantiates the need for a region-wide eradication program.

REFERENCES

1. HOOGSTRAAL, H. 1956. African Ixodoidea. Vol. 1. Ticks of the Sudan (with special reference to Equatoria Province and with preliminary reviews of the genera *Boophilus, Margaropus,* and *Hyalomma*). Research Report NM 005 050.29-07. United States Department of the Navy, Bureau of Medicine and Surgery. Washington, DC.

2. NOWAK, R. M. & J. L. PARADISO. 1983. Walker's Mammals of the World. 4th Edit. Johns Hopkins University Press. Baltimore, MD.

3. FAABORG, J. R. & W. J. ARENDT. 1985. Wildlife Assessments in the Caribbean. Institute of Tropical Forestry, Southern Forest Experiment Station, Forest Service, United States Department of Agriculture. Rio Piedras, Puerto Rico.

4. CHRISTIAN, A. 1989. Dominica's wildlife. *In* Proceedings of the Fourth Meeting of Caribbean Foresters, Wildlife management in the Caribbean Islands: 19-23. Institute of Tropical Forestry and Caribbean National Forest, United States Department of Agriculture. Rio Piedras, Puerto Rico.

5. LUDEKE, A. K., G. VINCENT, R. WALKER & D. DUMOND. 1989. What is the status of wildlife in Grenada today. *In* Proceedings of the Fourth Meeting of Caribbean Foresters, Wildlife Management in the Caribbean Islands: 24-31. Institute of Tropical Forestry and Caribbean National Forest, United States Department of Agriculture. Rio Piedras, Puerto Rico.

6. MILLS, H. N. 1989. Wildlife management in St. Kitts. *In* Proceedings of the Fourth Meeting of Caribbean Foresters, Wildlife Management in the Caribbean Islands: 53-62. Institute of Tropical Forestry and Caribbean National Forest, United States Department of Agriculture. Rio Piedras, Puerto Rico.

7. VALLANCE, M. 1989. Wildlife protection and wildlife management in Guadeloupe. *In* Proceedings of the Fourth Meeting of Caribbean Foresters, Wildlife Management in the Caribbean Islands: 32-36. Institute of Tropical Forestry and Caribbean National Forest, United States Department of Agriculture. Rio Piedras, Puerto Rico.

8. BOND, J. 1985. Birds of the West Indies. 4th edit. Houghton Mifflin Company. Boston, MA.

9. BARRÉ, N., G. I. GARRIS, G. BOREL & E. CAMUS. 1988. Host and population dynamics of *Amblyomma variegatum* (Acari: Ixodidae) on Guadeloupe, French West Indies. J. Med. Entomol. **25:** 111-115.

10. CORN, J. L., N. BARRÉ, B. THIEBOT, T. E. CREEKMORE, G. I. GARRIS & V. F. NETTLES. 1993. Potential role of cattle egrets, *Bubulcus ibis* (Ciconiformes: Ardeidae), in the

dissemination of *Amblyomma variegatum* (Acari: Ixodidae) in the eastern Caribbean. J. Med. Entomol. **30:** 1029-1037.

11. CORN, J. L., D. M. KAVANAUGH, T. E. CREEKMORE & J. L. ROBINSON. 1994. Wildlife as hosts for ticks (Acari) in Antigua, West Indies. J. Med. Entomol. **31:** 57-61.

12. GARRIS, G. I. 1987. *Amblyomma variegatum* (Acari: Ixodidae): population dynamics and hosts used during an eradication program in Puerto Rico. J. Med. Entomol. **24:** 82-86.

13. HOURRIGAN, J. L., R. K. STRICKLAND, O. L. KELSEY, B. E. KNISELY, C. C. CRAGO, S. WHITTAKER & D. J. GILHOOLY. 1969. Eradication efforts against tropical bont tick, *Amblyomma variegatum,* in the Virgin Islands. J. Am. Vet. Med. Assoc. **154:** 540-545.

14. CORN, J. L., B. D. SNYDER, N. BARRÉ & G. I. GARRIS. 1994. Effect of acaricide treatment of cattle on *Amblyomma variegatum* (Acari: Ixodidae) infestation of mongooses (Carnivora: Viverridae: *Herpestes auropunctatus*) in Guadeloupe, French West Indies. J. Med. Entomol. **31:** 490-495.

15. BARRÉ, N. & G. I. GARRIS. 1990. Biology and ecology of *Amblyomma variegatum* (Acari: Ixodidae) in the Caribbean: implications for a regional eradication program. J. Agric. Entomol. **7:** 1-9.

16. GRAHAM, O. H. & J. L. HOURRIGAN. 1977. Eradication programs for the arthropod parasites of livestock. J. Med. Entomol. **6:** 629-658.

17. GARRIS, G. I., B.H. BOKMA, R. K. STRICKLAND & G. P. COOMBS. 1989. Evaluation of the eradication program for *Amblyomma variegatum* (Acari: Ixodidae) on Puerto Rico. Exp. Appl. Acarol. **6:** 67-76.

18. NORVAL, R. A. I., B. D. PERRY, M. I. MELTZER, R. L. KRUSKA & T. H. BOOTH. 1994. Factors affecting the distributions of the ticks *Amblyomma hebraeum* and *A. variegatum* in Zimbabwe: implications of reduced acaricide usage. Exp. Appl. Acarol. **18:** 383-407.

19. HOOGSTRAAL, H. & M. N. KAISER. 1961. Ticks from European-Asiatic birds migrating through Egypt into Africa. Science **133:** 277-278.

20. HOOGSTRAAL, H., M. N. KAISER, M. A. TRAYLOR, S. GABER & E. GUINDY. 1961. Ticks (Ixodidae) on birds migrating from Africa to Europe and Asia. Bull. World Health Org. **24:** 197-212.

21. HOOGSTRAAL, H., M. N. KAISER, M. A. TRAYLOR, E. GUINDY & S. GABER. 1963. Ticks (Ixodidae) on birds migrating from Europe and Asia to Africa, 1959-61. Bull. World Health Org. **28:** 235-262.

22. HOOGSTRAAL H., M. A. TRAYLOR, S. GABER, G. MALAKATIS, E. GUINDY & I. HELMY. 1964. Ticks (Ixodidae) on migrating birds in Egypt, spring and fall 1962. Bull. World Health Org. **30:** 355-367.

23. KAISER, M. N., H. HOOGSTRAAL & G. E. WATSON. 1974. Ticks (Ixodoidea) on migrating birds in Cyprus, fall 1967 and spring 1968, and epidemiological considerations. Bull. Entomol. Res. **64:** 97-110.

24. BRINCK, P., A. SVEDMYR & G. VON ZEIPEL. 1965. Migrating birds at Ottenby Sweden as carriers of ticks and possible transmitters of tick-borne encephalitis virus. Oikos **16:** 88-99.

25. JAENSON, T. G. T., L. TÄLLEKLINT, L. LUNDQVIST, B. OLSEN, J. CHIRICO & H. MEJLON. 1994. Geographical distribution, host associations, and vector roles of ticks (Acari: Ixodidae, Argasidae) in Sweden. J. Med. Entomol. **31:** 240-256.

26. WALTER, V. G. & R. MASSA. 1987. Ein beitrag zur ektoparasitenfauna der zugvögel in Norditalien. J. Appl. Entomol. **103:** 523-527.

27. BARRÉ, N., G. UILENBERG, P. C. MOREL & E. CAMUS. 1987. Danger of introducing heartwater onto the American mainland: potential role of indigenous and exotic *Amblyomma* ticks. Onderstepoort J. Vet. Res. **54:** 405-417.

28. HEATWOLE, H. 1965. Some aspects of the association of cattle egrets with cattle. Anim. Behav. **13:** 79-83.

29. DINSMORE, J. J. 1973. Foraging success of cattle egrets, *Bubulcus ibis.* Am. Midl. Nat. **89:** 242-246.

30. ARENDT, W. J. 1988. Range expansion of the cattle egret (*Bubulcus ibis*) in the Greater Caribbean Basin. Colonial Waterbirds **11:** 252–262.
31. HANCOCK, J. & J. KUSHLAN. 1984. The Herons Handbook. Harper and Row, Publishers, Inc. New York, NY.
32. BLAKE, E. R. 1939. African cattle heron taken in British Guiana. Auk **54:** 470–471.
33. PALMER, R. S. 1962. Handbook of North American birds. Vol. 1. Yale University Press. New Haven, CT.
34. CROSBY, G. T. 1972. Spread of the cattle egret in the Western Hemisphere. Bird-Banding **43:** 205–212.
35. SPRUNT, A. 1954. The spread of the cattle egret. Smithson. Inst. Annu. Rep. **1954:** 259–276.
36. USDA. 1993. Bont ticks on St. Croix and in Puerto Rico. Foreign Animal Disease Report. Animal and Plant Health Inspection Service, United States Department of Agriculture. No. 21 (2/3): 1.
37. GARRIS, G. I. 1984. Colonization and life cycle of *Amblyomma variegatum* (Acari: Ixodidae) in the laboratory in Puerto Rico. J. Med. Entomol. **21:** 86–90.
38. BARRÉ, N. 1989. Biologie et ecologie de la tique *Amblyomma variegatum* (Acarina: Ixodina) en Guadeloupe (Antilles Francaise). Thése Docteur es-sciences. Universite de Paris-Sud. Paris, France.

Efficacy of Tail-Tag Decoys Impregnated with Pheromone and Acaricide for Control of Bont Ticks on Cattle[a]

S. A. ALLAN, R. A. I. NORVAL,[c] D. E. SONENSHINE,[b]
AND M. J. BURRIDGE

Department of Pathobiology
College of Veterinary Medicine
University of Florida
Post Office Box 110880
Gainesville, Florida 32611-0880

[b]Department of Biological Sciences
College of Sciences
Old Dominion University
Norfolk, Virginia 23529-0266

INTRODUCTION

The southern African bont tick, *Amblyomma hebraeum,* and the tropical bont tick, *Amblyomma variegatum,* are closely related species important as major vectors of heartwater disease of ruminants, caused by *Cowdria ruminantium*.[1] This disease is of considerable economic importance in the areas of sub-Saharan Africa where one or both species are found, and on several Caribbean islands, where *A. variegatum* is present.[2-4]

Both of these bont tick species belong to a small group of *Amblyomma* ticks that form feeding aggregations in response to aggregation-attachment pheromones produced by males that have fed for several days on a host.[5,6] Unfed males and females of both bont tick species and unfed nymphs of *A. hebraeum* are stimulated to attach to hosts in the presence of the pheromone.[7-10] Additionally, host location and selection by *A. hebraeum* are facilitated by attraction to the aggregation-attachment pheromone emitted by attached males.[11] Ticks rarely attach to a host not previously infested by feeding male ticks.[11,12] Extracts of blood-fed male *A. hebraeum* placed on potential hosts dramatically increases the attractiveness of the hosts to unfed *A. hebraeum* males and females.[11,13] This concept was utilized by Norval and associates[14] in the development of a slow-release device or attractant decoy that releases phero-

[a] This study was supported by cooperative agreement no. AFR-0435-A-00-9084-00 between the US Agency for International Development and the University of Florida and the Florida Dairy Checkoff fund.

[c] Deceased.

mone components to elicit long- and short-range attraction to hosts as well as stimulating aggregation and attachment of ticks on hosts. In addition to pheromones, decoys also release acaricide to control the ticks on the host. Attractant decoys were developed for attachment to the tails of cattle to maximize their effectiveness for acaricide delivery to body regions most frequently infested by bont ticks.[15-17] In this study, we report on the efficacy of the tail-tag decoy technology for control of bont ticks on cattle from field trails in Zimbabwe. Comparisons were made between the relative efficacy of the acaricides cyfluthrin, flumethrin, and alphacypermethrin.

MATERIALS AND METHODS

The field efficacy trials were conducted on Fallowfield Farm, Zimbabwe (40 km south of Plumtree) which is naturally infested with *A. hebraeum*. Treatments were compared during two trials; the first conducted from October 7 to December 29, 1992; and the second from February 11 to May 5, 1993. For each trial, 300 mixed-breed cattle were divided among treatments and housed in separate paddocks. In the first trial, cattle were treated with either pheromone/flumethrin tail tags (100 animals), pheromone/cyfluthrin tail tags (100 animals), tail tags with only pheromone (25 animals), or left as untreated controls (75 animals). Cattle were not treated with acaricide to remove ticks for the first trial. In the second trial, cattle were treated with either pheromone/flumethrin tail tags (100 animals), pheromone/cyfluthrin tail tags (100 animals), pheromone/alphacypermethrin tail tags (75 animals), or with tail tags containing only pheromone (15 animals). Prior to the second trial, all cattle were treated with a synthetic pyrethroid to remove any ticks.

Tail-tag decoys were prepared at the UF/USAID/SADC Heartwater Research Project facilities in Harare, Zimbabwe using polyvinylchloride(PVC) (Goodrich Geon 139 Resin), plasticizers (dioctylphthalate and epoxidized soybean oil) and stabilizers (organo-barium-zinc) following methods described by Norval.[11] Pheromone components consisted of compounds previously shown to be effective long-range attractants and those that elicited aggregation and attachment responses. These compounds included (by weight of the PVC mixture) 1% *o*-nitrophenol, 2% methyl salicylate, 0.2%, 2,6-dichlorophenol, and 0.1% phenylacetaldehyde (Aldrich Chemical Co.). Acaricides added were either 7.5% flumethrin, cyfluthrin [Bayer Zimbabwe (Pvt.) Ltd.], or alphacypermethrin (FMC Corporation). Tail-tag decoys were made with the pheromone components alone or in combination with one of the three acaricides. The rectangular tail-tag decoys (6.5 cm × 9.0 cm) weighed approximately 10 grams each and had two slits, 3 cm apart, cut in the long axis of the tags. Decoys were attached to cattle by wrapping a strip of filament adhesive tape (3M Corporation)(30 cm long) around the tail, threading the tape through the slits in the tail tag, and, finally, wrapping it around the outside of the tail tag and tail. The filament tape deteriorated under wet conditions and was replaced with a waterproof cloth tape (Servistar Corporation) which also rapidly deteriorated, and another waterproof tape (Can-Do National Tape Co., Inc.) was used to finish trial 1. Tail tags were initially attached during trial 2 using the waterproof Can-Do tape which subsequently became brittle during dry weather, and the trial was completed using the original filament tape.

Whole-body tick counts were conducted prior to tail-tag placement and at biweekly intervals until week 12. In the first trial, tick counts were made on all animals in each treatment herd for the first six weeks, then on about 50 cattle for each of the test groups containing 75 or 100 animals and all animals in the group of 25 animals. In the second trial, tick counts were made on 70 of the animals in each of the three acaricide-treated herds for the first six weeks and then on at least half of the animals for the remainder of the trial. All of the 15 animals in the control herd were counted at biweekly intervals. Tail-tag retention was noted for all animals when tick counts were made, and missing tail tags were replaced as needed.

Control of ticks was determined by using the formula $C = 100 - (T/U \times 100)$ where U is the mean number of ticks present on control cattle and T is the mean number of ticks present on treated cattle.[18] There were no differences between numbers of male or female bont ticks on untreated cattle compared to those on cattle with pheromone tail tags (ANOVA, $p < 0.05$), and all calculations of control were made using tick counts on cattle with pheromone tail tags as controls. All tick count data were transformed [$\log (x + 1)$] prior to analysis by ANOVA.

In view of the problems with tape deterioration and occasional complications due to constriction of the tail, effort was made to improve the attachment of tail tags. These trials were conducted at the University of Florida Dairy Research Unit using dairy cows (Holstein). Tail tags, which were prepared as described above but without pheromone components or acaricide, were rectangular (6 cm × 8 cm) and without slits. Tail tags were fastened to tails by applying cyanoacrylate adhesive (Bondini Corp.) to one side of the tag and folding the tag in half around a stand of hair from the tail switch. To facilitate bonding of the tag to the hair and onto itself, sheet metal pliers were clamped over the folded tag and hair for approximately 10 seconds. Heavy duty staples were also used to provide additional strength. Tail-tag retention was checked each week for 4 weeks with no replacement of tags if they fell off.

RESULTS

Retention of tail-tag decoys was similar between treatments within trial 1 ($p < 0.01$) and trial 2 ($p > 0.01$), and data for all treatments within a particular trial were combined. Retention in trial 1 was high with 92% of all animals retaining decoys (TABLE 1). By week 4 of trial 2, however, retention was quite low (66.5%) due to the degradation of the waterproof tape with exposure to sunlight. With use of the original filament tape, retention increased to above 80% for the remainder of the trial.

Numbers of male and female *A. hebraeum* increased throughout both trials 1 and 2 on untreated cattle and on cattle with pheromone tail tags (FIG. 1). By the end of trial 1, there were significantly more ticks (male or female) on untreated cattle and on cattle with pheromone tail tags than on cattle with cyfluthrin or flumethrin tail tags ($p < 0.05$). By the end of both trials 1 and 2, there were significantly fewer male bont ticks present on cattle with cyfluthrin tail tags compared to cattle with flumethrin tail tags ($p < 0.05$). By the end of the trial 2, significantly more male and female bont ticks were present on cattle with alphacypermethrin tail tags compared to those with either cyfluthrin or flumethrin tail tags ($p < 0.05$). Fewer males and females were present on cattle with alphacypermethrin tail tags compared to the pheromone or untreated controls ($p < 0.05$).

TABLE 1. Retention of Tail-Tag Decoys on Cattle throughout Two Field Trials in Zimbabwe

	Percent of Tail-Tag Decoys Retained	
Week No.	Trial 1	Trial 2
0	100.0	100.0
2	100.0	100.0
4	99.0	68.5
6	94.4	82.3
8	91.7	86.3
10	96.6	86.3
12	92.3	87.6

Other tick species found consistently on cattle during trials 1 and 2 included *Rhipicephalus evertsi, R. zambeziensis,* and several *Hyalomma* species. Effects of tail-tag decoys on numbers of these species on cattle are presented in FIGURE 2. Cattle with cyfluthrin- or flumethrin-impregnated tail tags were infested with fewer adult *R. evertsi, R. zambeziensis,* and *Hyalomma* spp. than untreated control cattle or those

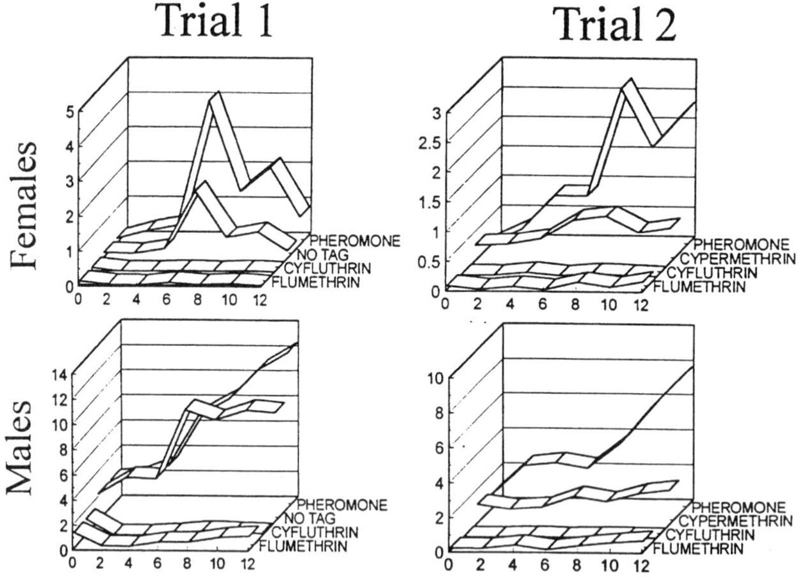

FIGURE 1. Mean numbers of *Amblyomma hebraeum* on cattle from weekly tick counts after treatment with tail-tag decoys containing pheromone alone, pheromone and cyfluthrin, flumethrin or alphacypermethrin, or untreated (no tags) during two field trials in Zimbabwe.

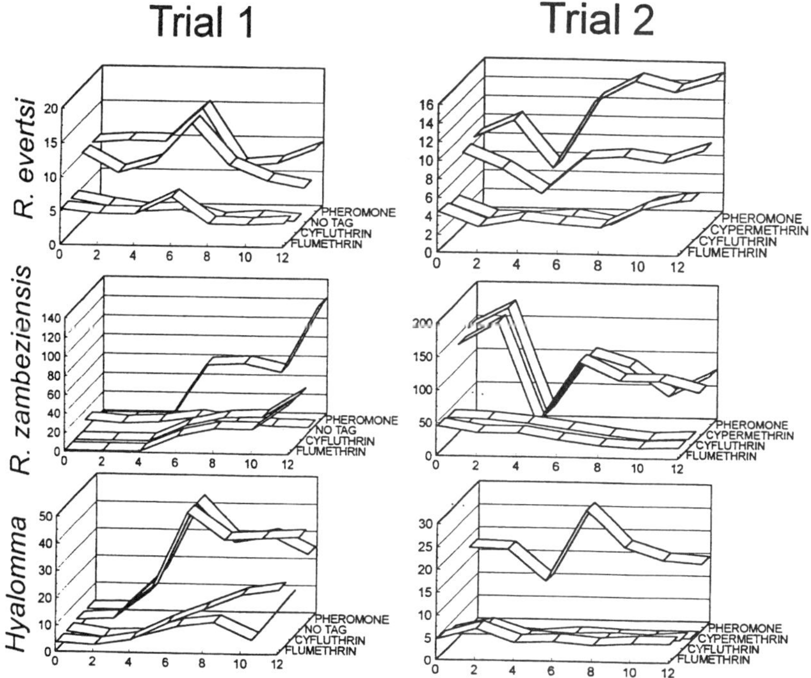

FIGURE 2. Mean numbers of *Rhipicephalus evertsi, Rhipicephalus zambeziensis* and *Hyalomma* spp. on cattle from weekly tick counts after treatment with tail-tag decoys containing pheromone alone, pheromone and cyfluthrin, flumethrin or alphacypermethrin, or untreated (no tags) during two field trials in Zimbabwe.

with tail tags containing only pheromone (Fig. 2). Numbers of ticks of all species on cattle treated with either flumethrin or cyfluthrin tail tags were similar at the end of both trials 1 and 2 ($p < 0.05$). Numbers of *R. evertsi* and *Hyalomma* spp. on untreated cattle were similar to those on cattle with pheromone tail tags ($p < 0.05$). Numbers of *R. zambeziensis,* however, were inexplicably lower on cattle with pheromone tags than on untreated cattle. Use of alphacypermethrin tail tags resulted in lower numbers of *R. evertsi* and *Hyalomma* spp. on cattle compared to those on cattle with pheromone tail tags ($p < 0.05$); however, there was no difference in numbers of *R. zambeziensis.* By the end of trial 2 there were no differences among acaricide treatments in numbers of *Hyalomma* on cattle ($p > 0.05$).

Excellent control of *A. hebraeum* males and females (91.8-99.5%) was obtained using cyfluthrin- or flumethrin-impregnated tail-tag decoys (TABLE 2). Control with use of alphacypermethrin-impregnated tail-tag decoys was moderate (49.9-74.3%). Control of other tick species was moderate, ranging from 40.1-73.2% for *R. evertsi,* 4.3-75.2% for *R. zambeziensis,* and from 33.5-82.2% for *Hyalomma* spp. (TABLE 2).

Retention of tail tags on 49 dairy cattle ranged from 97.9% (week 1) to 84.6% (week 4). No apparent aggravation or damage was seen on tails with attached decoys.

TABLE 2. Control of Ticks on Cattle Using Tail-Tag Decoys Impregnated with Pheromone and Cyfluthrin, Flumethrin, or Alphacypermethrin during Two Field Trials in Zimbabwe[a]

Species	Treatment	Present Control[b]	
		Trial 1	Trial 2
A. hebraeum Females	Cyfluthrin	97.8	98.8
	Flumethrin	92.0	91.8
	Alphacypermethrin	—	74.3
Males	Cyfluthrin	99.5	99.1
	Flumethrin	98.2	96.7
	Alphacypermethrin	—	49.9
R. evertsi	Cyfluthrin	69.4	73.2
	Flumethrin	57.5	65.3
	Alphacypermethrin	—	40.1
R. zambeziensis	Cyfluthrin	59.2	75.2
	Flumethrin	60.9	74.5
	Alphacypermethrin	—	4.3
Hyalomma	Cyfluthrin	69.2	82.2
	Flumethrin	46.0	72.2
	Alphacypermethrin	—	33.5

[a] Data for alphacypermethrin were only obtained for trial 2.
[b] Percent control calculated by the formula of Mount and associates.[18]

Moisture, high temperatures, and sunlight did not appear to affect retention. Tags that were not retained on tails were recovered from pens and all were still attached to strands of hair indicating that the loss was due to loss of hair.

DISCUSSION

Tail-tag decoys provided excellent consistent control of bont ticks on cattle (94.9% in trial 1; 99.3% in trial 2) for three months during both trials. Analysis by gas chromatography indicated that loss of pheromone components such as methyl salicylate and *o*-nitrophenol from the PVC matrix was slow with detectable levels present for 14 weeks (Sonenshine, D. E., unpublished data). The effective control of ticks throughout the trials, particularly with the cyfluthrin and flumethrin tail tags, indicated that acaricides were released from the PVC matrix at effective levels for the duration of the trials.

Initial studies incorporating acaricide with the aggregation-attachment pheromone of *Amblyomma maculatum*[19] or *A. hebraeum*[13] indicated the potential of this approach; however, effective control was short-lived. Our study provides clear evidence that slow-release devices containing acaricide in conjunction with pheromone components of bont ticks capable of eliciting short- and long-range attraction, aggregation, and attachment are a viable alternative to conventional control strategies for bont ticks.

Cyfluthrin and flumethrin provided greater control than alphacypermethrin. Control of other tick species on the cattle was considerably lower than for bont ticks, possibly because bont ticks were attracted to the pheromone source and were treated with a lethal dose of acaricide as they moved towards the source, or because preferred attachment sites for these species were not effectively treated with acaricide. In this study, control sufficient to reduce overall tick infestations to below the economic threshold was provided; however, all ticks were not killed. This may be beneficial as the presence of low numbers of ticks may be sufficient to maintain endemic stability of heartwater in areas where the disease is maintained with wild reservoir hosts.[20,21]

Attractant decoy tail tags have many advantages over conventional control methods in that they are relatively inexpensive, do not cause pollution, do not require specialized training or apparatus for application, are long lasting (3 mo.) and utilize bont-tick-specific pheromones to attract the ticks to areas of high acaricide concentration thereby requiring less acaricide to effectively treat an animal. The low cost and reduced amounts of acaricide required would reduce the drain on foreign currency reserves for large-scale control efforts in many countries. Tail tags can be adapted for tick eradication by placing a decoy on the anterior of a host by attaching it to a collar. Use of tail-tag decoys on some animals in conjunction with conventional control methods would allow selective removal of ticks remaining on pastures. While this technology was targeted for tick control, it has considerable potential as a surveillance method by placement of tail tags containing only pheromone on selected animals that can be routinely checked for the presence of ticks. Continued release of pheromone from tags would mimic hosts infested with feeding males, and host-seeking ticks would preferentially attach to these hosts.

The concept of attaching devices to the tails of cattle has previously been documented for fly[22,23] and tick control;[24,25] however, loss of devices from tails has been a consistent problem in these studies. During our study, deterioration under various climatic conditions of the tape attaching decoys to the tails was a problem as were the occasional complications arising from the presence of a constrictive device on the tail. The adaptation of tail tags for attachment to hair in tail switch overcomes these problems and greatly improves the potential application of this technology.

The pheromone components for *A. hebraeum* incorporated in the tail-tag decoy in this study also elicit similar responses in *A variegatum*.[26–28] A field trial in the Caribbean is planned to evaluate the efficacy of this technology, both as tail tags and on collars, for localized eradication of *A. variegatum*. Additionally, the potential for this technology as a surveillance method will also be evaluated.

ACKNOWLEDGMENTS

We thank P. L. Donovan, C. Smith, P. Dube, J. Phillips, K. Prine, and M. Stevenson for assistance, Fallowfield Farm in Zimbabwe and the University of Florida Dairy Research Unit for use of cattle and facilities, and Bayer Zimbabwe (Pvt.) Ltd, FMC Corporation, Witco Corporation, and B. F. Goodrich Company for technical information and supplies.

REFERENCES

1. PETNEY, T. N., I. G. HORAK & Y. RECHAV. 1987. The ecology of the African vectors of heartwater, with particular reference to *Amblyomma hebraeum* and *Amblyomma variegatum*. Onderstepoort J. Vet. Res. **54:** 381–395.

2. UILENBERG, G. 1983. Heartwater (*Cowdria ruminantium* infection): current status. Adv. Vet. Sci. Comp. Med. **27:** 427–480.

3. CAMUS, E., D. MARTINEZ, L. BEAUPERTHUY, A. BENDERDOUCHE, S. COISNE, C. CORBETTE, N. DENORMANDIE, G. GARRIS, E. HARRIS, T. KING, L. LANNOY, A. LOUVET, E. LUNEL, M. MONTROSE, B. NISBETT, B. NYACK, J. ROBINSON, P. ROUCHOSSE, G. B. SWANSTON, B. THIEBOT & G. D. THYE. 1993. Heartwater in Guadeloupe and in the Caribbean. Rev. Elev. Med. Vet. Pays Trop. **46:** 109–114.

4. BURRIDGE, M. J. 1985. Heartwater invades the Caribbean. Parasitol. Today **1:** 1–3.

5. NORVAL, R. A. I. & Y. RECHAV. 1979. An assembly pheromone and its perception in the tick *Amblyomma variegatum* (Acarina: Ixodidae). J. Med. Entomol. **16:** 507–511.

6. RECHAV, Y. 1977. Evidence for an assembly pheromone(s) produced by males of the bont tick *Amblyomma hebraeum* (Acarina, Ixodoidea). J. Med. Entomol. **14:** 71–78.

7. NORVAL, R. A. I., J. F. BUTLER & C. E. YUNKER. 1989. Use of carbon dioxide and natural or synthetic aggregation-attachment pheromone of the bont tick, *Amblyomma hebraeum*, to attract and trap unfed adults in the field. Exp. Appl. Acarol. **7:** 171–180.

8. NORVAL, R. A. I., T. PETER, C. E. YUNKER, D. E. SONENSHINE & M. J. BURRIDGE. 1991. Responses of the ticks, *Amblyomma hebraeum* and *A. variegatum* to known or potential components of the aggregation-attachment pheromone. II. Attachment stimulation. Exp. Appl. Acarol. **13:** 19–26.

9. NORVAL, R. A. I., T. PETER, D. E. SONENSHINE & M. J. BURRIDGE. 1992. Responses of the ticks *Amblyomma hebraeum* and *A. variegatum* to known or potential components of the aggregation-attachment pheromone. III. Aggregation. Exp. Appl. Acarol. **16:** 237–245.

10. NORVAL, R. A. I., T. PETER, D. E. SONENSHINE & M. J. BURRIDGE. 1992. Responses of the ticks, *Amblyomma hebraeum* and *A. variegatum* to known or potential components of the aggregation-attachment pheromone. IV. Attachment stimulation of nymphs. Exp. Appl. Acarol. **16:** 247–253.

11. NORVAL, R. A. I., H. R. ANDREW & C. E. YUNKER. 1989. Pheromone-mediation of host selection in bont ticks (*Amblyomma hebraeum* Koch). Science **243:** 364–365.

12. NORVAL, R. A. I., J. F. BUTLER & C. E. YUNKER. 1989. Use of carbon dioxide and natural or synthetic aggregation-attachment pheromone of the bont tick, *Amblyomma hebraeum*, to attract and trap unfed adults in the field. Exp. Appl. Acarol. **7:** 171–180.

13. RECHAV, Y. & G. B. WHITEHEAD. 1978. Field trials with pheromone-acaricide mixtures for control of *Amblyomma hebraeum*. J. Econ. Entomol. **71:** 149–151.

14. NORVAL, R. A. I., D. E. SONENSHINE, M. I. MELTZER & M. J. BURRIDGE. 1994. Attractant decoy for controlling bont ticks. United States Patent No. 5,296,227, US Patent Office. Washington, DC.

15. BEADLES, M. L., A. R. GINGRICH & J. A. MILLAR. 1977. Slow-release devices for livestock insect control: cattle body surfaces contracted by five types of devices. J. Econ. Entomol. **70:** 72–75.

16. WALKER, J. B. 1974. The ixodid ticks of Kenya. Commonwealth Institute of Entomology. London, England.

17. MCLEOD, J., M. H. COLBO, M. H. MADBOULY & B. MWANAUMO. 1977. Ecological studies of ixodid ticks (Acari: Ixoididae) in Zambia. III. Seasonal activity and attachment sites on cattle with notes on other hosts. Bull. Entomol. Res. **67:** 161–173.

18. MOUNT, G. A., R. H. GROTHAUS, J. T. REED & K. F. BALDWIN. 1976. *Amblyomma americanum:* area control with granules or concentrated sprays of diazinon, propoxur and chlorpyrifos. J. Econ. Entomol. **69:** 257–259.

19. GLADNEY, W. J., R. R. GRABBE, S. E. ERNST & D. D. OEHLER. 1974. The Gulf Coast tick: evidence of a pheromone produced by males. J. Med. Entomol. **11:** 303–306.

20. NORVAL, R. A. I., H. R. ANDREW, C. E. YUNKER & M. J. BURRIDGE. 1992. Biological process in the epidemiology of heartwater. *In* Tick Vector Biology. Medical and Veterinary Aspects. B. Fivaz, T. Petney & I. Horak, Eds.: 71-86. Springer-Verlag. Berlin, Germany.

21. NORVAL, R. A. I., B. D. PERRY, M. I. MELTZER, R. L. KRUSKA & T. H. BOOTH. 1994. Factors affecting the distributions of the tick *Amblyomma hebraeum* and *A. variegatum* in Zimbabwe: implications of reduced acaricide usage. Exp. Appl. Acarol. **18:** 383-407.

22. HOGSETTE, J. A., & J. P. RUFF. 1987. Control of stable flies and horn flies (Diptera: Muscidae) with permethrin tapes applied to tails of beef and dairy cattle. J. Econ. Entomol. **80:** 417-420.

23. TAYLOR, S. M., C. T. ELLIOTT & W. J. BLANCHFLOWER. 1987. A comparison of cypermethrin distribution on cattle hair after application of impregnated ear and tail tags. Pest Sci. **21:** 39-43.

24. SCHRODER, J. & P. C. VAN SCHALKWYK. 1988. Die doeltreffendheid van alfametriendeurweekte oorplaatjies teen beesbosluise. J. S. Afr. Vet. Assoc. **60:** 79-82.

25. MILLER, J.A. & J. E. GEORGE. 1994. Efficacy of a combination neckband and tailtag containing amitraz or cyhalothrin K against lone star ticks on pastured cattle. J. Agric. Entomol. **11:** 165-176.

26. SCHONI, R., E. HESS, W. BLUM & K. RAMSTEIN. 1984. The aggregation-attachment pheromone of the bont tick *Amblyomma variegatum* Fabricius (Acari: Ixodidae). Isolation, identification, and action of its active components. J. Insect Physiol. **30:** 613-618.

27. LUSBY, W. R., D. E. SONENSHINE, C. E. YUNKER, R. A. NORVAL & M. J. BURRIDGE. 1991. Comparison of known and suspected pheromonal constituents in males of the African ticks, *Amblyomma hebraeum* Koch and *Amblyomma variegatum* (Fabricius). Exp. Appl. Acarol. **13:** 143-152.

28. PRICE, T. L., JR., D. E. SONENSHINE, R. A. I. NORVAL, C. E. YUNKER & M. J. BURRIDGE. 1994. Pheromonal composition of two species of African *Amblyomma* ticks: similarities, differences and possible species specific components. Exp. Appl. Acarol. **18:** 37-50.

Puerto Rico Tick Program

Potential Conversion to an Integrated Pest Management Program

BOB H. BOKMA

*United States Department of Agriculture
Animal and Plant Health Inspection Service
Veterinary Services
Post Office Box 71355
San Juan, Puerto Rico 00936*

Infestations by ticks and infections by tick-borne pathogens are serious limiters of livestock production in tropical regions of the world.[1-3] This is potentially true in Puerto Rico, which is affected with the southern cattle tick *Boophilus microplus* (Canestrini) and the hemoparasitic pathogens of cattle *Babesia bigemina,* the protozoan agent of babesiosis (also known as piroplasmosis and Texas fever), and *Anaplasma marginale,* the rickettsial agent of anaplasmosis. Puerto Rico's cattle tick eradication program, which is conducted by the US Department of Agriculture (USDA) and the Puerto Rico Department of Agriculture, is directed specifically at *B. microplus.*

As is true for many animal disease eradication and control programs, the tick program has been under increasing scrutiny for continued funding. Recent Presidential references to this program increase uncertainty for future funding and participation by USDA, Animal and Plant Health Inspection Service, Veterinary Services (VS). Program management must plan for alternate strategies for tick control. This paper presents some integrated pest management (IPM) approaches that may be useful if funding is lost before the tick is eradicated.

BACKGROUND

B. microplus reappeared in Puerto Rico late in 1977, following a 24-year absence after the successful eradication in the early 1950s.[4,5] The current program is based on repeated whole-body spraying of all cattle on farms that are found to be infested when cattle are inspected. The current pesticide of choice is amitraz, at a concentration of 0.025% active ingredient. Injectable and oral bolus-form pesticides and those that are applied as pour-on or spot-on products have not yet been used. These products do not offer the levels of control that are required for success in eradication, their use requires difficult withdrawal times, or they lack registry for use in the United States. Dip-vat formulations have not been used due to environmental considerations related to spent pesticide disposal, as well as logistical problems, maintenance costs, and the lack of safe, cost-effective, and efficacious formulations.

94

Treatments of infested cattle are repeated every 21 days. This interval has been found to be effective under ideal situations, due to a combination of the residual effect after treatment and the tick's life cycle. Treatments extend for at least 231 days and until ticks are no longer detected on prerelease inspections. Occasional cattle herds presenting with acute outbreaks of anaplasmosis and babesiosis are treated more frequently. Under United States regulations, the chemotherapeutic product imidocarb dipropionate, which is widely available outside of the United States, is not approved for use in food animals.

Following hurricane Hugo in September 1989, all treatments in the eastern portion of Puerto Rico were discontinued for a period of about five months. Within a period of about one month, many cattle herd owners, especially of dairy heads, reported alarming numbers of acute deaths due to piroplasmosis. There was a general clamor for the renewal of treatment activities. Several illegal importations of imidocarb dipropionate were investigated by the Food and Drug Administration.

Surveillance for infestation is routinely conducted on all known cattle herds considered noninfested.

On the main island, there are a total of 17,400 herds with cattle and a total bovine population of roughly 346,000 head. Since 1990, the entire main island has been under active eradication. The program has been extended to the island municipality of Culebra, but not to Vieques, the other island municipality.

In general terms, the program's successes include an effective control of tick infestations and essentially all clinical disease due to babesiosis and anaplasmosis. Over 68% of cattle herds, varying from 25% to 100% per municipality, and over 50% of all cattle are free of infestation. In 7 of Puerto Rico's 78 municipalities, less than 10% of the total herds are considered infested. A total of 15 municipalities have less than 20% infested herds. Culebra has been found free of *B. microplus* repeatedly. Vieques is considered infested.

Unfortunately, the program has not been able to definitively curtail cattle movements that are not in compliance with regulations. Also, the program has been plagued by a significant reinfestation rate in herds released from treatments after completing a treatment cycle. While these problems occur in all areas, they are particularly important in the dairy milk-shed area of the northern coast.

IPM PRINCIPLES

IPM is achieving considerable acclaim for insect pest management and offers interesting solutions for tick-related management. In the tick realm, IPM may be defined as the simultaneous use of compatible techniques to manage tick burdens, diseases caused by tick-borne agents, and other losses caused by ticks. In Puerto Rico's case, these losses are primarily due to *B. microplus*. They can also include *Anocentor nitens,* the tropical horse tick, and *Amblyomma variegatum,* the tropical bont tick. IPM methods must act at a tick/pathogen/cattle interface[1,3,6–9] and can be classified as chemical, host resistance, and host management.[3,6,9] Another author focused on mechanisms that affect tick feeding and biology, with the consequent reduction in tick populations being secondary, or that affect the hemoparasite directly by reducing infection, development, or the transmission by ticks, or a combination of these.[1]

IPM measures may be so varied as herd management and pasture makeup, herd breeding, vaccination augmenting natural animal immunity, and strategic use of different acaricides and methods of application.[1,3]

A successful IPM program should effectively reduce the numbers of ticks feeding on cattle and the concomitant losses in animal production. It should enhance the level of host immunity to *B. microplus* and to *Anaplasma* and *Babesia*. Ideally, the IPM program will allow significant marketability of animals and animal products not affected because of imposed quarantines, withdrawal periods due to products used, or illness caused by ticks.[1]

ACARICIDE, PHEROMONE, AND OTHER TICK-MANAGEMENT METHODS

Acaricides are mainstays for tick management and are used on animals via body dips, sprays, pour-on or spot-on formulations, back rubbers, injectable products, boluses, and self-medication techniques. They may also be directly applied on pastures.[1,3]

As part of IPM, acaricides and the different application technologies that are or may become available offer different levels of control and interesting solutions. Acaricides can be applied strategically over time to minimize herd infestations and production losses, hide damage, residues in meat and milk, and the transmission of hemoparasitic disease, especially *Babesia*.[1,3] Strategic acaricide treatments will be important to protect the current immunologically naive population while gradually improving the immunological status. Frequent acaricide treatments have been particularly beneficial for acute outbreaks of babesiosis.

Demonstration projects and experimental protocols for dairy herd use of sprayable lambda cyhalothrin and/or cyfluthrin are a distinct possibility. Also, an injectable encapsulated formulation of avermectin has been offered for a demonstration project. Self-treatment methods for cattle as they feed or pass through alleyways are of interest. Finally, various acaricides on baits may also provide specific opportunities for use.

Pheromone methodology is not likely to become available for one-host ticks such as *B. microplus*.[10]

While sterile hybrid technology has also been tried to manage *Boophilus* tick populations, the numbers of sterile hybrids produced and the seeding of these on pastures remain limiting factors for effective use.[10,11]

The use of methodology to manage tick populations using predator species of arthropods and nematodes will also be of interest.[1,3]

PASTURE AND HOST MANAGEMENT

Water balance is critical to nonfeeding ticks.[2,12] Tillage and irrigation methods that achieve the lowest possible surface humidity can be useful in reducing tick longevity. Access of cattle on pastures used for grazing will make a difference in the buildup of tick populations. Pasture spelling or vacation, especially in the more arid areas, and the use of brush clearing, as well as pasture mowing, can be effective methods in reducing numbers of larval ticks.[2,3,6,13]

On grazed pastures, species of forage can include tick-killing varieties.[2,12,13]

Pasture management experts and the agricultural extension service can provide information on managing both cattle and pastures. These sources can also provide valuable advice on forage harvesting methods such as green chop, silage, and hay.

HOST RESISTANCE

Cattle breeding influences susceptibility to ticks and disease, with some European, *Bos indicus* (Zebu breeding or Brahman-type), and Creole breeds generally more resistant to both. Dairy and beef producers may also wish to increase use of specific *Bos taurus* or *B. indicus* breeding to improve tick resistance. These breeds may not have the genetic traits desired by producers who are used to specific European breeding.[1,6]

In many tropical countries, dairy producers are advized to cull out animals that have more ticks.[3,11] This technique does not seem to give clear results of reduced tick burdens in the offspring of animals retained for breeding purposes. Generally, the use of tick-control programs, combined with breeding for increased genetic potential for milk or beef production, is considered more profitable than breeding for tick or disease resistance.[1]

Vaccination for increased tick resistance has been tried with limited success in different countries.[1-3] If new biogenetic vaccines become available, VS can support this technology.

Low grade infection with *Anaplasma* and *Babesia*, particularly if cattle are infected as young animals, protects from acute disease. For it to be effective, tick control must not be exercised indiscriminately. Acute disease can easily follow periods of lax control. The cattle population in Puerto Rico is immunologically naive; thus continued tick control will be necessary until immunity can be enhanced. The governmental veterinary services will need to advise caution on relaxing tick control until this immunity can be achieved.

Different forms of vaccination aimed at *Babesia* have been commonly used in most tropical countries. Premunition, the transfer of infection via infected blood from older animals to the young, is the oldest and to our knowledge has never been used in Puerto Rico. This method is generally not considered satisfactory due to the potential role in the uncontrolled transmission of diseases. Projects to secure commercial vaccines licensed in the United States will also be explored and/or extended to Puerto Rico.

CHEMOTHERAPEUTIC TREATMENTS

Susceptible adult cattle that become acutely infected with anaplasmosis or babesiosis must be treated with chemotherapeutic agents such as oxytetracycline, diamidine derivatives, and imidocarb dipropionate. Puerto Rico may be an appropriate site for trials to further licensure of chemotherapeutic compounds.

CONCLUSION

Given the modern-day reality of potentially reduced funding for eradication of ticks and tick-borne diseases affecting livestock, new solutions for managing these are necessary. In Puerto Rico, where traditional eradication protocols have not yet been successful in eliminating ticks, there may be a need to use a variety of methods for the control of ticks and tick-borne diseases. The federal and commonwealth offices that are currently involved in the eradication program will be active partners in preparing for this approach.

ACKNOWLEDGMENTS

The author sincerely appreciates the assistance and ideas of Raymond Carruthers, PhD, Research Leader, Biological Pest Control Research, ARS, Weslaco, TX; Jose R. Diez, Director, Tick Eradication Program, San Juan, PR; Glen I. Garris, PhD, Senior Staff Entomologist, VS, Riverdale, MD; John George, PhD, Research Leader, Knipling-Bushland US Livestock Insects Research Laboratory, Kerrville, TX; Dario Gonzalez, Area Epidemiology Officer, Tick Eradication Program, San Juan, PR.

REFERENCES

1. KOCAN, K. M. 1994. Focusing on ticks for control of hemoparasitic diseases. *In* Proceedings of the United States Animal Health Association: **98:** 338–348. Prowriter Communications, Inc. Richmond, VA.
2. SHONENSHINE, D. E. 1991. Biology of Ticks. Vol. **1.** Oxford University Press. New York, NY.
3. SHONENSHINE, D. E. 1993. Biology of Ticks. Vol **2.** Oxford University Press. New York, NY.
4. GRAHAM, O. H. & J. L. HOURIGAN. 1977. Eradication programs for the arthropod parasites of livestock. J. Med. Entomol. **13:** 629–658.
5. SUTHERN, C. B. & G. P. COMBS. 1984. The Puerto Rico tick eradication program. *In* Proceedings of the United States Animal Health Association **88:** 406–412. Carter Printing Company. Richmond, VA.
6. GEORGE, J. E. 1990. Summing-up of strategies for the control of ticks in regions of the world other than Africa. Parassitologia **32:** 203–209.
7. SUTHERST, R. W., G. A. NORTON, N. D. BARLOW, G. R. CONWAY, M. BIRLEY & H. N. COMINS. 1979. An analysis of management strategies for cattle tick (*Boophilus microplus*) control in Australia. J. Appl. Ecol. **16:** 359–382.
8. NORTON, G. A., R. W. SUTHERST & G. F. MAYWALD. 1984. Implementation models: the case of the Australian Cattle Tick. *In* Pest and Pathogen Control: Strategic, Tactical, and Policy Models. G. R. Conway, Ed. Wiley. Chichester, West Sussex, England.
9. NORTON, G. A., R. W. SUTHERST & G. F. MAYWALD. 1983. A framework for integrating control methods against the cattle tick, *Boophilus microplus* in Australia. J. Appl. Ecol. **20:** 489–505.
10. GARRIS, G. I. 1995. Personal communication.
11. GEORGE, J. E. 1994. Personal communication.

12. NORVAL, R. A. I. 1992. The Epidemiology of Theileriosis in Africa. Academic Press
     Limited. London, England.
13. DESQUESNES, M. 1995. The Cattle Tick: *Boophilus microplus*. S. Votaky, Ed. (English
     edition.) Centre de Cooperation Internationale en Recherce Agronomique pour le Devel-
     oppement/Elevage et Medicine Veterinaire des pays Tropicaux and International Insti-
     tute for Cooperation in Agriculture. Cayenne, French Guiana & Georgetown, Guyana.

Effect of Management and Inherent Factors on Birth Seropositivity and Seroconversion to *Babesia bovis* in Calves in Costa Rica

ENRIQUE PÉREZ,[a,c] MARCELINO JAÉN,[b] AND
MARCO V. HERRERO [a,d]

[a]*Regional Graduate Program in Tropical Veterinary Sciences
School of Veterinary Medicine
Universidad Nacional
Costa Rica*

[b]*Instituto de Investigaciones Agropecuarias
Panamá, Panamá*

[c]*Herd Health Project
Costa Rican-Dutch Interuniversity Cooperation Program*

[d]*Tropical Disease Research Program
School of Veterinary Medicine
Universidad Nacional
Costa Rica*

INTRODUCTION

Babesia bovis (Babes, 1888) is the most pathogenic species of *Babesia* in several areas of the world where it affects cattle.[1] It is known to be present in Costa Rica.[2] It was reported to invade the central nervous system in a young animal.[3] Two areas of nonrandom seroprevalence of *B. bovis* were detected in Costa Rica. One focus is located mainly in the province of Guanacaste (northwestern and dry pacific region), with an annual precipitation of 1000 to 2000 mm, and an average temperature of 24°C. The other focus is located in the lowlands of the provinces of Alajuela, Heredia, and Limon (tropical moist forest, southern Caribbean coast) with an annual precipitation, of 2000 to 4000 mm and an average temperature higher than 24°C.[4] Another study in this endemic area (Guanacaste: Tilarán) showed that the number of young animals (less than 1 year old) seropositive to *B. bovis* fluctuates during the year from 24.1% to 80.4%.[5]

The objective of this study was to investigate the relationships among managerial factors, host factors, ecological conditions, and the immunological response of newborn and older calves to *B. bovis*.

MATERIAL AND METHODS

Dual Purpose and Beef Farms

The Tilaran area was selected as a study area because the Herd Health Project of the School of Veterinary Medicine, Universidad Nacional, was conducting a pilot project in that area, as part of a livestock information system and a serum bank was available. Eleven farms were selected from a list of 23. Herds were eligible for inclusion if the farmer was willing to collaborate. The Tilaran region is part of a tropical moist forest area with an average rainfall of 2000 to 4000 mm, 0.5 to 1 evapotranspiration index, and ambient temperature between 24° and 30°C.

The extensive beef cow-calf production system on medium and some large farms is the main livestock enterprise in the area with an average farm size of 170 Ha (range 47-871 Ha), with a stocking rate of 1.06 animals per hectare (range 0.38-2.80). This type of production system is traditional in the area, with farmers having on average 23 years of experience. Labor used was one man per each 64 cows (range 18-298). Hiring of laborers was not uncommon to the system, particularly if the farm was too big (relative to family labor). On average, farms had 18 paddocks (range 3-60) mainly with Star Grass (*Cynodon plectostachyus*) and Jaragua (*Hyparrhenia rufa*). Supplementation with mineral salt was common on all farms. The culling rate was 11.78% (range 1-32), and the calving rate 65% (range 24-80).

The dual purpose production system in the region comprises medium-scale farms, with an average of 75 Ha (range 5 to 252 Ha). The average farming experience was 16 years, and on average one man of labor was used for each 44 cows (7-164). Family labor was primarily used for the routine management activities such as milking, calf caring, and feeding practices. Nutrition was pasture based with Star Grass combined with native grass and Jaragua. The average number of paddocks was 17 (2-83) with a stocking rate of 1.82 animals per hectare (0.61-2.34). Supplementation with molasses and mineral salt often occurred during the dry period (May to December). Few producers applied a cut and carry system with stall feeding. The annual culling rate was 8.82% (range 4-19), and the average calving rate was 46% (range 38-57). Milk production levels of 1500 to 2000 kg per lactation (270 days) were obtained during the study period.

Samples

A total of 3624 samples from a serum bank collected by the Herd Health Project technicians and stored at the School of Veterinary Medicine was used for this study. Each sample has animal and farm identification as well as the date of collection. The sera samples were from 467 animals, bled every 30 days from birth to weaning.

ELISA Assay

The assay used was the ELISA kit provided by the International Atomic Energy Agency,[6] with some modifications described as follows: 100 µl of the *Babesia bovis* antigen was added to each well in a 96 microplate and incubated at 4°C overnight.

The antigen (Ag) was donated by Dr. Don Berrie (Long Pocket Laboratories, CSIRO, Australia), and it was diluted 1:500 in a 0.05 M carbonate/bicarbonate buffer at pH 9.6. The next morning, the Ag was discarded and the microplate washed. As blocking agent, skim milk (5%) diluted in 0.05 M carbonate/bicarbonate buffer pH = 9.6 was used; the plate was incubated for 1 h at 37°C and washed. The unknown sera, diluted 1:300 in PBS, pH 7.4 plus Tween 20 (0.05%) plus skim milk (5%), was added; and the plate was incubated again under the same conditions and washed. The IgG anti-bovine conjugate, whole molecule conjugated with peroxidase, was diluted 1:11000 in PBS at pH 7.4 plus Tween 20 (0.05%) plus skim milk (5%) and added. The incubation and wash steps were repeated. Finally, 3.7 mM OPD (*o*-phenylenediamine) plus 3.5 mM hydrogen peroxide were added, and the plate was incubated for five minutes at room temperature. The reaction was stopped with 2 M sulfuric acid and read at 492 nm in an ELISA reader (Titertek Multiskan MK II).

Controls

Positive and negative controls were also donated by Dr. Don Berrie and used in each plate.

Cutoff Line

Two hundred Costa Rican sera negative for hemoparasite from a tick-free area were run. The cutoff line used was the optical density mean plus three times the standard deviation (0.229). All sera with OD higher than the cutoff line were considered positive. When compared with the IAEA positivity criteria, we found that our positive sera were within the kit's range.

Statistical Analysis

The unit of analysis was the individual animal. Standardized definitions for management practices were discussed with the calf caretakers and farm managers. Response variable was seropositivity. Exposure variables were: gender (male, female), breed (*Bos indicus,* crossbred), type of production system (dual purpose, beef), parity, calving history (normal, dystocia), place of birth (corral, paddock), time of birth (night, day), colostrum intake (first 6 hours of life, more than 6 hours) colostrum administration (dam, teat bottle), life zone (five zones), calving season (dry, rainy) and farm (11 farms).

A primary screening was performed using $2 \times K$ contingency tables of exposure variables with outcomes (PROC FREQ, SAS 1991). All covariates associated with outcome variables with a p value ≤ 0.25 were considered in the secondary steps of the screening procedure. A stratified analysis of host variables by farm level variables was done. When required (if the null hypothesis of homogeneity was rejected, $p <$ 0.05), an interaction term for multivariate analysis was created. All main effects and interaction terms selected in the screening procedures were used in the multivariate analyses. A backward elimination process was employed. An examination of the

Wald statistic for each variable and assessment of the odds ratios estimates were done at each step. All nonsignificant terms were eliminated, and a new model fitted. A proportionate change of 5% in the estimates was used as a criterion to force the covariate to remain in the model as a potential confounder. Models were compared using the likelihood ratio test. To control for possible herd effects, a logistic-binomial regression model for distinguishable data was used.[7] This model was selected because the data had variables measured both at herd level (herd effects) and animal level (host effects) associated with the individual responses rather than just with the herds.

The following general model was used:

$$\pi_j = \frac{e^{z'_j\beta}}{1 + e^{z'_j\beta}} \, j = 1, \ldots, J$$

where z_j is the covariate vector for the jth proportion in the set, and β is a vector of parameters. The success probability is given by:

$$logit(\pi_j) = z'_j\beta + \sigma\nu$$

where μ is the standardized binomial distribution, and $\sigma \geq 0$.

Time to Event Analysis

Time to event analysis was performed using the Cox Proportional Hazards model. The Proportional Hazards Model[8] was used to analyze the effect of management factors on time to seropositivity. In this model, the probability that some event occurred (hazard function), including covariate information, was calculated. The hazard function represented the risk of failure (seropositive) for a calf who had to time t. This is conditional on survival up to the beginning of the interval. The variables used were chosen by the screening procedure described earlier and refitted with the Cox Proportional model stratified in production system and ecological life zone. The proportional hazard assumption was examined, graphing the log [−log S (t, ecological life zone)], log [−log S (t, type of farm)] for various periods and the levels of covariate used in the stratification. Differences between survival curves were assessed using the Wilcoxon test.

In notational form:

$$h(t) = \lim_{\Delta t \to 0^+} \frac{P(t \leq T \leq t + \Delta t | t \leq T)}{\Delta t}$$

RESULTS

Two hundred and seventy-six of 467 (59.1%) of newborn animals were positive at 4.4 days old. Seropositivity risk factors at birth were shown by univariate (TABLE

TABLE 1. Univariate Analysis of Factors Associated with Seropositivity at Birth, Tilaran, Costa Rica, 1990–1991

Factor	OR[c]	95% CI[d]	p Value
Females vs. males[a]	1.11	0.75-1.63	0.593
B. indicus vs. *B. taurus*	1.53	1.03-2.26	0.028
Dual purpose vs. beef	1.48	1.00-2.19	0.040
Cows vs. first parity heifers	0.37	0.75-1.63	0.001
Normal calving vs. dystocia	0.55	0.19-1.54	0.209
Birth place (corral vs. paddock)	6.40	0.82-1.36	0.045
Time of birth (night vs. day)	1.60	1.04-2.45	0.023
Colostrum intake (after 6 h vs. first 6 h)	0.25	0.14-0.45	0.001
Colostrum intake (help vs. voluntary)	0.28	0.16-0.61	0.001
Colostrum source (teat bottle vs. dam)	0.28	0.06-1.30	0.061
Tropical moist forest premountain transition[b]	3.24	1.55-6.85	0.001
Premountain wet forest	8.57	3.61-20.6	0.001
Premountain wet forest basal transition	5.49	2.31-13.4	0.001
Premountain wet forest, rain forest transition	6.19	2.11-18.7	0.001

[a] Second factor used as base category.
[b] Tropical moist forest used as base category.
[c] OR 1:1 odds ratio.
[d] 95% CI = 95% confidence interval for the odds ratio.

1) and logistic with random effects (TABLE 2) analysis. Calves that drank colostrum within 6 hours had odds of being seropositive 2.9 times greater than calves drinking colostrum after 6 hr. Also, calves born in a corral had odds of being seropositive 38.8 times greater than calves born in paddocks. In addition, calves from normal delivery had odds 2.56 times greater than calves born from distocya. Calves from heifers were less likely to be seropositive than calves from multiparous cows. Moreover, there were differences among ecological zones.

Once the animals were born, a decrease in the number of seropositive animals during the next two months was observed. It decreased from 59.1% to 33.8% (31.9 days) to 28.9% (62.2 days). However, there was an increase in antibody prevalence to *B. bovis* in the following months. An increase was observed in the number of seropositive animals to 50% at 145 days and to 80% at 244 days. Finally, at weaning, most of the animals were seropositive (FIG. 1).

Cox regression model analysis indicated that seroconversion began at 28 days and increased as animals aged. However, it was found that the greater odds for seropositivity to *B. bovis* were by 119 days after birth; up to this period, 60% of calves seroconverted. It was possible to detect a few seroconversions until 300 days postbirth.

In this model, the pattern of survival was different by production system. (FIG. 2). In the cow-calf farms, the number of seroconversions by age was lower than in the dual-purpose farms. At 94 days, 50% of the beef calves had seroconverted, and 90% by day 148. On the other hand, by day 60, 50% of the dual-purpose calves had seroconverted, and 90% by day 88.

TABLE 2. Logistic Binomial Multiple Regression for Seropositivity to *B. bovis* at Birth, Costa Rica, 1991–1992

Factor	β	Standard Error	*p* Value	OR[c]	95% CI[d]
Colostrum intake (after 6 h vs. first 6 h)[a]	−1.070	0.306	0.001	0.34	0.18–0.68
Birth place (corral vs. paddock)	3.641	1.32	0.006	38.1	2.8–504
Dystocia vs. normal calving	−0.935	0.381	0.014	0.39	0.18–0.82
First parity heifer vs. older cows	−1.016	0.324	0.002	0.36	0.19–0.68
Tropical moist forest[b] premountain transition	1.128	0.502	0.025	3.09	1.10–8.2
Premountain wet forest[b]	2.037	0.570	0.001	7.66	2.5–23.4
Premountain wet forest basal transition[b]	1.839	0.577	0.001	6.29	2.0–19.4
Premountain wet forest, rain forest transition[b]	1.835	0.693	0.008	6.26	1.6–24.3

[a] First category used as risk factor.
[b] Tropical moist forest used as base category.
[c] OR = odds ratio.
[d] 95% CI = 95% confidence interval for the odds ratio.

DISCUSSION

Tilarán is an area endemic for *B. bovis* where it is expected that many cows in their reproductive age will be seropositive. Previous data showed a seroprevalence between 53.4% and 66.2% in animals between 1 and 2 years old and between 44.2% and 77% in animals older than 2 years old in the same area.[5]

In this study, all the host factors associated with seropositivity were related to colostrum intake: colostrum from first parity heifers (odds ratio = 0.36) could contain less immunoglobulins than that from older herd-mates.[9] Moreover, calves born in corrals had higher odds of being seropositive than those born in paddocks. Perhaps time to access to colostrum could be an explanation, because calves born in corrals are supervised for colostrum intake. Dams that suffered from calving difficulties can have calves that suffered from exhaustion, weakness, a delay in time to rise, and as a consequence less chance of having colostrum in the proper time. In calves, absorption of colostrum continues until 24 hours, but maximum absorption occurs within the first 6 hours after birth.[10]

The number of seropositive animals decreased in the first two months postbirth from 59.1% to 28.9%. Seropositive cows are able to transmit *Babesia* antibodies to their newborns by colostrum,[11–13] but these antibodies will decrease in time after birth.[12,14,15] In the absence of a sustained field booster stimulus, irregular seroconversions will take place after this period; for example, in this study, the number of seropositive animals changed in one exceptional cow-calf farm from 54.4% (at birth) to 30.3% (31.9 days), then from 2.6% (62.2 days) to 0% (91.8 days). A slight increase to 14.2% (152.5 days), a decrease again from 11.1% (182.4 days) to 3.3% (212.1

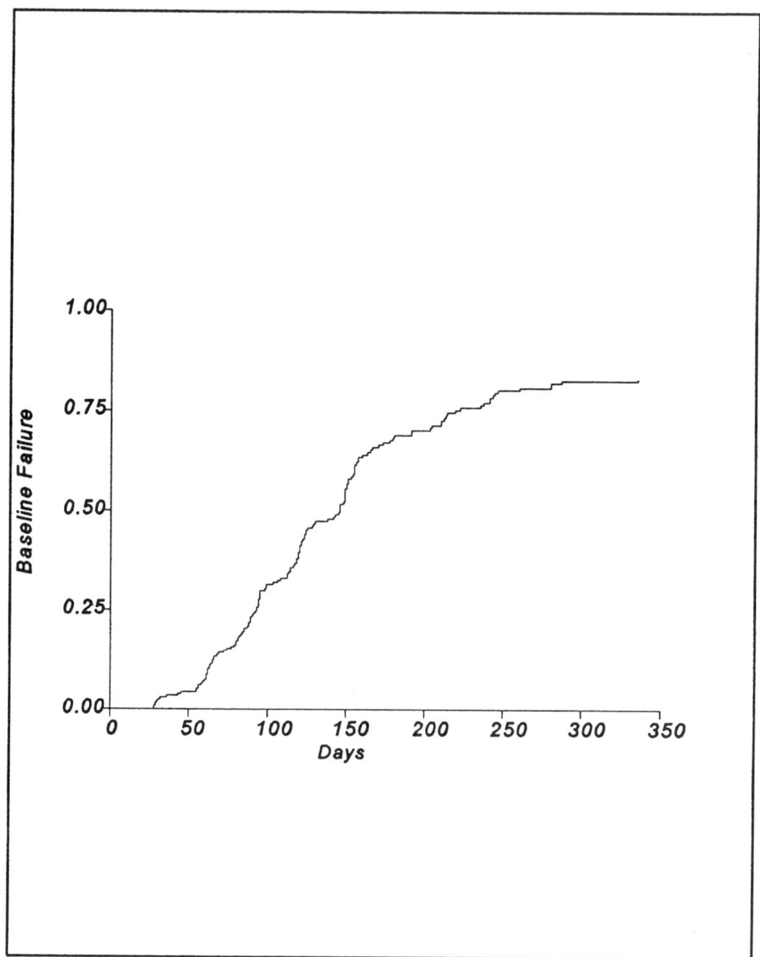

FIGURE 1. Proportional Hazards model for seropositivity to *B. bovis,* Tilarán, Costa Rica, 1991–1992.

days) and to 5% (240.1 days). A higher incidence of clinical cases could be present in this type of farm.

The *B. bovis* life cycle requires a well-established vector population to accomplish biological transmission; the known vector for this parasite in Costa Rica is *Boophilus microplus.*[2] Infection of the tick population is based only on alimentary infection of engorging female ticks followed by transovarial infection of a minor proportion of their eggs; only the larval stage will transmit this parasite to the bovine host.[16] Preliminary work with *B. microplus* determined 55%–84.4% of infection rates in engorged females in Tilarán.[17] In this study, we found differences among ecological

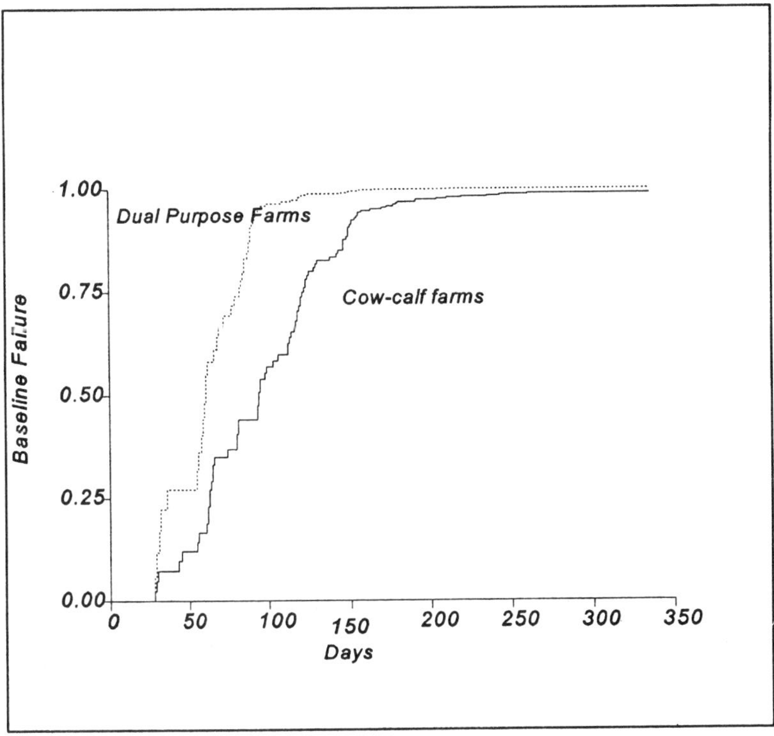

FIGURE 2. Proportional Hazards model for seropositivity to *B. bovis* by production system, Tilarán, Costa Rica, 1991–1992.

zones, suggesting that abiotic conditions such as temperature, precipitation, humidity, and evapotranspiration may interfere with the inoculation thresholds for *B. bovis,* determined by tick activity. *Boophilus microplus* unfed larvae showed an increase longevity at constant temperatures of 15° to 22°C than at 29°C or higher.[18] Tropical moist forests have an average temperature of 24° to 30°C. On the other hand all the premountain life zones (TABLES 1 and 2) have an average temperature ranging from 12° to 24°C. It is possible that larva development was affected in our study by the temperature, explaining the increased odds to seroconversion in the premountain life zones.

Vector-host contact is needed for the exposure of the host to the parasite and consequently for the development of an immune response. The increase in number of seropositive animals over time could be explained by an active transmission cycle taking place in the study area. However, the absence of clinical cases is explained by a high number of seropositive animals. Under these conditions, it is expected that most animals will become seropositive in time and clinical cases will not be frequent. Subclinical infections and carrier animals are required to maintain the transmission cycle because there is no vertical transmission of *B. bovis.*

Finally, the ratification of the hazard function on the production system showed that calves from beef farms were less likely to develop seropositivity than dual-purpose calves. The management factor could be the explanation for this difference: beef farms had an extensive pasture-based management (stocking rate of 1.06 animals per hectare), and dual-purpose farms had a more intensive management similar to dairy farms, keeping the calves with the cows during day hours in a small paddock near the milking parlor and indoors during the night, thus increasing the possibilities of exposure and transmission.

SUMMARY

The relationship between calf characteristics, farm management, seropositivity, and age at seroconversion to *Babesia bovis* was assessed. A total of 3624 samples obtained from 494 animals on 11 farms was analyzed by an indirect ELISA assay, and individual results classified as positive or negative. The animals were sampled from birth to weaning, beginning in April 1990 and ending in April 1992. We found 59.1% ($n = 467$) of seropositive neonatal calves (mean age 4.4 days). Using a logistic binomial regression model as the analytical method for predicting the likelihood of calf seropositivity, we found that the following factors were related to seroconversion at birth: time to colostrum consumption, place of birth, calving history, parity (number of lactations), and ecological life zone. To analyze the effect of management factors on time to seroconversion, we used the Proportional Hazards method. In this model the pattern of seroconversion was different for the type of production system. Calves from dual-purpose farms seroconverted earlier than calves from cow-calf farms.

ACKNOWLEDGMENTS

We thank the staff of the Herd Health Project and the Laboratory of Entomology at the School of Veterinary Medicine for their help and technical support during this study. Marcelino Jaén was an MSc student during this project; supported by the Panamá Government and the DAAD (German Academic Exchange Program). This work was partially funded by the University of Missouri and the Dutch Inter-University Cooperation Program (NUFFIC/DGIS). Support was also obtained from the International Agency of Atomic Energy, which provided the ELISA kit and technical advice. We thank Dr. Don Berrie (Long Pocket Laboratories, CSIRO, Australia), who donated the antigen and control sera for the ELISA assay.

REFERENCES

1. KUTTLER, K. L. 1988. World-wide impact of babesiosis. *In* Babesiosis of Domestic Animals and Man. M. Ristic, Ed.: 5. CRC Press Inc. Boca Raton, FL.
2. MAG. 1980. Ministerio de Agricultura y Ganaderia. Dirección de Sanidad Animal. Informe Final. Proyecto de Estudio de Factibilidad para el control de la garrapata. Departamento de Publicaciones MAG. San José, Costa Rica.
3. SOLANO, M. 1986. Babesiosis cerebral bovina. Ciencias Vet. **7:** 9-11.

4. PÉREZ, E., M. V. HERRERO, C. JIMÉNEZ, T. E. CARPENTER & G. BUENING. 1994. Epidemiology of bovine anaplasmosis and babesiosis in Costa Rica. Prev. Vet. Med. **20:** 23-31.

5. PÉREZ, E., M. V. HERRERO, C. JIMÉNEZ, D. HIRD & G. BUENING. 1994. Effect of management and host factors on seroprevalence of bovine anaplasmosis and babesiosis in Costa Rica. Prev. Vet. Med. **20:** 33-46.

6. IAEA/FAO JOINT PROGRAM. 1991. Animal Production and Health. Babesiosis ELISA kit. Indirect immunoassay for detection of antibody *Babesia bovis*. December, 1991.

7. MAURITSEN, R. H. 1984. Logistic regression with random effects. PhD Dissertation. Dept. of Biostatistics. University of Washington. Seattle, WA.

8. COX, D. R. 1972. Regression model and life tables. J. R. Stat. Soc. B **34:** 187-202.

9. LeBLANC, M. M., T. THANG, J. L. BALDWIN & E. L. PRITCHARD. 1981. Factors that influence passive transfer of immunoglobulins in Poás. JAVMA **200**(2): 179-183.

10. STOTT, G. H. 1980. Immunoglobulin absorption in calf neonates with special consideration of stress. J. Dairy Sci. **63:** 681.

11. CALLOW, L. L. 1984. Animal Health in Australia. Protozoal and Rickettsial Diseases **5:** 123-160. Australian Government Publishing Service. Watson, Ferguson and Co. Brisbane, Canberra, Australia.

12. MADRUGA, C. R., E. AYCARDI, R. H. KESSLER, S. M. A. MOREIRA, G. R. DeFIGUEIREDO & J. B. E. CURVO. 1984. Niveis de anticorpos anti-*Babesia bigemina* e *Babesia bovis* en bezerros de raza Nelore. Ibagé e cruzamientos de Nelore. Pesq. Agropec. Bras. **19:** 163-168.

13. DE LA RIOS, L. G. 1987. Immunidad pasiva y activa contra *Babesia bovis* en temeros de un área enzoótica del noroeste argentino. Rev. Ibér. Parasitol. **47:** 237-245.

14. JAMES, M. A., A. CORONADO, W. LÓPEZ, R. MELÉNDEZ & M. RISTIC. 1985. Seroepidemiology of bovine anaplasmosis and babesiosis in Venezuela. Trop. Anim. Health. Prod. **17:** 9-18.

15. DE LA RIOS, L. G., H. D. AGUIRRE & B. A. GALDO. 1989. Infección natural por *Babesia bovis* y *Babesia bigemina* en dos rodeos bovinos con diferentes niveles de infestación por *Boophilus microplus*. Rev. Latin. Microbiol. **31:** 39-43.

16. FRIEDHOFF, K. T. 1988. Transmission of *Babesia*. *In* Babesiosis of Domestic Animals and Man. M. Ristic, Ed.: 23-52. CRC Press Inc. Boca Raton, FL.

17. HERMANNS, P., R. H. DWINGER, G. M. BUENING & M. HERRERO. 1994. Seasonal incidence and haemoparasite infection rates of ixodid ticks Acaari: Ixodidae detached from cattle in Costa Rica. Rev. Biol. Trop. **42:** 623-632.

18. HITCHCOK, L. F. 1955. Studies on the non-parasite stages of the cattle tick, *Boophilus microplus*. Aust. Vet. Journal. **3:** 295-311.

Epizootiologic Instability of Bovine Populations against *Babesia bovis* (Piroplasmida: Babesiidae) in the Region of Poas, Costa Rica

VICTOR ALVAREZ,[a] ENRIQUE PÉREZ,[b,c] AND
MARCO V. HERRERO [b,d,e]

[a]*Programa de Reactivación Ganadera*
Ministerio de Agricultura y Ganaderia
Costa Rica

[b]*Regional Graduate Program in Tropical Veterinary Sciences*
School of Veterinary Medicine
Universidad Nacional
Costa Rica

[c]*Herd Health Project*
Costa Rican-Dutch Interuniversity Cooperation Program
School of Veterinary Medicine
Universidad Nacional
Costa Rica

[d]*Tropical Disease Research Program*
School of Veterinary Medicine
Universidad Nacional
Costa Rica

INTRODUCTION

Bovine babesiosis is a tick-borne disease transmitted by *Boophilus microplus* in Costa Rica. *Babesia bigemina* and *B. bovis,* both parasites of bovine erythrocytes, have been found in this country.[1]

The disease caused by *B. bovis* is characterized by high fever, anemia, and hemoglobinuria. This parasite may invade the central nervous system, producing convulsions, incoordination, and coma. It often causes bovine mortality. When the host survives, it may lose weight, decrease milk production, abort, and have a long term recover.[2] Cerebral babesiosis has also been reported in Costa Rica.[3]

Host-acquired immunity is mainly mediated by antibodies, but some cellular intervention has been studied.[4,5] Serological techniques are used to determine whether or not an individual has been exposed to an antigen and has built up an acquired

[e] Author to whom correspondence should be addressed at: Laboratorio de Entomología, Escuela de Medicina Veterinaria, Universidad Nacional, Apdo 304-3000 Heredia, Costa Rica.

humoral immunity response. In a population study, this could be applied to determine areas of epizootiologic instability.[6]

The objective of this study was to test the hypothesis that the bovine population in the region of Poás has a great proportion of susceptible individuals to *B. bovis* because of epizootiologic instability. In addition, the distribution of seropositive animals was studied regarding ecological zones.

MATERIALS AND METHODS

Study Area

We define the region of Poás as the geographic area where the Veterinary Automated Management and Production Program (VAAMP) of the School of Veterinary Medicine, Universidad Nacional, has its experimental units. These units are a group of 23 farms, specialized in dairy production.

The herds are composed of *Bos taurus* breeds, Holstein (75%) had Jersey (25%). Median farm size is 80 hectares, with a median number of paddocks of 60, and median herd size of 107. The farm topography is irregular with grazing areas not uniformly distributed; crops or forested areas alternate in some places of each farm.

This geographic region is located in Poas (latitude: 10° 06′ 03″ N; longitude: 84° 14′ 45″ W), Santa Barbara (latitude: 10° 04′ 36″ N; longitude: 84° 09′ 24″ W) and Vara Blanca (latitude: 10° 10′ 25″ N; longitude: 84° 09′ 35″ W). The altitude range goes from 1600 meters above sea level to 2100 meters above sea level.

Sampling

From a total population of 2650 animals, we selected 335 animals from 20 farms; 193 animals in lower mountain rain forest (12 farms, bp-MB), 37 from premountain wet forest (2 farms, bh-MB), and 105 from lower mountain wet forest (6 farms, bmh-MB).[7] In the area, there is not a system for active epidemiological surveillance; however, reports from 8 farms indicated a prevalence of 3.22% of hemoparasite clinical cases. Another problem associated with available information is that no differential diagnosis is practiced.

We collected blood from the jugular vein of each animal, twice in each dry season, for two consecutive years. For the purpose of this study, only animals bled three or four times were included. Blood was collected in vacutainer tubes without anticoagulant. The samples were transported to the laboratory, centrifuged at 3000 rpm, and stored at −20°C until their use.

ELISA Essay

All sera were analyzed following the procedure described by the International Atomic Energy Agency.[8] We compared our cutoff line in 200 negative sera obtained from a tick-free area in Costa Rica; we designated that optical densities over the mean plus three times the standard deviation were positive. This value is similar to

TABLE 1. Percentage of Seropositive and Seronegative Bovines to *Babesia bovis* (Piroplasmida: Babesiidae) in the Region of Poás

Percentage of Animals	Sampling Period[a]			
	I	II	III	IV
Seropositive	21.20	21.49	19.50	19.79
Seronegatives	78.80	78.51	80.50	80.21
n	273	321	328	283

[a] I = Dec. 1990–Jan. 1991; II = Apr. 1991–May, 1991; III = Dec. 1991–Jan. 1992; IV = Mar. 1992.

the one proposed by the IAEA. Working dilution of unknown sera was 1:200. All positive sera were diluted 1:400 and 1:800 to detect antibody titers.

Data Analysis

The unit of analysis was the individual animal. Responses to the test were positive or negative. Seropositivity results were expressed as seroprevalence. Seroprevalence was defined as $P = a/b$ where a is the number of positive animals and b the number of animals tested.

A test of hypothesis for proportions from independent groups was used to compare between ecological zones (Microstat. 1984. Ecosoft. Inc.). An additional multivariate analysis using logistic binomial regression was conducted to analyze the effect of ecological zone comparing with premountain wet forest and correcting by production system. Each farm systems was classified according to Solano.[9]

RESULTS

In Poás, a mean of over 79.5% of the animals sampled were seronegative; with a standard deviation among periods of 0.99% and a variation coefficient of 1.24%. The mean seroprevalence was 20.49%, with a standard deviation among sampling periods of 0.99% and a variation coefficient of 4.83% (TABLE 1).

A small number of animals seroconverted between sampling periods. The number of seroconversions, from negative to positive, was 7 (between sampling periods I–II), 11 (between sampling periods II–III) and 14 (between sampling periods III–IV) suggesting field booster stimuli to the immune system of the animals. Some animals changed from positive to negative between periods; 7 between I and II, 14 between II and III, and 8 between III and IV. The seropositive animals' titers are shown in TABLE 2; higher titers than 1:800 were not tested.

The distribution of seropositive animals by ecological zone is shown in TABLE 3. No significant differences ($p > 0.05$) were observed using univariate analysis; however, lower mountain rain forest showed the greatest odd ratios for seropositivity

TABLE 2. Bovine Antibody Titers to *Babesis bovis* (Piroplasmida, Babesiidae) in the Poás Region

Antibody Titer	Sampling Period[a]			
	I	II	III	IV
1:200	21	29	28	22
1:400	12	20	23	17
1:800	25	20	12	17
n	58	69	64	56

[a] I = Dec. 1990–Jan. 1991; II = Apr. 1991–May 1991; III = Dec. 1991–Jan. 1992; IV = Mar. 1992.

compared with the others when multivariate analysis was used (TABLE 4). Although the prevalence of antibodies to *B. bovis* was low in the region, 80% of the farms had seropositive animals. Of 20 farms sampled, 4 do not have seroreactors, 8 show less than 7%, 6 had between 13% and 41%, and 2 over 50% of seroreactors. The last two farms were located in two different ecological zones, bmh-MB and bp-MB.

DISCUSSION

Babesiosis is a disease where several biological populations intervene. Its transmission cycle included four components: A vector, *B. microplus,* that transmitted the pathogen and assured its persistence in the field due to transovarial transmission.[10,1] An etiological agent, *B. bovis,* that must complete its biological cycle using a vector and a host.[11,12] A host, which could be *Bos taurus* among other vertebrate animals.[13,4] And, a specific type of environment where all populations should survive.

The host responded to the presence of *B. bovis* by developing an immunological response.[14,15] In our region studied, most animals do not show acquired immunity

TABLE 3. Distribution of Seropositive Bovines (percentage) to *Babesia bovis* (Piroplasmida: Babesiidae) by Ecological Zone

Ecological Zone[a]	Sampling Period[b]			
	I	II	III	IV
bmh-P	31.0	28.5	27.0	20.0
bmh-Mb	16.7	21.0	20.6	20.2
bp-Mb	20.9	20.4	17.5	19.5

[a] bmh-P = premountain wet forest; bmh-Mb = lower mountain wet forest; bp-Mb = lower mountain rain forest.
[b] I = Dec. 1990–Jan. 1991; II = Apr. 1991–May 1991; III = Dec. 1991–Jan. 1992; IV = May 1992.

TABLE 4. Logistic Binomial Regression Analysis for Seropositivity of Bovines against *Babesis bovis* (Piroplasmida: Babesiidae) in Poás (by Ecological Zone)

Ecological Life Zone[a]	Odds Ratio	Confidence Interval (95%)	p Value
bp-Mb			
1[b]	28.41	4.40–183.60	<0.001
2	5.91		0.07
3	5.41	1.57–18.62	0.07
4	13.13	2.36–73.02	0.03
bmh-Mb			
1	3.43	0.91–12.94	0.069
2	1.51	1.14–18.26	0.032
3	1.70	0.55– 5.47	0.340
4	2.60	0.58–11.57	0.030

[a] bmh-P = premountain wet forest; bmh-Mb = lower mountain wet forest; bp-Mb = lower mountain rain forest.

[b] 1 = Dec. 1990–Jan. 1991; 2 = Apr. 1991–May 1991; 3 = Dec. 1991–Jan. 1992; 4 = May 1992.

which means that exposure to the parasite by intervention of an infected vector is lacking.[16,4]

Few animals show significant antibody titers to *B. bovis*. The ELISA test used has very high sensitivity and specificity; however, the antibodies detected do not necessarily protect the host from infection.[17] The occurrence of seropositive animals may be due to exposure during animal movement between Poas and other endemic areas in the country. In the other areas, the animals could be exposed to *B. bovis* due to their contact with an infected vector. Another explanation is that this contact may occur in the same region; in this case, *B. microplus* populations must be infected. A preliminary study[18] shows high infection rates in ticks from one farm; however, sample size in this study was small (only five animals from one farm), reduced to a period of six months, and it was not clear if the ticks were infected or if the parasite was in the blood obtained from fully engorged females. In the same farm, we found a seroprevalence of 41%.

In this study, differences among ecological zones were observed. This could mean that abiotic conditions such as temperature, precipitation, humidity, and evapotranspiration may play a differential role in tick distribution or in infection rates in these areas.

Production systems, cattle breeds, and management are not uniform in the area. Therefore, seroprevalence could be related to farm characteristics. Based on the serological characteristics of the herds and the frequency of clinical cases, it is suggested that the cattle situation in the area is epidemiologically unstable for *B. bovis*.

SUMMARY

In Poás (Costa Rica), more than 78% of the cattle population is susceptible to *Babesia bovis* (Babes, 1888) which indicates that care should be taken during animal movement to avoid tick exposure. Seroprevalence is less than 22%; the frequency distribution of antibody titers is presented. Significant differences ($p < 0.05$) were observed in the distribution of seropositive animals between ecological zones. Lower mountain rain forest and lower mountain wet forest presented higher risk for seropositivity. Based on the serological characteristics of the herds and the frequency of clinical cases, it is suggested that the cattle situation in the area is epidemiologically unstable for *B. bovis*.

ACKNOWLEDGMENTS

We are thankful for the technical and professional help provided by the Tropical Disease Research Program staff. Funds for this work were partially obtained from the University of Missouri. During this study, Victor Alvarez was supported by the National Council on Science and Technology (CONICIT) and the Ministery of Agriculture (MAG) to conduct his MSc studies. This paper was presented as a partial requirement to complete an MSc degree in Tropical Veterinary Sciences at Universidad Nacional. The ELISA assay, reagents, and equipment were provided by the International Atomic Energy Agency.

REFERENCES

1. MAG. DIRECCIÓN DE SANIDAD ANIMAL. 1980. Informe Final Proyecto Estudio de Factibilidad para el control de la Garrapata. San José, Costa Rica.
2. KUTTLER, K. L. 1988. World-wide impact of babesiosis. *In* Babesiosis of Domestic Animals and Man. M. Ristic, Ed.: CRC Press Inc. Boca Raton, FL.
3. SOLANO, M. 1986. Babesiosis cerebral bovina. Cienc. Vet. **8:** 9-11.
4. JAMES, M. A., A. CORONADO, W. LOPEZ, R. MELENDEZ & M. RISTIC. 1985. Seroepidemiology of bovine Anaplasmosis and Babesiosis in Venezuela. Trop. Anim. Health Prod. **17:** 9-18.
5. TRAMBALLO, L. J., G. M. BUENING & R. M. MCLAUGHLIN. 1992. The effect of neutrophils, tumor necrosis factor, and granulocyte macrophage/colony stimulating factor on *Babesia bovis* and *Babesia bigemina* in culture. Vet. Parasitol **43:** 177-188.
6. MARTIN, S. W., A. H. MEEK & P. WILLEBERG. 1988. Veterinary Epidemiology. Principles and Methods. First Edit. Iowa State University Press.
7. HOLDRIDGE, L. 1978. Ecologia basada en zonas de vida. Ed. IICA. San José, Costa Rica.
8. IAEA/FAO. 1991. Joint FAO/IAEA programme. Babesiosis. Indirect ELISA kit. Vienna, Austria.
9. SOLANO, C. 1993. La crianza de novillas de reemplazo en fincas lecheras de altura de la zona de Poás, Costa Rica: el crecimiento preparto y la subsecuente producción de leche. Tesis M. Sc. en Producción Animal Tropical. Universidad Autónoma de Yucatán, México.
10. SOULSBY, E. J. L. 1982. Helminths, arthropods and Protozoa of Domestic Animals. Seventh Ed. Lea and Febiger.
11. MAHONEY, D. F. & D. F. ROSS. 1972. Epizootiological factors in the control of bovine babesiosis. A.V.A. Conference Paper. Aust. Vet. J. **48:** 292-298.
12. GUGLIELMONE, A. A. 1992. Epizootiologia de las enfermedades hemoparasitarias de los vacunos. Rvta. Cub. Cienc. Vet. **24**(2-3): 74-107.
13. HALL, W. T. K. 1963. The immunity of calves to tick-transmitted *Babesia argentina* infection. Aust. Vet. J. **39**(Oct.): 386-389.

     ANNALS NEW YORK ACADEMY OF SCIENCES

14. MAHONEY, D. F. 1962. Bovine babesiosis. Diagnosis of infection by a complement fixation test. Aust. Vet. J. (Feb.): 40-45.
15. GOFF, W. L., G. G. WAGNER, T. M. CRAIG & R. F. LONG. 1984. The role of specific immunoglobulins in antibody-dependent cell-mediated cytotoxicity assays during *Babesia bovis* infection. Vet. Parasitol. **14:** 117-128.
16. KUDAMBA, C., R. S. CAMPBELL, N. I. PAULL & R. G. HOLROYD. 1982. Serological studies of babesiosis and anaplasmosis of cattle. Aust. Vet. J. **59:** 101-104.
17. WALTISBUHL, D. J., B. V. GOODGER, I. G. WRIGHT, M. A. COMMINS & D. F. MAHONEY. 1987. An enzyme linked immunosorbent assay to diagnose *Babesia bovis* infection in cattle. Parasitol. Res. **73:** 126-131.
18. HERMANNS, P., R. H. DWINGER, G. M. BUENING & M. HERRERO. 1994. Seasonal incidence and hemoparasite infection rates of ixodid ticks (Acari: Ixodidade) detached from cattle in Costa Rica. Rev. Biol. Trop. **42**(3): 623-634.

Comparative Sensitivity of Two Tests for the Diagnosis of Multiple Hemoparasite Infection of Cattle[a]

J. V. FIGUEROA,[b] J. A. ALVAREZ, G. J. CANTO, J. A. RAMOS, J. J. MOSQUEDA, AND G. M. BUENING[c]

CENID-PAVET, INIFAP
Jiutepec, Morelos, 62550 Mexico

[c]Department of Veterinary Microbiology
UM/C
Columbia, Missouri 65211

INTRODUCTION

Bovine babesiosis caused by *Babesia bigemina* and/or *B. bovis* is considered one of the most economically important diseases of cattle in tropical and subtropical regions of the world.[1,2] Both *Babesia* species share the tick *Boophilus* spp. as the vector for their biological transmission to cattle.[1,2]

Clinical manifestations of the disease include fever, anorexia, dullness, weakness, ataxia, hemoglobinuria, icterus, anemia, and presence of intraerythrocytic parasites.[1,2] Under field conditions, it is often the case that both organisms are found together as well as in association with the hemorickettsia *Anaplasma marginale*.[1-3]

Laboratory diagnosis of *Babesia* and *Anaplasma* infection in cattle is based on the microscopic examination of Giemsa-stained blood smears for the presence of intraerythrocytic bodies. This, however, depends upon the ability of the microscope technologist to determine the *Babesia* species involved.[1,2]

A feature of bovine babesiosis and anaplasmosis is that animals that recover from a primary acute disease become carriers.[1,2] Carrier cattle may serve as a reservoir of infection for the vectors and are difficult to differentiate from uninfected cattle as the hemoparasites usually cannot be readily demonstrated by light microscopy.[1,2] The present study was designed to compare the diagnostic capability of a PCR-based assay to detect cattle simultaneously infected with either *Babesia bovis* and *Babesia bigemina* or *B. bovis*, *B. bigemina*, and *Anaplasma marginale*.

[a] This study was partially financed by USAID Grant No. DAN 4178-A-00-7056-00, and by INIFAP-SARH, Mexico.

[b] Address correspondence to: CENID-PAVET, INIFAP, Apartado Postal No. 206, Civac, Morelos, C.P. 62500, Mexico.

MATERIALS AND METHODS

Source of Hemoparasites

Attenuated Mexican strains of *Babesia bigemina* and *B. bovis* were maintained in culture essentially as previously described.[4,5] Blood containing *A. marginale*-infected erythrocytes was obtained from an intact calf which had been experimentally inoculated with a Mexican (Texcoco) isolate.

Experimental Animals

Sixteen intact, 1–2 year old, Holstein-Friesian bulls (animal identification numbers 1-16) from a tick-free area in Central Mexico were simultaneously inoculated with 1×10^9 culture-derived, *B. bigemina*-infected erythrocytes and 1×10^8 culture-derived *B. bovis*-infected erythrocytes. Sixteen more bulls (animal identification numbers 17-32) were inoculated with approximately 1×10^8 *B. bovis*- and *B. bigemina*-infected erythrocytes of Mexican field isolates. These bulls had previously been inoculated with 10^7 *A. marginale*-infected erythrocytes. Blood samples were collected prior to and after the inoculation of cattle using evacuated tubes containing EDTA. Blood samples were processed as described,[6,7] and packed erythrocytes for PCR analysis were kept at $-20°C$ until analyzed as one batch.

PCR Primers

Six sets of primers derived from reported DNA sequences[6–9] were synthesized in an Applied Biosystems-DNA synthesizer (Model 380B). Group I primers (3 sets) comprised BiIA/IB, BoF/R, and 9/10. Group II primers (3 sets) comprised BiIAN/IBN, BoFN/RN, and 11/12. Primer sequences have been previously reported.[7] By using group I primers in the PCR assay, DNA fragments of 278 bp, 356 bp, and 200 bp were amplified from blood containing *B. bigemina*-, *B. bovis*-, or *A. marginale*-infected erythrocytes, respectively;[6–8] the three fragments were coamplified from artificially mixed blood containing the three hemoparasite species.[7]

PCR DNA Amplification

To amplify hemoparasite DNA from blood samples, 20-μl aliquots of sedimented erythrocytes from *in vitro* culture (positive controls) and from the experimental animals were placed in 200 μl of lysis buffer (0.015% saponin, 35 mM NaCl, 1 mM EDTA) and vortexed gently.[6,8] After 10 min centrifugation at $12,000 \times g$, the hemoglobin was removed and the pellets resuspended in 250 μl of PCR buffer[10] [10 mM Tris-HCl (pH 8.9), 50 mM KCl, 1.5 mM $MgCl_2$, 0.01% gelatin] and centrifuged again. Pellets were then resuspended in 99.5 μl of PCR buffer containing 200 μM of each dNTP and 1 μM of each group I primer, and incubated at 100°C for 10 min. After a brief centrifugation, 2.5 U of Taq polymerase (Promega Co., Madison, WI) was added and the PCR tubes were placed in a TempCycler 50 (Coy, Ann Arbor,

MI) for thermocycling (35 cycles) using the following temperature profile: template denaturation, 1 min at 95°C (2 min for the first cycle); primer annealing, 1 min at 60°C; and primer extension, 1.5 min at 73°C; with a final extension at 73°C for 13.5 min.

Preparation of PCR-Labeled Probes

Purified pBbi55 DNA,[6] purified pBv60 DNA,[9] and pStCroix DNA[7] served as template for the synthesis of probe via PCR using group II primers BiIAN/IBN (*B. bigemina*), BoFN/RN (*B. bovis*), and 11/12 (*A. marginale*). The procedure was carried out essentially as described previously[6–8] where dTTP is replaced at about 35% with Digoxigenin-dUTP.[11] By using group II primers in the PCR assay, DNA fragments of 170 bp, 291 bp, and 150 bp were amplified and used as parasite-specific probes for *B. bigemina, B. bovis,* and *A. marginale,* respectively.[6–8]

Analysis of PCR Products

Twenty-μl aliquots of each amplification reaction were spotted onto each of three different nylon membranes using a dot-blot apparatus. After DNA denaturation, neutralization, and fixation,[12] the membranes were each hybridized with the species-specific nonradioactive probe (100 ng/ml of hybridization solution). DNA hybrids were detected by the alkaline phosphatase-based chemiluminescent reaction as described earlier.[6,7]

RESULTS AND DISCUSSION

Blood samples from the inoculated animals analyzed by the Multiplex PCR assay and hybridized with the *B. bigemina*-specific probe showed positive reactions in 30 out of 31 (blood sample from bull no. 32 was not available to test for *Babesia* sp.) tested bulls (FIG. 1). The first day on which they were detected positive varied among the animals from 6-9 days postinoculation (DPI). However, the majority of tested bulls were *B. bigemina* PCR/DNA probe positive between days 9 and 14 PI.

Out of the 31 blood samples tested by the *B. bovis* PCR/DNA probe system, 29 resulted positive (FIG. 2), although very weak hybridization signals were observed in some animals. Similar to what was found with the *B. bigemina* system, the day on which the animals were first detected positive to *B. bovis* varied among the bulls. Positive signals were detected from day 7 to 23 PI. None of the samples collected from the 31 cattle previous to inoculation showed hybridization signal when tested by the PCR/DNA probe assay.

Except for one animal (no. 21), all *A. marginale*-infected bulls showed a positive hybridization signal after PCR/DNA probe analysis (FIG. 3). Some of the bulls were detected positive from day 17 and remained so up to day 72 PI.

Overall, light microscopy examination of peripheral blood samples collected from the inoculated animals confirmed the presence of *B. bigemina* in all 31 bulls examined. However, the number of animals being detected positive in a particular day PI varied

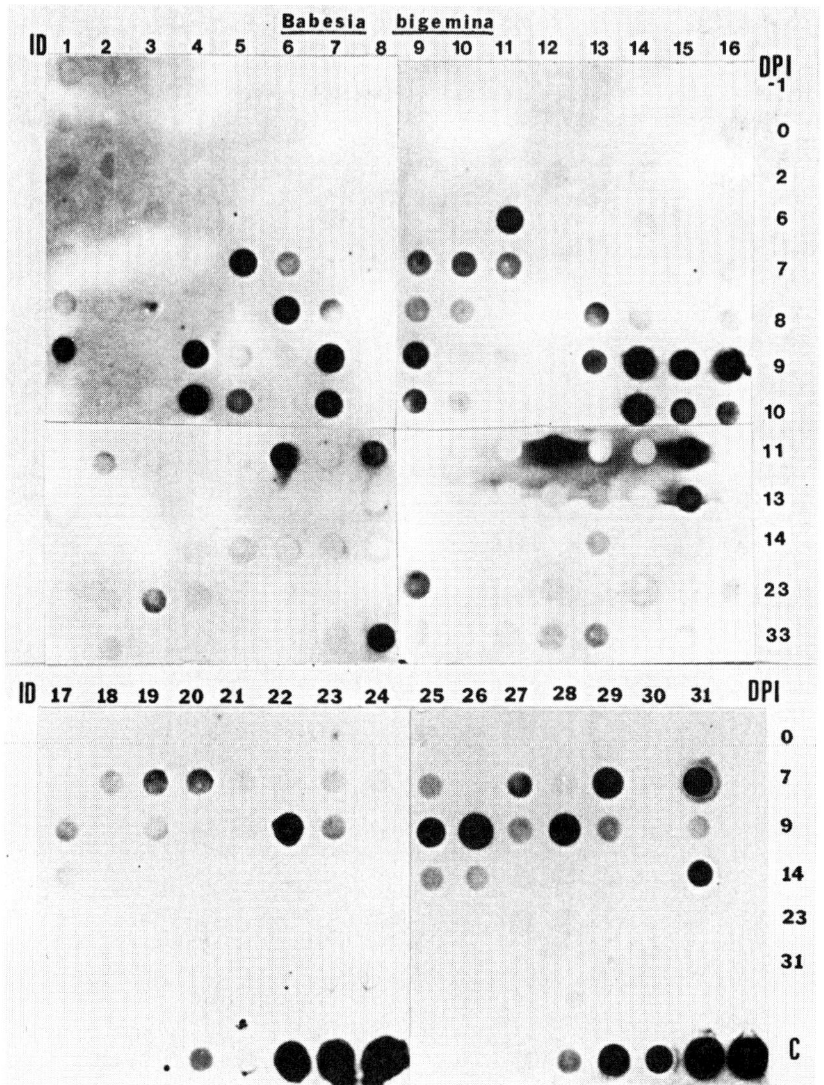

FIGURE 1. Analysis of PCR products by dot-blot nucleic acid hybridization with *B. bigemina* DNA probe. ID no. 1-16: samples corresponding to bulls inoculated with culture-derived *Babesia*-infected erythrocytes. ID no. 17-31: samples corresponding to bulls infected with the field isolate of *B. bigemina.* DPI: days postinoculation. C: control samples.

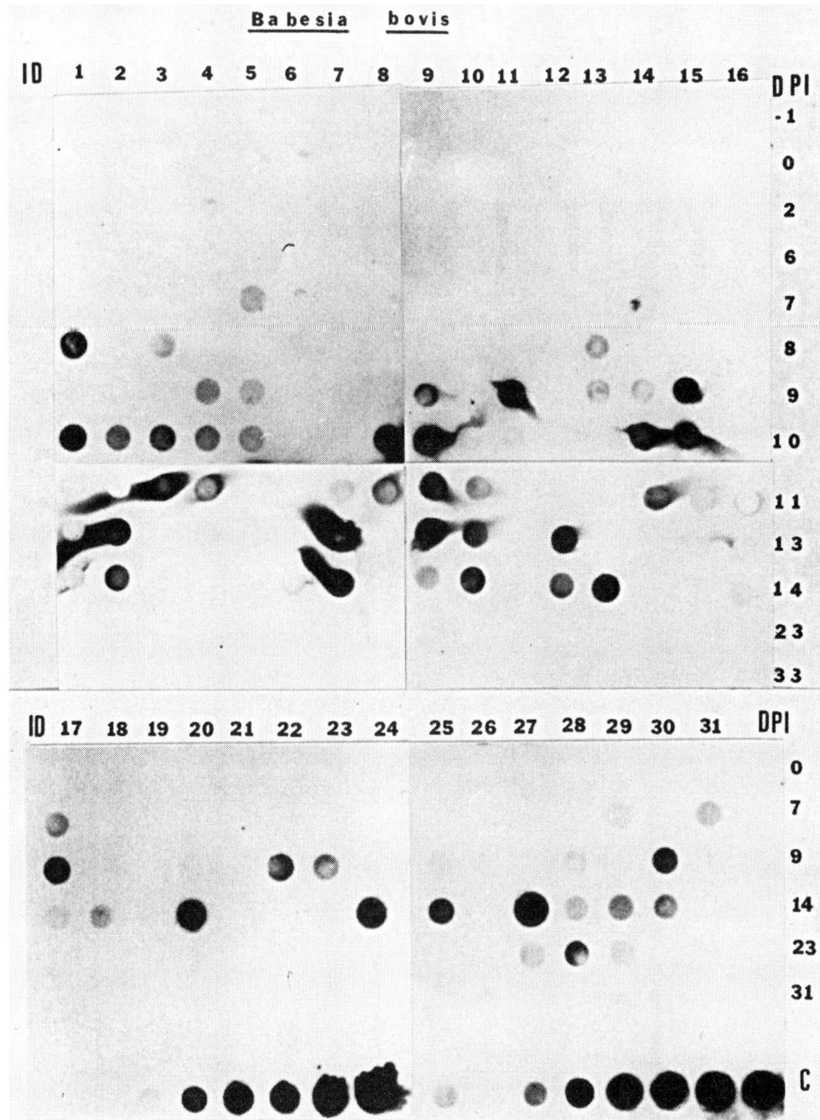

FIGURE 2. Analysis of PCR products by dot-blot nucleic acid hybridization with *B. bovis* DNA probe. ID no.1-16: samples corresponding to bulls inoculated with culture-derived *Babesia*-infected erythrocytes. ID no. 17-31: samples corresponding to bulls infected with the field isolate of *B. bovis*. DPI: days postinoculation. C: control samples.

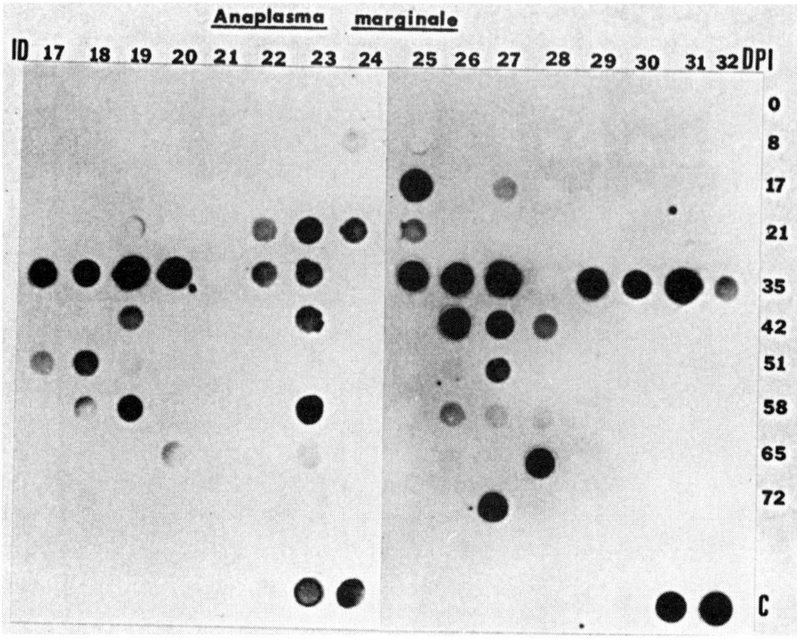

FIGURE 3. Analysis of PCR products by dot-blot nucleic acid hybridization with *A. marginale* DNA probe. ID 17–32: samples corresponding to 16 out of the 32 bulls inoculated with calf-derived *Anaplasma*-infected erythrocytes. DPI: days post-*A. marginale* inoculation. C: control samples.

from 23 to 28 throughout the study (TABLES 1 and 2). However, the parasitemias were too low to be quantitated (<0.01%, see TABLE 1), and for the most part, it took up to two and a half hours of microscopic examination to find infected erythrocytes, particularly the *B. bovis*-infected ones. *B. bovis* was detected in peripheral blood smears in 22 of the 31 inoculated bulls. The low parsitemias observed in the animals inoculated with high doses of infected erythrocytes (10^9 *B. bigemina*- and 10^8 *B. bovis*-infected red cells, respectively) is a characteristic that may be attributed to the fact that the *Babesia* strains utilized in this study had been maintained continuously in culture for long periods of time. Prolonged cultivation may have given rise to the apparent attenuation of the parasites as the experimental animals showed very mild clinical manifestations of disease after inoculation (unpublished data). In contrast, animals no. 17 to no. 31, which were inoculated with the field isolates, showed relatively higher numbers of parasitized erythrocytes. *A.marginale*-infected erythrocytes were detected in 10 inoculated bulls by day 35 PI (TABLE 3), and in 15 bulls on day 40 PI (not shown). TABLE 4 shows the calculations of sensitivity, specificity, and predictive values for the MPCR assay to diagnose the hemoparasite infection of cattle.

Conclusive evidence for an accurate diagnosis of bovine babesiosis and/or anaplasmosis is by demonstrating the presence of the infectious agent.[13] Current proce-

TABLE 1. Light Microscopy Results of Cattle Inoculated with Culture-Derived *Babesia* sp.[a]

DPI:	6		7		8		9		10		11		13		14		Overall Positive	
ID	BB	Bv	BB	Bv	BB	Bv	BB	Bv	BB	Bv	BB	Bv	BB	Bv	BB	Bv	BB	Bv
16	ND	ND	(+)	(−)	(+)	(−)	(+)	(−)	(+)	(−)	(−)	(−)	(−)	(−)	(−)	(−)	(+)	(−)
15	(+)	(−)	(−)	(−)	(+)	(−)	(+)	(−)	(+)	(−)	(+)	(+)	(−)	(−)	(−)	(−)	(+)	(+)
14	(−)	(−)	(−)	(−)	(+)	(−)	(+)	(+)	(+)	(−)	(+)	(−)	ND	ND	(−)	(−)	(+)	(+)
13	(−)	(−)	(+)	(−)	(+)	(−)	(+)	(−)	(+)	(−)	(−)	(−)	(−)	(+)	(−)	(−)	(+)	(+)
12	(−)	(−)	(−)	(−)	(−)	(−)	(−)	(−)	(−)	(−)	(+)	(+)	(−)	(−)	(+)	(−)	(+)	(+)
11	(−)	(−)	(+)	(−)	(+)	(+)	(+)	(−)	(−)	(+)	(−)	(−)	(−)	(+)	(−)	(−)	(+)	(+)
10	(−)	(−)	(+)	(−)	(+)	(−)	(−)	(−)	(+)	(+)	(+)	(+)	(−)	(−)	ND	ND	(+)	(+)
9	(−)	(−)	(+)	(−)	(+)	(−)	(+)	(−)	(+)	(−)	(−)	(+)	(−)	(+)	(−)	(−)	(+)	(+)
8	(+)	(−)	(+)	(+)	(−)	(−)	(−)	(−)	(−)	(−)	(−)	(−)	(−)	(−)	(−)	(+)	(+)	(+)
7	(−)	(−)	(−)	(−)	(+)	(−)	(+)	(−)	(+)	(+)	(+)	(−)	(−)	(−)	(+)	(−)	(+)	(+)
6	(+)	(−)	(+)	(−)	(+)	(+)	(+)	(+)	(+)	(−)	(+)	(−)	ND	ND	(−)	(−)	(+)	(+)
5	(−)	(−)	(−)	(−)	(+)	(−)	(+)	(−)	(+)	(−)	(+)	(−)	ND	ND	(−)	(−)	(+)	(−)
4	(+)	(−)	(+)	(−)	(+)	(−)	(+)	(−)	(+)	(−)	(+)	(−)	(−)	(−)	(−)	(−)	(+)	(−)
3	(−)	(−)	(+)	(−)	(−)	(−)	(−)	(−)	(−)	(−)	(−)	(−)	(−)	(−)	(−)	(−)	(+)	(−)
2	(+)	(−)	(+)	(−)	(+)	(−)	(+)	(−)	(−)	(−)	(+)	(+)	(−)	(−)	(+)	(−)	(+)	(+)
1	(+)	(−)	(−)	(+)	(+)	(−)	(+)	(−)	(−)	(+)	(+)	(−)	(−)	(−)	(−)	(−)	(+)	(+)
Total	6	0	10	2	13	2	12	2	10	4	10	5	0	3	3	1	16	12

[a] DPI = days postinoculation of blood stabilates. ID = animal identification. BB = *Babesia bigemina.* Bv = *Babesia bovis.* ND = not done. (+) = 2–10 infected erythrocytes found in blood smears. At least 30 minutes search per slide. (−) = up to two and a half hours search.

TABLE 2. Light Microscopy Results of Animals Inoculated with Field Isolates of *Babesia* sp.[a]

DPI:	7 PPE		9 PPE		14 PPE		Overall Positive	
ID	BB	Bv	BB	Bv	BB	Bv	BB	Bv
32	ND	ND	ND	ND	ND	ND	ND	ND
31	(+)	(−)	(+)	0.02	(−)	(−)	(+)	(+)
30	(+)	(−)	(+)	(−)	(+)	(+)	(+)	(+)
29	(−)	(−)	0.05	(−)	(+)	(+)	(+)	(+)
28	(+)	(−)	(+)	(+)	(−)	(−)	(+)	(+)
27	(+)	(−)	0.02	(−)	0.018	(+)	(+)	(+)
26	(+)	(−)	0.2	(+)	0.011	(−)	(+)	(+)
25	(+)	(−)	0.07	(−)	(−)	(−)	(+)	(−)
24	(+)	(−)	0.06	(−)	(+)	(−)	(+)	(−)
23	(+)	(+)	0.05	(−)	(−)	(−)	(+)	(+)
22	(+)	(−)	(+)	(−)	(−)	(−)	(+)	(−)
21	(+)	(−)	(+)	(−)	0.02	(−)	(+)	(−)
20	(+)	(−)	(+)	(−)	0.013	0.013	(+)	(+)
19	(+)	(−)	(+)	(−)	(−)	(−)	(+)	(−)
18	(+)	(−)	(+)	(+)	(+)	(−)	(+)	(+)
17	(−)	(−)	(+)	(+)	(+)	(+)	(+)	(+)
Total	13	1	15	5	8	5	15	10

[a] DPI = days postinoculation of blood stabilates. ID = animal identification. PPE = percentage of parasitized erythrocytes. BB = *Babesia bigemina*. Bv = *Babesia bovis*. ND = not done. (+) = 2–20 infected erythrocytes found in blood smears (PPE < 0.01%). At least 30 minutes examination per slide. (−) = up to two and a half hours search.

dures utilized to show that an animal is chronically infected include the thick blood smear technique[14] immunosuppression with corticosteroids; splenectomy of suspected animals; brain biopsy (*B. bovis* infection); blood subinoculation from a suspected carrier to a susceptible, splenectomized recipient[13] and more recently, culture isolation of *Babesia* spp.[6,8] However, most of these procedures are tedious, time consuming, and expensive. A highly sensitive method able not only to detect the presence of *Babesia* spp. and *A. marginale* in a suspected carrier animal, but also to specifically differentiate animals infected solely with *B. bovis* or *B. bigemina* from those concurrently infected with both *Babesia* and *Anaplasma* species, is needed. In this study we showed that such analysis can be carried out by simultaneous PCR amplification of parasite DNA and detection of PCR products using species-specific DNA probes.

To validate the Multiplex PCR assay, an experiment with a larger number of cattle having a mixed *Babesia* infection had to be carried out, as the blood samples previously analyzed using a multiplex PCR system had been mixed artificially.[7] Since the animals inoculated with the attenuated strains of *Babesia* showed low parasitemias and very mild clinical reactions, it could be argued that they resembled what may usually occur in carrier cattle and served the purpose for validation of the test. This

TABLE 3. Light Microscopy Results of Animals Inoculated with *A. marginale*[a]

Days PI:	8	17	21	35	Overall Positive
ID	PPE	PPE	PPE	PPE	PPE
32	(−)	(−)	(−)	(+)	(+)
31	(−)	(−)	(−)	0.31	(+)
30	(−)	(−)	(−)	1.87	(+)
29	(−)	(−)	ND	0.08	(+)
28	(−)	(−)	(−)	0.42	(+)
27	(−)	(+)	(−)	0.06	(+)
26	(−)	(−)	(−)	0.034	(+)
25	(−)	(+)	(−)	ND	(+)
24	(−)	(−)	(−)	(−)	(−)
23	(−)	(−)	(−)	(+)	(+)
22	(−)	(−)	(−)	(−)	(−)
21	(−)	(−)	(−)	(−)	(−)
20	(−)	(−)	(−)	(−)	(−)
19	(−)	(−)	(−)	(+)	(+)
18	(−)	(−)	(−)	(−)	(−)
17	(−)	(−)	(−)	(−)	(−)
Total	0	2	0	9	10

[a] PI = days postinoculation of blood stabilate containing *A. marginale*. (+) = 2–10 infected erythrocytes found by light microscopy of Giemsa-stained blood smears. PPE <0.01%. At least 30 minutes examination per slide.

(−) = up to two and a half hours search. ND = not done.

TABLE 4. Ability of MPCR and LM Performed on Blood Samples to Detect Hemoparasite Infection

Assay	Species	Percent Sensitivity	Percent Specificity	Percent Positive PV[a]	Percent Negative PV
	B. bigemina	96.7	100	100	97
MPCR	*B. bovis*	93.5	100	100	94
	A. marginale	93.8	100	100	97
	B. bigemina	100	100	100	94
LM	*B. bovis*	70.9	100	100	58.8
	A. marginale	93.8	100	100	100

[a] PV: predictive value.

was demonstrated by detecting almost all bulls positive to *B. bigemina* and *B. bovis*. However, whether the bulls inoculated with the culture-derived parasites actually become carriers remains to be tested.

In conclusion, the Muliplex PCR assay is proposed as a valuable tool with high sensitivity and specificity for the detection of *B. bovis*- and/or *B. bigemina*- and/or *A. marginale*-infected animals. The MPCR assay can be utilized to process large numbers of samples (batch sizes as high as 400 in a couple of days) in a research laboratory to conduct, for example, epidemiological studies in order to determine the distribution of the hemoparasite species in carrier cattle.

SUMMARY

The purpose of this study was to evaluate the efficacy of light microscopy (LM) examination of blood smears and a multiplex polymerase chain reaction (MPCR) assay, in terms of their ability to detect cattle experimentally infected with *Babesia bovis, Babesia bigemina,* and *Anaplasma marginale*. Blood samples were collected from 32 intact, 1–2 year old, Holstein bulls, previous to and after simultaneous inoculation of culture-derived or field isolates of *B. bovis*- and *B. bigemina*-infected erythrocytes. To establish the triple hemoparasite infection, 16 of the bulls were also inoculated with a calf-derived isolate of *A. marginale*. The results showed that both tests had 100% specificity. In contrast, the sensitivities of the MPCR assay against the LM test were 93.5% and 70.9%; 96.7% and 100%; and 93.8% and 93.8% for *B. bovis, B. bigemina,* and *A. marginale* infection, respectively. The advantages and disadvantages of the MPCR assay to differentially diagnose cattle with multiple hemoparasite infection are discussed.

ACKNOWLEDGMENTS

Plasmid pBv60 was kindly provided by Dr. T. F. McElwain, Department of Veterinary Microbiology and Pathology, WSU, Pullman, WA 99163. Provision of Plasmid pStCroix is acknowledged to Dr. R. Aboytes, INIFAP-SARH, Mexico.

REFERENCES

1. CALLOW, L. L. 1984. Protozoan and rickettsial diseases. *In* Australian Bureau of Animal Health, Animal Health in Australia. **5:** 121–216. Australian Government Publishing Service. Canberra, Australia.
2. RISTIC, M. 1981. Babesiosis. *In* Diseases of Cattle in the Tropics. M. Ristic & I. McIntrye Eds.: 443–468. Martinus Nijhoff Publishers. The Hague, the Netherlands.
3. DALGLIESH, R. J. & N. P. STEWART. 1983. The use of tick transmission by *Boophilus microplus* to isolate pure strains of *Babesia bovis, Babesia bigemina* and *Anaplasma marginale* from cattle with mixed infections. Vet. Parasitol. **13:** 317–323.
4. PALMER, D. A., G. M. BUENING & C. A. CARSON. 1982. Cryopreservation of *Babesia bovis* for *in vitro* cultivation. Parasitology **84:** 567–572.
5. VEGA, C. A., G. M. BUENING, T. J. GREEN & C. A. CARSON. 1985. *In vitro* cultivation of *Babesia bigemina*. Am. J. Vet. Res. **46:** 416–420.

6. FIGUEROA, J. V., L. P. CHIEVES, G. S. JOHNSON & G. M. BUENING. 1992. Detection of *Babesia bigemina* infected carriers by polymerase chain reaction amplification. J. Clin. Microbiol. **30:** 2576-2582.

7. FIGUEROA, J. V., L. P. CHIEVES, G. S. JOHNSON & G. M. BUENING. 1993. Multiplex polymerase chain reaction based assay for the detection of *Babesia bigemina, Babesia bovis* and *Anaplasma marginale* DNA in bovine blood. Vet. Parasitol. **50:** 69-81.

8. FIGUEROA, J. V., L. P. CHIEVES, G. S. JOHNSON, W. L. GOFF & G. M. BUENING. 1994. Polymerase chain reaction-based diagnostic assay to detect cattle chronically infected with *Babesia bovis*. Rev. Lat. Microbiol. **36:** 47-55.

9. SUAREZ, C. E., G. H. PALMER, D. P. JASMER, S. A. HINES, L. E. PERRYMAN & T. F. MCELWAIN. 1991. Characterization of the gene encoding a 60-kilodalton *Babesia bovis* merozoite protein with conserved and surface exposed epitopes. Mol. Biochem. Parasitol. **46:** 45-52.

10. SAIKI, R. K., D. H. GELFAND, S. STOFFEL, S. J. SCHARF, R. HISUCHI, G. T. HORN, K. B. MULLIS & H. A. EHRLICH. 1988. Primer-directed enzymatic amplification of DNA with a thermostable DNA polymerase. Science **239:** 487-491.

11. EMANUEL, J. R. 1991. Simple and efficient system for synthesis of nonradioactive nucleic acid hybridization probes using PCR. Nucleic Acids Res. **19:** 2790.

12. SAMBROCK, J., E. F. FRITSH & T. MANIATIS. 1989. Molecular Cloning. A Laboratory Manual. 2nd edit.: 6.15; 9.31. Cold Spring Harbor Laboratory. Cold Spring Harbor, NY.

13. CALLOW, L. L., R. J. ROGERS & A. J. DE VOS. 1986. standard diagnostic techniques for tick-borne diseases (babesiosis and anaplasmosis) of cattle in Australia. *In* Australian Agricultural Council, Standard Diagnostic Techniques for Animal Diseases. **29:** 1-29. CSIRO. Canberra, Australia.

14. MAHONEY, D. F. 1962. The epidemiology of babesiosis in cattle. Aust. J. Sci. **24:** 327-335.

Characterization of Helper T Cell Responses against Rhoptry-Associated Protein 1 (RAP-1) of Babesial Parasites[a]

WENDY C. BROWN,[b,e] SERGIO D. RODRIGUEZ,[b,d]
ISIDRO HOTZEL,[c] BARBARA J. RUEF,[b]
CAROL G. CHITKO-McKOWN,[b]
TERRY F. McELWAIN,[c] AND GUY H. PALMER [c]

[b]*Department of Veterinary Pathobiology*
Texas A & M University
College Station, Texas 77843-4467

[c]*Department of Veterinary Microbiology and Pathology*
Washington State University
Pullman, Washington 99164-7040

[d]*Centro National de Investigaciones Forestales y Agropecurias*
Apdo. Postal no. 206 Civac C.P. 62500
Edo. de Morelos, Mexico

INTRODUCTION

Babesiosis in domestic livestock is an important disease of many tropical and subtropical countries. Considerable progress has been made towards the identification of babesial protein antigens that stimulate humoral and cellular immunity that may be useful as components of a subunit vaccine.[1-3] Among these are a multigene family of 58-60 kDa proteins, designated rhoptry associated-protein(s) 1, or RAP-1, that are associated with both the merozoite surface and rhoptries of the apicomplexan protozoan parasites *Babesia bovis* and *B. bigemina*.[4-11] Proteins produced by specialized organelles of the merozoite apical complex that include rhoptries, micronemes, spherical bodies, and dense granules are believed to play a major role in host erythrocyte invasion or possibly egression.[12] Many of these apical-complex-associated proteins have been targeted for vaccine development against human malarial parasites.[13] The native form of the 58-kDa RAP-1 protein of *B. bigemina* was shown to confer partial protective immunity in cattle.[5] Similarly, the native form of *B. bovis* RAP-1,

[a]This work was supported in part by the United States Department of Agriculture-BARD grant 1855-90R, the National Institutes of Health grant AI30136, the USDA NRICGP grants 92-37204-8180 and 93-37206-9075, and a Fullbright Research Scholarship awarded to SDR.

[e]Address correspondence to: Department of Veterinary Microbiology and Pathology, Washington State University, Pullman, WA 99164-7040.

which was originally described as a 60–70 kDa protease designated 21B4, also induced protective immunity against *B. bovis* challenge in calves, as did a recombinant glutathione-*S*-transferase fusion protein consisting of a fragment of the *B. bovis* RAP-1 protein.[1] Together, these studies indicate the feasibility of developing RAP-1 epitope-based vaccines.

The serological response against RAP-1 proteins has been partially characterized. Cattle naturally or experimentally infected with *B. bovis* or *B. bigemina* develop RAP-1 specific antibodies that recognize surface-exposed epitopes.[4,14] Merozoite surface exposed B cell epitopes are conserved among all geographical strains of *B. bovis* and *B. bigemina*, respectively, but not between these species.[5,15] Comparison of *B. bovis* and *B. bigemina* RAP-1 sequences revealed a broadly conserved 300 amino acid bloc located at the amino terminus that contains 45% sequence identity and a strictly conserved 14-amino acid residue.[7] However, it was subsequently found that surface-exposed, antibody epitopes were not conserved between species, and the highly conserved sequence was very poorly immunogenic for cattle.[16] Therefore the conserved sequences among these apical complex proteins that are apparently required for function do not appear to contain conserved B-cell epitopes.[3,16] A neutralization-sensitive, surface-exposed, nonlinear epitope on *B. bigemina* RAP-1 was localized to the amino terminal region with monoclonal antibodies; however, there was a poor correlation between titer of antibody to this epitope and protective immunity in both *B. bigemina*-infected and RAP-1 immunized cattle, suggesting that antibody against this epitope alone is not sufficient to provide protective immunity.[17] The studies summarized in this paper were therefore designed to characterize the cellular immune response against RAP-1 proteins and to identify RAP-1 T helper (Th) cell epitopes for potential inclusion in a subunit vaccine.

Considerable evidence has accumulated in favor of a requirement for Th1-like responses in induction of protective immunity against intracellular protozoan parasites, including *Babesia* spp. (reviewed in Reference 2). Because the immunogenicity of RAP-1 proteins for T cells and the nature of the Th cell response against these antigens had not been previously determined, studies were undertaken to characterize Th cell responses against RAP-1 proteins of *B. bovis* and *B. bigemina*. RAP-1 genes of both *B. bovis* (Bv60) and *B. bigemina* (p58) were subcloned into the pMAL expression vector (New England Biolabs, Beverly, MA) and expressed as fusion proteins with maltose binding protein (MBP). This system permits rapid purification by affinity to amylose resin. Recombinant proteins were then assessed for the ability to induce recall responses *in vitro* from *B. bovis*- or *B. bigemina*-infected and immune cattle, and the Th cell response against *B. bovis* RAP-1 was characterized. In a second set of experiments, native *B. bigemina* RAP-1 protein was used to immunize calves, and the Th cell responses to native and recombinant forms of *B. bigemina* RAP-1 were also characterized.

RESULTS

RAP-1 Is Immunodominant for Th Cells from B. bovis*–Immune Cattle*

Our experiments demonstrated that the *B. bovis* RAP-1 protein is strongly immunogenic for Th cells from immune cattle, and that Th cell epitopes on RAP-1 are shared among different *B. bovis* isolates but not with the *B. bigemina* homologue.[18] Peripheral blood mononuclear cells (PBMC) from two *B. bovis*-immune cattle, C97 and C15,[19] proliferated in a dose-dependent manner against recombinant *B. bovis* RAP-1 (cloned from the Mexico isolate)[6] but not to control MBP. Similarly, T cell lines established by stimulating lymphocytes with recombinant *B. bovis* RAP-1 for two or three weeks responded against RAP-1 fusion protein, but not the control fusion protein, MBP, and responded to crude merozoite antigen prepared from different isolates (Mexico, Texas, and Australia) of *B. bovis*, but not *B. bigemina*. T cell lines established by stimulating lymphocytes with *B. bovis* membrane antigen for two to seven weeks proliferated strongly against crude membrane (CM) and soluble high speed supernatant (HSS) merozoite antigen of *B. bovis*, and also proliferated against CM prepared from *B. bigemina*, but did not respond to control uninfected red blood cell (URBC) antigen. Importantly, these *B. bovis*-induced oligoclonal T cell lines responded to *B. bovis* RAP-1, for as long as seven weeks following their initiation and continuous culture with crude *B. bovis* merozoite membrane antigen, indicating the very immunodominant nature of the RAP-1 protein. These CD4[+] T cell lines were specific for *B. bovis* RAP-1, and did not recognize either MPB or *B. bigemina* RAP-1. Together, these results demonstrated that the immunodominant T cell epitopes are conserved between different geographical isolates of *B. bovis*, but not with *B. bigemina*.

These conclusions were confirmed at the clonal level. T cells were cloned by limiting dilution from RAP-1 specific T cell lines in the presence of antigen [either RAP-1 or crude merozoite membrane antigen prepared from *B. bovis* (Mexico)], antigen presenting cells, and bovine T cell growth factor, as described for other recombinant antigens.[20,21] RAP-1 specific CD4[+] T cell clones proliferated against Mexico, Texas, Australia, and Israel isolates of *B. bovis*, but neither *B. bigemina* merozoites, *B. bigemina* RAP-1, nor the fusion protein, MBP. All clones were MHC-restricted. Importantly, we were able to induce proliferation of a RAP-1 specific T cell clone with *B. bovis*-infected erythrocytes, demonstrating the feasibility of inducing an anamnestic Th cell response in RAP-1 immunized animals when the parasite is encountered in its natural state. These Th cell clones will be used to identify the immunodominant T cell epitopes by testing proliferation against a panel of truncated fusion proteins.[20]

B. bovis RAP-1 specific Th cell clones expressed different profiles of cytokines when examined by reverse-transcription polymerase chain reaction (RT-PCR) using cytokine-specific primers.[22] Although the levels of cytokine message for IL-2, IL-4, and IL-10 in the RAP-1 specific Th cell clones were either undetectable or low, as compared with results obtained in the same experiment when RNA was prepared from Con A-stimulated PBMC and with results observed previously with *B. bovis* and *Fasciola hepatica*-stimulated clones,[23,24] all clones expressed IFN-γ. The clones were classified as either Th1 or Th0 cells, and as previously found for *B. bovis*-specific T cell clones,[23] no Th2 clones were identified. IFN-γ was also measured in supernatants by ELISA (IDEXX, Portland, ME).[22]

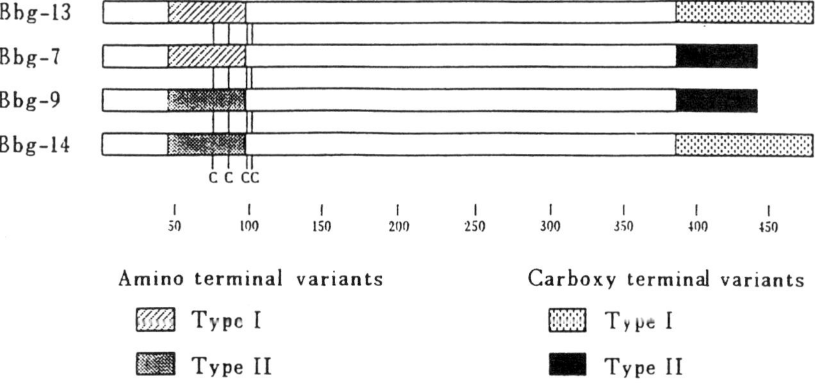

FIGURE 1. Schematic representation of RAP-1 gene copies. The relative position of the four RAP-1 gene copies, designated Bbg-13, Bbg-7, Bbg-9, and Bbg-14 and the dimorphic amino and carboxy terminal variant types is indicated. Four conserved cystein residues are indicated by a C.

Identification of Th Cell Epitopes on B. bigemina *RAP-1*

Four genomic DNA copies of RAP-1 were identified in a cloned Mexico isolate (JG-29) of *B. bigemina*.[9] These were designated clones Bbg 7, Bbg 9, Bbg 13, and Bbg 14. At least three of the genes appear to be expressed.[9] The four genes express two regions of sequence dimorphism at the 5' and 3' termini.[9] The 5' dimorphic sequence has two variants, referred to as amino terminal variants type I and II, and the 3' dimorphic region also has two variants, referred to as carboxy terminal variants type I and II, respectively (FIG. 1). Thus, clones 7 and 13 share the amino terminal type I variant, whereas clones 9 and 14 share the amino terminal type II variant. Clones 13 and 14 share the carboxy terminal type I variant, whereas clones 7 and 9 share the carboxy terminal type II variant. In addition, clones 7 and 13 share a B cell epitope recognized by neutralizing monoclonal antibodies 14/16.1.7 and C2F3G3.[5,17,25,26] The recombinant RAP-1 protein used for the majority of *in vitro* T cell assays was derived from clone 13, unless indicated otherwise.

The *B. bigemina* RAP-1 homologue was characterized for the induction of memory T cell responses and for immunogenicity.[27] As found with *B. bovis*-infected and immune cattle, one *B. bigemina*-infected cow (G2) was tested for recall proliferative responses against recombinant (clone 13) *B. bigemina* RAP-1. PBMC proliferated vigorously against both *B. bovis* and *B. bigemina* CM antigens, but not against control URBC. When tested against recombinant RAP-1 proteins, the response was specific for *B. bigemina* RAP-1, with no proliferation detected against either the *B. bovis* RAP-1 homologue or control MBP fusion protein, again indicating the lack of conservation among Th cell epitopes on RAP-1 proteins from these two babesial species.

Four calves were immunized with native RAP-1 protein obtained by affinity purification with neutralizing monoclonal antibody 14/16.1.7 that recognized both clone 7 and clone 13.[4,5] Sera from all four animals reacted specifically with a 58-

kDa protein on immunoblots of crude *B. bigemina* merozoite lysate. PBMC from two of these animals (2216 and 2234) responded specifically in proliferation assays against *B. bigemina* parasite extracts (CM and HSS fractions), but did not respond to *B. bovis* or control URBC. PBMC from these animals also responded specifically to recombinant (clone 13) *B. bigemina* RAP-1. RAP-1 specific T cell lines were established by stimulation of PBMC with either recombinant (clone 13) RAP-1 or with *B. bigemina* CM antigen and RAP-1. A panel of nine CD4[+] T cell clones were obtained and characterized extensively. All clones proliferated specifically against *B. bigemina* merozoite antigen and recombinant (clone 13) *B. bigemina* RAP-1; no clones responded to either *B. bovis* parasites or *B. bovis* RAP-1, again demonstrating the nonconserved nature of the immunodominant Th cell epitopes.

A differential proliferative response by the Th cell clones against different Latin American isolates, including Mexico (immunizing), Texcoco, Puerto Rico, and St. Croix, was observed. One group of clones (group I) responded poorly to Texcoco, Puerto Rico, and St. Croix parasites, whereas the second group of Th cell clones (group II) responded more vigorously to these isolates. This result suggested that the *B. bigemina* RAP-1 specific Th clones recognized at least two different epitopes. This hypothesis was confirmed when the proliferative responses of the two groups of Th cell clones directed against the four expressed gene copies of RAP-1 (FIG. 1) were compared. Two patterns of response were observed. The T cell epitope recognized by seven group I clones mapped to the carboxy variable region shared by clones 13 and 14, whereas a second T cell epitope recognized by two group II clones mapped to the conserved region in the amino terminal half of the RAP-1 protein. Although the level of expression of the different gene copies in different parasite isolates has not been determined, these data indicate a differential expression of the two T cell epitopes in the different geographical isolates. As one might predict, the carboxy terminal type I variant epitope appears to be less frequently expressed in the heterologous isolates than does the T cell epitope conserved among the four RAP-1 genes.

The panel of *B. bigemina*-specific Th cell clones was also characterized for cytokine mRNA expression and IFN-γ production. Following antigenic stimulation, all Th cell clones produced IFN-γ and expressed IFN-γ mRNA. Undetectable to low levels of IL-2 and/or IL-4 transcripts were observed, whereas the majority of clones expressed IL-10 mRNA. As with *B. bovis* RAP-1 specific Th cells, these clones were also classified as Th1- or Th0-like.

CONCLUSIONS

In conclusion, *B. bovis* and *B. bigemina* RAP-1 proteins are strongly immunogenic for Th cells from *B. bovis*- or *B. bigemina*-infected and immune cattle, and are immunodominant antigens when presented to T cells in the form of a crude membrane prepared from homogenized merozoites.[18,27] Studies with Th cell clones specific for either *B. bovis* or *B. bigemina* RAP-1 demonstrated that the Th cell epitopes are shared among geographically different isolates of these respective parasites, but are not shared across species. Immunization of calves with affinity purified, native *B. bigemina* RAP-1 protein induced an antigen-specific CD4[+] T cell response that could be recalled *in vitro* with either native parasite antigen or recombinant RAP-1 when PBMC, T cell lines, and Th cell clones from these calves were examined. Studies measuring proliferation of nine Th cell clones against different geographical isolates of *B. bigemina* and against four unique *B. bigemina* RAP-1 gene products derived from a cloned Mexico isolate revealed a heterogeneity in the specificity of the Th cell clones that indicated the presence of at least two Th cell epitopes. Two Th cell epitopes on the different *B. bigemina* RAP-1 proteins were identified. One epitope mapped to a region conserved among all four gene copies, and one epitope mapped to the dimorphic carboxy terminal variable region present on the native RAP-1 gene products used for immunization. Characterization of the cytokines expressed by *B. bovis* RAP-1- and *B. bigemina* RAP-1 specific Th cell clones showed that the majority of clones produced IFN-γ and expressed a Th0 cytokine profile. Additional Th cell epitopes are currently being identified on both *B. bovis* and *B. bigemina* RAP-1 proteins. The immunodominant nature of RAP-1 epitopes for IFN-γ producing Th cells, the antigenic conservation of at least one Th cell epitope, and the ability to induce recall responses against whole or fractionated merozoites provide a rationale for developing RAP-1 Th cell epitopes as components, together with neutralizing antibody epitopes, of a subunit vaccine for bovine babesiosis.

REFERENCES

1. WRIGHT, I. G., R. CASU, M. A. COMMINS, B. P. DALRYMPLE, K. R. GALE, B. V. GOODGER, P. W. RIDDLES, D. J. WALTISBUHL, I. ABETZ, D. A. BERRIE, Y. BOWLES, C. DIMMOCK, T. HAYES, H. KALNINS, G. LEATCH, R. MCCRAE, P. E., MONTAGUE, I. T. NISBET, F. PARRODI, J. M. PETERS, P. C. SCHEIWE, W. SMITH, K. RODE-BRAMANIS & M. A. WHITE. 1992. The development of a recombinant *Babesia* vaccine. Vet. Parasitol. **44:** 3-13.
2. BROWN, W. C. & A. C. RICE-FICHT. 1994. Use of Th cells to identify potential vaccine antigens of *Babesia bovis*. Parasitol. Today **11:** 145-149.
3. PALMER. G. H. & T. F. MCELWAIN. 1995. Molecular basis for vaccine development against anaplosmosis and babesiosis. Vet. Parasitol. **57:** 233-253.
4. MCELWAIN, T. F., G. H. PALMER, W. L. GOFF & T. C. MCGUIRE. 1987. Identification of *Babesia bigemina* and *Babesia bovis* merozoite proteins with isolate- and species-common epitopes recognized by antibodies in bovine immune sera. Infect. Immun. **56:** 1658-1660.
5. MCELWAIN, T. F., L. E. PERRYMAN, A. J. MUSOKE & T. C. MCGUIRE. 1991. Molecular characterization and immunogenicity of neutralization-sensitive *Babesia bigemina* merozoite surface proteins. Mol. Biochem. Parasitol. **47:** 213-222.
6. SUAREZ, C. E., G. H. PALMER, D. P. JASMER, S. A. HINES, L. E. PERRYMAN & T. F. MCELWAIN. 1991. Characterization of the gene encoding a 60-kilodalton *Babesia bovis* merozoite protein with conserved and surface exposed epitopes. Mol. Biochem. Parasitol. **46:** 45-52.
7. SUAREZ, C. E., T. F. MCELWAIN, E. B. STEPHENS, V. S. MISHRA & G. H. PALMER. 1991. Sequence conservation among merozoite apical complex proteins of *Babesia bovis*, *Babesia bigemina* and other apicomplexa. Mol. Biochem. Parasitol. **49:** 329-332.
8. MISHRA, V. S., E. B. STEPHENS, J. B. DAME, L. E. PERRYMAN, T. C. MCGUIRE & T. F. MCELWAIN. 1991. Immunogenicity and sequence analysis of recombinant p58: a neutralization-sensitive, antigenically conserved *Babesia bigemina* merozoite surface protein. Mol. Biochem. Parasitol. **47:** 207-212.

9. MISHRA, V. S., T. F. MCELWAIN, J. B. DAME & E. B. STEPHENS. 1992. Isolation, sequence, and differential expression of the p58 gene family of *Babesia bigemina*. Mol. Biochem. Parasitol. **53:** 149-158.

10. MACHADO, R. Z., T. F. MCELWAIN, C. E. SUAREZ, S. A. HINES & G. H. PALMER. 1993. *Babesia bigemina:* isolation and characterization of merozoite rhoptries. Exp. Parasitol. **77:** 315-325.

11. DALRYMPLE, B. P., R. E. CASU, J. M. PETERS, C. M. DIMMOCK, G. R. GALE, R. BOESE & I. G. WRIGHT. 1993. Characterisation of a family of multi-copy genes encoding rhoptry protein homologues in *Babesia bovis, Babesia ovis,* and *Babesia canis*. Mol. Biochem. Parasitol. **57:** 181-192.

12. PERKINS, M. E. 1992. Rhoptry organelles of apicomplexan parasites. 1992. Immunol. Today **8:** 28-32.

13. HOWARD, R. J. & B. L. PASLOSKE. 1993. Target antigens for asexual malaria vaccine development. Parasitol. Today **9:** 369-372.

14. GOFF, W. L., W. C. DAVIS, G. H. PALMER, T. F. MCELWAIN, W. C. JOHNSON, J. F. BAILEY & T. C. MCGUIRE. 1988. Identification of *Babesia bovis* merozoite surface antigens by using immune bovine sera and monoclonal antibodies. Infect. Immun. **56:** 2363-2368.

15. PALMER, G. H., T. F. MCELWAIN, L. E. PERRYMAN, W. C. DAVIS, D. R. REDUKER, D. P. JASMER, V. SHKAP, E. PIPANO, W. L. GOFF & T. C. MCGUIRE. 1991. Strain variation of *Babesia bovis* merozoite surface-exposed epitopes. Infect. Immun. **59:** 3340-3342.

16. SUAREZ, C. E., G. H. PALMER, S. A. HINES & T. F. MCELWAIN. 1993. Immunogenic B-cell epitopes of *Babesia bovis* rhoptry-associated protein 1 are distinct from sequences conserved between species. Infect. Immun. **61:** 3511-3517.

17. USHE, T. C., G. H. PALMER, L. SOTOMAYOR, J. V. FIGUEROA, G. M. BUENING, L. E. PERRYMAN & T. F. MCELWAIN. 1994. Antibody response to a *Babesia bigemina* rhoptry-associated protein 1 surface-exposed and neutralization-sensitive epitope in cattle. Infect. Immun. **62:** 5698-5701.

18. BROWN, W. C., W. B. TUO, S. A. HINES, T. F. MCELWAIN & G. H. PALMER. 1994. T cell epitopes of *Babesia bovis* rhoptry-associated protein 1 (RAP-1) are immunodominant for Th cells of immune cattle and conserved among geographically different isolates. Abstract 131, 75th Annual Conference for Research Workers in Animal Diseases (CRWAD), Chicago, IL.

19. BROWN, W. C., K. S. LOGAN, G. G. WAGNER & C. L. TETZLAFF. 1991. Cell-mediated immune responses to *Babesia bovis* antigens in cattle following infection with tick-derived or cultured parasites. Infect. Immun. **59:** 2418-2426.

20. BROWN, W. C., S. ZHAO, V. WOODS, C. A. TRIPP, C. L. TETZLAFF, V. T. HEUSSLER, D. A. E. DOBBELAERE & A. C. RICE-FICHT. 1993. Identification of two Th1 cell epitopes on the 77 kDa *Babesia bovis* merozoite protein (Bb-1) by use of truncated recombinant fusion proteins. Infect. Immun. **61:** 236-244.

21. BROWN, W. C., G. H. PALMER, T. F. MCELWAIN, S. A. HINES & D. A. E. DOBBELAERE. 1993. *Babesia bovis:* characterization of the T helper cell response against the 42 kilodalton merozoite surface antigen (MSA-1) in cattle. Exp. Parasitol. **77:** 97-110.

22. BROWN, W. C., V. M. WOODS, C. G. CHITKO-MCKOWN, S. M. HASH & A. C. RICE-FICHT. 1994. IL-10 is expressed by bovine type 1 helper, type 2 helper and unrestricted parasite-specific T cell clones, and inhibits proliferation of all three subsets in an accessory cell-dependent manner. Infect. Immun. **62:** 4697-4708.

23. BROWN, W. C., V. M. WOODS, D. A. E. DOBBELAERE & K. S. LOGAN. 1993. Heterogeneity in cytokine profiles of *Babesia bovis*-specific bovine CD4[+] T cell clones activated in vitro. Infect. Immun. **61:** 3273-3281.

24. BROWN, W. C., W. C. DAVIS, D. A. E. DOBBELAERE & A. C. RICE-FICHT. 1994. CD4[+] T cell clones obtained from cattle chronically infected with *Fasciola hepatica* and specific for adult worm antigen express both unrestricted and Th2 cytokine profiles. Infect. Immun. **62:** 818-827.

25. FIGUEROA, J. V., G. M. BUENING, D. A. KINDEN & T. J. GREEN. 1990. Identification of common surface antigens among *Babesia bigemina* isolates using monoclonal antibodies. Parasitology **100:** 161–175.
26. FIGUEROA, J. V., G. M. BUENING, V. S. MISHRA & T. F. MCELWAIN. 1992. Screening a *Babesia bigemina* cDNA library with monoclonal antibodies directed to surface antigens. Ann. N.Y. Acad. Sci. **653:** 122–130.
27. RODRIGUEZ, S. D., T. F. MCELWAIN, G. H. PALMER & W. C. BROWN. 1994. Bovine T cell responses against *Babesia bigemina* rhoptry associated (RAP-1) protein. Abstract 130, 75th Annual Conference for Research Workers in Animal Diseases (CRWAD), Chicago, IL.

Reactive Oxygen and Nitrogen Intermediates and Products from Polyamine Degradation Are Babesiacidal *In Vitro*[a]

W. C. JOHNSON, C. W. CLUFF, W. L. GOFF,[b] AND
C. R. WYATT[c]

Animal Disease Research Unit
USDA-ARS
Pullman, Washington 99164-7030

[c]*Veterinary Microbiology and Pathology*
Washington State University
Pullman, Washington 99164-7040

INTRODUCTION

Babesiosis is a hemoparasitic disease of cattle that is responsible for significant economic losses throughout tropical and subtropical regions of the world.[1] A characteristic of animals that survive babesial infection is the appearance of abnormal morphologic intraerythrocytic forms, which have been termed crisis forms.[2] The development of crisis forms is antibody independent[2] and is associated with the presence of mononuclear aggregations in the spleen.[3] Crisis forms appear to result from a cell-mediated event involving soluble mediator(s) released from macrophages.[4]

One mechanism that results in the production of reactive intermediates is the catabolism of polyamines by polyamine oxidase (PAO). Bovine serum contains significant levels of PAO,[5] which generates reactive oxygen intermediates (ROI), aminoaldehydes, and acrolein[6,7] from polyamines such as spermine or spermidine. Microbicidal activity has been demonstrated from polyamine degradation products[8-11] and, in the case of *Plasmodium falciparum,* results in the development of crisis forms.[8]

Generation of ROI and generation of reactive nitrogen intermediates (RNI) are two mechanisms whereby a macrophage releases microbicidal products. Each mechanism has been suggested to play a role in the resolution of a variety of parasitic diseases including *Babesia,*[12,13] *Plasmodium,*[14-20] *Theileria,*[21] *Francisella,*[22] *Trypanosoma,*[23,24] *Schistosoma,*[25] *Toxoplasma,*[26] *Listeria,*[26] *Leishmania,*[27] and *Entamoeba.*[28] The induction of one or both of these reactive pathways for control of an infection depends on the species of the parasite. *Plasmodium yoelli, P. chabaudi,* and *P. falciparum*

[a] This work was supported by USDA-ARS-ADRU Babesia project CWU-5348-34000-010-00D.

[b] Author to whom correspondence should be addressed at: USDA-ARS-ADRU, 337 Bustad Hall, Pullman, WA 99164-7030.

are sensitive to ROI,[4–16,18] but *P. berghei* is insensitive.[16] Similarly, species of *Babesia*[13] and *Trypanosoma*[24] differ in their sensitivities to, or ability to elicit an ROI response. The developmental stage of a parasite can also influence the organism's sensitivity to reactive intermediates.[29]

The release of ROI from bovine mononuclear cells is well characterized,[30] and recently, RNI have been elicited from bovine macrophages activated with interferon-γ in conjunction with either a phagocytic stimulus or tumor necrosis factor-α.[31,32] The elucidation of mechanisms involved in the development of crisis forms of *Babesia sp.* is important for our understanding of the cellular mechanisms involved in an immune response *in vivo*. In this report, we investigate the effect of products from RNI, ROI, or polyamine catabolism, added exogenously or generated enzymatically, on the *in vitro* growth of *Babesia bovis* and the development of crisis forms.

MATERIALS AND METHODS

In Vitro *Growth of* Babesia bovis

Asynchronous *in vitro* microaerophilous stationary phase (MASP) cultures of a Mexican isolate of *B. bovis* were established from a steer with a tick-induced infection. Cultures were maintained as previously described.[33,34] The culture medium consisted of 40% normal bovine serum in media M199 (Gibco Laboratories, Grand Island, NY) supplemented with 10 mg/ml glucose, 1 mM calcium chloride dihydrate (Fisher Scientific, Fair Lawn, NJ), 100 units/ml each of Amphotericin-B and streptomycin (Gibco Laboratories, Grand Island, NY) and buffered with 10 mM 3-(*N*-tris-[hydroxymethyl]-methyl-amino)-2-hydroxypropane sulfonic acid, pH 7.35.

Reactive Oxygen Intermediates

Exogenous H_2O_2 or acrolein (Sigma, St. Louis, MO), diluted in culture medium with a 5% suspension of normal bovine erythrocytes, was added to replicate wells of an equal volume of *B. bovis* culture, such that a 0.5% parasitemia resulted. ROI, including H_2O_2, were also produced enzymatically by adding xanthine oxidase (XO) (Sigma, St. Louis, MO), diluted in culture medium, to replicate cultures of *B. bovis* with a 0.5% parasitemia containing 5 mM xanthine (Sigma, St. Louis, MO). All cultures were incubated in the presence of 1 mg/ml catalase (Sigma, St. Louis, MO), overnight at 37°C, 5% CO_2. Growth of the parasite was evaluated by hydroethidine viability staining and subsequent flow cytometry (FC) as previously described.[35] Briefly, 200 μl samples were diluted with 150 μl room temperature phosphate-buffered saline (PBS), centrifuged at 1500 × *g* at room temperature for 10 minutes, and incubated in PBS containing 20 μg/ml hydroethidine (Polysciences, Inc., Warrington, PA) at 37°C for 5 minutes. Samples were then diluted with an equal volume of PBS and centrifuged as before. The supernatants were removed, and the samples suspended in 200 μl PBS and analyzed immediately using a Beckton Dickinson FACScan. Development of crisis forms was determined by examining Giemsa stained thin smears of material from pooled replicate wells.

Polyamine Catabolism

Generation of H_2O_2 and acrolein by the catabolism of spermine or spermidine was accomplished by incubation of *B. bovis* cultures, 0.5% parasitemia, supplemented with either spermine or spermidine (Sigma, St. Louis, MO) in the presence or absence of 1 mg/ml catalase. Additionally, control cultures were made of 1 mM aminogaunidine (Sigma, St. Louis, MO), which neutralizes the activity of PAO,[5] prior to the addition of exogenous polyamine. Dilutions of acrolein were added directly to replicate cultures to examine the effect of acrolein without H_2O_2. All cultures were incubated overnight and evaluated for growth and crisis form development as described above.

Reactive Nitrogen Intermediates

Babesia bovis cultures with a parasitemia of 0.5% were cultivated in the presence of dilutions of sodium nitrite (Sigma, St. Louis, MO), *S*-nitroso-*N*-acetyl-penicillamine (SNAP) (Sigma, St. Louis, MO), a spontaneous emitter of nitric oxide (NO) with a six-hour half-life, or DL-penicillamine (PEN) (Sigma, St. Louis, MO). Concentrations of nitrite formed in cultures incubated with SNAP were measured by evaluating supernatants from duplicate SNAP or PEN cultures without bovine erythrocytes by the Griess reaction as previously described.[19] For washout experiments, parasite cultures were incubated in Puck's Saline G (PSG) (Gibco Laboratories, Grand Island, NY) containing doubling dilutions of either PSG, potassium ferricyanide (FCN), or sodium nitroprusside (NP) (Sigma, St. Louis, MO), for 30 minutes at 37°C, 5% CO_2, and 3% O_2. NP is a compound that rapidly decays and emits NO, or FCN, a stable breakdown product of NP. Cultures were then suspended, diluted in PSG, and centrifuged at $1500 \times g$ at 4°C for 10 minutes. Supernatants were aspirated, and the erythrocyte pellets suspended in culture medium, plated in 24-well tissue-culture plates (Costar, Boston, MA), and incubated overnight at 37°C, 5% CO_2, and 3% O_2. Evaluation of growth by hydroethidine staining and the development of crisis forms was performed as described above.

Student *t*-tests were employed for statistical analysis.

RESULTS

Polyamine Catabolism

The effect of overnight incubation of *B. bovis* with spermine is shown in FIGURE 1. Compared to control cultures, maximal babesiacidal effects were achieved by the addition of 300 μM spermine ($p < 0.005$). Erythrocytes incubated in higher concentrations became crenated and developed high autofluorescence, rendering them unsuitable for evaluation by FC. Addition of 1 mg/ml catalase to cultures containing dilutions of spermine did not ameliorate the babesiacidal effects of polyamine catabolism ($p < 0.005$). Addition of 1 mM aminoguanidine, an inhibitor of serum PAO, abrogated the effect of spermine catabolism on the growth of the parasite, and prevented erythrocyte changes that occurred in high concentrations of spermine (data not shown).

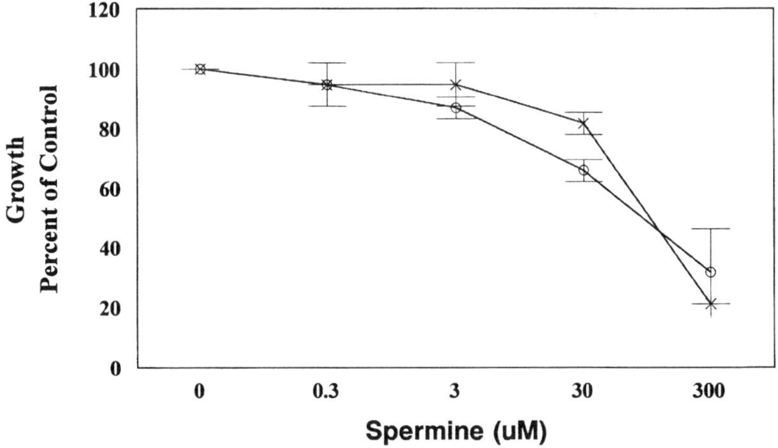

FIGURE 1. Inhibition of 24-hour *Babesia bovis* growth *in vitro* in the presence of the polyamine, spermine, and the failure of catalase to reverse the effects. Spermine, crosses; spermine plus catalase (1 mg/ml), circles.

Xanthine Catabolism

FIGURE 2 illustrates that *B. bovis* growth was compromised, compared to controls, by incubation of the parasite with XO ($p < 0.001$). Xanthine catabolism by XO yields H_2O_2, but addition of 1 mg/ml catalase did not prevent the appearance of crisis forms

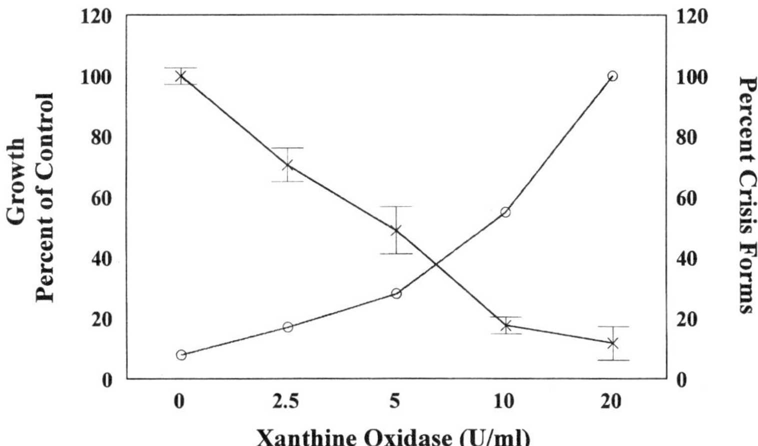

FIGURE 2. Inhibition of 24-hour *Babesia bovis* growth *in vitro* and the development of parasite crisis forms in the presence of xanthine (5 mM) and xanthine oxidase. Percent parasitized erythrocytes, crosses; percent crisis forms, circles.

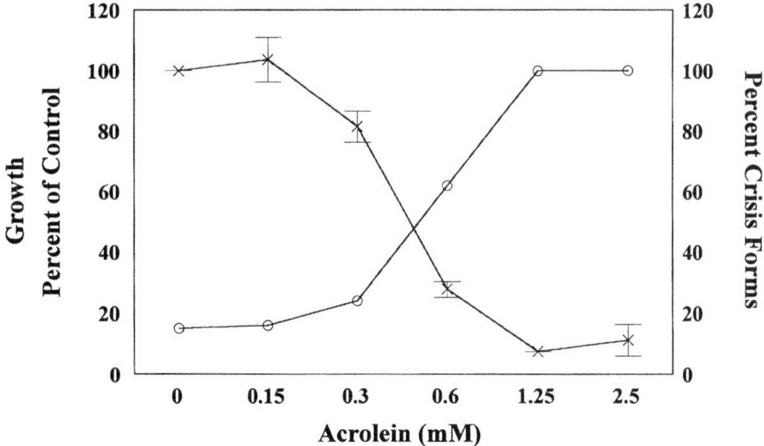

FIGURE 3. Inhibition of 24-hour *Babesia bovis* growth *in vitro* and the development of parasite crisis forms in the presence of acrolein. Percent parasitized erythrocytes, crosses; percent crisis forms, circles.

(data not shown). Growth of the parasite with XO + catalase was partially improved as compared to XO cultures (38.8% of the normal control vs. 16.6%). Increasing concentrations of catalase were tested but proved to be toxic for the parasite irrespective of the addition of XO.

Hydrogen Peroxide

To determine directly if H_2O_2 was babesiacidal, sufficient H_2O_2 was added to cultures to make a final concentration of 100 μM. Evaluation by FC after 24 hours revealed parasite growth that was not significantly affected by the presence of H_2O_2 (96.3% of control cultures). Additionally, crisis forms in the H_2O_2 cultures comprised between 5 and 15%, a range that was equivalent to control cultures.

Acrolein

To evaluate the possibility that products other than H_2O_2 were responsible for the babesiacidal effects of spermine catabolism, acrolein was serially diluted in a set of cultures. A dose-dependent effect of acrolein on *B. bovis* growth ($p < 0.001$) was demonstrated (FIG. 3), with an associated increase in parasite crisis forms. Incubation of cultures in 5 mM acrolein produced high erythrocyte autofluorescence similar to incubation of erythrocytes with 3 mM spermine. Removal of acrolein after incubation of cultures for one hour did not reverse its babesiacidal effect or prevent erythrocyte changes that occurred in higher concentrations of acrolein (data not shown).

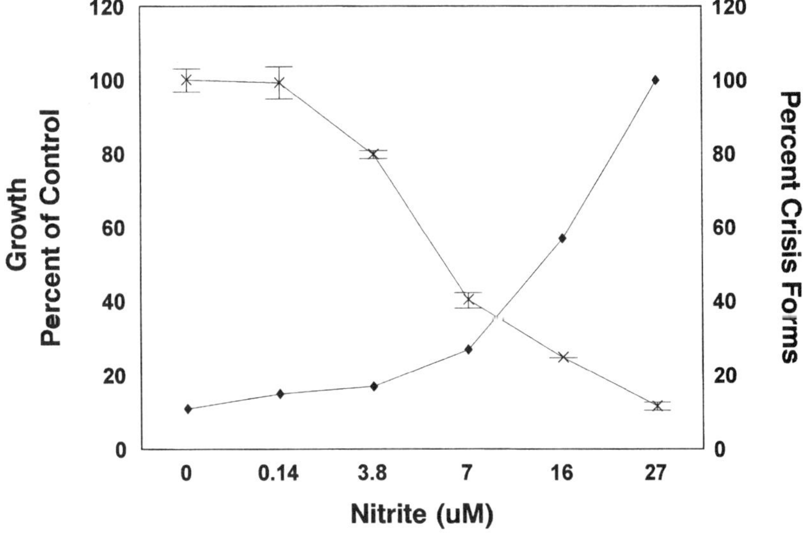

FIGURE 4. Inhibition of 24-hour *Babesia bovis* growth *in vitro* and the development of parasite crisis forms in the presence of *S*-nitroso-*N*-acetyl-penicillamine sufficient to generate nitric oxide and subsequent measurable sodium nitrite at the indicated concentrations. Percent parasitized erythrocytes, crosses; percent crisis forms, circles.

Nitric Oxide

Spontaneous generation of NO by SNAP was used to evaluate the effect of NO exposure on *B. bovis* in continuous culture. All experiments utilizing SNAP were set up in parallel with erythrocyte-free culture conditions in order to accurately measure nitrite concentrations by the Griess reaction. Overnight incubation in the presence of a concentration of SNAP resulting in 7 μM nitrite was sufficient to reduce the parasitemia by 50% (FIG. 4). Cultures incubated for an additional 24 hours were composed of 100% crisis forms (data not shown). Concentrations of SNAP resulting in 27 μM nitrite were capable of producing cultures with 100% crisis forms within 24 hours ($p < 0.005$). To insure that NO and not nitrite was responsible for the babesiacidal effects, cultures containing as much as 20 μM nitrite were also examined. The presence of nitrite in the culture medium did not affect the growth of *B. bovis*.

To determine the necessary period of exposure to NO, washout experiments using NP were performed (FIG. 5). In contrast to SNAP, NP rapidly degrades to emit NO. Thirty-minute exposures to dilutions of NP between 0.6–2.5 μM were babesiacidal ($p < 0.005$). Incubation with equal concentrations of FCN did not significantly alter *B. bovis* growth.

To determine if NO exposure exerted an effect on the host erythrocyte, which could prevent subsequent parasite growth, normal erythrocytes in PSG were exposed to 100 μM NP for one hour, washed with PSG, and then used as the source of

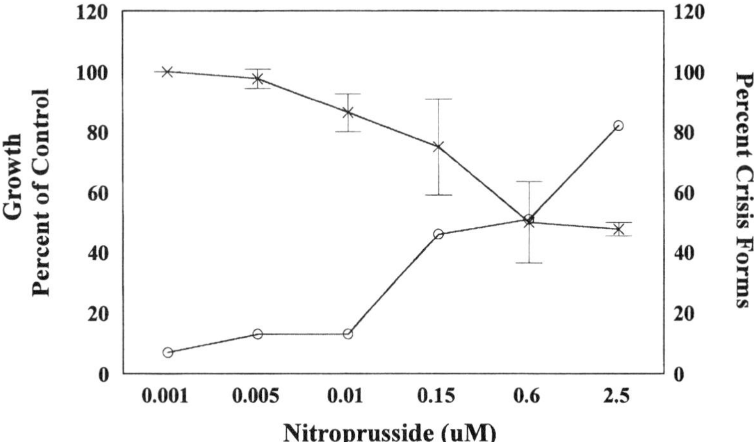

FIGURE 5. Inhibition of 24-hour *Babesia bovis* growth *in vitro* and the development of parasite crisis forms resulting from a 30-minute exposure to sodium nitroprusside at the indicated concentrations. Percent parasitized erythrocytes, crosses; percent crisis forms, circles.

erythrocytes for subculturing high parasitemic 48-hour cultures. The presence of NO-exposed erythrocytes had no effect as compared to cultures divided with erythrocytes similarly treated with either PSG or FCN.

DISCUSSION

Cell-free generation of reactive intermediates has been used to examine the role of products of cell-mediated immunity in the resolution of infection by a variety of pathogens.[17,18,27,28] With regard to the obligate intraerythrocytic Apicomplexa, particular attention has been paid to those products that induce the formation of intraerythrocytic crisis forms.[3,8,14,18] In this study, crisis forms were induced by the catabolism of polyamines or xanthine and by NO.

Hydrogen peroxide is a product common to both xanthine and polyamine catabolism; thus, a role for this ROI would appear likely. However, catalase was ineffective in completely reversing the deleterious effect on babesial growth of catabolism of either substrate. Additionally, direct addition of H_2O_2, in concentrations exceeding that produced by 10^7 phorbol myristate acetate stimulated neutrophils/ml,[32] failed to compromise the growth of *B. bovis in vitro,* suggesting that babesiacidal effects were not due to H_2O_2. While exposure to ROI results in the formation of crisis forms in rodent *Babesia,*[3] the apparent insensitivity of *B. bovis* to H_2O_2 is in agreement with rodent adapted cattle *Babesia* that have been studied *in vivo.*[13] Reactive oxygen intermediates might still play a synergistic role with RNI as has been suggested in other systems.[28]

The babesicidal effects of the catabolism of the compounds examined can not be fully attributable to the generation of ROI. Xanthine oxidase utilizes xanthine and converts hypoxanthine to xanthine. *Babesia* are incapable of *de novo* purine synthesis and require exogenous hypoxanthine for growth.[34,36] The babesiacidal effects of the XO reaction might reflect the consumption of hypoxanthine from the media as exogenous xanthine was not required (data not shown).

Secondary products of polyamine catabolism by serum PAO include acrolein and aminoaldehydes. We have demonstrated that the addition of acrolein, an inhibitor of nucleic acid and protein synthesis,[37] results in the induction of crisis forms and is babesiacidal. Experiments using spermidine in place of spermine demonstrated that this polyamine was less effective as a babesiacidal agent (data not shown). The enhanced toxicity of spermine relative to spermidine might reflect the stoichiometry of the interaction between the polyamine and serum PAO, since twice as much acrolein is produced from spermine catabolism as from spermidine catabolism.[7]

Two characteristics of plasmodial and babesial infection are splenomegaly[3,38,39] and a transient immunosuppression.[40,41] A dramatic increase in polyamine biosynthesis occurs in the spleens of mice infected with *P. berghei* as a consequence of increased proliferation of spleen cells.[38] As a result, increased concentrations of putrescine, spermidine, and spermine are confined within spleen cells, and not associated with the parasites or increased in the blood. The eventual catabolism of polyamines by intracellular PAO differs from serum PAO in that aminoaldehydes, and not acrolein, are formed.[5] Smith *et al.* have suggested that these aminoaldehydes, rather than acrolein, are responsible for inhibiting proliferation of lymphocytes exposed to polyamine degradation products[7] and may similarly affect other cell types. Gahl and Pitot suggest that acrolein formed by serum PAO and polyamines is rapidly complexed with serum proteins, preventing host cell damage.[5] Whether acrolein or aminoaldehydes are responsible for babesiacidal effects of polyamine catabolism *in vivo* cannot be determined from our results because we used extracellular polyamines and serum PAO. Incubation of erythrocytes in high concentrations of extracellular polyamines and serum PAO results in profound changes in the erythrocyte, as revealed by alteration of the erythrocytes' autofluorescent characteristics in FC, similar to incubation with high concentrations of acrolein.

Crisis forms of *B. bovis* developed in the presence of NO. Exposure to as little as 0.6 μM NO for 30 minutes was sufficient to irreversibly compromise intraerythrocytic parasites. Given the asynchronous nature of *B. bovis* MASP cultures, the sensitivity of the parasite to NO, within this short period of time, is probably not restricted to a single developmental stage, as has been suggested for other intracellular parasites.[21] Interestingly, the effect of NO appeared to be restricted to interactions with the parasite directly, as NO-exposed erythrocytes were fully capable of supporting normal parasite growth. Thus, while NO can interact with erythrocyte proteins, converting oxyhemaglobin to nitrosothiolhemoglobin or methemoglobin and H_2O_2,[19] altering cell morphology and down-regulating cellular functions such as potassium efflux,[42] these potential host cell modifications did not adversely affect the ability of parasites to invade and divide in erythrocytes. Furthermore, this effect is restricted to NO and not nitrite, as continuous incubation with the oxidized product had no effect on parasite growth. In fact, bovine erythrocytes effectively scavenged exogenously added nitrite from the media, removing 80% of the extracellular nitrite in one hour (data not shown).

Young calves are, to some degree, naively resistant to babesial infection,[43] as are adult animals infected with BCG, *Corynebacterium parvum,* or *Brucella abortus* S19.[3] Naive immunity requires the presence of the spleen[44] and mature T cells,[44–46] does not require B cells,[44] is unaffected by the down-regulation of a humoral response by the administration of cyclophosphamide,[3] and is abolished by down-regulating cell-mediated responses by cortisone treatment.[45] Macrophages, once activated to release reactive intermediates, appear to control the intraerythrocytic parasite infection. If NO represents an effector molecule *in vivo* for *B. bovis,* then the release of interferon-γ by Th1 $\alpha\beta$T cells or $\gamma\delta$T cells will be required to activate the macrophage.[31,32] Interestingly, young naively resistant calves, unlike adult cattle, have splenic lymphocytes that proliferate in response to babesial antigens and contain a high proportion of $\gamma\delta$T cells.[47,48] Finally, our results with *B. bovis* indicate that nitric oxide might be well suited for a role as an effector molecule for control of the organism *in vivo*. The amount of NO that induces crisis forms and is babesicidal is within the capacity of bovine mononuclear cells and splenic macrophages.[21,31,32] Furthermore, nitric oxide appears to be without the associated host cell damage of other crisis form mediators such as acrolein. We are currently investigating antigen/adjuvant systems that promote Th1 T cell responses to babesial antigens.

SUMMARY

Products released from activated macrophages have been demonstrated to have microbicidal activity against a variety of microorganisms. Reactive oxygen intermediates (ROI) and reactive nitrogen intermediates (RNI) have been shown to affect the induction of degenerate (crisis) forms of *Plasmodium* spp. Polyamines are degraded into acrolein which has also been shown to be toxic to *Plasmodium* spp. We have investigated the possibility that these products act similarly with *Babesia bovis*. Crisis forms of *B. bovis* developed in erythrocyte cultures after the introduction of supernatants containing ROI, RNI, and acrolein. Xanthine degradation by xanthine oxidase leads to the formation of superoxide anion, hydrogen peroxide, and hydroxyl radicals. The degradation in the presence of *B. bovis* was toxic to the parasite. The toxicity was partially reversed by the addition of the ROI scavenger catalase. However, H_2O_2 added directly had little effect, suggesting a role for the other ROI products. Spermine degradation by polyamine oxidase and direct addition of acrolein was toxic in a dose-dependent manner. Finally, spontaneous generation of nitric oxide from sodium nitroprusside or *S*-nitroso-*N*-acetyl-penicillamine was also toxic in a dose-dependent manner. These data lead us to suggest a role for activated macrophages in the primary immune response against *B. bovis*.

ACKNOWLEDGMENTS

We wish to thank Ernie Smith, Pete Steiner, and Gary Johnson for the maintenance of animals and the collection of biological materials necessary for these experiments.

REFERENCES

1. McCosker, P. J. 1981. The global importance of babesiosis. *In* Babesiosis. M. Ristic & J. P. Kreier, Eds.: 1-24. Academic Press. New York, NY.

2. Clark, I. A., J. E. Richmond, E. J. Wills & A. C. Allison. 1977. Intra-erythrocytic death of the parasite in mice recovering from infection with *Babesia microti*. Parasitology **75:** 189-196.

3. Clark, I. A. 1979. Protection of mice against *Babesia microti* with cord factor, COAM, zymosan, glucan, *Salmonella* and *Listeria*. Parasite Immunol. **1:** 179-196.

4. Allison, A. C. & I. A. Clark. 1977. Specific and non-specific immunity to haemoprotozoa. Am. J. Trop. Med. Hyg. **26:** 216-222.

5. Gahl, W. A. & H. C. Pitot. 1982. Polyamine degradation in foetal and adult bovine serum. Biochem. J. **202:** 603-611.

6. Alarcon, R. A. 1964. Isolation of acrolein from incubated mixtures of spermine with calf serum and its effects on mammalian cells. Arch. Biochem. Biophys. **106:** 240-242.

7. Smith, C. J., J. I. Hussain & J. C. Allen. 1985. Inhibition of cell proliferation by polyamines does not depend on the cytotoxicity of acrolein. Biochem. Soc. Trans. **13:** 326-329.

8. Ferrante, A., C. M. Rzepczyk & A. C. Allison. 1983. Polyamine oxidase mediates intra-erythrocytic death of *Plasmodium falciparum*. Trans. R. Soc. Trop. Med. Hyg. **77**(6): 789-791.

9. Foldes, J. & A. Matyi. 1994. The immunomodulating effect of a new polyamine (the MAP-1987) administered with chloroquine in plasmodia infected mice. Acta Microbiol. Immunol. Hun. **41:** 73-82.

10. Morgan, D. M. L., J. R. Christense & A. C. Allison. 1981. Polyamine oxidase and the killing of intracellular parasites. Biochem. Soc. Trans. **9:** 563-564.

11. Rzepczyk, C. M., A. Saul & A. Ferrante. 1984. Polyamine oxidase-mediated intra-erythrocytic killing of *Plasmodium falciparum*: evidence against the role of reactive oxygen metabolites. Infect. Immun. **43**(1): 238-244.

12. Montealegre, F., M. G. Levy, M. Ristic & M. A. James. 1985. Growth inhibition of a *Babesia bovis* in culture by secretions from bovine mononuclear phagocytes. Infect. Immun. **50:** 523-526.

13. Langley, R. J. & J. S. Gray. 1989. Non-specific resistance to *Babesia divergens* in the mongolian gerbil (*Meriones unguiculatus*). Int. J. Parasitol. **19:** 265-269.

14. Clark, I. A., N. H. Hunt, G. A. Butcher & W. B. Cowden. 1987. Inhibition of murine malaria (*Plasmodium chabaudi*) in vivo by recombinant interferon-gamma or tumor necrosis factor, and its enhancement by butylated hydroxyanisole. J. Immunol. **139**(10): 3493-3496.

15. Dockrell, H. M. & J. H. L. Playfair. 1984. Killing of *Plasmodium yoelii* by enzyme-induced products of the oxidative burst. Infect. Immun. **43**(2): 451-456.

16. Dockrell, H. M. & J. H. L. Playfair. 1983. Killing of blood-stage murine malaria parasites by hydrogen peroxide. Infect. Immun. **39:** 456-459.

17. Malhotra, K., D. Salmon, J. Le Bras & J. L. Vilde. 1988. Susceptibility of *Plasmodium falciparum* to a peroxidase-mediated oxygen-dependent microbicidal system. Infect. Immun. **56**(12): 3305-3309.

18. Ockenhouse, C. F., S. Schulman & H. L. Shear. 1984. Induction of crisis forms in the human malaria parasite *Plasmodium falciparum* by gamma-interferon-activated, monocyte-derived macrophages. J. Immunol. **133**(3): 1601-1608.

19. Rockett, K. A., M. M. Awburn, W. B. Cowden & I. A. Clark. 1991. Killing of *Plasmodium falciparum in vitro* by nitric oxide derivatives. Infect. Immun. **59:** 3280-3283.

20. Seguin, M. C., F. W. Klotz, I. Schneider, J. P. Weir, M. Goodbary, M. Slayter, J. J. Raney, J. U. Aniagolu & S. J. Green. 1994. Induction of nitric oxide synthase protects against malaria in mice exposed to irradiated *Plasmodium berghei* infected

mosquitoes: involvement of interferon gamma and CD8+ T cells. J. Exp. Med. **180:** 353-358.

21. VISSER, A. E., A. ABRAHAM, L. J. B. SAKYI, C. G. D. BROWN & P. M. PRESTON. 1995. Nitric oxide inhibits establishment of macroschizont-infected cell lines and is produced by macrophages of calves undergoing bovine tropical theileriosis or East Coast fever. Parasite Immunol. **17:** 91-102.

22. ANTHONY, L. S. D., P. J. MORRISSEY & F. E. NANO. 1992. Growth inhibition of *Francisella tularensis* live vaccine strain by IFN-gamma-L-arginine-activated macrophages is mediated by reactive nitrogen intermediates derived from L-arginine metabolism. J. Immunol. **148:** 1829-1834.

23. GAZZINELLI, R. T., I. P. OSWALD, S. HIENY, S. L. JAMES & A. SHER. 1992. The microbicidal activity of interferon-gamma-treated macrophages against *Trypanosoma cruzi* involves an L-arginine-dependent, nitrogen oxide-mediated mechanism inhibitable by interleukin-10 and transforming growth factor-beta. Eur. J. Immunol. **22:** 2501-2506.

24. VRAY, B., P. DE BAETSELIER, A. OUAISSI & Y. CARLIER. 1991. *Trypanosoma cruzi* but not *Trypanosoma brucei* fails to induce a chemiluminescent signal in a macrophage hybridoma cell line. Infect. Immun. **59:** 3303-3308.

25. JAMES, S. L. & J. GLAVEN. 1989. Macrophage cytotoxicity against schistosomula of *Schistosoma mansoni* involves arginine-dependent production of reactive nitrogen intermediates. J. Immunol. **143**(12): 4208-4212.

26. LANGERMANS, J. A. M., M. E. B. VAN DER HULST, P. H. NIBBERLING & R. VAN FURTH. 1992. Endogenous tumor necrosis factor alpha is required for enhanced antimicrobial activity against *Toxoplasma gondii* and *Listeria monocytogenes* in recombinant gamma interferon-treated mice. Infect. Immun. **60:** 5107-5112.

27. LIEW, F. Y., S. MILLOTT, C. PARKINSON, R. M. J. PALMER & S. MONCADA. 1990. Macrophage killing of *Leishmania* parasite in vivo is mediated by nitric oxide from L-arginine. J. Immunol. **144:** 4794-4797.

28. LIN, J. Y. & K. CHADEE. 1992. Macrophage cytotoxicity against *Entamoeba histolytica* trophozoites is mediated by nitric oxide from L-arginine. J. Immunol. **148:** 3999-4005.

29. SCOTT, P., S. JAMES & A. SHER. 1985. The respiratory burst is not required for killing of intracellular and extracellular parasites by a lymphokine-activated macrophage cell line. Eur. J. Immunol. **15:** 553-558.

30. BOUNOS, D. I. 1992. Comparison of oxidant production by bovine neutrophils and monocyte-derived macrophages stimulated with *Brucella abortus*. Inflammation **16:** 215-225.

31. ADLER, H., E. PETERHANS, J. NICOLET & T. W. JUNGI. 1994. Inducible L-arginine-dependent nitric oxide synthase activity in bovine bone marrow-derived macrophages. Biochem. Biophys. Res. Commun. **198:** 510-515.

32. GOFF, W. L., W. C. JOHNSON, C. R. WYATT & C. W. CLUFF. Assessment of bovine mononuclear phagocytes and neutrophils for induced L-arginine-dependent nitric oxide production. Vet. Immunol. Immunopath. (In press.)

33. LEVY, M. G. & M. RISTIC. 1980. *Babesia bovis:* continuous cultivation in a microaerophilus stationary phase culture. Science **207:** 1218-1220.

34. GOFF, W. L. & C. E. YUNKER. 1986. *Babesia bovis:* increased percentage parasitized erythrocytes in cultures and assessment of growth by incorporation of [3H] hypoxanthine. Exp. Parasitol. **62:** 202-210.

35. WYATT, C. R., W. L. GOFF & W. C. DAVIS. 1995. A flow cytometric method for assessing viability of intraerythrocytic hemoparasites. J. Immunol. Methods **140:** 23-30.

36. IRVIN, A. D., E. R. YOUNG & R. E. PURNELL. 1978. The *in vitro* uptake of tritiated nucleic acid precursors by *Babesia* spp. of cattle and mice. Infect. Immun. **8:** 19-24.

37. KIMES, B. W. & D. R. MORRIS. 1971. Inhibition of nucleic acid and protein synthesis in *Escherichia coli* by oxidized polyamines and acrolein. Biochim. Biophys. Acta **228:** 235-244.

38. HIBASAMI, H., T. TSUKADA, Y. NISHIGUCHI, M. SAKURAI, S. SHIRAKAWA & K. NAKASHIMA. 1994. Remarkable activation of polyamine biosynthesis in hematopoieses and hyperpla-

sia of spleen in mice with hemolytic anemia caused by infection with *Plasmodium berghei.* Int. J. Parasitol. **24:** 213-217.

39. HILDEBRANDT, P. K. 1981. The organ and vascular pathology of babesiosis. *In* Babesiosis. M. Ristic & J. P. Kreier, Eds.: 459-473. Academic Press. New York, NY.

40. ROCKETT, K. A., M. M. AWBURN, E. J. ROCKETT, W. B. COWDEN & I. A. CLARK. 1994. Possible role of nitric oxide in malarial immunosuppression. Parasite Immunol. **16:** 243-249.

41. PURVIS, A. C. 1977. Immunodepression in *Babesia microti* infections. Parasitology **75:** 197-205.

42. CARAMELO, C., A. RIESCON, J. OUTEIRINO, I. ILLAS, G. BLUM, B. MONZU, V. CASADO, L. SANCHEZ, J. R. MOSQUERA, S. CASADO & A. L. FARRE. 1995. Effects of nitric oxide on red blood cells: changes in erythrocyte resistance to hypotonic hemolysis and potassium efflux by experimental maneuvers that decrease nitric oxide. Biochem. Biophys. Res. Commun. **199:** 447-454.

43. TRUEMAN, K. F. & G. W. BLIGHT. 1981. The effect of age on resistance of cattle to *Babesia bovis.* Aus. Vet. J. **54:** 301.

44. STEVENSON, M. M., M. F. TAM & D. RAE. 1990. Dependence on cell-mediated mechanisms for the appearance of crisis forms during *Plasmodium chabaudi* AS infection in C57BL/6 mice. Microbial Pathogen. **9:** 303-314.

45. ALLISON, A. C. 1978. Macrophage activation and nonspecific immunity. Int. Rev. Exp. Pathol. **18:** 303-345.

46. KNOWLES, D. P., L. S. KAPPMEYER & L. E. PERRYMAN. 1994. Specific immune responses are required to control parasitemia in *Babesia equi* infection. Infect. Immun. **62:** 1909-1913.

47. WYATT, C. R., C. W. CLUFF, W. C. DAVIS & W. L. GOFF. 1992. Differential response of bovine peripheral blood and splenic lymphocytes to *Babesia bovis* antigens *in vitro.* (abstract.) 73rd Conference of Research Workers in Animal Diseases 17.

48. WYATT, C. R., C. MADRUGA, C. CLUFF, S. PARISH, M. J. HAMILTON, W. GOFF & W. C. DAVIS. 1994. Differential distribution of gamma delta T-cell receptor lymphocyte subpopulations in blood and spleen of young and adult cattle. Vet. Immunol. Immunopathol. **40:** 187-199.

Incidence and Intensity of *Babesia* spp. Sporokinetes in Engorged *Boophilus microplus* from a Dairy Herd in Venezuela

R. D. MELENDEZ AND M. FORLANO

Universidad Centroccidental Lisandro Alvarado
Decanato de Ciencias Veterinarias
Area de Parasitología
Apartado Postal 665
Barquisimeto, Lara, Venezuela

INTRODUCTION

Bovine babesiosis in Latin America is caused by the intraerythrocytic parasites *Babesia bigemina* and *Babesia bovis,* which constitute a significant limiting factor in the development and improvement of the cattle industry throughout the subcontinent.[1] The transmission of bovine babesiosis is carried out by parasitic stages of the one-host tick *Boophilus microplus*[2,3] (Acarina: Ixodoidea), which is the only known vector of both hemoparasitic protozoa in South America. Bovine babesiosis causes high economic losses in cattle herds in this region, directly due to mortality occurring mainly in *Bos taurus* susceptible animals, and indirectly as a consequence of (1) reduced weight gain and milk production, (2) the cost of therapeutic treatment for disease animals, and (3) the cost of acaricide treatments for tick control.

Actually, these expenses were reported as high as US $20 million per year in Puerto Rico, a tropical island,[4] an example that reflects the importance of the negative impact of babesiosis in the cattle industry in developing nations.

The life cycle of *Babesia* spp. is a complex system of interactive factors that, combined or separated, act upon the three elements of the system: cattle, *Babesia,* and *Boophilus* tick. The weakest point of this system is the protozoan *Babesia,* and cattle are the strongest one.[1] Some of these factors affect directly or indirectly the transmission process of *B. bovis* and *B. bigemina* from the vertebrate host to the vector tick. On the other hand, other factors affect the transovarial transmission for these babesias, including the mechanism of vertical transmission found only in *B. bigemina.*[5]

Climatic factors in a region such as rain and temperature play a key role in limiting or favoring the life cycle of babesiosis of domestic animals. Thus, some weather factors can cause larval mortality, stop the oviposition process of engorged females (EF) of *B. microplus,* and alter the normal embryonic development of tick eggs.[6] Specifically, in Venezuela there are two seasons: rainy (May through November) and dry (December through April).[7] The population of cattle ticks decreases during the rainy season, and it reaches high levels along the dry season.[8] A seasonality for *B. microplus* has also been reported from other countries.[9,10] and this parameter is of great importance in epizootiological studies about bovine babesiosis.[11,12] Thus, changes in the size of the cattle tick population will affect the inoculation rate of *Babesia* spp. into its vertebrate host. Besides, for *B. bigemina* and *B. bovis,* a seasonal incidence has also been found in tick hemolymph, which closely follows that of the protozoan infection rate in cattle, either grazing or at experimental paddocks.[13]

Finally, another factor that cannot be overlooked in the epizootiology of bovine babesiosis in a region is the efficacy of the acaricide treatments applied in the field against *B. microplus.*

Thus, tick infestation in a herd might be reduced by chemical control to levels that are insufficient to maintain the transmission of *Babesia* spp.[11] Nevertheless, it is well documented that ticks develop resistance to acaricides,[14,15] in particular when these chemical weapons are improperly used, which is often seen in countries like Venezuela.[16]

Consequently, a better understanding of bovine babesiosis in a region ought to be based on a comprehensive study of the interactive factors that determine its epizootiology in a given geographic area.

The main objectives of this paper were to assess during 14 months some factors that affect the transmission of bovine babesiosis in the west central region of Venezuela. In particular, we study the incidence and the mean intensity of *Babesia* spp. sporokinetes in hemolymph of EF *Boophilus microplus* detached from Carora breed heifers, which were located in two farms, one with efficient and the other with inefficient methods of tick control.

MATERIALS AND METHODS

Experimental Animals

Two ranches (A and B) dedicated to raising Carora breed cattle were selected for this experiment. Ranches were located close to Carora town (10° 35′N and 70° 15′W), and they were 45 km apart. The Carora breed is a local dairy bovine developed in the region since 1930, and officially approved as a cattle race in 1989 by the Venezuelan Ministry of Agriculture and Livestock. This local dairy breed is the product of empirical genetic crosses among Brown Swiss cattle, indigenous Creole animals, and *Bos indicus* to a minor degree.

The Carora breed cattle shows two main characteristics: (a) it has an average milk yield/cow per day of 12 l, and (b) it shows some resistance to tropical diseases and ticks. A group of 80 to 90 heifers was checked fortnightly for *B. microplus* ticks in ranch A up to 14 consecutive months, and a second group of 16 to 22 half to one-year-old animals was also evaluated for ticks at ranch B.

Tick Collection

Only EF *Boophilus microplus* were collected from the skin of experimental cattle which were checked for ixodids at early hours (9:00-10:00 AM), twice a month, and during 20-30 min each time. Collected ticks were transferred to the laboratory in vials stoppered with cotton plugs. There, ticks were incubated under conditions explained elsewhere,[17] and they were kept in an incubator for 8-10 days, when hemolymph samples were obtained.[18]

Examination of Ticks

To assess the level of infection of EF *B. microplus* with *Babesia* spp. sporokinetes, hemolymph microdrops were collected from incubated ticks in a glass slide after cutting one leg of each.[19] Microdrops were fixed with methanol, stained with Giemsa, and examined under oil immersion using a ×100 objective and a ×10 ocular. The morphology of *Babesia* spp. sporokinetes searched for in stained hemolymph smears was that described previously,[2,3,20] the detection of any sporokinete in a smear was taken as a positive result. We preferred to use the term sporokinete for the *Babesia* spp. stage in the hemolymph tick after the life cycle description for this protozoan given by Young and Morzaria.[21] No attempt was made to differentiate between *B. bigemina* and *B. bovis* sporokinetes. Two parameters were calculated with the accumulated data after 14 months of evaluation: (a) the incidence and (b) the intensity of *Babesia* spp. sporokinetes in hemolymph samples of EF *B. microplus*. The incidence was defined as the percentage of *Babesia* spp. positive ticks related to the total number of ticks collected each time per ranch. Intensity was defined as the mean number of *Babesia* spp. sporokinetes counted in a stained microdrop of hemolymph, divided by the total number of microscopic fields observed in the hemolymph, and divided by the total number of positive ticks.[22] Each monthly number for incidence and intensity per ranch represents the mean value of two observations.

Tick Control

Ranches A and B had put in effect programs for tick control (TC) on their herds, but with different management and efficacy. Ranch A had a year-round plan of acaricide treatments, spraying cattle every 15 days with organophosphate (OP) solutions, which were alternated for a few months with pyrethroid compounds. On the other hand, ranch B occasionally used acaricide treatments with OP solutions and without a particular frequency; spraying was based on the visual finding of ticks on the herd.

Climatic Factors

Two weather factors were recorded from the experimental area: temperature and precipitation. These parameters were monitored daily at "La Pastora Sugarcane Factory," hydrology and meteorology section, located between ranches A and B. The mean monthly precipitation in the region for the last 10 years (1983-1992) is presented.

TABLE 1. Incidence (%) of *Babesia* spp. Sporokinetes in EF[a] *B. microplus* Collected from Carora Bred Cattle

Ranch	Total No. *B. microplus*	Total Hemolymphs Tested	Microscopic Results			Incidence (%)
			+[b]	−[c]	w/s[d]	
A	76	76	05	71	0	6.5
B	493	461	103	358	32	22.3[e]
Totals	569	537	108	429	32	20.1

[a] Engorged females.
[b] Positive.
[c] Negative.
[d] Without sample
[e] Significant difference at $p < 0.01$.

Statistical Analysis

The differences detected in incidence at both ranches were analyzed by chi-square analysis using one degree of freedom. The results for intensity at both ranches were statistically compared at the dry and rainy season using the χ^2 test.

RESULTS

The total number of EF *B. microplus* detached from the experimental groups of Carora breed cattle per ranch plus the results of the microscopical examination of hemolymph smears are presented in TABLE 1. The number of collected EF *B. microplus* was 6.5 times greater at ranch B than at ranch A. *Amblyomma cajennense* was another tick species seen on these herds, but it was not recorded.

The total incidence of *Babesia* spp. sporokinetes in EF *B. microplus* for both ranches was 20.1%, and the figure detected for incidence at ranch B (22.3%) was highly significant ($p < 0.01$) compared with that found at ranch A (6.5%) TABLE 1. All ticks collected from ranch A yielded hemolymph microdrops for examination after the incubation time, whereas samples were not collected from 32 ticks (6.5%) from ranch B, because these ticks often became dark or were dead after the incubation time (8-10 days).

The monthly variation of incidence of *Babesia* spp. sporokinetes in its vector tick is presented in FIGURE 1. This shows peaks of incidence at both ranches which are approximately coincident with the months of the dry season for the region; however, the values of mean incidence detected at both ranches during the rainy season are very low or negative (July through December). The mean incidence of *Babesia* spp. sporokinetes in its vector tick at ranch A always had values below 18%, and lower than at ranch B.

The precipitation data gathered for this region showed that the rainy season has 2 peaks, in May and in October, and the dry season spans from December through

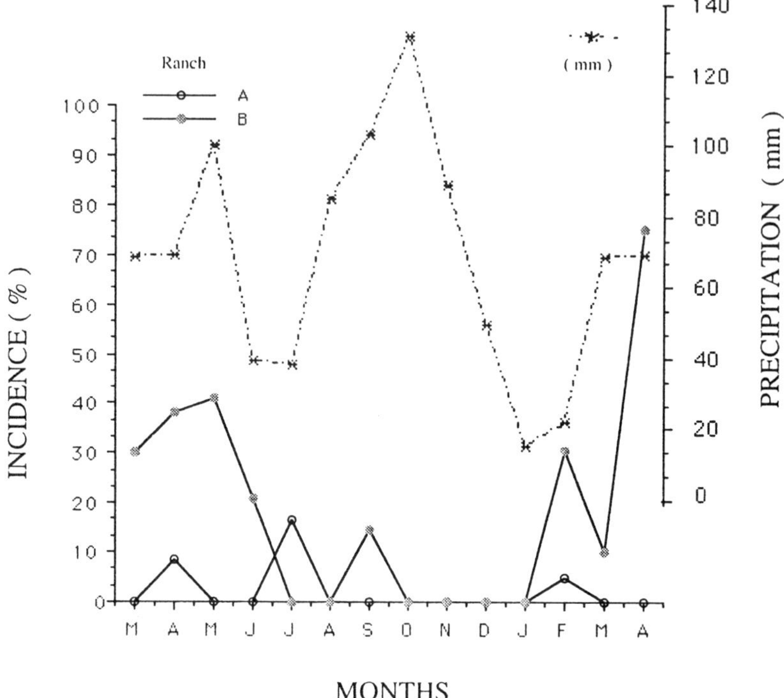

FIGURE 1. Incidence (%) of *Babesia* spp. sporokinetes in *B. microplas* during the dry and rainy seasons at two ranches with different methods of tick control.

April (Fig. 1). The mean monthly precipitation for this region in the last 10 years (1983-1992) was 70.1 mm. The mean temperature for the studied area in the same period was 26°C (mean range: 19.9°-32.1°C).

The mean intensity of *Babesia* spp. sporokinetes in EF *B. microplus* for both ranches was assessed. This factor was calculated counting the total number of *Babesia* spp. sporokinetes in hemolymph microscopic fields. The mean number of microscopic fields at ×100 objective was 29.7 with a range of 5 to 42 fields. The number of sporokinetes per microscopic field ranged from 4 to 27 at ranch A, and from 1 to 56 at ranch B.

The mean intensity of *Babesia* spp. in its vector tick during the time of this research (14 months) at ranch B was 2.33 compared with 0.19 at ranch A during the same period. Proportionally, the number of *B. microplus* positive to *Babesia* spp. sporokinetes from ranch A was 3 times lower than at ranch B, where *B. microplus* was found all year round.

In general, the mean intensity detected at the studied area seems not associated with a particular climatic season. Nevertheless, the mean intensity for ranch B during the rainy season was significant ($p < 0.05$) compared with the value at the dry season (TABLE 2). In contrast, nonsignificant differences were seen for ranch A during both seasons. The values of mean intensity for ranch B were higher and significant compared with those at ranch A, either during the dry ($p < 0.05$) or the rainy season ($p < 0.01$).

The programs of tick control worked with different efficacy at both ranches. At A, the acaricide control has been kept systematically for several years, putting great pressure on the tick population, decreasing the size of it, and cattle were even free of ticks during several months (5 out of 14). On the other hand, ranch B showed ticks (*Boophilus* and *Amblyomma*) all year round, with a larger population since their tick control program was mismanaged.

DISCUSSION

The incidence and the mean intensity of *Babesia* spp. sporokinetes in EF *B. microplus* collected from two herds of Carora breed cattle were studied during 14 consecutive months. Higher numbers of *Babesia* spp. infected EF ticks were detected from ranch B than from ranch A, which directly indicates a higher tick infection rate (TIR) at ranch B, and its suggests indirectly that the parasitemia of *Babesia* spp. should have a greater incidence at this ranch. The fact that 32 ticks from ranch B (6.5%) died during the incubation time in this experiment, and none from ranch A, seems to support the second statement. Besides, it is well known that a significant percentage of EF *B. microplus* died after feeding on cattle with a parasitemia over 20% for *Babesia bigemina,* and over 5% for *B. bovis*[2,3].

The monthly variation of incidence of *Babesia* spp. sporokinetes in their tick vector showed peaks of incidence during the dry season at both ranches, and very low or negative results during the rainy season. This finding ratifies a seasonality for *Babesia* spp. infected ticks, which would lead to greater chances of transmission during the dry season. Previous studies for this region and for Carora breed cattle[23] have showed that blood incidence of *Babesia bigemina* peaks also during the dry season, and it shows very low or negative results during the rainy season. These results of incidence in the vertebrate host are closely coincident with the data of incidence for *Babesia* spp. sporokinetes in their vector tick found in this research. These epizootiological results about cattle babesiosis and its vector in a tropical country with two seasons, should be helpful to develop an efficient and strategic program of control for this hemoparasitic disease.

Since ranch A had established a suitable program of TC through acaricide treatments, a low population of *B. microplus* was expected at this ranch, and a reduced TIR by *Babesia* spp. which according to Mahoney and Ross[11] leads to low inoculation rate (h) of *Babesia* spp., and to insufficient transmission of these cattle hemoparasites. Few cases (3) of clinical babesiosis were diagnosed at ranch A during this experiment. On the other hand, cases of clinical babesiosis were unreported at ranch B since it should have a greater TIR, more chances of transmission for *Babesia* spp., and probably a situation of enzootic stability.

The mean intensity of *Babesia* spp. sporokinetes in their tick vector monitored at both ranches showed high and significant differences at ranch B compared with ranch A (TABLE 2).

Several studies for South America have proved that *B. bigemina* has a higher prevalence of antibodies in cattle than does *B. bovis*[12] actually, clinical babesiosis

TABLE 2. Mean Intensity of *Babesia* spp. Sporokinetes in EF *B. microplus* Related with the Season of the Year

	Season	
Ranch	Dry (Dec.-Apr.)	Rainy (May-Nov.)
A[a]	0.07	0.32
B[b]	1.32[c]	3.43[d]

[a] Nonsignificant difference when compared dry vs. rainy.
[b] Significant difference at $p < 0.05$, dry vs. rainy.
[c] Significant difference at $p < 0.05$, ranch B vs. A.
[d] Significant difference at $p < 0.01$, ranch B. vs A.

caused by *B. bovis* is occasionally seen in Venezuela. The main routes of transmission for *B. bigemina* to the tick vector are by alimentary and by combined routes (transstadial and alimentary). Since ranch A had an efficient system of TC which appears to reduce the transmission of *Babesia* spp. to its vector, lower levels of mean intensity were detected in ticks from this ranch. In contrast, at ranch B, ticks should have greater chances of alimentary and combined infection by *Babesia* spp. merozoites from infected cattle because the population of ticks was greater at this ranch. This may explain the differences detected for mean intensity.

Although the levels of monthly incidence of *Babesia* spp. sporokinetes in their tick vector were low or negative during the rainy season at both ranches (FIG. 1) the levels of intensity for this protozoan in *B. microplus* were greater during the same season. It is stated that there is not a correlation between parasitemia and infectivity for ticks at parasitemias above 0.1%, and parasitemias above 1% may cause high mortality rates in EF ticks and eggs.[24] In addition, the phase of parasitemia does affect alimentary infection significantly, as tick infectivity decreases rapidly in the course of parasitemia.[24] Friedhoff and Smith[24] concluded that extremely low parasitemias that may not be detectable by blood transmission can induce alimentary infection in ticks. According to James *et al.*,[23] the levels of *B. bigemina* parasitemia for Carora breed cattle are very low or negative during the rainy season in this region. Therefore, this low level of *B. bigemina* parasitemia should be one factor that helps to increase the levels of intensity of *Babesia* spp. sporokinetes in their tick vector during the rainy season. The infection of *B. microplus* by merozoites of *Babesia* spp. is a parasitic process, and this protozoan may be overdispersed on the tick population. In other words, the distribution of *Babesia* spp. population throughout its tick host would not be random, but the majority of parasites are found in just a few host individuals.[25] This mechanism of overdispersion is well known in parasitic nematode populations.[26] Overdispersion may also participate in the differences seen in intensity in this experiment.

SUMMARY

The objective of this study was to determine the incidence and intensity of *Babesia* spp. sporokinetes in hemolymph of engorged females (EF) *Boophilus microplus* monthly detached from two herds of a local dairy breed of cattle named Carora breed. The experimental herds were located in 2 ranches, 45 km apart, having different management of tick control (TC). Proper methods of TC were commonly used at ranch "A," whereas ineffective methods were used at ranch "B." *Babesia* spp. sporokinetes were diagnosed in EF *B. microplus* by microscopic exam of tiny smears of hemolymph stained with Giemsa. A total of 537 smears were checked, and 108 were positive to *Babesia* spp. (20.1%). The incidence of *Babesia* spp. was higher ($p < 0.01$) in ranch "B," 22.3%, compared with the value detected in ranch "A," 6.5%. Incidence of *Babesia* spp. was greater in the dry season (Dec-Apr) compared with the values detected during the rainy season (May-Nov). The intensity of *Babesia* spp. was 3.43 at ranch "B" ($p < 0.01$), and 0.32 at ranch "A" during the rainy season. The values detected for intensity kept no correlation with those two seasons, although it is known that the incidence and seroprevalence of bovine babesiosis are greater during the dry season. The importance of these parameters in the epizootiology of bovine babesiosis in Venezuela is discussed.

ACKNOWLEDGMENTS

The authors thank the Consejo de Desarrollo Científico Humanístico y Tecnológico (CDCHT) de la Universidad Centroccidental Lisandro Alvarado, Barquisimeto, Venezuela, for the financial support given to this project. The cooperation of owners and employees of "Sicarigua" and "Libertad" ranches is acknowledged. The manuscript was reviewed by Dr. Hugo Leyva, Endocrinology Laboratory, Barquisimeto, and we thank him for his suggestions.

REFERENCES

1. FAO. 1991. Epizootiología de las enfermedades hemoparasitarias de los vacunos: 1-58. RLAC/91/31 GAN-35. Oficina regional de la FAO para América Latina y el Caribe. Santiago, Chile.
2. RIEK, R. F. 1964. The life cycle of *Babesia bigemina* (Smith & Kilborne, 1893) in the tick vector *Boophilus microplus* (Canestrini) (Acarina: Ixodoidea). Austr. J. Agric. Res. **15:** 802-821.
3. RIEK, R. F. 1966. The life cycle of *Babesia argentina* (Lignieres, 1903) (Sporozoa: Piroplasmidae) in the tick vector *Boophilus microplus* (Canestrini). Austr. J. Agric. Res. **17:** 247-254.
4. GERI, L. F., F. VOLLMERHAUSEN, D. WILSON, G. HARRIS & T. POPHAM. 1989. Puerto Rico tick erradication program benefit-cost analysis. USDA, APHIS. Hyattsville, MD.
5. HODGSON, J. L. 1992. Biology and transmission of *Babesia bigemina* in *Boophilus microplus*. *In* Tropical Veterinary Medicine: Current Issues and Perspectives. Ann. N.Y. Acad. Sci. **653:** 42-51.
6. NARI, A., H. CARDOZO, J. BERDIE, F. CANABEZ & R. BAWDEN. 1979. Estudio preliminar sobre la ecología de *Boophilus microplus* en Uruguay; ciclo no parasitario en un área considerada poco apta para su desarrollo. Veterinaria Montevideo **15:** 25-31.
7. MINISTERIO DE AGRICULTURA Y CRIA, FONAIAP, Venezuela. 1976. Zonas de Vida de Venezuela. Memoria explicativa sobre el mapa ecológico: 265. Edit. Sucre. Caracas, Venezuela.
8. POWER, L. A., R. SILVESTRI, & J. C. CHACON. 1985. Incidencia de *B. microplus* y *A. cajennense* en explotaciones bovinas de los estados Barinas, Falcón, Lara y Yaracuy. Rev. Fac. Ciencias Vet. **32**(1-4): 21-29.

9. CORDOVES, C. O., M. FERNANDEZ, H. POLANCO, O. BLAYA, M. SEDANE, A. CAMACHO & P. A. SANCHEZ. 1991. El control de la garrapata en la República de Cuba. Simposium Internacional de Garrapatas. Paxtepex, Morelos, México: 1-3.

10. WHARTON, R. H. 1974. Ticks with special emphasis on *Boophilus microplus*. *In* Control of Arthropods of Medical and Veterinary Importance. R. Pal & R. H. Wharton, Eds. Plenum Publishing Co. New York, NY.

11. MAHONEY, D. F. & D. R. ROSS. 1972. Epizootiological factors in the control of bovine bebesiosis. Aust. Vet. J. **48:** 292-298.

12. YOUNG, A. S. 1988. Epidemiology of babesiosis. *In* Babesiosis of Domestic Animals and Man. M. Ristic, Ed.: 81-98. CRC Press. Boca Raton, FL.

13. JOHNSTON, L. A. Y. 1967. Epidemiology of bovine babesiosis in Northern Queensland. Aust. Vet. J. **44**(6): 265-267.

14. NUÑEZ, J. L., M. E. MUÑOZ-COBEÑAS & H. L. MOLTEDO. 1982. *Boophilus microplus:* the Common Cattle Tick: 181-200. Springer-Verlag. Berlin & Heidelberg.

15. SHAW, R. D., M. COOK & R. E. CARSON. 1968. Developments in the resistant status of the southern cattle tick to organophosphorus and carbamate insecticides. J. Econ. Entomol. **61:** 1590-1594.

16. CORONADO, A. 1994. Acaricide resistance in the cattle tick *Boophilus microplus:* an update on the resistance situation in Venezuela. Proceedings of Workshop FAO/UN/IPVDF. Porto Alegre, RS, Brazil: 1-3.

17. DALGLIESH, R.J. & N. P. STEWART. 1982. Some effects of time, temperature and feeding on infection rates with *Babesia bovis* and *Babesia bigemina* in *Boophilus microplus* larvae. Int. J. Parasitol. **12**(4): 323-326.

18. GUGLIELMONE, A., A. J. MANGOLD, A. C. BERMUDEZ & A. HADANI. 1985. Detección de merozoitos grandes (vermículos) de *Babesia* en teleoginas de *Boophilus microplus* alimentadas sobre terneros con distintos niveles de parasitemia de *Babesia bigemina* y *Babesia bovis* (= *Babesia argentina*). Rev. Ibér. Parasitol. **45**(4): 303-311.

19. BURGDORFER, W. 1970. Hemolymph test. A technique for detection of *Rickettsiae* in ticks. Am. J. Trop. Med. Hyg. **19**(6): 1010-1014.

20. SMITH, R. D. 1978. Ciclo biológico de *Babesia* en la garrapata. Ciencia Veterinaria (México) **2:** 233-264.

21. YOUNG, A.S. & S. P. MORZARIA. 1986. Biology of *Babesia.* Parasitol. Today. **2**(8): 211-219.

22. MARGOLIS, L., G. W. ESCH, J. C. HOLMES, A. M. KURIS & G. A. SCHAD. 1982. The use of ecological terms in parasitology (report of an ad hoc committee of the American Society of Parasitologists). J. Parasitol. **68**(1): 131-133.

23. JAMES, M., A. CORONADO, W. LOPEZ, R. D. MELENDEZ & M. RISTIC. 1985. Seroepidemiology of bovine anaplasmosis and babesiosis in Venezuela. Trop. Anim. Health. Prod. **17:** 9-18.

24. FRIEDHOFF, K. T. & R. D. SMITH. 1981. Transmission of Babesia by ticks. *In* Babesiosis. M. Ristic & J. Kreier, Eds.: 267-321. Academic Press. New York, NY.

25. KENNEDY, C. R. 1975. Ecological Animal Parasitology. 1st Edit. Blackwell Scientific Publications. Oxford, England.

26. ANDERSON, R. M. & R. M. MAY. 1982. Population dynamics of human helminth infections: control by chemotherapy. Nature **297:** 557-563.

Strategies to Interrupt the Development of *Anaplasma marginale* in Its Tick Vector

The Effect of Bovine-Derived Antibodies[a]

KATHERINE M. KOCAN, EDMOUR F. BLOUIN,
GUY H. PALMER,[b] INGA S. ERIKS,[b] AND
WANDA L. EDWARDS

Department of Veterinary Pathology
College of Veterinary Medicine
Oklahoma State University
Stillwater, Oklahoma 74078

[b]*Department of Veterinary Microbiology and Pathology*
College of Veterinary Medicine
Washington State University
Pullman, Washington 99164

INTRODUCTION

Anaplasma marginale is a rickettsia transmitted by ticks to cattle that invades and multiplies in bovine erythrocytes resulting in mild to severe anemia.[1-4] The disease, anaplasmosis, is the only major tick-borne disease of cattle in the United States and is endemic in many geographic areas throughout the world.[1] Annual mortality and morbidity due to this disease among US beef cattle have been estimated at 50,000 to 100,000 head and a cost of $300 million. The resultant loss of cattle to the export market has been estimated at $45 million.[5] Anaplasmosis is transmitted to cattle biologically by ticks or mechanically by blood-contaminated fomites or mouthparts of biting flies. Approximately 20 species of ticks have been incriminated as vectors worldwide.[2-3] Transstadial transmission (larvae to nymph and nymph to adult) has been demonstrated repeatedly.[6-10] Transmission also occurs by adults infected as larvae without reexposure as nymphs[11] and by male ticks transferred from infected to susceptible cattle.[12-18] Transovarial transmission of *A. marginale* was reported by Howell[19] but has not been confirmed by others.[3,11] Several isolates of *A. marginale* are infective for ticks (Idaho, Virginia, Washington isolates) but others apparently do not infect ticks (Illinois, Florida isolates).[15,20-21] A complex development cycle[10,22-23] occurs within ticks that begins in midgut cells, with

[a]Supported by Oklahoma Agricultural Experiment Station Projects 1669 and 1443; SmithKline Beecham Animal Health; and the Oklahoma Center for the Advancement of Science and Technology (OCAST), Project ARO-021.

157

subsequent infection in gut muscle cells. Final development occurs in salivary glands from where the rickettsia is transmitted to the vertebrate host. At each of these sites of development, *A. marginale* multiplies within membrane-bound inclusions called colonies. Two major stages have been recognized in ticks. The reticulated form is the first stage to be seen and appears to be a vegetative stage that divides by binary fission within the inclusion. The reticulated stage condenses into the electron dense stage that is posited to survive outside cells and be infective for others.

Attempts to control anaplasmosis thus far have focused on cattle and have included vaccination with killed or attenuated vaccines, prophylactic treatment of calves with tetracyclines, tick control, and reduction of carrier cattle in herds.[24–32] The killed vaccine induces immune response reducing the clinical syndrome, but cattle challenge-exposed thereafter can acquire infection and become carriers. Their blood can, therefore, serve as an infective source for ticks and/or mechanical transmission via biting flies and blood-contaminated instruments. Although vaccination of cattle reduces the number of clinical cases of anaplasmosis, more complete control of this disease will occur when immunized cattle are prevented from acquiring infection via exposure to ticks, thus reducing the number of carriers in the field.

New strategies for control of anaplasmosis are being focused on the tick vector. Potential control methods for hemoparasites in ticks may target pathogens during their development cycle using drugs or pathogen-specific immunoglobulins or may interrupt tick function using drugs or tick-specific immunoglobulins that may have a secondary inhibitory effect on hemoparasites in ticks.

DEVELOPMENT OF VACCINES AGAINST HEMOPARASITES IN TICKS

Development of vaccines against hemoparasites in ticks appears to be feasible because small amounts of vertebrate host immunoglobulins cross the midgut epithelium of invertebrates and enter the hemolymph without breakdown. Therefore, antibodies produced in cattle may affect hemoparasites when they are taken into ticks with the bloodmeal. Ticks feed for long periods as they concentrate the bloodmeal and, therefore, should be exposed to antibodies present in the serum.

A review of the evidence for passage of intact immunoglobulins from vertebrates to invertebrates can be found in Sauer *et al.* 1994.[33] Briefly, hemolysins produced in rabbits against sheep erythrocytes crossed the midgut epithelium of *Ixodes ricinus* via the blood meal. They were detected by a complement fixation (CF) test, and were shown to retain their immunological properties in tick hemolymph.[34] Since the CF test requires formation of antigen-antibody complexes, these results demonstrated that *I. ricinus* hemolymph contained intact and reactive antibodies. Antibodies ingested during normal or artificial feeding retained their antigenicity in the hemolymph of *D. variabilis*.[35] Ben-Yakir *et al.* used an ELISA to measure concentrations of antiovalbumin IgG in hemolymph of *Amblyomma americanum* and *D. variabilis* while they were feeding on several different hosts.[36–37] In further studies, a radioimmunoassay detected intact host IgG to ovalbumin in hemolymph of *Hyalomma excavatum, Rhipicephalus sanguineus,* and *Otobius moubata*.[38] Minoura *et al.* detected intact IgG in hemolymph of *Ornithodoros moubata* fed on rabbits or fed ^{125}I-labeled IgG through an artificial membrane.[39] The concentration of IgG in tick hemolymph increased as feeding progressed. Therefore, cattle immunized against ticks and/or against stages of hemoparasites within ticks would be expected to produce antibodies that might affect the vector and/or the parasite and aid in control of both.

Development of vaccines against tick stages of cattle hemoparasites will depend, in part, on whether the various developmental stages in ticks express unique antigens. If so, antigens could be targeted for use in a recombinant vaccine that might interrupt parasite development, resulting in reduced challenge doses for cattle or elimination of pathogen transmission. Precedent for this approach comes from research on malarial parasites as reviewed by Kocan 1995.[40]

Stage-specific proteins have been demonstrated on the protozoan hemoparasites of cattle, *Theileria* and *Babesia*. Unique antigens were demonstrated on *Theileria* sporozoites, including those of *T. parva*[41–44] and *T. annulata*.[45–47] The sporozoite stage of *T. parva* was neutralized with a monoclonal antibody directed against a sporozoite-specific antigen, suggesting that it would be a good candidate for development of a vaccine to prevent infection of bovine lymphocytes.[48] Furthermore, sporozoites from different stocks were neutralized with one monoclonal antibody against this sporozoite surface protein.[48,49] A recombinant antisporozoite vaccine, currently being developed for *T. parva*, induces protection in cattle challenged with tick-transmitted sporozoites.[50,51] The effect of antisporozoite antibodies on the development and transmission of *T. parva* by ticks has not been reported. However, because sporozoite antigen is first expressed after 2–3 days of tick feeding as schizogony commences,[42] this vaccine may also interrupt parasite development and transmission by ticks. This potential second vaccine target in ticks, as well as the primary one in cattle, might be exploited to reduce the sporozoite challenge dose by ticks while avoiding the likelihood of overwhelming the bovine immune system, and thus may enhance the performance of the vaccine.

Although *A. marginale* has a complex developmental cycle in ticks, unique antigens on tick stages, differing from those expressed by organisms within erythrocytes, have not been investigated. Proteins common to erythrocytic and tick midgut stages were identified using antisera of cattle immunized with infected tick midguts.[52] The impact of cyclical *A. marginale* rickettsemia in carrier cattle on tick infection rates was described by Eriks *et al.* using an *A. marginale*-specific DNA probe.[53] A positive correlation was found between rickettsemia levels in cattle and resulting infections in ticks. It did not appear from these studies that antibodies present in carrier cattle acted on tick stages of *A. marginale* to reduce tick infections. Presumably these antibodies in cattle were produced against the erythrocytic-derived *A. marginale*. *Anaplasma* in tick salivary glands may prove to have antigens that are unique from those on the erythrocytic stage.

EXPERIMENTAL DESIGN FOR TESTING THE EFFECT OF BOVINE-DERIVED ANTIBODIES ON DEVELOPMENT AND TRANSMISSION OF *A. MARGINALE* BY TICKS

Recent research in our laboratory has focused on attempts to interrupt the development and transmission of *A. marginale*, in its tick vectors, *Dermacentor* spp.[40,54]

EXPERIMENTAL DESIGN

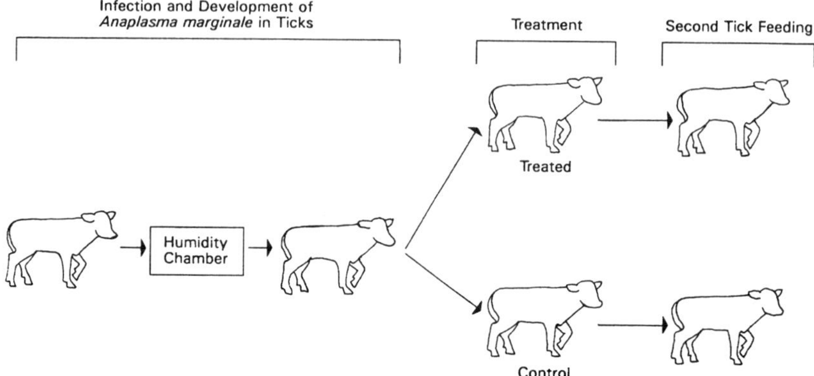

FIGURE 1. Experimental design for testing the effect of treatments given to cattle on *Anaplasma marginale* in ticks. Male *Dermacentor andersoni,* infected as adults, were allowed to feed on treated and control cattle to test the effect of the treatment (drugs or antibodies) on the ability of ticks to transmit *A. marginale.* The ticks were then transferred to susceptible cattle to test whether exposure to the treatment affected their ability to transmit *A. marginale* a second time. Infection of ticks was monitored after feeding by determining infection rates in tick salivary glands using microscopy and a DNA probe.

Interruption of the life cycle of *A. marginale* may result in prevention of transmission of the rickettsiae to cattle or reduce the number of organisms in the tick challenge. We developed an experimental design (FIG. 1) for testing the effect of drugs and antibodies ingested via the bloodmeal on development and transmission of *A. marginale* by ticks. Male ticks infected as adults are used in this experimental design because they develop high infection rates, remain persistently infected, and can feed and transmit *A. marginale* multiple times. The salivary glands of male ticks become heavily infected with *A. marginale* and can be easily monitored for infection. In this experimental design, male *D. andersoni* are allowed to feed (acquisition feeding) for seven days on a splenectomized calf with an ascending parasitemia of *A. marginale* in order to acquire *A. marginale* infection. The ticks are then removed and held in a humidity chamber for 5 to 7 days. This holding period is necessary to avoid the possibility of mechanical transmission of *A. marginale* by transfer of infective blood on tick mouthparts and to allow for development of *A. marginale* in tick gut cells. In previous studies, we determined that development of *A. marginale* in tick midgut cells does not begin until the ticks are removed from the infected calf used for acquisition feeding. After acquisition feeding, the ticks are divided into two groups, and half are allowed to transmission feed for seven days on calves treated with drugs or immunized with *A. marginale* while the other half are allowed to feed on untreated control calves. At the end of the feeding period, 20 ticks are removed, the salivary glands are dissected and examined for infection with *A. marginale.* One salivary gland from each tick is processed for light microscopy (LM) for determination of colony densities, while the other is tested with an *A. marginale*-specific DNA probe.

The calves are monitored for infection with *A. marginale* by determination of the packed cell volume, parasitemia, and by serologic tests. In this experimental design, the exposed and control ticks can be transferred to an additional set of susceptible calves for a second feeding. This second transmission feeding can test whether *A. marginale* infection and transmission by ticks are altered by previous exposure to the treatment. As in the first transmission feeding, after a feeding period of 7 days, infection in ticks is monitored by microscopy and a DNA probe. In summary, this experimental design allows for infection of ticks (acquisition feeding), exposure through feeding (transmission feeding) of ticks to cattle that have received treatment (drugs or vaccines), and, if warranted, an additional feeding (repeat feeding) of the treatment-exposed ticks to determine whether the treatment altered the ability of these ticks to transmit *A. marginale* to untreated, susceptible cattle.

EFFECT OF BOVINE-DERIVED ANTIBODIES ON DEVELOPMENT AND TRANSMISSION OF *A. MARGINALE* IN TICKS

We have completed two preliminary experiments using the experiment design described to test whether bovine-derived antibodies are inhibitory to development and transmission of *A. marginale* by ticks.[55] Bovines were immunized with purified outer membrane proteins of erythrocyte-derived *A. marginale* as described previously.[56,57] The effect of immunization of bovines was monitored by determination of *A. marginale*-specific antibodies and clinical signs. In the first trial, two splenectomized calves were used in treated and control groups. The immunized calves developed high titers to *A. marginale* as determined with immunoblots (1:80,000 and 1:160,000). In the second trial, four intact yearling cattle were immunized and developed titers ranging from 1:100 to 1:10. Even though bovines in both studies developed titers to *A. marginale,* tick transmission of *A. marginale* was not prevented in either trial by ingestment of *Anaplasma*-specific antibodies by ticks; the immunized and control cattle developed anaplasmosis as a result of being fed on by the infected male ticks. *Anaplasma* infection in ticks did not appear to be affected by the ticks feeding on immunized bovines because the infection rates in tick salivary glands were similar to those in the controls. It is probable that tick stages of *A. marginale* have an antigenic composition that differs from that of the bovine erythrocytic stage. Conserved antigens on tick stages, if present, did not appear to be affected by the bovine-derived antibodies, which presumably are to the erythrocytic stage of *A. marginale.* In order to predict the effect of immunoglobulins on *A. marginale* in ticks, the antigenic composition of the tick salivary gland stage will need to be clearly defined and passage of immunoglobulins against tick stages from cattle to ticks will need to be confirmed.

SUMMARY

Anaplasma marginale is a rickettsia transmitted by ticks that invades and multiplies in bovine erythrocytes causing the disease anaplasmosis. A complex developmental cycle occurs within ticks that begins in midgut cells, with subsequent infection in gut muscle cells. Final development occurs in salivary glands from where the rickettsia is transmitted to the vertebrate host. At each site of development, *A. marginale* multiplies within membrane-bound inclusions. Attempts to control anaplasmosis have focused on cattle and have included immunization and prophylactic treatment with tetracyclines. New strategies for control of anaplasmosis are being focused on the tick vector. Development of vaccines against hemoparasites in ticks may be feasible because vertebrate host immunoglobulins appear to cross the midgut epithelium of invertebrates and enter the hemolymph without breakdown. We tested the effect of *A. marginale* antibodies ingested by ticks with the bloodmeal on infections in ticks. Cattle were immunized with purified outer membrane proteins of erythrocytic-derived parasites. Infections in ticks exposed to the immunized cattle were determined using an *Anaplasma*-specific DNA probe, light and electron microscopy, and tick transmission studies. Vaccine-derived antibodies did not appear to affect the development and transmission of *A. marginale* in ticks. Further studies are needed to determine if bovine antibodies remain intact within ticks and whether the tick stage of *A. marginale* has unique surface antigens from the erythrocytic stage.

REFERENCES

1. BRAM, R. A. 1975. Tick-borne livestock diseases and their vectors. 1. The global problem. World Anim. Rev. **6:** 1–5.
2. DIKMAN, G. 1950. The transmission of anaplasmosis. Am. J. Vet. Res. **11:** 5–16.
3. EWING, S. A. 1981. Transmission of *Anaplasma marginale* by arthropods. *In* Proceedings of the Seventh National Anaplasmosis Conference. 395–423. Mississippi State University.
4. RISTIC, M. 1968. Anaplasmosis. *In* Infectious Blood Diseases of Man and Animals. D. Weinman & M. Ristic, Eds.: 478–542. Academic Press. New York, NY.
5. McCALLON, B. R. 1973. Prevalence and economic aspects of anaplasmosis. *In* Proceedings of the Sixth National Anaplasmosis Conference. E. W. Jones, Ed.: 1–3. Heritage Press. Stillwater, OK.
6. REES, C. W. 1934. Transmission of anaplasmosis by various species of ticks. USDA Tech Bull. 134, No. 418. Washington, D.C.
7. ANTHONY, D. W. & T. O. ROBY. 1962. Anaplasmosis studies with *Dermacentor variabilis* (Say) and *Dermacentor andersoni* Stiles (*D. venustus* Marx) as experimental vectors. *In* Proceedings, Fourth National Anaplasmosis Conference: 78–81. Reno, Nevada.
8. KOCAN, K. M., J. A. HAIR & S. A. EWING. 1980. Ultrastructure of *Anaplasma marginale* Theiler in *Dermacentor andersoni* Stiles and *Dermacentor variablis* (Say). Am. J. Vet. Res. **41:** 1966–1976.
9. KOCAN, K. M., J. A. HAIR, S. A. EWING & L. G. STRATTON. 1981. Transmission of *Anaplasma marginale* Theiler by *Dermacentor andersoni* Stiles and *Dermacentor variabilis* (Say). Am. J. Vet. Res. **42:** 15–18.
10. KOCAN, K. M. 1986. Development of *Anaplasma marginale* in ixodid ticks: coordinated development of a rickettsial organism and its tick host. *In* Morphology, Physiology and Behavioral Biology of Ticks. J. R. Sauer & J. A. Hair, Eds.: 472–505. Ellis Horwood, Ltd. Chichester, England.
11. STICH, R. W., K. M. KOCAN, G. H. PALMER, S. A. EWING, J. A. HAIR & S. J. BARRON. 1989. Transstadial and attempted transovarial transmission of *Anaplasma marginale* by *Dermacentor variabilis*. Am. J. Vet. Res. **50:** 1377–1380.
12. POTGIETER, F. T. 1979. Epizootiology and control of anaplasmosis in South Africa. J. South African Vet. Assoc. **50:** 367–372.
13. STILLER, D., K. M. KOCAN, W. EDWARDS, S. A. EWING, J. A. HAIR & S. J. BARRON. 1989. Detection of colonies of *Anaplasma marginale* Theiler in salivary glands of three *Dermacentor* spp. infected as nymphs or adults. Am. J. Vet. Res. **50:** 1386–1391.

14. STILLER, D., M. E. COAN, W. L. GOFF, L. W. JOHNSON & T. C. McGUIRE. 1989. The importance and putative role of *Dermacentor* spp. males in anaplasmosis epidemiology: transmission of *Anaplasma marginale* to cattle by *ad libitum* interhost transfer of *D. andersoni* males under seminatural conditions Abstract. *In* Proceedings, Eighth National Veterinary Hemoparasite Disease Conference, 206. St. Louis, MO.

15. ZAUGG, J. L., D. SILLER, M. E. COAN & S. D. LINCOLN. 1986. Transmission of *Anaplasma marginale* Theiler by males of *Dermacentor andersoni* Stiles fed on an Idaho-infected chronic carrier cow. Am. J. Vet. Res. **47:** 2269-2271.

16. COAN, M. E., J. L. ZAUGG & D. STILLER. 1987. Probable anaplasmosis transmission by natural intrastadial interhost transfer of adult 3-host ticks. *In* Proceedings, Eighth Annual Western Conference Food Animal Disease Research: 34. University of Idaho. Boise, Idaho.

17. STILLER, D. & W. L. JOHNSON. 1983. Experimental transmission of *Anaplasma marginale* Theiler by males of *Dermacentor albipictus* (Packard) and *Dermacentor occidentalis* Marx (Acari:Ixodidae). *In* Proceedings, 87th Meeting of the US Animal Health Association: 59-65. Las Vegas, NV.

18. POTGIETER, F. T. 1981. Tick transmission of anaplasmosis in South Africa. Rhodes University Tick Research Unit: 53-56. Grahmstown, South Africa.

19. HOWELL, D. E., G. W. STILES & J. H. MOE. 1941. The hereditary transmission of anaplasmosis by *Dermacentor andersoni* Stiles. Am. J. Vet. Res. **2:** 165-166.

20. SMITH, R., M. G. LEVY, M. S. KUHLENSCHMIDT, J. H. ADAMS, D. G. RZECHULA, T. A. HARDT & K. M. KOCAN. 1986. Isolate of *Anaplasma marginale* not transmitted by ticks. Am. J. Vet. Res. **47:** 127-129.

21. WICKWIRE, K. B., K. M. KOCAN, S. J. BARRON, S. A. EWING, R. D. SMITH & J. A. HAIR. 1987. Infectivity of three *Anaplasma marginale* isolates for *Dermacentor andersoni*. Am. J. Vet. Res. **48:** 96-99.

22. KOCAN, K. M., D. STILLER, W. L. GOFF, P. L. CLAYPOOL, W. EDWARDS, S. A. EWING, T. C. McGUIRE, J. A. HAIR & S. J. BARRON. 1992. Development of *Anaplasma marginale* in male *Dermacentor andersoni* transferred from parasitemic to susceptible cattle. Am. J. Vet. Res. **53:** 499-507.

23. KOCAN, K. M., W. L. GOFF, D. STILLER, P. L. CLAYPOOL, W. EDWARDS, S. A. EWING, J. A. HAIR & S. J. BARRON. 1992. Persistence of *Anaplasma marginale* (Rickettsiales:Anaplasmataceae) in male *Dermacentor andersoni* (Acari:Ixodidae) transferred from infected to susceptible calves. J. Med. Entomol. **29:** 657-668.

24. RICHEY, E. F. 1981. Bovine anaplasmosis. *In* Current Veterinary Therapy—Food Animal Practice: 767-772. W. B. Saunders. Philadelphia, PA.

25. KUTTLER, K. L. & R. A. TODOROVIC. 1973. Techniques of premunization for the control of anaplasmosis. *In* Proceedings of the Sixth National Anaplasmosis Conference. E. W. Jones, Ed.: 106. Heritage Press. Stillwater, OK.

26. CORRIER, D. R., O. VIZCAINO, C. A. CARSON, M. RISTIC, K. L. KUTTLER, G. S. TREVINO & A. J. LEE. 1980. Comparison of three methods of immunization against bovine anaplasmosis: an examination of post-vaccinated effects. Am. J. Vet. Res. **41:** 1062.

27. NORMAN, B. B. 1973. Current status of vaccination for control of bovine anaplasmosis. *In* Proceedings of the Sixth National Anaplasmosis Conference. E. W. Jones, Ed.: 103. Heritage Press, Stillwater, OK.

28. CARSON, C. A., D. M. SELLS & M. RISTIC. 1976. Cell-mediated immunity in bovine anaplasmosis and correlation with protection induced by vaccination. Vet. Parasitol. **2:** 75.

29. KUTTLER, K., J. L. ZAUGG & L. W. JOHNSON. 1984. Serologic and clinical responses of premunized, vaccinated, and previously infected cattle to challenge exposure by two different *Anaplasma marginale* isolates. Am. J. Vet. Res. **45:** 2223-2226.

30. DENNIS, R. W., P. J. O'HARA, M. F. YOUNG & K. D. DORRIS. 1970. Neonatal immunohemolytic anemia and icterus of calves. J. Am. Vet. Med. Assoc. **156:** 1961-1964.

31. HENRY, E., B. B. NORMAN, D. FLY, R.W. WICHMANN & S. M. YORK. 1983. Effects and use of a modified live *Anaplasma marginale* vaccine in beef heifers in California. J. Am. Vet. Med. Assoc. **183:** 66-69.

32. PALMER, G. H. 1989. Anaplasmosis vaccines. *In* Veterinary Protozoan and Hemoparasite Vaccines. I. G. Wright, Ed.: 2-25. CRC Press. Boca Raton, FL.

33. SAUER, J. R., J. L. MCSWAIN & R. C. ESSENBERG. 1994. Cell membrane receptors and regulation of cell function in ticks and bloodsucking insects. Int. J. Parasitol. **24:** 33-52.

34. BROSSARD, J. & O. RAIS. 1984. Passage of hemolysins through the midgut epithelium of female *Ixodes ricinus* L. fed on rabbits infested or reinfested with ticks. Experientia **40:** 561-563.

35. ACKERMAN, S., F. B. CLARE, T. W. MCGILL & D. E. SONENSHINE. 1981. Passage of host serum components, including antibody, across the digestive tract of *Dermacentor variabilis* (Say). J. Parasitol. **67:** 737-740.

36. BEN-YAKIR, D., J. C. FOX, J. T. HOMER & R. W. BARKER. 1986. Quantitative studies of host immunoglobulin G passage into the hemocoel of the ticks, *Amblyomma americanum* and *Dermacentor variabilis. In* Morphology, Physiology, and Behavioral Biology of Ticks. J. R. Sauer & J. A. Hair, Eds.: 329-360. Ellis Horwood. Chichester, England.

37. BEN-YAKIR, E., J. C. FOX, J. T. HOMER & R. W. BARKER. 1987. Quantification of host immunoglobulin in the hemolymph of ticks. J. Parasitol. **73:** 669-671.

38. BEN-YAKIR, D. 1989. Quantitative studies of host immunoglobulin G in the hemolymph of ticks (Acari). J. Med. Entomol. **26:** 243-246.

39. MINOURA, H., Y. CHINZEI & S. KITAMURA. 1985. *Ornithodorus moubata:* host immunoglobulin G in tick haemolymph. Exp. Parasitol. **60:** 355-363.

40. KOCAN, K. M. 1995. Targeting ticks for control of selected hemoparasitic diseases of cattle. Vet. Parasitol. **57:** 121-151.

41. MUSOKE, A. J., V. M. NANTULYA, G. BUSCHER, R. A. MASAKE & B. OTIM. 1982. Bovine immune response to *Theileria parva:* neutralizing antibodies to sporozoites. Immunology **45:** 663-668.

42. DOBBELAERE, D. A. E., P. WEBSTER, B. L. LEITCH, W. P. VOIGT & A. D. IRVIN. 1985. *Theileria parva:* expression of a sporozoite surface coat antigen. Exp. Parasitol. **60:** 90-100.

43. DOBBELAERE, D. A. E., S. A. SHAPIRO & P. WEBSTER. 1985. Identification of a surface antigen on *Theileria parva* sporozoites by monoclonal antibody. Proc. Natl. Acad. Sci. USA **82:** 1771-1775.

44. NENE, V., K. P. IAMS, E. GOBRIGHT & A. J. MUSOKE. 1992. Characterisation of the gene encoding a candidate vaccine antigen of *Theileria parva* sporozoites. Mol. Biochem. Parasitol. **51:** 17-28.

45. MIRANPURI, G. S. 1986. Evaluation of purified *Theileria annulata* sporozoite antigen from the tick *Hyalomma anatolicum anatolicum.* Vet. Parasitol. **22:** 9-19.

46. WILLIAMSON, S., A. TAIT, D. BROWN, A. WALKER & P. BECK. 1989. *Theileria annulata* sporozoite surface antigen expressed in *Escherichia coli* elicits neutralizing antibody. Proc. Natl. Acad. Sci. USA **86:** 4639-4643.

47. KACHANI, M., R. A. OLIVER, C. D. G. BROWN, H. OUHELLI & R. L. SPOONER. 1992. Common and stage-specific antigens of *Theileria annulata.* Vet. Immunol. Immunopathol. **34:** 221-234.

48. DOBBELAERE, D. A. E., P. R. SPOONER, W. C. BARRY & A. D. IRVIN. 1984. Monoclonal antibody neutralizes the sporozoite stage of different *Theileria parva* stocks. Parasite Immunol. **6:** 361-370.

49. MUSOKE, A. J., V. M. NANTULYA, F. R. RURANGIRWA & G. BUSCHER. 1984. Evidence for a common protective antigenic determinant on sporozoites of several *Theileria parva* strains. Immunology **52:** 231-238.

50. MUSOKE, A., S. MORZARIA, C. NKONGE, E. JONES & V. NENE. 1992. A recombinant sporozoite surface antigen of *Theileria parva* induces protection in cattle. Proc. Natl. Acad. Sci. USA **89:** 514-518.

51. Musoke, A., V. Nene & S. P. Morzaria. 1993. A sporozoite-based vaccine for *Theileria parva*. Parasitol. Today **9:** 385–388.
52. Palmer, G. H., K. M. Kocan, S. J. Barron, J. A. Hair, A. F. Barbet, W. C. Davis & T. C. McGuire. 1985. Presence of common antigens including major surface protein epitopes between cattle and tick stages of *Anaplasma marginale*. Infect. Immun. **50:** 881–886.
53. Eriks, E. S., D. Stiller & G. H. Palmer. 1993. Impact of persistent *Anaplasma marginale* rickettsemia on tick infection and transmission. J. Clin. Microbiol. **31:** 2091–2096.
54. Kocan, K. M. 1995. Strategies for control of *Anaplasma marginale* focused on the tick vector. *In* Current Topics in Veterinary Research. Council of Scientific Information. India. (In press.)
55. Kocan, K. M., E. F. Blouin, G. H. Palmer & W. E. Edwards. 1994. The effect of *Anaplasma marginale* antibodies ingested by ticks with the bloodmeal on infections in salivary glands. Abstract No. 127, 75th Conference of Research Workers in Animal Diseases.
56. Tebele, N., T. C. McGuire & G. H. Palmer. 1991. Induction of protective immunity by using *Anaplasma marginale* initial body membranes. Infect. Immun. **59:** 3199–3204.
57. Tebele, N. & G. H. Palmer. 1991. Crossprotective immunity between the Florida and a Zimbabwe stock of *Anaplasma marginale*. Trop. Anim. Health Prod. **23:** 197–202.

TRYPNET

New Hemoparasite Information Network[a]

S. VOKATY,[b] M. DESQUESNES,[c] L. APPLEWHAITE,[d]
J. FAVRE,[e] R. LIEUW-A-JOE,[f] M. PARRIS-AARON,[d]
AND L. BANSSE-IISA[f]

[b]*Inter-American Institute for Cooperation on Agriculture (IICA)*
Georgetown, Guyana

[c]*Centre de Cooperation Internationale en Recherche Agronomique*
pour le Developpement—Elevage et Medecine Veterinaire des Pays
Tropicaux (CIRAD-EMVT)
Institut Pasteur
Cayenne, French Guiana

[d]*Ministry of Agriculture*
Georgetown, Guyana

[e]*Services Veterinaires Departementaux*
Cayenne, French Guiana

[f]*Ministerie Van Landbouw*
Veeteelt & Visserij
Paramaribo, Suriname

INTRODUCTION

The "Guianas" are three little-known countries located in the humid tropical region on the north coast of South America. These three countries share borders and historic and geographic similarities, but have maintained little contact, due to language, political, administrative, and economic differences. Guyana, the former British Guiana, is an independent democratic country, whose national language is English. It has a human population of 750,000 and a bovine population of almost 300,000. Suriname, the former Dutch Guiana, is also an independent democratic country, whose national language is Dutch. It has a human population of approximately 400,000 and a bovine population of just under 100,000. French Guiana (or Guyane Francaise) is an overseas department of France, therefore not a sovereign nation. It has a human population of 115,000 and a bovine population of 7000. Guyana and Suriname are considered to be developing countries. As a French overseas department, French Guiana enjoys social and economic benefits similar to those in metropolitan France. In all three countries, the great majority (80-90%) of human and livestock populations are concentrated along the Atlantic coastline.

[a]This project received funding from the European Community and French Interministerial Fund for the Caribbean (FIC).

Due to their geographic proximity, similar climates, and ecosystems, these countries share various livestock pests and diseases. The difference lies in their ability to respond to such problems. Due to its significant technical, financial, and human resources, as well as its relatively small livestock population, French Guiana has been able to conduct animal disease research, monitoring, and control programs. As a consequence of economic crisis and structural adjustment programs, the veterinary services of Guyana and Suriname have been confronted with scarce technical, material, financial, and human resources. Guyana's veterinary diagnostic laboratory ceased operations in 1990. Suriname's laboratory continues to operate on a limited basis, but has great difficulty with material supply and equipment maintenance, due to a shortage of foreign exchange. These limitations make it difficult or impossible for the veterinary services of Guyana and Suriname to implement livestock disease monitoring and control programs.

The livestock hemoparasitic diseases anaplasmosis and trypanosomiasis and babesiosis and their arthropod vectors constitute important constraints to animal health and livestock production in tropical South American countries. In South America, *Anaplasma marginale* is transmitted by biting insects and ticks. This disease causes fever, anemia, weight loss, jaundice, abortion, and mortality in cattle.[1]

The epidemiology, clinical significance, and economic impact of non-tsetse transmitted trypanosomes are less clearly understood in the Americas. In South America, the trypanosomes *Trypanosoma vivax and Trypanosoma evansi* are mechanically transmitted by biting insects including *Tabanus*,[2,3] *Cryptotylus*,[4] and *Stomoxys* species. Vampire bats (*Desmodus rotundus*),[5] ticks,[6,7] and triatomine bugs[8] have also been suggested as vectors in the transmission of *T. vivax*. In French Guiana, a temporal association has been found between dry season (November to January), high Tabanid density, and *Trypanosoma vivax* clinical outbreaks.[9]

Serological evidence for *Trypanosoma vivax* has been found in the Americas from El Salvador to Paraguay.[10] The organism is believed to have been imported with a shipment of cattle from Senegal to French Guiana, Martinique, and Guadeloupe in 1830.[11] Previous studies in Guyana have found high seroprevalence rates to *T. vivax* in sheep in coastal Guyana. Applewhaite found a seroprevalence rate of 63.4% in sheep on ELISA (enzyme linked immunosorbent assay).[12] Vokaty *et al.* found 64% seroprevalence to *Trypanosoma* sp. in sheep in coastal Guyana on indirect fluorescent antibody test.[13] In French Guiana, a survey of 3000 cattle yielded a seroprevalence rate of 22% on ELISA testing for *T. vivax* antibody.[9] *T. vivax* has been found to be prevalent in cattle, water buffalo, and sheep in Colombia, but outbreaks of severe disease are sporadic. In Colombia and French Guiana, these outbreaks in adult cattle are characterized by fever, anemia, decreased milk yields, rapid weight loss, abortion, and occasional mortalities.[14,15] However, calves may show high parasitemias without clinical signs. The sporadic clinical outbreaks have been attributed to decreased herd immunity caused by stress such as poor nutrition, intercurrent infection, or vaccination.[15]

Trypanosoma evansi has been found in South America from Panama to Argentina.[10] It causes clinical disease in horses, donkeys, buffalo, mules, and dogs. The disease is known as Surra in Africa and Asia, and as "Mal de Caderas" in Brazil and Argentina. This disease is characterized by intermittent fever, anemia, dependent edema, lethargy, weigh loss, nervous signs, and eventually death.[16] Epidemics of

"Mal de Caderas" have caused significant mortality in horses in the Pantanal region in southern Brazil, thereby limiting development of the cattle ranching industry.[17] Significant *T. evansi* epidemics have also been described in Colombia and Venezuela.[10] Natural infections have been detected in several South American wildlife species including the capybara (*Hydrochoerus hydrochaeris*),[10] coatis (*Nasua nasua*),[18] and vampire bats (*Desmodus rotundus*).[5] Serological evidence exists for *T. evansi* in sheep in coastal Guyana, but clinical disease has not been reported.[13]

HEMOPARASITE NETWORK

The project "Hemoparasite Network for the Guianas" is a collaborative effort between the Inter-American Institute for Cooperation on Agriculture (IICA) in Guyana and Suriname, the Centre de Cooperation Internationale en Recherche Agronomique pour le Developpement–Elevage et Medecine Veterinaire Tropicale (CIRAD-EMVT), which has a specialized laboratory for research into arthropod-borne diseases in Cayenne, and the government veterinary services of Guyana, Suriname, and French Guiana. The general objective of the project is "to increase knowledge of the epidemiology, clinical and economic importance of hemoparasites, particularly trypanosomes, in Guyana, Suriname and French Guiana in order to develop effective control methods and thus improve the health and productivity of livestock."

The specific objectives are:

i. To exchange information on hemoparasites, especially Trypanosomes, with researchers in Latin America and the Caribbean by formation of a Hemoparasite Information Network (TRYPNET)

ii. To strengthen the diagnostic capabilities of the veterinary services of Guyana and Suriname by:
    a. Creation of a Hemoparasite Reference Laboratory for the Guianas at CIRAD-EMVT in Cayenne.
    b. Training Guyanese and Surinamese technicians in hemoparasite diagnostic techniques.
    c. Purchase of materials and regents for the veterinary diagnostic laboratories in Suriname and Guyana.
    d. Laboratory support for other livestock disease investigations whenever possible.

iii. To increase information on hemoparasites, particularly Trypanosomes, in South America, by conducting an epidemiological study in cattle in Guyana, Suriname, and French Guiana.

TRYPNET: HEMOPARASITE INFORMATION NETWORK

The Hemoparasite Information Network was initiated in 1994 and is ongoing. Two CIRAD-EMVT publications were translated from French to English and distributed to livestock production and animal health professionals in Guyana and Suriname. The English versions were titled "Horseflies of the Guyanas: Biology, Veterinary Significance & Control Methods" and "The Cattle Tick, *Boophilus microplus*." A quarterly

TABLE 1. Preliminary Results of Hemoparasite Survey in Guyana, Suriname, and French Guiana

	n	+CCT[a]	Adj. + CCT[b]	%	+ELISA TRP AB[c]	%	Anaplasma n	+[d]	%
French Guiana	507	13	0	2.6	33	6.5	469	106	22.6
Suriname	503	5	0	1	163	32.4	259	131	50.6
Guyana	408	10	3	0.7	168	41.2	308	78	25.3

[a] Positive for *Trypanosome* sp. on capillary centrifuge technique (WOO).[19]
[b] Adjusted for *T. vivax* positive.
[c] Positive on ELISA for *Trypanosome* sp. antibody—Ferenc method.[20]
[d] Positive for *Anaplasma* on stained blood smear.

newsletter called *Trypnews* has published two issues in English and Spanish. These are sent to animal health professionals and researchers in North, Central, South America, the Caribbean, Africa, and Europe. Initial response has been enthusiastic, and an expansion of the network to include Venezuela and Brazil is planned. A small technical meeting on hemoparasites, especially New World Trypanosomes, is planned in Georgetown, Guyana in 1996.

EPIDEMIOLOGIC STUDY IN FRENCH GUIANA, GUYANA, AND SURINAME

Materials and Methods

Bovine blood samples and demographic data are collected by the veterinary services in Guyana and Suriname, in both field and abattoir. Two thin blood smears are made for each sample and stained with RAL stain.[g] One is read locally and the other at the Hemoparasite Reference Laboratory. Packed cell volume (PCV) and capillary centrifuge technique (WOO test) are conducted within 3 hours of sampling. Blood is centrifuged, and serum samples are split into two; one is stored in a local frozen serum bank and the other half is sent to the Hemoparasite Reference Laboratory, where ELISA tests for trypanosome antigen and antibody are performed.

Preliminary Results

Survey results to date appear in TABLE 1.

Discussion

The relatively high seroprevalence rates on ELISA test suggest that trypanosomes, most likely *T. vivax,* are endemic in cattle in Guyana and Suriname. This finding agrees

[g] Hematovet Kit RAL, Rhone Merieux, contains methanol, eosin, and methylene blue stains.

with previous small ruminant studies in Guyana done by Vokaty and Applewhaite. *T. evansi* has yet to be conclusively identified in the Guianas. However, there is anecdotal evidence for equine trypanosomiasis in the Rupununi, the ranching area in southern Guyana that borders on Brazil.[21] Further studies are necessary to determine the epidemiology, clinical significance, and economic impact of *T. vivax* and *T. evansi* infection in livestock in South America. However, improved laboratory tests would be necessary to conduct such research. Serologic tests developed for African trypanosomes do not always work well for South American strains of *T. vivax* and *T. evansi*.[22] By putting New World trypanosome researchers in contact to share technical information, it is hoped that TRYPNET will help to solve such challenging questions.

SUMMARY

Although Guyana, Suriname, and French Guiana share borders and climatic and geographic similarities, the countries have maintained little contact, due to language, political, and administrative differences. In 1993, two international organizations involved in the improvement of animal health, the Inter-American Institute for Cooperation on Agriculture (IICA) and CIRAD-EMVT (Centre de Cooperation Internationale en Recherche Agronomique pour le Developpement–Elevage et Medecine Veterinaire des Pays Tropicaux), jointly developed a collaborative project between the veterinary services of the three countries entitled ''Hemoparasite Network for the Guianas.''

This project seeks to pool livestock, laboratory, and technical resources between the three countries in order to generate and exchange information on hemoparasites of livestock. A Hemoparasite Reference Laboratory for the Guianas has been created at the CIRAD-EMVT laboratory in Cayenne, French Guiana. Besides processing ruminant serum samples from the three countries, specialists from this organization conduct training in hemoparasite diagnostic techniques for laboratory personnel from Guyana and Suriname. A large-scale epidemiologic study of hemoparasites of cattle in the three countries is under way, to determine the prevalence, distribution, and clinical and economic significance of hemoparasites in the three countries, particularly *Trypanosoma vivax* and *T. evansi*. Preliminary results are presented and discussed. A Hemoparasite Information Network (TRYPNET) has been initiated, including a quarterly hemoparasite newsletter (TRYPNEWS), published in English and Spanish and disseminated to researchers in the Americas, Europe, and Africa. In 1995/96, it is proposed to expand the network's scope to include Venezuela and Brazil.

REFERENCES

1. BLOOD, D. C. & O. M. RADOSTITS, Eds. 1989. Veterinary Medicine. 7th Edit. Bailliere Tindall. London, England.
2. RAYMOND, H. L. 1990. Tabanus importunus, vecteur mecanique experimental de *Trypanosoma vivax* en Guyane Francaise. Ann. Parasitol. Hum. Comp. **65**(1): 44–46.
3. OTTE, M. J. & J. Y. ABUABARA, 1991. Transmission of South American *Trypanosoma vivax* by the neotropical horsefly *Tabanus nebulosus*. Acta Tropica **49**: 73–76.
4. FERENC, S., H. L. RAYMOND & R. LANCELOT. 1988. Mechanical transmission of South American *Trypanosoma vivax* by the *Tabanid Cryptotylus unicolor*. Proceedings of the 18th Congress of Entomology: 295.

5. HOARE, C. A. 1965. Vampire bats as vectors and hosts of equine and bovine trypanosomes. Acta Trop. **22:** 204.
6. MORZARIA, S. P., A. A. LATIF, F. JONGEJAN & A. R. WALKER, 1986. Transmission of a *Trypanosoma* sp. to cattle by the tick *Hyalomma anatolicum anatolicum*. Vet. Parasitol. **19:** 13-21.
7. LOPEZ, G. V., K. C. THOMPSON & H. BAZALAR. 1979. Transmission experimental de *Trypanosoma vivax* por la garrapata *Boophillus microplus*. Revista ICA Bogota (Colombia) **14:** 93-96.
8. GARDINER, P. R. 1989. Recent studies of the biology of *Trypanosoma vivax*. Adv. Parasitol. **28:** 229-316.
9. DESQUESNES, M. & P. GARDINER. 1993. Epidemiologie de la trypanosomose bovine (*Trypanosoma vivax*) en Guyane francaise. Rev. Elev. Med. Vet. Pays Trop. **46**(3): 163-170.
10. WELLS, E. A. 1984. Animal trypanosomiasis in South America. Prev. Vet. Med. **2:** 31-41.
11. CURASSON G. 1943. Traité de protozoologie veterinaire et comparée. I. Trypanosomes. Vigot Freres. Paris, France.
12. APPLEWHAITE, L. 1990. Small Ruminant Trypanosomiasis in Guyana. Proceedings of XVII Biennial Caribbean Veterinary Congress, Nov. 7-10, 1990. Antigua.
13. VOKATY, S., V. O. M. MCPHERSON, E. CAMUS & L. APPLEWHAITE. 1993. Ovine trypanosomosis: a seroepidemiological survey in coastal Guyana. Rev. Elev. Med. Vet. Pays Trop. **46**(1-2): 57-59.
14. CAMUS, E. & A. MARTRENCHAR. 1990. Infection experimentale de zebus guyanais avec *Trypanosoma vivax*. Revue Elev. Med. Vet. Pays Trop. **43**(4): 467-472.
15. BETANCOURT, A. E. 1978. Studies on the epidemiology and economic importance of Trypanosoma vivax Ziemann, 1905, in Colombia. PhD Thesis. Texas A&M University. College Station, Texas.
16. IKEDE, B. O., I. FATIMAH, W. SHARIFUDDIN & T. A. BONGSO. 1983. Clinical and pathological features of natural *Trypanosoma evansi* infection in ponies in West Malaysia. Trop. Vet. **1**(3): 151-157.
17. WILCOX, R. 1992. Cattle and environment in the Pantanal of Mato-Grosso, Brazil, 1870-1970. Agricultural History **66**(2): 232-256.
18. NUNES, V. L. B. & E. T. OSHIRO. 1990. *Trypanosoma* (Trypanozoon) *evansi* in the coati from the Pantanal region of Mato Grosso do Sul State, Brazil. Trans. R. Soc. Trop. Med. Hyg. **84:** 692.
19. WOO, P. Y. K. 1970. The haematocrit centrifuge technique for the diagnosis of African trypanosomiasis. Acta Tropica **27:** 384-386.
20. FERENC, S. A., V. STOPINSKI & C. H. COURTNEY. 1990. The Development of an enzyme-linked immunosorbent assay for *Trypanosoma vivax* and its use in a seroepidemiological survey of the Eastern Caribbean Basin. Int. J. Parasitol. **20**(1): 51-56.
21. APPLEWHAITE, L. (Personal communication.)
22. DESQUESNES, M. 1995. Proceedings of the 3rd Biennial Society for Tropical Veterinary Medicine, Costa Rica. N.Y. Ann. Sci. (In press.)

Evaluation of Three Antigen Detection Tests (Monoclonal Trapping ELISA) for African Trypanosomes, with an Isolate of *Trypanosoma vivax* from French Guyana

M. DESQUESNES

CIRAD-EMVT
c/o Institut Pasteur
BP 6010
97306 Cayenne, French Guyana

INTRODUCTION: REPORT OF PRELIMINARY RESULTS OBTAINED AT ILRAD AND IN FRENCH GUYANA WITH TRYPANOSOME-ANTIGEN-ELISA

Three species-specific monoclonals developed by Nantulya *et al.*[1] were used to set up antigen trapping-ELISA detection tests (AgDT) for African trypanosomes, *Trypanosoma vivax, T. brucei,* and *T. congolense.*[2] In Africa, the authors recorded a sensitivity from 80 to 96%[3,4] and a specificity of 100%.[1]

In 1991–1992, an epidemiological survey was carried out on cattle in French Guyana, with these AgDT for *T. vivax* and *T. brucei* (the latter was used to detect *T. evansi* antigens). Half of the 2953 sera were tested at ILRAD, half at CIRAD-EMVT-Guyane. The prevalence of *T. vivax* antigens was 12%.[5] The prevalence of positivity observed with the *T. brucei* assay was 20% (unpublished data). However, *T. evansi* has never been observed in French Guyana. Also, the presence of this parasite has never been demonstrated. Furthermore, despite mice inoculation with blood samples from cattle or wild pigs positive to *T. brucei* AgDT, the parasite has never been isolated. A prevalence of 20% for this parasite therefore appeared doubtful.

The *T. brucei* AgDT was thus suspected to have a specificity of approximately 80%.

Even more doubtful were the 18% positive results obtained with *T. congolense* AgDT when further evaluation was pursued in French Guyana with local cattle sera (unpublished data).

To reevaluate these tests, in French Guyana, two sheep were experimentally infected[6] with a stock of *T. vivax* isolated in French Guyana by Lancelot in 1988.[7] Before inoculation, the animals were negative to all the *Trypanosoma* spp. detection tests that were used. One sheep was bled daily for 270 days, the other for 130

Figure 1a: parasitaemia and optical densities with *T. vivax* AgDT.

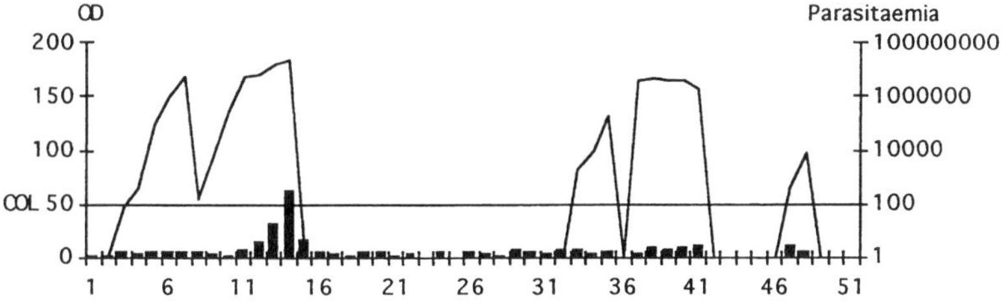

Figure 1b: parasitaemia and optical densities with *T. brucei* AgDT

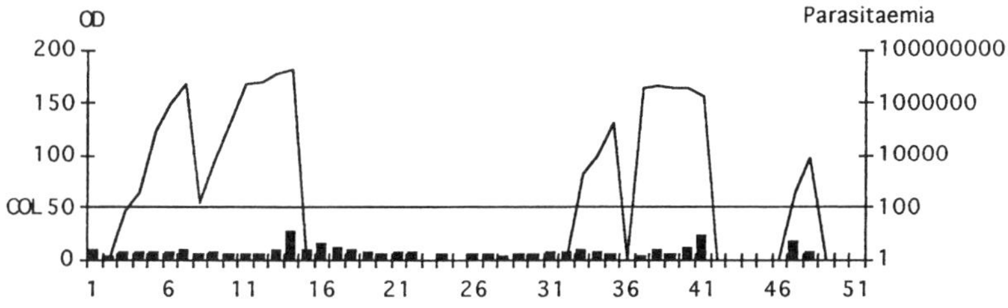

Figure 1c: parasitaemia and optical densities with *T. congolense* AgDT

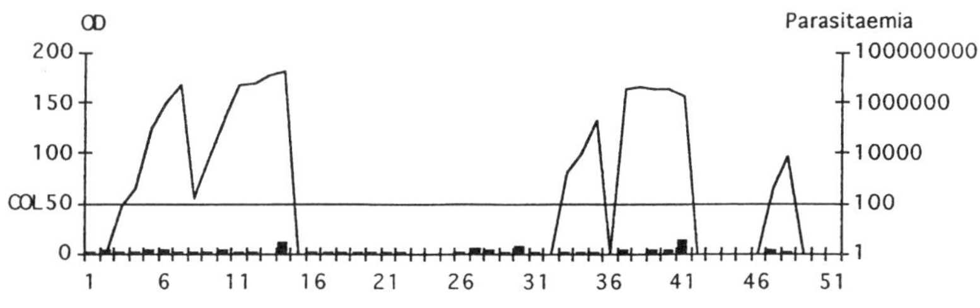

FIGURE 1. Parasitemia and ODs of calf 1.

days. Parasitological techniques such as the microhematocrit centrifugation technique (MHCT),[8] blood buffy-coat technique (BCT)[9] stained blood smears, and AgDT were processed. A summary of the results is presented in TABLE 1 (complementary information is available in FIGURE 5: graphs showing parasitemia and optical densities for the three tests). Only *T. vivax* was identified by observation of the stained blood or buffy coat smears, on the basis of morphology and size of the parasites, as described by Hoare.[10]

Using the cutoff line (COL) indicated by ILRAD: 0.05 optical density (OD), the sensitivity of the *T. vivax* AgDT was 2%, far below that of the MHCT, 52%.

Using the same COL, the specificities of *T. brucei* and *T. congolense* AgDT were respectively 74% and 78%.

Figure 2a: parasitaemia and optical densities with *T. vivax* AgDT.

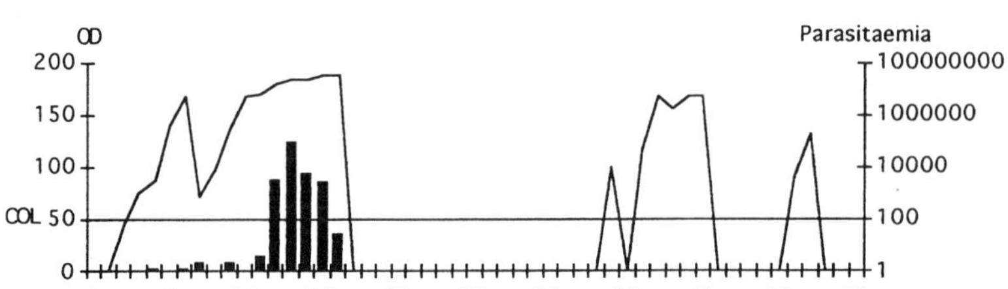

Figure 2b: parasitaemia and optical densities with *T. brucei* AgDT

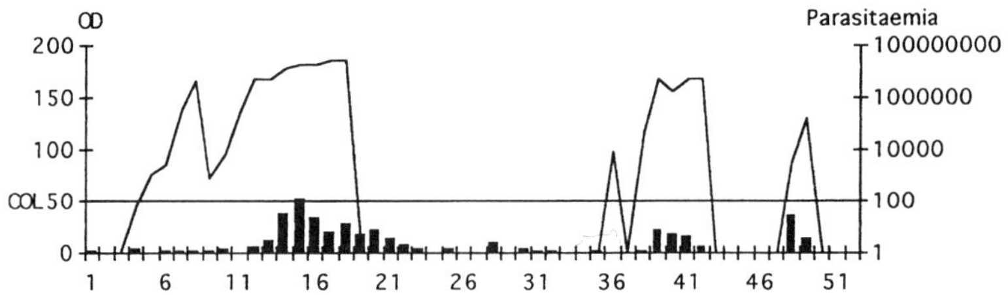

Figure 2c: parasitaemia and optical densities with *T. congolense* AgDT

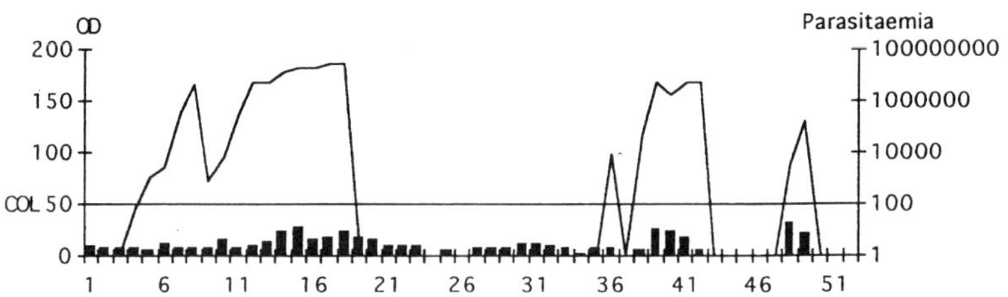

FIGURE 2. Parasitemia and ODs of calf 2.

To confirm the results obtained in French Guyana, blind tests were performed at ILRAD, and CRTA, Burkina Faso, with samples from these sheep. Results did not show a significant difference between the three laboratories.[6]

Statistical analysis of the results was made according to the methods described by Schwartz:[11]

- confidence intervals were calculated with a probability of 95%;

- quantitative analysis of the OD values were made using Student's *t*-test applied between 2 independant means (mean ODs) at a 5% level of significance;

Figure 3a: parasitaemia and optical densities with *T. vivax* AgDT.

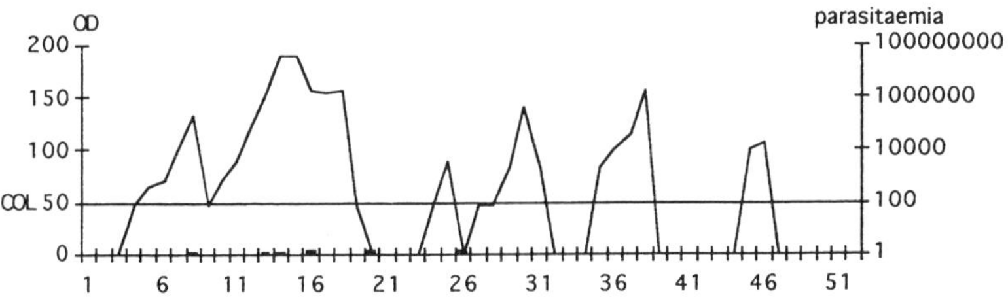

Figure 3b: parasitaemia and optical densities with *T. brucei* AgDT

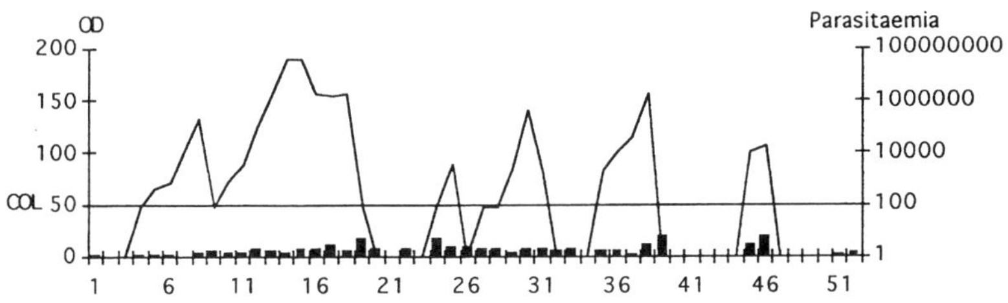

Figure 3c: parasitaemia and optical densities with *T. congolense* AgDT

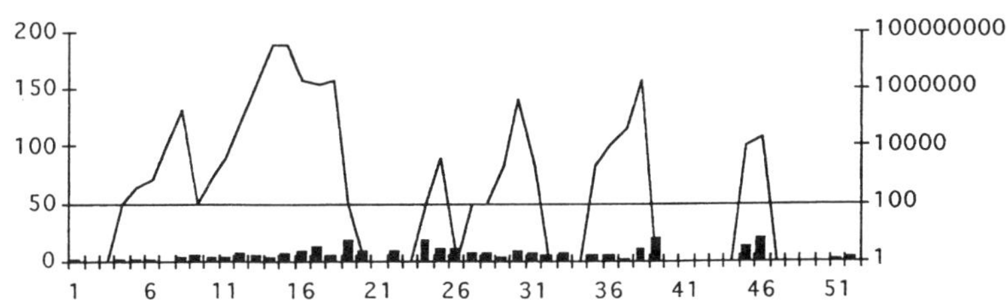

FIGURE 3. Parasitemia and ODs of calf 3.

- qualitative analysis of the results of the AgDT for correlated samples was made using the Yates-corrected chi-square at a 5% level of significance.

EXPERIMENTAL INFECTION OF CALVES AT ILRAD

To confirm these preliminary results, 4 Boran (*Bos indicus*) calves, 5 months of age, raised in a trypanosome-free area, were kept in a fly-proof stable and inoculated via the jugular vein with 10^5-10^6 parasites of the stock of *T. vivax* from French Guyana (IL 4007). Animals were bled daily from the jugular vein, using vacutainer tubes (with and without heparin) during the first 51 days of infection. Sera were separated by spinning, and frozen at $-20°C$ until the ELISA tests were processed.

Figure 4a: parasitaemia and optical densities with *T. vivax* AgDT.

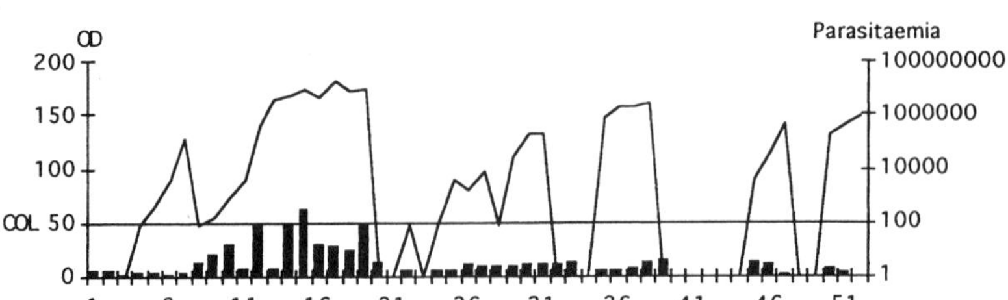

Figure 4b: parasitaemia and optical densities with *T. brucei* AgDT.

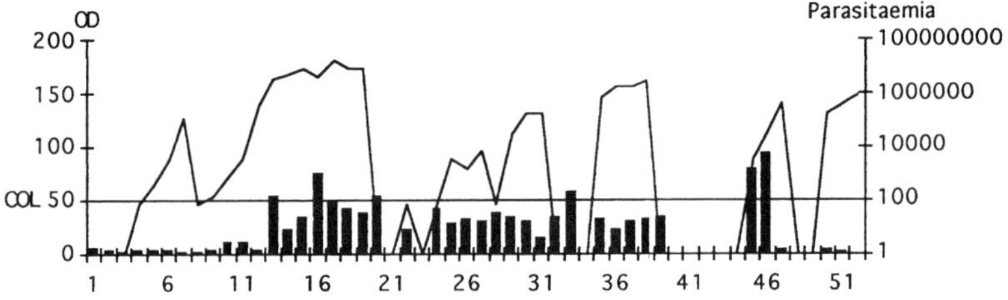

Figure 4c: parasitaemia and optical densities with *T. congolense* AgDT.

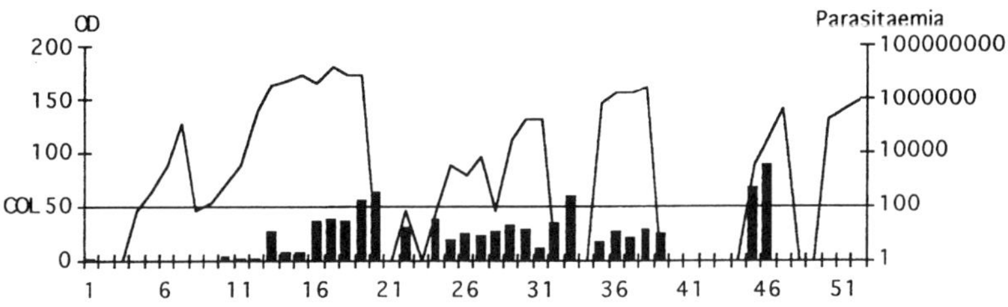

FIGURE 4. Parasitemia and ODs of calf 4.

Prior to the infection, the animals were shown to be negative to all *Trypanosoma* spp. detection tests (MHCT, BCT, and AgDT for the three species). The AgDT were carried out with Dynatech Microtiter® plates, according to ILRAD's protocol.

The OD readings recorded during the infection are indicated in FIGURES 1 to 4; FIGURES a indicate the results with *T. vivax* AgDT, FIGURES b with *T. brucei* AgDT, and FIGURES c with *T. congolense* AgDT. The values indicated are 1000 × the values given by the ELISA-reader. The ODs are indicated by the histograms with a linear scale from 0 to 200. The COL of AgDT (0.05 × 1000 = 50) and the minimum level of detection of parasites by the MHCT (about 100 trypanosomes/ml) are represented in such a way that they appear at the same level on the figures.

TABLE 1. Numbers and Percentages of Positive Results to MHCT and AgDT for *Trypanosoma* spp. in 2 Sheep Experimentally Infected with a Stock of *T. vivax* from French Guyana

Test:	No. of Tests Performed	AgDT for *T. vivax*	AgDT for *T. brucei*	AgDT for *T. congolense*	MHCT
Sheep 1: (270 days of infection)	216	4 1.8%	77 35.5%	66 30.5%	87 40%
Sheep 2: (130 days of infection)	122	3 2.5%	11 9%	7 6%	89 73%
Total: Mean %	338	7 2%	88 26%	73 22%	176 52%

The mean ODs of the 40 tests performed for each animal during infection are indicated in TABLE 2 (value of the ELISA reader × 1000), with mean values of 4 positive and 4 negative controls, and mean values of the 4 calves prior to the infection (confidence intervals are indicated at a risk of 5%).

Quantitative analysis does not show significant differences between the mean ODs of the 4 samples collected before infection and the 160 samples collected during the infection.

Quantitative analysis does not show significant differences between the mean ODs of infected animals (160 samples) and negative reference controls (4 samples).

The COL indicated by ILRAD was used to determine positive and negative samples; the results are indicated in TABLE 3, along with results of parasitological tests. Qualitative analysis does not show significant difference between the results of the ELISA tests for the 3 species (6, 7, and 5 positive results). Mean percentages of positive and negative results to the tests are indicated in TABLE 4 with confidence intervals (at a risk of 5%).

In this experimental infection with *T. vivax,* it should be noted that the animals remained infected throughout the experiment. True positive results are obtained with *T. vivax* AgDT and MHCT (confirmed with the BCT for species identification). Positive results with *T. brucei* or *T. congolense* AgDT must be considered as false-positive results since these parasites appear not to be present in the isolate (see discussion).

DISCUSSION

T. vivax Isolate IL 4007

For information, the following controls were carried out to support the work described here:

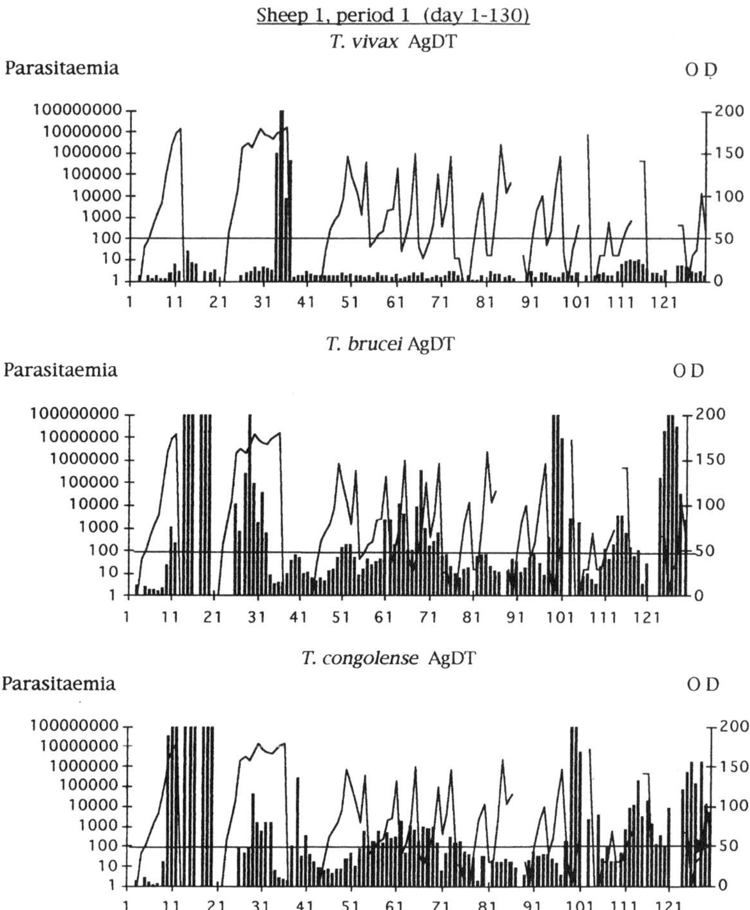

FIGURE 5. Parasitemia and results of AgDT with two sheep experimentally infected with the isolate of *T. vivax* from French Guyana.

- 20 mice inoculations with the *T. vivax* isolate have been done, and parasites have never been shown to grow in mice; *T. evansi* would have grown;

- stained blood and buffy-coat smears were regularly checked, no trypanosome other than *T. vivax* were observed;

- PCR for *vivax* has been carried out on these serum samples, showing positive results with two sets of specific primers for *T. vivax:* TVW 1 and 2, giving a 140-150 bp product.[12] Similarly, new *T. vivax*-specific primers developed by R. Masake at ILRAD gave a characteristic 400bp product whose specificity was confirmed through DNA-probe (personal communication).

Sheep 1   period 2 (day 131-270)

T. vivax Ag-ELISA

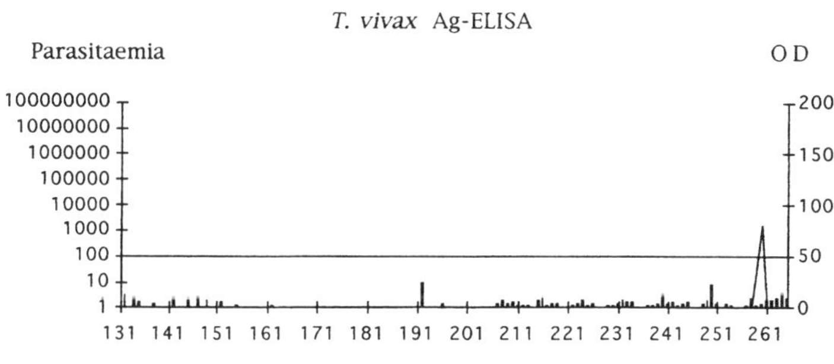

T. brucei Ag-ELISA

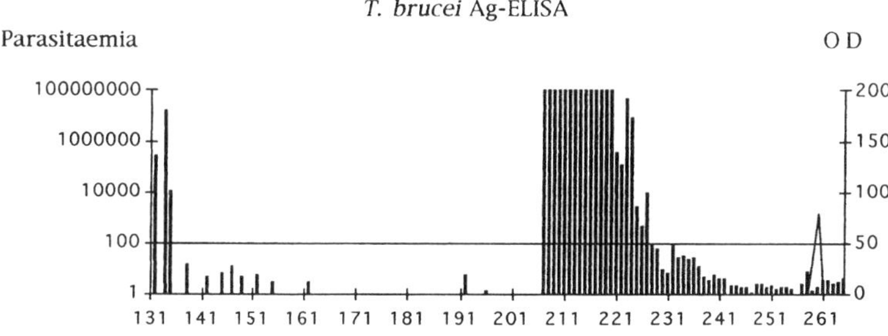

T. congolense Ag-ELISA

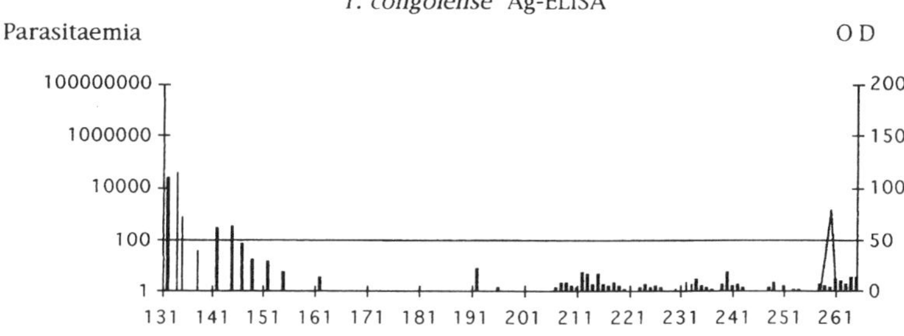

FIGURE 5. *Continued.*

- PCR for other trypanosomes has been processed on these serum samples, showing negative results with the specific primers for *T. brucei* group (TBR 1 and 2),[12] and minicircle primers for *T. evansi.*[13]

From all these results we conclude that the isolate contains only *T. vivax.*

Specificity of **T. brucei** *and* **T. congolense** *AgDT*

The specificities of *T. brucei* AgDT and *T. congolense* AgDT were respectively 95.6% ± 3.2% and 96.9% ± 2.7%. The detection of *T. vivax* antigens with *T. brucei* AgDT and *T. congolense* AgDT should be explored in further studies.

<u>Sheep 2 (day 1-130)</u>

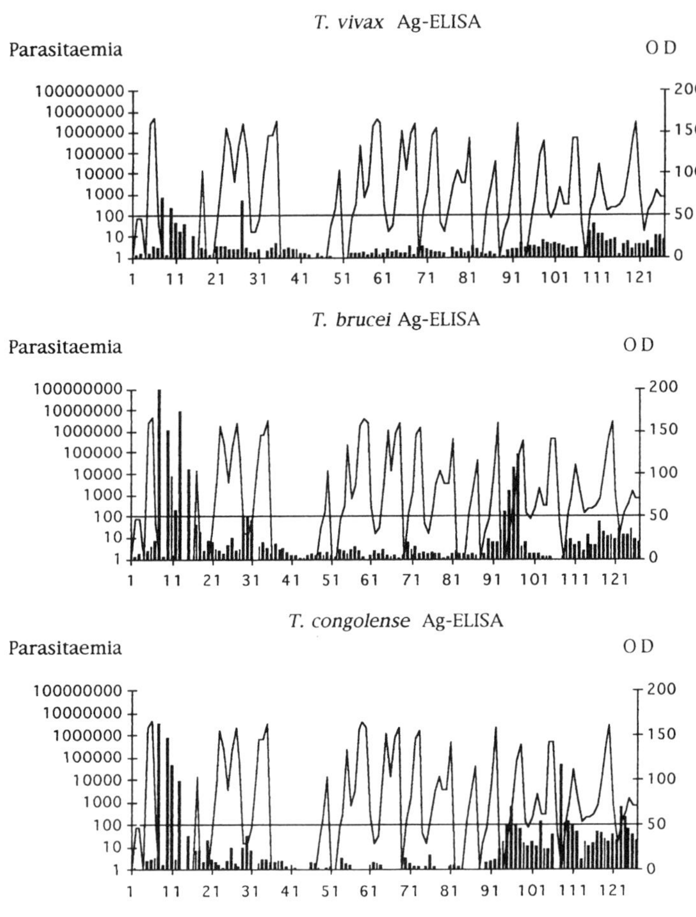

FIGURE 5. *Continued.*

Sensitivity of T. vivax *AgDT*

The mean sensitivity of the *T. vivax* AgDT was low, 3.8% ± 3%. Moreover, the ODs obtained with *T. vivax* AgDT were often so close to zero (mean OD: 8.5 ± 4.2) that it is not only impossible to differentiate a negative from a positive sample, but the sensitivity limit of the ELISA reader is reached.

Similar Results in Africa?

Some researchers working in Africa have also recorded poor sensitivity of the *T. vivax* AgDT, and suspected poor specificity of *T. brucei* and *T. congolense* AgDT:

TABLE 2. Mean OD of the AgDT for *T. vivax, T. brucei,* and *T. congolense* Obtained with 4 Calves Experimentally Infected with a Stock of *T. vivax* from French Guyana

Calf Number	*T. vivax* AgDT	*T. brucei* AgDT	*T. congolense* AgDT
1	7.8	8.4	2.4
2	11.3	10.5	13.2
3	0.3	6.5	5.7
4	14.6	27.3	21.4
Mean OD of the 4 calves (160 tests) during the infection	8.5 ± 4.2	13.2 ± 2.6	10.7 ± 7.1
Mean OD of the 4 calves (4 tests) prior to the infection	2.0 ± 4.5	3.2 ± 2.5	4.0 ± 6.9
Mean OD of the 4 negative controls	14.5 ± 4	3.5 ± 4	3.2 ± 3
Mean OD of the 4 positive controls	437 ± 110	163 ± 10	134 ± 8

- Cuisance,[14] in North Cameroun, in a tsetse-free area, suspected cross-reactivity of *T. vivax* antigens with *T. brucei* and *T. congolense* AgDT.

- Kanwe *et al.,*[15] in Burkina Faso, have recorded a sensitivity of the *T. vivax* AgDT of 9.9% in sheep and goats experimentally infected with *T. vivax.*

- Similarly, Faye,[16] in Gambia, mentioned a sensitivity of the *T. vivax* AgDT of 11% for animals found infected with *T. vivax.* These samples were generally also positive to *T. brucei* and *T. congolense* AgDT.

TABLE 3. Positive Results of Parasitological Tests and AgDT for *T. vivax, T. brucei,* and *T. congolense* with 4 Calves Experimentally Infected with a Stock of *T. vivax* from French Guyana

Calf No.	No. of Tests Performed	MHCT	*T. vivax* AgDT	*T. brucei* AgDT	*T. congolense* AgDT
1	40	22	1	0	0
2	40	23	4	1	0
3	40	29	0	0	0
4	40	35	1	6	5
Total	160	109	6	7	5

TABLE 4. Mean Percentages of Results Obtained with MHCT and AgDT for *T. vivax, T. brucei,* and *T. congolense*

Test:	MHCT	*T. vivax* AgDT	*T. brucei* AgDT	*T. congolense* AgDT
Positive:	66%	3.8%	4.4%	3.1%
	±7.3%	±3%	±3.2%	±2.7%
Negative:	34%	96.2%	95.6%	96.9%
	±7.3%	±3%	±3.2%	±2.7%

The parasitological pictures described by these authors are very similar to those described here, with 4 calves and 2 sheep experimentally infected with an isolate of *T. vivax.*

CONCLUSION

With the South American *T. vivax* isolate we have used, low levels of sensitivity were obtained with the *T. vivax* AgDT. It is important to note that *T. brucei* AgDT and *T. congolense* AgDT presented 3-4% false-positive results (never described before).

When new tests are developed that might be used throughout the world, numerous isolates of parasites should be tested before one can consider the test reliable. The ELISA techniques are suitable for large-scale diagnosis and are thus a useful tool for trypanosome epidemiological surveys, but still encounter important problems. New immunological reagents may need to be developed to improve the accuracy of the tests in circumstances where parasites occur that are similar antigenically to the isolate of *T. vivax* used here.

SUMMARY

Preliminary studies in French Guyana with sheep experimentally infected with a local isolate of *Trypanosoma vivax* tended to show poor sensitivity and/or specificity of the monoclonal antibodies used in kits for the antigen-detection (Ag) ELISA for *T. vivax, T. brucei,* and *T. congolense.*

To reevaluate these kits, 4 calves were infected at ILRAD, Nairobi, Kenya, with the same isolate. Blood samples were taken daily for 51 days, and examined directly on blood smears and buffy coat, and using Ag-ELISA for the three species. For the 4 calves, on 158 tests performed over the first 51 days of infection, the percentages of positive results were 66% on buffy coat; on Ag-ELISA 3.8% for *T. vivax,* 4.4% for *T. brucei,* and 3.1% for *T. congolense.* Blood smears showed only *T. vivax.*

These results confirm those previously obtained in French Guyana: the test for *T. vivax* (at least in the initial stage) shows a very low sensitivity, far below that of parasitological techniques, and the specificity of the *T. brucei* and *T. congolense* tests is low. Certain surprising results obtained in Africa might also be due to a poor sensitivity and specificity of the monoclonals used.

As ELISA is the technic of choice for epidemiological surveys, and antigen detection a logical way to confirm whether an animal is actively infected by trypanosomes, the Ag-ELISA remains a necessary tool for epidemiological surveys of trypanosomes; new monoclonals are required to develop more specific and sensitive tests.

ACKNOWLEDGMENTS

I am very thankful to Mr. S. Minja (research associate) for processing the ELISA tests, Dr. R. Masake for providing reagents, and Mr. J. Thuo (technician) for PCR and DNA-probe reactions, and to Mr. S. Kemei, for technical assistance. I am also very grateful to Dr. A. Peregrine and Dr. P. Lessard for revising the manuscript, and to Prof. G. Uilenberg who presented the paper at the STVM-95 meeting.

REFERENCES

1. NANTULYA, V. M., A. J. MUSOKE, F. R. RURANGIRWA, N. SAIGAR & S. H. MINJA. 1987. Monoclonal antibodies that distinguish *Trypanosoma congolense, T. vivax* and *T. brucei.* Parasite Immunol. **9:** 421–431.
2. NANTULYA, V. M. & K. J. LINDQVIST. 1989. Antigen-detection enzyme immunoassays for the diagnosis of *Trypanosoma vivax, T. congolense* and *T. brucei* infections in cattle. Trop. Med. Parasit. **40:** 267–272.
3. NANTULYA, V. M. 1990. Trypanosomiasis in domestic animals: the problem of diagnosis. Rev. Sci. Tech. Off. Int. Epiz. **9:** 357–367.
4. NANTULYA, V. M. 1991. Molecular diagnosis of parasites. Experientia **47:** 142–145.
5. DESQUESNES, M. & P. R. GARDINER. 1993. Épidémiologie de la trypanosomose bovine (*Trypanosoma vivax*) en Guyane Française. Rev. Élev. Méd. Vét. Pays Trop. **46:** 463–470.
6. DESQUESNES, M. & S. LA ROCQUE. Comparaison de la sensibilité du tests de WOO et d'un test de détection des antigènes de *Trypanosoma vivax* chez deux moutons expérimentalement infectés avec une souche guyanaise du parasite. (Submitted for publication.)
7. LANCELOT, R. 1988. La trypanosomose bovine à *Trypanosoma vivax* en Guyane Française, contribution à l'étude clinique et épidémiologique. Thèse de Doctorat vétérinaire. Alfort.
8. WOO, P. T. K. 1970. The haematocrit centrifuge technique for the diagnosis of African trypanosomiasis. Acta Trop. **27:** 384–386.
9. MURRAY, M., P. K. MURRAY & W. I. M. MCINTYRE. 1977. An improved parasitological technique for the diagnosis of African trypanosomiasis. Trans. R. Soc. Trop. Med. Hyg. **71:** 325–326.
10. HOARE, C. A. 1972. The Trypanosomes of Mammals: 125–141. Blackwell Scientific Publications, Edinburgh. Western Printing Services Ltd., Bristol.
11. SCHWARTZ, D. 1963. Méthodes statistiques à l'usage des médecins et des biologistes: 211–227; 283–287. Editions médicales Flammarion. Paris, France.
12. MASIGA, D. K., A. J. SMYTH, P. HAYES, T. J. BROMIDGE & W. C. GIBSON. 1992. Sensitive detection of trypanosomes in tsetse flies by DNA amplification. Int. J. Parasitol **22:** 909–918.
13. ARTAMA, W. T., M. W. AGEY & J. E. DONELSON. 1992. DNA comparisons of *Trypanosoma evansi* (Indonesia) and *Trypanosoma brucei* spp. Parasitology **104:** 67–74.

14. CUISANCE, D. 1994. Etude de la trypanosomose animale sans glossines dans le Nord-Cameroun. Ministère de la Coopération/CIRAD-EMVT, 20 rue P. Curie, 94704 Maisons-Alfort, France.
15. KANWE, A. B., Z. BENGALY, D. SAULNIER & G. DUVALLET. 1992. Évaluation du test de détection des antigènes circulants de trypanosomes à l'aide d'anticorps monoclonaux. Infections expérimentales et naturelles. Rev. Élev. Méd. Vét. Pays Trop. **45:** 265-271.
16. FAYE, J. 1993. The use of an antigen-detection ELISA for the diagnosis of trypanosome infections in N'Dama cattle in Gambia. *In* Improving the Diagnosis and Control of Trypanosomiasis and Other Vector-Borne Diseases of African Livestock Using Immunoassay Methods. IAEA-TECDOC-707: 51-57.

An Amino Acid Sequence Coded by the Exon 2 of the BoLA DRB3 Gene Associated with a BoLA Class I Specificity Constitutes a Likely Genetic Marker of Resistance to Dermatophilosis in Brahman Zebu Cattle of Martinique (FWI)[a]

J. C. MAILLARD,[b] D. MARTINEZ,[b] AND A. BENSAID[c]

*Centre de Cooperation Internationale en Recherche Agronomique
pour le Developpement (CIRAD)
Departement d'Elevage et de Medecine Veterinaire des Pays
Tropicaux (EMVT)*

*[b]Mission Guadeloupe
Domaine Duclos
BP 515
97165-Pointe à Pitre Cedex
Guadeloupe (FWI), France*

*[c]Laboratoire Pathotrop
10 rue Pierre Curie
94704-Maisons Alfort Cedex, France*

INTRODUCTION

Dermatophilosis is a world-wide infectious skin disease caused by infection with the actynomycete *Dermatophilus congolensis.*[1] In tropical and subtropical regions, it dramatically reduces ruminant productivity and impedes the introduction of improved breeds, which are highly susceptible. The disease presents in two forms. The chronic disease affects variable and sometimes extensive areas of the skin (presence of scabs) and is associated with low productivity. The severe form is lethal and can lead to the destruction of entire herds of cattle.

The epidemiology of dermatophilosis in the Caribbean has been extensively studied.[2,3] Several major environmental risk factors including management, climate (warm and wet conditions), and the presence of the tick *Amblyomma variegatum*[4]

[a]This work was financially supported by grants from the European Community DGXII (TS3-CT91-0012), the General Council of Martinique, and the French Ministry of Overseas (Cordet 91-D105).

185

are well recognized and determine the severity of this disease. The significance of the tick is as a factor triggering severe infections, and it has been suggested that tick saliva acts as an immunosuppressive agent favoring the disease. However, mechanisms underlying this effect are not well understood and studies are hampered because the severe, lethal disease cannot be reproduced experimentally. Control measures are based on the use of antibiotics to treat affected animals, and acaricides to reduce tick pressure. Both measures present difficulties in the eradication of this disease, and they are expensive even for wealthy stock breeders. Vaccine development remains a remote hope, but thus far, all attempts have failed.

Varying degrees of susceptibility to dermatophilosis among ruminants have been reported since the beginning of the century when importation of European cattle was attempted in Africa[5] and in the French West Indies.[2] It was found that while local cattle populations, living for a long time in the presence of risk factors, displayed a low prevalence of dermatophilosis, exotic cattle, generally imported, were severely infected. Considering the failures of vaccination, it seems that a protective immunity is not generated. Hence, the reasons why local populations are more resistant to the disease probably derive rather from genetic factors than acquired immunity. Indeed, field observations tend to favor the hypothesis that within a particular breed, lineages of individuals are more resistant than others to dermatophilosis.[6]

The epidemiological situation in Martinique, an island of the French West Indies (FWI), provides an environment in which dermatophilosis can be conveniently studied. The Brahman zebu population, introduced in 1952 from Texas, displays a wide variety of animals ranging from the relatively resistant to the relatively susceptible. To confirm that this resistance to dermatophilosis is under genetic control, we have conducted basic and well-controlled experiments.

We chose to study the molecules of the major histocompatibility complex (MHC) termed by the bovine leucocyte antigens (BoLA) which are recognized for their important role in mechanisms controlling the immune response and resistance to pathogens.[7,8] These molecules are essential for the presentation of antigens and their recognition by T cells. The class I molecules present the antigenic endopeptides to CD8+ T cells controlling the cytotoxic response. The class II molecules present the antigenic exopeptides to CD4+ T cells which control both cellular and humoral immune response by means of cytokine release, lymphocyte proliferation, specific antibody production, or cell-mediated cytolysis.[9–11] In the class II region, we studied the DR group.

MATERIAL AND METHODS

Epidemiological Survey and Animal Material

One hundred and twenty-seven animals within herds located in different parts of the island were studied. Several herds could be surveyed for long periods encompassing several rainy and dry seasons. For most of the animals, the duration of the survey was over 3 years, and once a month, the animals were checked for the presence of characteristic lesions of dermatophilosis. When lesions were observed, generally at the end of the rainy season, animals were treated with antibiotics. Individuals with extensive clinical signs or needing multiple treatments were classified as susceptible; many of them died during the survey. Tick pressure was evaluated by tick counts on animals. Resistant cattle were selected on the basis that they were reared in the same herds and under the same conditions as susceptible animals but did not display clinical signs during the survey. They never needed antibiotic treatment.

In all animals, a specific serological monitoring by ELISA[12] was performed to demonstrate contact with the pathogen.

Major Histocompatibility Complex Typing

Class I Antigens

Leucocytes of the 127 animals were collected from buffy coats and tested in a classical microlymphocytotoxicity assay using the international panel of antisera defining BoLA class I allo specificities.[13]

Class II Gene

The MHC class II DR molecules are heterodimers composed of an alpha and a beta chain. Each chain is coded by a gene, DRA for the alpha chain and DRB for the beta one. The terminal parts of these 2 chains contain the antigen recognition site (ARS). In the beta chain, this terminal first domain is coded by the exon 2 composed of a 90 amino acid sequence. Fourteen of them display polymorphisms and constitute the ARS.[14,15]

This exon 2 of the BoLA-DRB3 gene was analyzed by using the polymerase chain reaction (PCR)-restriction fragment length polymorphism (RFLP) method.[16-19]

DNA (100 ng) prepared from each of the 30 selected animals was subjected to 30 cycles of PCR in the following conditions: a denaturation step at 94°C for 1 min, an annealing step at 50°C for 30 s, and an extension step at 72°C for 1 min. At the end of the PCR, an additional 5 min at 72°C was performed in all samples. The PCR buffer was composed of 50 mM Tris-HCl pH = 9, 20 mM ammonium sulfate and 0.2 mM of dNTPs in a final reaction volume of 50 µl to which 1 U of *Thermus flavus* polymerase was added. Primers, BOD 31 (5′ GATGGATCCTCTCTCTGCAG-CACATTTCCT 3′) and BOD 32 (5′ CTTGAATTCGCGCTCACCTCGCCGCTG 3′) were used at a concentration of 300 ng each/50 µl. An aliquot of each PCR product was run in agarose gel in order to check for PCR specificity.

The restriction map of exon 2 of the DRB3 gene is shown in FIGURE 1. PCR products were further cleaved with restriction enzymes chosen for their ability to cut DNA at polymorphic sites of this exon 2. These restriction enzymes were Dpn II (↓GATC), Hae III (GG ↓ CC), Fnu 4HI (GC ↓ NGC) and Rsa I (GT ↓ AC) used in single and double combinations in order to determine the two alleles for each animal. Patterns were obtained by electrophoresis in agarose gel. Amino acid residues were deduced from the specific DNA sequences that are cut by each restriction enzyme type. Ten polymorphic amino acid positions of which 5 were located in ARS were studied.

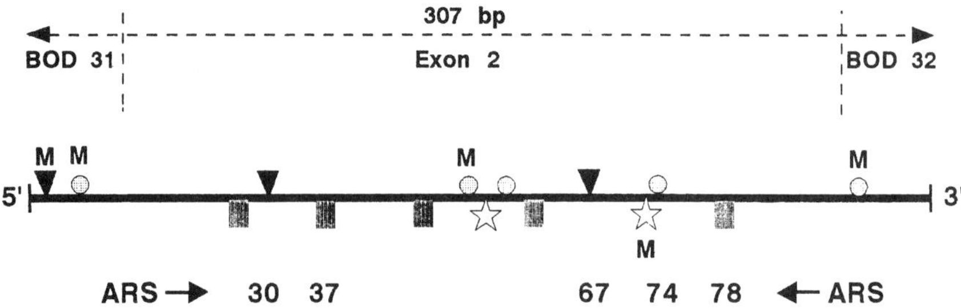

FIGURE 1. Restriction map of exon 2 of the BoLA-DRB3 gene with the four endonucleases Dpnll (triangles), Fnu 4 HI (circles), Rsal (squares) Haelll (stars). ARS: polymorphic positions of the antigen recognition site. M: monomorphic restriction site. BOD 31 and BOD 32 are primers for PCR.

Genetic Informative Families

In view of the results obtained with unrelated animals studied, and the potential markers identified in the BoLA system, it was necessary to investigate their genetic transmission. Within a single herd, 36 animals belonging to three different informative families were followed. All the individuals of these families were checked for the same management and epidemiological conditions as previously described. Because of management conditions, the potential sires were not well known and three or four bulls were available in each herd. Using the classical electrophoretic markers in blood protein systems (hemoglobin, albumin and transferrins), plus MHC class I typing, exclusion of paternity was possible. In the same way, maternal haplotype compatibilities were confirmed. So, for each individual, knowing haplotypes of the dam and her offspring, it was possible to deduce the haplotype of the probable sire (data not shown).

Because of the lower susceptibility to dermatophilosis reported in young animals under one-year-old and not weaned, their epidemiological status was established only in the second year of life.

RESULTS

Epidemiological Data

Data obtained from the epidemiological survey are summarized in TABLE 1.

Thirty unrelated individuals were chosen among the 127 checked animals and classified; 12 as highly resistant and 18 as more susceptible to dermatophilosis.

MHC Typing

Class I Antigens

From a panel of 46 international specificities, only one displayed a significant result. It was the ED13 antiserum that defines the BoLA-A8 specificity present in 83% of the resistant animals and in 28% of the susceptibles, with a *p* value <0.01.

TABLE 1. Epidemiological Data on Dermatophilosis in Zebu Brahman Cattle of Martinique[a]

Herds	Num. Sampl.	% Herd	Ticks	Dipping Frequency	% Seropositive	Clinic. Cases	Survey
Cerise	6	100	+++	0	100	83	1 obs.
Trinité	7	100	+++	0	100	71	1 obs.
Pimantade	41	34	+++	1/week	95	68	3 years
3 Rivières	17	9	++	1/week	94	33	3 years
Ceron	8	8	++	1/week	89	39	3 years
Assclin	27	9	++	1/2 weeks	85	70	2 years
Desportes	21	8	+	1/week	67	33	2 years

[a] Num. Sampl.: number of individuals sampled in the herd. % Herd: percentage of sampled animals in the herd. % Seropositive: percentage of seropositive animals, by ELISA. Clinic. Cases: percentage of sick animals in each herd. Ticks: % of animals having ticks in each herd (+<20, 20<++<50, 50<+++<100).

Class II DRB3 Gene

FIGURE 2 shows electrophoretic patterns of 2 sample animals.

TABLE 2 shows the A8 typing and both alleles of exon 2 of the DRB3 gene for each of the 30 animals in both resistant and susceptible groups.

We studied 10 polymorphic amino acid residues deduced from restriction sites considered. Five of them are ARS positions: 30, 37, 67, 74, and 78, marked by a (+).

In the resistant group, a particular amino acid sequence E (glutamic acid), I (isoleucine), A (alanine), Y (tyrosine) in ARS position 66, 67, 74, and 78 respectively, was found in all animals, either in homozygotic or in heterozygotic forms. Ten out of 12 also displayed the BoLA class I A8 specificity.

In the susceptible group, 10 out of 18 animals showed this EIAY sequence, but only 3 of them had the BoLA class II DRB3 EIAY sequence and the BoLA class I A8 specificity simultaneously.

Interestingly, a serine residue (S) at position 30 of the ARS was found in 8 of the susceptible animals (44%) and was completely absent from the resistant group. Furthermore, this serine at position 30 and the EIAY sequence were never found simultaneously on the same haplotype for one animal.

Statistical analysis (TABLE 3) indicates a good correlation between these BoLA class I and class II markers and the characters of resistance or susceptibility to dermatophilosis. The A8 specificity and the EIAY sequence are significant markers of resistance, whereas the serine residue is a marker of susceptibility.

The best correlation with resistance character (p value <0.001) was obtained by the association *in a same animal,* of the class I A8 specificity, the class II EIAY sequence, and the lack of the amino acid serine in ARS position 30 (TABLE 3).

To validate these results and to study the genetic transmission of these markers, we followed three different families.

TABLE 2. Amino Acid Residues in the 10 Studied Positions Coded by Exon 2 of DRB3 gene and Class I AB Typing for both Haplotypes of Each Resistant and Susceptible Animal[a]

Resistant Animals (n = 12)

A.A. posi. ARS	26	30	37	47	56	60	66	67	74	78	A8 specif.
ARS		+	+					+	+	+	
RA:	—	—	—	—	—	—	E	I	A	Y	+
RA:	—	—	—	—	—	—	E	I	A	Y	
2453:	—	—	—	—	—	—	E	I	A	Y	+
2453:	—	—	—	—	—	—	E	I	A	Y	
69:	—	—	—	—	—	—	E	I	A	Y	+
69:	—	—	—	Y	P	Y	E	I	A	Y	
481:	—	—	—	—	—	—	E	I	A	Y	+
481:	—	—	—	Y	P	Y	E	I	A	Y	
X7:	Y	—	—	Y	P	—	E	I	A	Y	—
X7:	—	—	Y	—	P	Y	E	I	A	Y	
50:	—	—	—	—	—	—	E	I	A	Y	+
50:	—	—	—	—	—	—	—	—	—	Y	
X4:	—	—	—	—	—	—	E	I	A	Y	+
X4:	—	—	—	—	—	—	—	—	—	Y	
2721:	—	—	—	—	—	—	E	I	A	Y	+
2721:	—	—	—	—	—	—	—	—	—	Y	

Susceptible Animals (n = 18)

A.A. posi. ARS	26	30	37	47	56	60	66	67	74	78	A8 specif.
ARS		+	+					+	+	+	
2399:	—	S	—	Y	P	Y	—	—	—	Y	—
2399:	—	S	—	Y	P	Y	—	—	—	Y	
42:	—	S	—	Y	P	Y	—	—	—	Y	—
42:	—	—	—	—	—	—	—	—	—	Y	
B25:	—	S	—	Y	P	Y	—	—	—	Y	—
B25:	—	—	—	—	P	Y	—	—	A	Y	
2331:	—	S	—	Y	P	Y	—	—	—	Y	+
2331:	—	—	—	—	—	—	E	I	—	Y	
Z:	—	S	—	—	P	—	E	I	—	Y	—
Z:	—	—	—	Y	P	Y	—	—	—	—	
2341:	—	S	—	—	P	—	—	—	—	—	—
2341:	—	—	—	—	P	—	E	I	—	—	
2243:	—	S	—	Y	P	Y	—	—	—	Y	+
2243:	—	—	—	—	—	—	E	I	A	Y	
A:	—	(S)	—	Y	P	Y	E	I	—	—	—
A:	—	(S)	—	Y	P	Y	E	I	A	Y	

MAILLARD *et al.*: A GENETIC MARKER

Resistant Animals (*n* = 12)

											ARS
44:	—	—	—	—	—	—	E	I	A	Y	+
44:	—	—	—	—	—	—	E	I	—	Y	
F:	—	—	—	—	—	—	E	I	A	Y	+
F:	—	—	—	—	P	—	—	—	—	—	
96:	—	—	—	—	—	—	E	I	A	Y	+
96:	—	—	—	—	P	Y	—	—	A	—	
47:	—	—	—	Y	P	Y	E	I	A	Y	—
47:	—	—	—	—	P	—	—	—	—	Y	

											ARS
117:	—	—	—	—	(P)	—	E	I	A	Y	—
117:	—	—	—	—	(P)	—	E	I	—	Y	—
B:	—	—	—	—	P	—	—	—	—	Y	—
B:	—	—	—	—	—	—	E	I	A	Y	
2377:	—	—	Y	Y	P	—	E	I	—	—	—
2377:	—	—	—	—	—	—	E	I	A	Y	—
150:	—	—	—	—	P	Y	—	—	A	—	—
150:	—	—	Y	Y	P	—	E	I	A	Y	
213:	—	—	—	—	—	—	—	—	A	Y	+
213:	—	—	—	—	P	Y	E	I	A	Y	
X2:	—	—	Y	Y	P	—	E	I	A	Y	+
X2:	—	—	—	—	—	—	E	I	A	Y	
2638:	—	—	—	Y	P	Y	E	I	A	Y	—
2638:	—	—	—	Y	P	Y	E	I	A	Y	
2273:	—	—	—	Y	P	Y	E	I	A	Y	—
2273:	—	—	—	—	—	—	E	I	A	Y	
58:	—	—	Y	Y	P	—	—	—	—	—	—
58:	Y	—	—	—	P	—	—	—	—	—	
E:	—	—	—	—	P	Y	—	—	A	Y	—
E:	—	—	—	—	—	—	—	—	—	Y	—

[a] A.A. posi.: amino acid positions, ARS: antigen recognition sites, A: alanine, E: glutamic acid, I: isoleucine, F: proline, S: serine, Y: tyrosine.

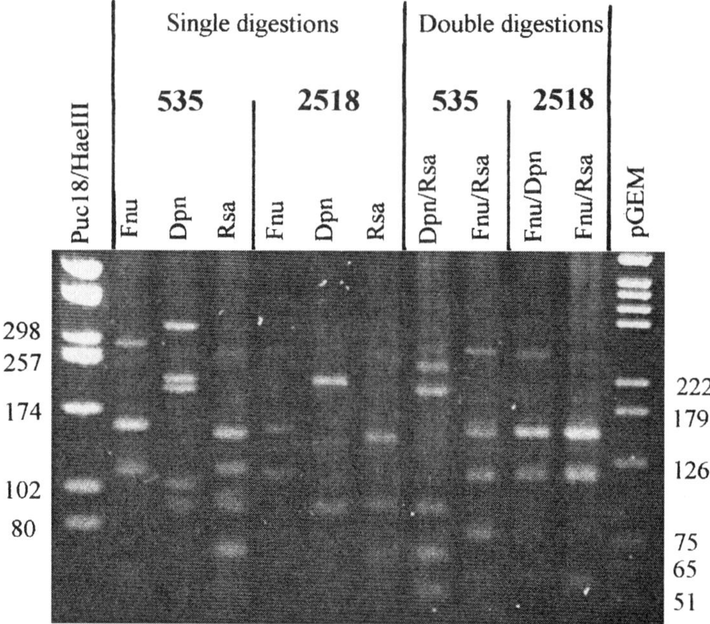

FIGURE 2. Electrophoretic patterns of single and double digestions on DRB3-exon 2 amplification products from two different individuals (535 and 2518). Puc 18 digested by HaeIII and pGEM are molecular weight markers.

Informative Families

Typing results in the three families are shown in FIGURES 3, 4, and 5, respectively.

Family 1

This family started from the nonstudied cow 113 which was the dam of animals 574 and 535. The 574 lineage is composed of animals having the EIAY sequence

TABLE 3. Statistical Results of BoLA Class I and Class II Markers in Groups of Resistant and Susceptible Animals

	Resistant $n = 12$	Susceptible $n = 18$	Prob. (1 dl)	Signif.
A8 +	0.83	0.28	<0.01	++
EIAY +	1.00	0.55	<0.1	+
S +	0.00	0.44	<0.1	+
A8 + /EIAY + /S−	0.83	0.11	<0.001	+++

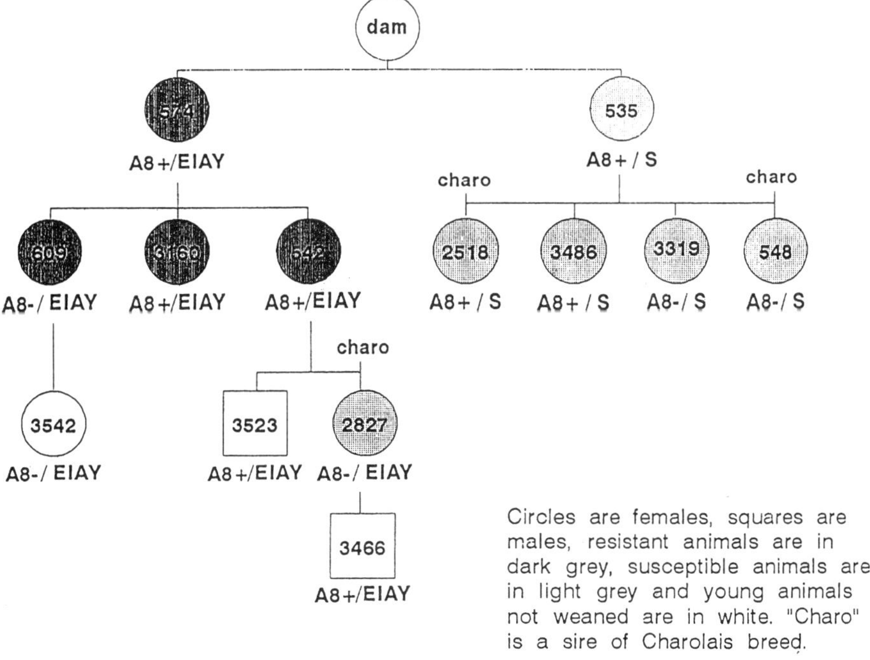

FIGURE 3. BoLA markers in family 1.

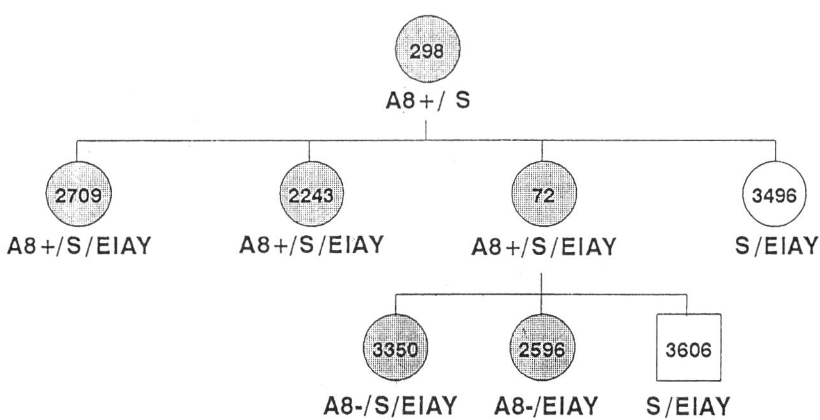

FIGURE 4. BoLA markers in family 2. All the individuals are susceptible (light grey) and two calves are in white.

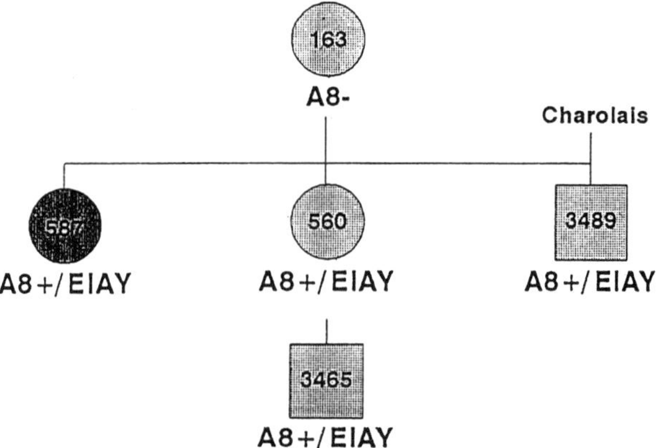

FIGURE 5. BoLA markers in family 3. Susceptible animals are in light grey whereas the resistant one is in dark grey.

marker more often associated with the A8 specificity. Among them, only animal 2827 was susceptible, but it was a Charolais crossbred. On the other side, all the animals of the 535 lineage were susceptible. None of them presented the resistance marker but all had the serine30 susceptible marker. Furthermore, two of them are Charolais crossbred like the animal 2827 of the resistant lineage. In this family, the results seem to confirm the markers identified and indicate that crossbreeding with exotic susceptible breeds seems to increase the susceptibility to the disease.

Family 2

All the animals were classified as susceptible whereas most of them had the EIAY sequence with or without the A8 specificity. But interestingly, they all had the serine30 susceptibility marker.

Family 3

In this family, four animals had exactly the same haplotypes showing (A8+/ EIAY), but three of them are classified susceptible, and one of which was a Charolais crossbred.

DISCUSSION AND CONCLUSIONS

In the epidemiological study, high prevalences of seropositivity by Elisa and of clinical cases indicate that infection is widespread. Disease prevalence increases with tick pressure and decreases with acaricide dipping.

Concerning identified markers, and regarding family 1, it is tempting to conclude that the haplotype formed by the combination of the EIAY BoLA-DRB3 allele and the BoLA-A8 specificity is a good marker for resistance to dermatophilosis in zebu Brahman cattle, but some caution should be taken. First, family studies indicate that these two MHC genes are not linked in all resistant animals. Second, the fact that susceptible animals have the DRB3 EIAY sequence and the A8 specificity indicates that probably MHC genes are not the only ones directly involved in conferring resistance to dermatophilosis. Other genes, within or nearby the MHC, could be implicated and must be studied. Third, the serine at position 30 was never found on the same haplotype as the EIAY sequence. It could be a good marker of susceptibility, and regarding the family 2, we could conclude that this susceptible marker is genetically dominant compared to resistant markers identified when they are present together in a same animal. Resistant animals showing the serine susceptible marker were never found, whereas we found susceptible animals with resistant markers. Several explanations are possible. The EIAY sequence is located in 66, 67, 74, 78 ARS positions, but in the middle of this sequence, there are two other ARS in positions 70 and 71 which we have not studied because we did not find endonucleases cutting in these two polymorphic positions. These results correspond to those obtained in bovine leucosis[20] where particular amino acid motifs at these same positions 70/71 and 74-78 displayed good correlation. To clarify this fact, cloning and sequencing of the entire exon 2 DRB3 alleles on all our experimental animals are in progress.

Another possibility is that as we worked on genomic DNA, it is possible that some animals having markers did not express the exon 2 of this DRB3 gene. A study on complementary RNA must be done to investigate expressed molecules.

In family 3, further study on the three animals 560, 587, and 3465, which have the same haplotypes with different epidemiological status, will be highly informative.

Finally, for this disease, we probably have a case of multigenic control where other genes could be involved. It remains to find potential candidate genes regarding the different physiological mechanisms of resistance or susceptibility presently known, and so to investigate their polymorphisms.

To get more genetic information on segregation of markers known now or in the future, our families will be followed and other started with known typed parents judiciously crossbred.

SUMMARY

One hundred and twenty-seven Brahman cattle from several locations in Martinique (FWI), reared under different environmental conditions, were followed over three years and checked for clinical signs of dermatophilosis. To confirm that these animals had been in contact with the pathogen *Dermatophilus congolensis,* their sera were tested by ELISA. On the basis of this epidemiological study, 12 animals were classified as resistant (seropositive without clinical signs), belonging to herds in which the prevalence of the disease ranged from 25 to nearly 98%. Eighteen animals classified as highly susceptible displayed severe characteristic skin lesions. These 30 selected animals were typed for class I antigens of the major histocompatibility complex (MHC). MHC class II genes were analyzed using the polymerase chain reaction (PCR) and restriction fragment length polymorphism (RFLP) techniques, on the exon 2 of the bovine leucocyte antigen (BoLA) DRB3 gene. Several alleles were found, according to patterns provided by the restriction enzymes used: Fnu 4HI, Dpn II, Hae III, and Rsa I. A particular sequence "EIAY" at amino acid positions 66/67/74/78 located in the antigen recognition sites (ARS) was found in the 12 animals classified as resistant, and 10 of them displayed also class I BoLA-A8 specificity. On the other hand, only 3 out of the 18 susceptible animals showed simultaneously the BoLA-DRB3 "EIAY" sequence and BoLA-A8 specificity. Interestingly, a serine residue at position 30 of the ARS was found in 8 of the susceptible animals and was completely absent from all resistant animals. Furthermore, in a same animal, the serine at position 30 and the EIAY sequence were never found simultaneously on the same haplotype.

These results show a strong correlation between the resistant character to dermatophilosis and the association of MHC haplotypes: the BoLA-A8 specificity and the BoLA-DRB3 "EIAY" sequence at ARS positions 66/67/74/78 with the lack of serine in position 30.

To confirm these results, family segregation studies are in progress and some interesting observations have been obtained.

ACKNOWLEDGMENTS

Herd breeders of Martinique, especially Y. Chene and E. Lecoeur, who have participated in this study are thanked for making their records accessible. Finally, the departmental Veterinary Service of Martinique and more particularly C. Palin are thanked for help in sampling cattle.

REFERENCES

1. VAN SACEGHEM, R. 1915. Dermatose contagieuse (impetigo contagieux). Bull. Soc. Pathol. Exo. **8:** 354–359.
2. BARRE, N., G. MATHERON, B. ROBEZ, F. ROGER, D. MARTINEZ & C. SHEIKBOUDOU. 1988. La dermatophilose des bovins dans les Antilles françaises. II. Facteurs de receptivité liés aux animaux. Rev. Elev. Med. Vet. Pays Trop. **41**(4): 339–345.
3. MARTINEZ, D., G. AUMONT, M. MOUTOUSSAMY, D. GABRIEL, J. C. TATAREAU, N. BARRE, F. VALLEE & B. MARI. 1993. Epidemiological studies on dermatophilosis in the Caribbean. Rev. Elev. Med. Vet. Pays Trop. **46**(1–2): 323–327.
4. MARTINEZ, D., N. BARRE, B. MARI, T. VIDALENC. 1992. Studies of the role of *Amblyomma variegatum* in the transmission of *Dermatophilus congolensis*. *In* Tick Vector Biology. B. Firaz, T. Petney & I. Horak, Eds.: 87–99. Springer-Verlag. Berlin, Germany.
5. LEROY, P. & P. MARCHOT. 1987. The resistance to dermatophilosis of Dinka cattle breed, Dinka crossbred and Boran, Friesian, Jersey, Sahiwal crossbreds. Ann. Rech. Vet. **18:** 107–109.
6. DUMAS, R., P. LHOSTE, N. CHABEUF & J. BLANCOU. 1971. Note sur la sensibilité héréditaire des bovins à la streptothricose. Rev. Elev. Med. Vet. Pays Trop. **24:** 349–353.
7. LEWIN, H. A. 1989. Disease resistance and immune response genes in cattle: strategies for their detection and evidence of their existence. J. Dairy Sci. **72:** 1334–1348.
8. OSTERGARD, H., B. KRISTENSEN & S. ANDERSEN. Investigations in farm animals of association between the MHC system and disease resistance and fertility. Livestock Prod. Sci. **22:** 49–67.

9. KLEIN, J. 1986. Natural History of the Major Histocompatibility Complex. John Wiley & Sons, Inc. New York, NY.

10. LUNDEN, A., S. SIGURDARDOTTIR, I. EDFORS-LILJA, B. DANELL, J. RENDEL & L. ANDERSSON. 1990. The relationship between bovine major histocompatibility complex class II polymorphism and diseases studied by use of bull breeding values. Anim. Genet. **21:** 221–232.

11. ROTHBARD, J. B. & M. L. GEFTER. 1991. Interactions between immunogenic peptides and MHC proteins. Annu. Rev. Immunol. **9:** 527–565.

12. MARTINEZ, D., B. MARI, G. AUMONT & T. VIDALENC. 1993. Development of a single dilution ELISA to detect antibody to *Dermatophilus congolensis* in goat and cattle sera. Vet. Microbiol. **34:** 47–62.

13. Different Joint Reports of the International bovine lymphocyte antigens (BoLA) workshops, held in 1979, 1982, 1986, 1990, 1992, 1994: Analysis of alloantisera against bovine lymphocytes. Special Issues of Anim. Blood Grps Bioch. Genet. and Anim. Genetics.

14. GROENEN, M. A. M., J. J. VAN DER POEL, R. J. M. DIJKHOF & M. J. GIPHART. 1990. The nucleotide sequence of bovine MHC class II DQB and DRB genes. Immunogenetics **31:** 37–44.

15. SIGURDARDOTTIR, S., C. BORSCH, K. GUSTAVSSON & L. ANDERSSON. 1991. Cloning and sequence analysis of 14 DRB alleles of the bovine major histocompatibility complex by using the polymerase chain reaction. Anim. Genet. **22:** 199–209.

16. SIGURDARDOTTIR, S., A. LUNDEN & L. ANDERSSON. 1988. Restriction fragment length polymorphism of DQ and DR class II genes of the bovine major histocompatibility complex. Anim. Genet. **19:** 133–150.

17. MAEDA, M., N. URYU, N. MURAYAMA, H. ISHII, M. OTA, K. TSJUJI & H. INOKO. 1990. A simple and rapid method for HLA-DP genotyping by digestion of PCR-amplified DNA with allele specific endonucleases. Hum. Immun. **27:** 111–121.

18. VAN EIJK, M. J. T., J. A. STEWART-HAYNES & H. A. LEWIN. 1992. Extensive polymorphism of the BoLA-DRB3 gene distinguished by PCR-RFLP. Anim. Genet. **23**(6): 483–496.

19. MAILLARD, J. C., C. PALIN, I. TRAP & A. BENSAID. 1993. An attempt to identify genetic markers of resistance or susceptibility to dermatophilosis in zebu Brahman population of Martinique. Revr. Elev. Med. Vet. Pays Trop. **46**(1–2): 291–295.

20. XU, A., M. J. T. VAN EIJK, C. PARK & H. A. LEWIN. 1993. Polymorphism in BoLA-DRB3 exon 2 correlates with resistance to persistent lymphocytosis caused by bovine leukemia virus. J. Immunol. **151**(12): 6977–6985.

Malignant Catarrhal Fever Virus

Characterization of a United States Isolate and Development of Diagnostic Assays

HONG LI, DAVID T. SHEN, DONAL O'TOOLE,[a]
WILLIAM C. DAVIS,[b] DONALD P. KNOWLES,
JOHN R. GORHAM, AND
TIMOTHY B. CRAWFORD [b,c]

Animal Diseases Research Unit
USDA Agricultural Research Service
Pullman, Washington 99164

[a]*University of Wyoming*
State Veterinary Laboratory
Laramie, Wyoming 82070

[b]*Department of Veterinary Microbiology and Pathology*
Washington State University
Pullman, Washington 99164

INTRODUCTION

Malignant catarrhal fever (MCF) is a dramatic, often lethal, infection of many species of bovidae and cervidae, caused by two or more distinct but closely related members of the subfamily *Gammaherpesvirinae*.[1-4] Based on the reservoir ruminant species from which the causative virus is transmitted, two major epidemiological entities have been described: wildebeest-associated (WA) and sheep-associated (SA) MCF. There are no marked differences in clinicopathologic features between the two.[3] The wildebeest agent is a relatively well-characterized lymphotrophic gamma-herpesvirus, designated as *Alcelaphine herpesvirus*-1 (AHV-1).[5] This virus is indigenous and nonpathogenic in the wildebeest (subfamily Alcelaphinae). Although the sheep-associated agent has not been isolated, there is considerable circumstantial evidence that domestic sheep are carriers of a virus closely related to AHV-1 and transmit it to susceptible ruminants such as domestic cattle, deer, and bison.[3,6-8] This is strengthened by the recent demonstration of SA-MCF viral DNA sequences with homology to Epstein-Barr virus (EBV) and *Herpesvirus saimiri* in leukocytes from normal domestic sheep and SA-MCF-affected cattle.[9]

The diagnosis and control of SA-MCF have traditionally been problematic, due to the lack of reliable serologic tests, the inability to cultivate the sheep-associated agent *in vitro,* and the lack of DNA probes for the viral genome. In this communication,

[c] Author to whom correspondence should be addressed.

we describe the *in vitro* growth characteristics of a US isolate of MCF virus (MCFV) and review (1) the identification of the major viral proteins of the US isolate using monoclonal antibodies (MABs),[10] (2) development of a competitive inhibition ELISA (CI-ELISA) for MCFV antibody,[8] (3) comparison of the CI-ELISA with a PCR assay for SA-MCFV DNA,[11] based on the previously reported primers,[9] for detection of the SA-MCFV infection in domestic sheep and in cattle and deer with clinical SA-MCF, and (4) the data generated by these assays on the prevalence of SA-MCFV to date.

MATERIALS AND METHODS

Viruses, Sera, and Peripheral Blood Lymphocyte (PBL) Samples

The Minnesota isolate of MCFV, the WC-11 (cell culture passaged strain of AHV-1), and fetal mouflon sheep kidney (FMSK) cells were kindly provided by Dr. W. Heuschele, Center for Reproduction of Endangered Species, Zoological Society of San Diego, CA. Bovine herpesvirus 1 (BHV-1) (LA strain, ATCC VR-188), bovine herpesvirus 4 (BHV-4) (DN-599, ATCC VR-631), and bovine turbinate (BT) cells (ATCC CRL-1390) were obtained from the American Type Culture Collection, Rockville, MD. Bovine viral diarrhea virus (BVDV), parainfluenza virus type 3 (PI-3), ovine respiratory syncytial virus (ORSV), bovine respiratory syncytial virus (BRSV), ovine adenoviruses types 5 and 6, ovine herpesvirus 1 (OHV-1), and caprine herpesvirus 1 were kindly supplied by Dr. J. Evermann, Washington Animal Disease Diagnostic Laboratory, Pullman, WA. Sera and PBLs were collected from sheep in 8 states. Sera and PBL were collected from field cases of MCF in cattle and deer from Idaho, Montana, Washington, and Wyoming. The diagnoses of MCF for these cases were based on clinical signs and confirmed by histopathology using established diagnostic criteria, such as the presence of disseminated necrotizing lymphoid vasculitis, lymphadenopathy, ulcerative enteritis, and panophthalmitis.[3] Lymphocytes were purified by isodensity centrifugation using Accu-Paque Lymphocytes (Accurate Chemical & Scientific Corp., Westbury, NY) and stored in Tris-EDTA buffer at $-20°C$. DNA was purified from PBL by proteinase K digestion and phenol/chloroform extraction,[12,13] dissolved in water, and quantified by OD_{260}.

One-Step Growth Curve and Kinetics of Appearance of Viral Antigens in Cell Culture

Viral infectivity was measured by a plaque assay in which dilutions of virus were inoculated onto FMSK monolayers in 6-well plates and overlayed with high glucose Dulbecco's Modified Eagle Medium (DMEM) containing 2% methylcellulose. After 5 days, plaques were stained with crystal violet solution and counted microscopically. To obtain the one-step growth curve, FMSK cell monolayers grown in 12-well plates were inoculated with the stock virus of Minnesota isolate of MCFV at a multiplicity of infection (MOI) of 0.5. After 1 h at 37°C, the monolayers were washed 3× with DMEM, and 2 ml of DMEM, 10% fetal bovine serum was added to each well. Samples were collected at intervals and stored at $-80°C$. The relative proportion of cell-associated and cell-free virus infectivity was examined using infected cultures in 25-cm² flasks. Cultures were harvested in duplicate at intervals. Cell-free virus titers were determined by titrating the supernatant following centrifugation at 1500 × g for 15 min. The cell-associated fraction was obtained by washing cell pellets 3×, resuspending in 5 ml of DMEM, and liberating virus by a single freeze-thaw cycle followed by vigorous agitation.

The kinetics of appearance of viral antigens in FMSK cells following infection with MN-MCFV were examined by sequential indirect immunofluorescence (IIF) and radioimmunoprecipitation. For IIF, MN-MCFV-infected cells were harvested sequentially by trypsinization, washed, spun onto glass slides, fixed with acetone, and reacted with rabbit anti-AHV-1 serum (Lee BioMolecular Research Lab Inc., San Diego, CA), followed by fluoresceinated anti-rabbit IgG. Radioimmunoprecipitation was performed as previously described.[10,14] Briefly, monolayers of FMSK cells were infected with MN-MCFV at an MOI of 3.0. in the presence of 40 μCi/ml of [^{35}S]methionine (New England Nuclear), and lysates harvested at 8, 12, 16, 20, and 72 h postinfection (PI). Labeled lysates were incubated with antiserum or monoclonal antibodies defined earlier.[10] Antigen-antibody complexes were precipitated with recombinant protein G beads (GENEX Co., Gaithersburg, MD), washed, solubilized, and subjected to SDS-PAGE. To identify a suitable time during the replication cycle to harvest mRNA preparatory to cDNA cloning efforts, preliminary experiments were conducted to confirm the presence of mRNA encoding the major viral proteins, using *in vitro* translation of mRNA as described previously for other organisms.[15] Total RNA was extracted from MCFV-infected cells 16 h after infection at MOI of 3.0. Purified mRNA was translated *in vitro* using 2 μg of polyadenylated mRNA per reaction and a rabbit reticulocyte lysate (Promega, Madison, WI). Translation products (10^6 CPM) were immunoprecipitated with antiserum and analyzed by SDS-PAGE.

Production of Monoclonal Antibodies and Identification of Major Viral Proteins

Hybridomas were produced by previously described methods,[10,16] screened for reactivity to MCFV antigens in ELISA and IIF, and further characterized by immunoprecipitation and immunoblot.[14,17] Viral neutralizations were performed to determine the neutralizing epitope(s). Briefly, volumes of medium containing 200 plaque-forming units of MN-MCFV were mixed with 100 μl of each MAb (20 μg/ml), incubated at 37°C for 1 h, and surviving virus titrated as above.

CI-ELISA

Monoclonal antibodies with high ELISA activity were evaluated for their ability to compete for their epitope with antibodies from sheep seropositive to MCFV as defined by IIF and radioimmunoprecipitation. CI-ELISA was performed using a modification of a described procedure.[15,18] Ninety-six-well polystyrene plates were coated with MN-MCFV antigens and control antigens, blocked with milk solution, and test sera and a MAb added and incubated for 1 h. Plates were then washed and developed as described in detail elsewhere.[11] Sera were run in triplicate. Negative control sera were defined by an absence of MCFV antibody detectable by IIF and immunoprecipitation. When the mean of the three replicate OD readings of a test serum was more than three standard deviations (SD) below the mean of a panel of six negative control sera (run in triplicate), the serum was considered positive for MCFV antibody.

Polymerase Chain Reaction (PCR)

A two-step PCR amplification was employed, using previously described primer sets derived from a lymphoblastoid cell line that originated from a cow with acute MCF:[9] no. 556 (5′-AGTCTGGGGTATATGAATCCAGATGGCTCTC-3′) and no. 775 (5′-AAGATAAGCACCAGTTATGCATCTGATAAA-3′) in the first step and no. 556 and no. 555 (5′-TTCTGGGGTAGTGGCGAGCGAAGGCTTC-3′) in the second step.

All amplification reactions were performed in a 100 ·μl volume containing 10 mM Tris-HCl (pH 8.0), 2 mM MgCl$_2$, 200 μM dATP, dCTP, dGTP, dTTP (Boehringer Mannheim Co., Indianapolis, IN), 20 pM of each primer, and 2 units *Taq* DNA polymerase (Boehringer Mannheim Co.). Thermal cycling conditions for both steps were 5 min at 95°C, followed by 34 cycles of 94°C (1 min), 60°C (1 min), and 72°C (2 min), followed by a final 7 min at 72°C for extension. Ten microliters of the first-step PCR reaction mixture was used as target for the second-step amplification. For product visualization, 10 μl of the amplified PCR products were analyzed by 2% agarose gel electrophoresis and ethidium bromide staining.

RESULTS

One-Step growth Curve and Viral Antigen Synthesis

As shown in FIGURE 1, the eclipse phase of the viral replication cycle was approximately 23 h. Maximum infectivity titers were attained at 96 to 120 h PI (FIG. 1). The earliest cytopathic effect was first observed at approximately 48 h (data not shown). The cell-associated fraction exceeded 75% of the total viral yield (data not shown). The first viral antigens were detected by IIF at 4 h PI. Prior to 16 h PI, antigens were apparent in the cytoplasm, but by 24 h PI, antigens were abundant in both cytoplasm and nuclei (data not shown). To further define the time of major viral protein synthesis, radiolabeled viral antigens were immunoprecipitated with polyclonal antiserum to MCFV or monoclonal antibodies. At 16 h PI, most major viral proteins[10] were detectable by radioimmunoprecipitation by either antiserum (FIG. 2A) or MAbs (data not shown). As shown in FIGURE 2B, most of the major viral proteins were detected by antiserum in the *in vitro* translation products (FIG. 2B, lane 3). A nonglycosylated 17-kDa protein was also prominent (FIG. 2B, lane 1), suggesting that abundant mRNA for this protein was present at 16 h PI. However, several larger viral proteins were not detected in the *in vitro* translation products by MAbs or antiserum, suggesting that many of the epitopes on the larger proteins are glycosylated or conformational, which would not be expressed by *in vitro* translation.

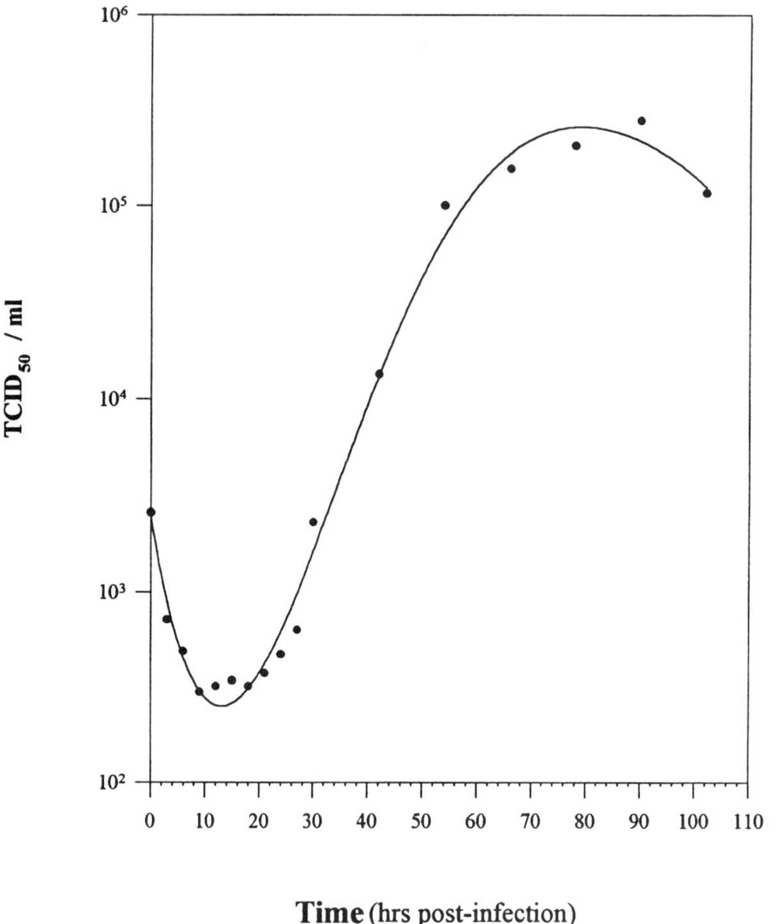

Time (hrs post-infection)

FIGURE 1. One-step growth curve of MN-MCFV. The value for each point was the average of duplicate samples.

Identification of Major Proteins by MAbs

The predominant proteins of MCFV were clustered into six major groups (TABLE 1), based on their MAb reactivity patterns. Group I was comprised of a 110-kDa protein that was labeled by both [^{35}S]methionine and [^{3}H]glucosamine. Group II was a 130-kDa protein that labeled with [^{35}S]methionine, but not with [^{3}H]glucosamine. Neither group I nor group II proteins were detected by any of the MAbs in immunoblot, indicating that the epitopes defined by the MAbs in the library were unstable to denaturation conditions. In group III, a protein complex containing 5 bands of 115/110/105/78/45 kDa was immunoprecipitated by the MAbs and was labeled with both

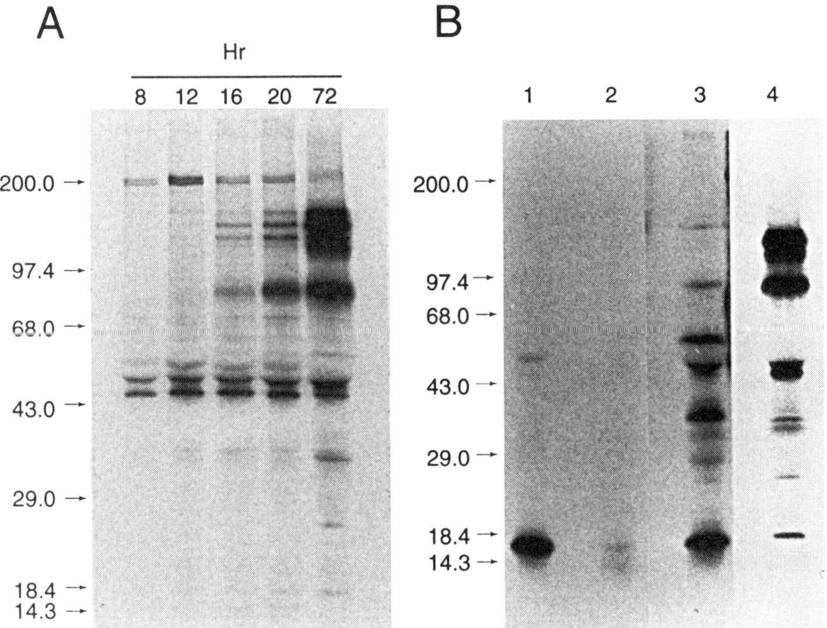

FIGURE 2. **A:** Immunoprecipitation of [³⁵S]methionine-labeled proteins of MN-MCFV infected cell lysate harvested at 8, 12, 16, 20, and 72 h postinfection with rabbit antiserum against the WC-11 strain of AHV-1. **B:** Immunoprecipitation of [³⁵S]methionine-labeled proteins translated *in vitro* from polyA mRNA purified from MN-MCFV infected cells at 16 h postinfection (lanes 1 to 3) with a group VI MAb (lane 1); a group II MAb (lane 2); and antiserum to AHV-1 (WC-11) (lane 3). Immunoprecipitation of [³⁵S]methionine-labeled proteins of MN-MCFV infected cell lysate harvested at 72 h postinfection with antiserum to AHV-1 (WC-11) (lane 4).

[³⁵S]methionine and [³H]glucosamine. Groups IV and V were comprised of a 36/34-kDa protein dimer and a 24-kDa protein, respectively, and these proteins were only recognized by their MAbs in immunoblot, but not in immunoprecipitation. Group VI comprised a nonglycosylated 17-kDa protein, the epitope of which survived in immunoblotting conditions. As also shown in TABLE 1, proteins in group I carried strong neutralizing determinants (over 80% plaque reduction by their MAbs). Of the three MAbs against group III proteins tested, one neutralized significantly, but at a lower level (53% reduction). The MAbs against the other viral proteins did not significantly neutralize infectivity. Comparison of immunoprecipitation patterns between MCFV isolates showed that the only difference observed in immunoprecipitation patterns between MN-MCFV and AHV-1 (WC-11) was that the single 110-kDa protein of AHV-1 (WC-11) was about 1 kDa smaller than its counterpart in MN-MCFV (data not shown). No differences were observed between the immunoprecipitation patterns of MN-MCFV and the putative sheep associated virus reported from Austria (Au-732).

TABLE 1. Characteristics of Major Proteins of Malignant Catarrhal Fever Virus

Group	GP[a]	Epitope[b] Glycosylation Dependent	Conformation Dependent	Protein Size (kDa)	Neutralizing Activity[c]
I	+	+	+	110	+++
II	–	–	+	130	–
III	+	+	–	115/110/105/78/45	+
IV	–	–	–	36/34	–
V	–	–	–	24	–
VI	–	–	–	17	–

[a] The protein(s) recognized by MAbs were glycosylated.

[b] Determined by available MAbs.

[c] Determined by available MAbs using neutralization test: +++ = over 80% of plaque reduction; + = variable reduction (0 to 53%); – = not significant (0 to 35%) reduction. The significance was based on upper limit to the 99% confidence interval calculated from control wells ($n = 7$), which was 42% plaque reduction.

Detection of MCFV Antibody in Sheep and Other Ruminants by CI-ELISA

A MAb against a conserved epitope was identified by a competition assay with sera from sheep that tested positive to MCFV antibody in IIF and immunoprecipitation. The MAb (15-A) bound an epitope located on the group III glycoprotein complex (FIG. 3). MAb 15-A was inhibited in CI-ELISA by rabbit antisera against MN-MCFV and the WC-11 strain of AHV-1, by serum from sheep inoculated with a US isolate of AHV-1 and by serum from a calf infected with Au-732 virus.[19] When tested by immunofluorescence and CI-ELISA, MAb 15-A did not react with any of 8 common sheep/goat or 5 common bovine viruses, including ovine herpesvirus 1, caprine herpesvirus 1, ovine progressive pneumonia virus, caprine arthritis encephalitis virus, ovine adenoviruses 5 and 6, ovine respiratory syncytial virus, ovine parainfluenza virus 3, bovine herpesviruses 1, 2, and 4, bovine viral diarrhea virus, and bovine respiratory syncytial virus. Of 149 serum samples collected from sheep associated with MCF outbreaks in 7 states, 88 were positive by CI-ELISA and 106 of the 149 were positive by IIF (TABLE 2). All sera positive by CI-ELISA were also positive by immunofluorescence. The converse, however, was not true: some immunofluorescence-positive sera were negative by CI-ELISA.

Comparison of CI-ELISA with PCR for Detection of MCFV Infection in Sheep and Other Ruminants

PCR amplified a strong band comprised of a 238 bp fragment PBL DNA of sheep, cattle, deer, bison, and other ruminants with clinical SA-MCF, using primers previously reported by Baxter *et al.*[9] The base sequence of the 238 bp DNA fragment

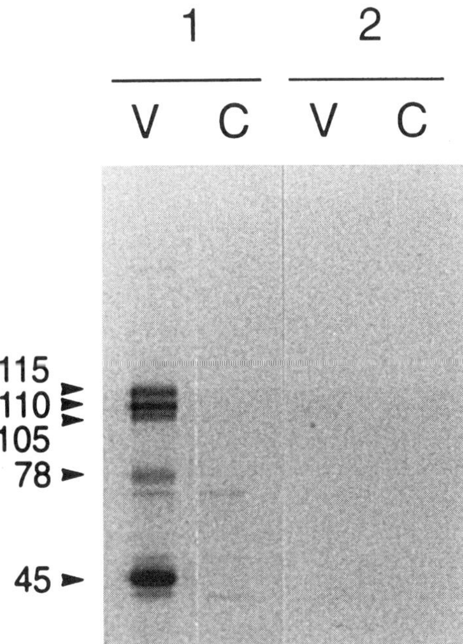

FIGURE 3. Immunoprecipitation of [^{35}S]methionine-labeled proteins of MN-MCFV infected cell lysate (V) or uninfected cell lysate (C) with MAb 15-A (lane 1) and an isotype control MAb (lane 2).

was identical to the sequence of the DNA cloned from a lymphoblastoid cell line derived from a cow with clinical SA-MCF.[9] Using ethidium bromide staining, the PCR detected 10 fg of DNA from the pUC18 construct containing the 238 bp fragment. The sensitivity of detection was enhanced approximately 25-fold by the use of Southern hybridization with the ^{33}P-labeled DNA probe rather than ethidium bromide staining (data not shown). Sera and PBL DNA from 144 adult sheep were examined by both the CI-ELISA (serum) and the PCR (PBL DNA). One hundred

TABLE 2. Comparison of CI-ELISA with HF and PCR for Detection of MCFV Infection in Sheep from Diverse Locations

Source of Samples[a]	No. of Tested	No. of Positive by		
		CI-ELISA	HF	PCR
Adult sheep and lambs	149	88	106	ND[b]
Adult sheep	144	136	ND	143

[a] Collected from the states of California, Georgia, Idaho, Kentucky, North Dakota, Montana, Tennessee, Washington, and Wyoming.

[b] Not done.

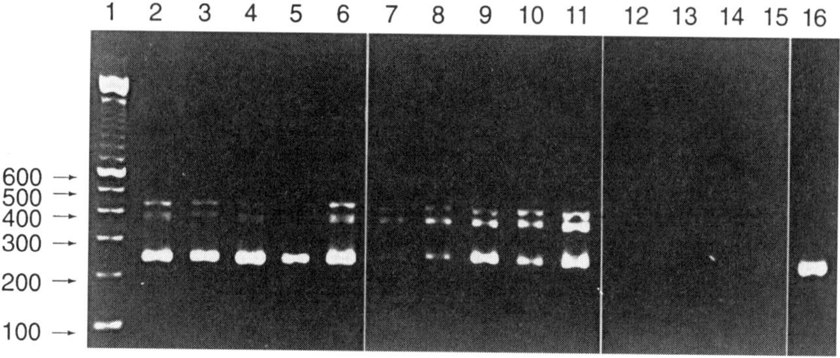

thirty-six (94%) were positive by the CI-ELISA and 143 (99%) were strongly positive by PCR (TABLE 2 and FIG. 4). Sera and PBL DNA samples were collected from seven cattle and two deer that were acutely ill with MCF. CI-ELISA detected antibody in the sera from 7 of 9 cattle and deer samples, and PCR detected viral DNA sequences in PBL of all 9 (TABLE 3 and FIG. 4). The PCR did not amplify the 238 bp band from DNA purified from the Minnesota or Austria isolates of MCFV, from the FMSK cell line, or PBL DNA from normal cattle.

TABLE 3. Comparison of CI-ELISA and PCR for Detection of MCFV Infection in Cattle and Deer with Clinical SA-MCF

Species[a]	Location	CI-ELISA	PCR
Cattle	Idaho	+	+
Cattle	Montana	+	+
Cattle	Montana	+	+
Cattle	Washington	−	+
Cattle	Wyoming	+	+
Cattle	Wyoming	+	+
Cattle	Wyoming	−	+
Deer	Washington	+	+
Deer	Washington	+	+

[a] All cases were diagnosed as MCF by clinical signs and histopathological lesions, and all the animals had a history of contact with sheep.

DISCUSSION

The *in vitro* virus replication study showed that the maximum titers and replication cycle of MN-MCFV are generally consistent with earlier studies of AHV-1 (WC 11 and C500).[20,21] While the MN-MCFV was isolated from clinical MCF cattle in contact with sheep,[22] no differences in the major protein profiles or DNA restriction fragment patterns (W. Heuschele, personal communication) were found between the MN-MCFV and the WC-11 strain of AHV-1. Moreover, a PCR using SA-MCFV primers did not amplify bands from the Minnesota or Austria isolates of MCFV. These findings are strong evidence that these two isolates (MN-MCFV and Au-732) do not represent the ovine herpesvirus 2 (OHV-2),[23] the common SA-MCFV. On other hand, these two isolates were readily detected by primers based on WC-11 DNA sequences,[24] suggesting that they probably belong to the AHV-1 group.

The 11 major viral proteins clustered into six major antigenic groups as defined by a panel of MAbs. The protein complex of 115/110/105/78/45 kDa identified in the present study is consistent with a previous report on MCFV.[25] Five additional unreported viral proteins with molecular masses of 130, 110, 36/34, 26, and 17 kDa were identified with MAbs. Although neutralizing epitopes on the protein complex were previously reported,[25] we found that the predominant neutralizing epitopes as defined by our MAb library was located on a single 110-kDa protein that was not identified in the earlier study.

Polyclonal antisera from several ruminant species infected with MCFV strongly immunoprecipitated a 130-kDa protein band. This protein in nonglycosylated, and no neutralizing epitopes were identified on the protein. It may represent a major structural component of the virion such as a capsid or tegument protein. Since this 130-kDa protein was strongly precipitated by a wide range of anti-MCFV sera and was present in all MCFV isolates tested, it should be considered as a potential candidate for diagnostic tests.

An antigenic epitope conserved among all known isolates of MCFV has been identified to develop a specific assay for identifying infected sheep and other infected ruminants. MAb 15-A defined a conserved carbohydrate-dependent epitope that was present on all isolates of MCFV tested but was not present on 13 common sheep and cattle viruses. Based on this MAb, a CI-ELISA was developed. In comparison of testing 149 sheep sera by CI-ELISA and IIF, the agreement between the two assays was 83%. No serum that was negative by IIF was positive by CI-ELISA. It has been documented that MCFV shares some antigens with other ruminant herpesviruses such as bovine herpesvirus 1, 2, 4, and caprine herpesvirus.[26,27] Cross-reactivity between MCFV and viruses such as these apparently is responsible for the greater number of sheep scoring positive in IIF than in CI-ELISA.

The PCR assay specifically amplified a 238 bp DNA fragment from PBLs of either inapparently infected adult sheep or cattle and deer with clinical SA-MCF. It would appear that the SA-MCF agent in North America is closely related at a base sequence level to the SA-MCF agents in Scotland and in Indonesia.[9,28] Parallel testing of 144 random samples from normal adult sheep revealed over 95% agreement between CI-ELISA and PCR. The relatively lower percentage (94%) of detection by CI-ELISA may be due to lower antibody production by some infected sheep. Of seven cattle and two deer with confirmed acute MCF, all nine were PCR-positive, but only seven of the nine were seropositive by CI-ELISA. This probably reflects a lack of sufficient time to mount an immune response in some rapidly fatal cases. Thus, the PCR appears to be more appropriate and reliable for recognizing acute clinical MCF than antibody detection by CI-ELISA. These latter studies thus have confirmed that (1) the two assays are useful for detecting sheep-associated MCFV infection; (2) the PCR assay is specific for SA-MCFV DNA; (3) SA-MCFV DNA consistently can be detected in peripheral blood mononuclear cells of cattle and deer during acute illness; and (4) virtually all domestic adult US sheep are carriers of the virus.

SUMMARY

Malignant catarrhal fever (MCF) is a severe lymphoproliferative disease of certain domestic and wild ruminants. Two distinct but closely related viruses cause clinically indistinguishable syndromes in susceptible ruminant species: wildebeest-associated MCF virus (WA-MCFV) and sheep-associated MCF virus (SA-MCFV). Neither the pathogenesis nor the epidemiology of SA-MCF is understood, primarily because of a lack of adequate detection methods for the etiologic agent or antibody against that agent. Work designed to develop these tests has been under way in our laboratory. To obtain basic information about the virus, the *in vitro* growth properties of a US isolate of MCF virus were studied and its major viral proteins identified and characterized by a panel of monoclonal antibodies generated against the isolate. A monoclonal antibody to a broadly conserved epitope of MCF virus was identified, and a competitive-inhibition ELISA (CI-ELISA) was developed for detection of anti-MCF antibody in sheep and other ruminants. The monoclonal antibody (15-A) reacted with an epitope located on a glycoprotein complex, which was present in all isolates of MCF virus examined. Antibody from a wide variety of ruminants infected with MCF virus of both sheep and wildebeest origin competed with the monoclonal antibody 15-A for the epitope, which was not present on 14 other common ruminant viruses. The assay detected antibody in inapparently infected sheep, and in cattle, deer, and bison with clinical MCF. A PCR assay for DNA of the sheep-associated virus was developed, based on previously reported primers. Comparative studies demonstrated that the CI-ELISA was specific for MCFV antibody and that the PCR was more reliable for diagnosis of clinical MCF.

ACKNOWLEDGMENTS

We thank Dongyue Zhuang, Dr. Leilani Wilson, Soonhee Kwon, and Lori Fuller for technical assistance and Teresa Harkins, Technology Transfer Laboratory, Washington State University, for synthesis of primers. The diligent efforts of Dr. Garry Snowder at USDA Sheep Experimental Station in Dubois, Idaho, Cody Molle at University of Wyoming, and Steve Maki at University of Idaho are greatly appreciated. The authors also acknowledge Drs. Katherine O'Rourke, Douglas Jasmer, and Travis McGuire for valuable discussions and suggestions.

REFERENCES

1. BRIDGEN, A., A. J. HERRING, N. F. INGLIS & H. W. REID. 1989. Preliminary characterization of the alcelaphine herpesvirus-1 genome. J. Virol. **70:** 1141-1150.
2. HEUSCHELE, W. P. 1988. Malignant catarrhal fever: a review of a serious disease hazard for exotic and domestic ruminants. Zool. Garten N. F. **58:** 123-133.
3. PLOWRIGHT, W. 1990. Malignant catarrhal fever virus. *In* Virus Infections of Ruminants. Z. Dinter & B. Morein, Eds.: 123-150. Elsevier Science Publishers B.V. New York, NY.
4. ROSSITER, P. B. 1985. Immunology and immunopathology of malignant catarrhal fever. Prog. Vet. Microbiol. Immun. **1:** 121-144.
5. REID, H. W., D. BUXTON, E. BERRIE, I. POW & J. FINLAYSON. 1984. Malignant catarrhal fever. Vet. Rec. **114:** 582-584.
6. PIERSON, R. E., D. THAKE, A. E. McCHESNEY & J. STORZ. 1973. An epizootic of malignant catarrhal fever in feed-lot cattle. J. Am. Vet. Med. Assoc. **163:** 349-350.
7. REID, H. W., D. BUXTON, W. CORRIGAL, A. R. HUNTER, D. A. McMARTIN & R. RUSHTON. 1979. An outbreak of malignant catarrhal fever in red deer (*Cervas elaphus*). Vet. Rec. **104:** 120-123.
8. LI, H., D. T. SHEN, D. P. KNOWLES, J. R. GORHAM & T. B. CRAWFORD. 1994. Competitive-inhibition enzyme-linked immunosorbent assay for antibody in sheep and other ruminants to a conserved epitope of malignant catarrhal fever virus. J. Clin. Microbiol. **32:** 1674-1679.
9. BAXTER, S. I. F., I. POW, A. BRIDGEN & H. W. REID. 1993. PCR detection of the sheep-associated agent of malignant catarrhal fever. Arch. Virol. **132:** 145-159.
10. LI, H., D. T. SHEN, W. C. DAVIS, D. P. KNOWLES, J. R. GORHAM & T. B. CRAWFORD. 1995. Identification and characterization of the major proteins of malignant catarrhal fever virus. J. Gen. Virol. **76:** 123-129.
11. LI, H., D. T. SHEN, D. O'TOOLE, D. P. KNOWLES, J. R. GORHAM & T. B. CRAWFORD. Investigation of sheep-associated malignant catarrhal fever virus infection in ruminants by PCR and competitive inhibition enzyme-linked immunosorbent assay. J. Clin. Microbiol. **33:** 2048-2053.
12. SEAL, B. S., R. B. KLIEFORTH, W. H. WELCH & W. P. HEUSCHELE. 1989. Alcelaphine herpesvirus 1 and 2 SDS-PAGE analysis of virion polypeptides, restriction endonuclease analysis of genomic dna and virus replication restriction in different cell types. Arch. Virol. **106:** 301-320.
13. MANIATIS, T., S. G. FISCHER, M. W. KIRSCHNER & U. K. LAEMMLI. 1982. Molecular Cloning, a Laboratory Manual. Cold Spring Harbor Lab. Cold Spring Harbor, NY.
14. LI, H., D. T. SHEN, D. BURGER, W. C. DAVIS & J. R. GORHAM. 1991. Analysis of bovine herpesvirus 4 (DN 599) major antigens with monoclonal antibodies and polyclonal immune serum. Arch. Virol. **119:** 225-238.
15. KNOWLES, D. P., JR., L. E. PERRYMAN, W. L. GOFF, C. D. MILLER, R. D. HARRINGTON & J. R. GORHAM. 1991. A monoclonal antibody defines a geographically conserved surface protein epitope of Babesa equi merozoites. Infect Immun. **59:** 2412-2417.
16. DAVIS, W. C. 1988. Enhancement of myeloma-B-cell hybridoma outgrowth in primary cultures with B cell mitogens. Period. Biol. **90:** 367-374.
17. SHEN, D. T., D. BURGER & J. R. GORHAM. 1991. Characterization of monoclonal antibodies to bovine herpesvirus type 1, Los Angeles strain. V. Micro. **28:** 25-37.
18. ANDERSON, J. 1984. Use of monoclonal antibody in a blocking ELISA to detect group specific antibodies to bluetongue virus. J. Immunol. Methods **74:** 139-149.
19. SCHULLER, W., S. CERNY-REITERER & R. SILBER. 1990. Evidence that the sheep associated form of malignant catarrhal fever is caused by a herpes virus. J. Vet. Med. **37:** 442-447.
20. MUSHI, E. Z. 1983. In Vitro growth properties of the herpesvirus of bovine malignant catarrhal fever. Bull. Anim. Hlth. Prod. Afr. **31:** 295-299.
21. HARKNESS, J. W. & D. M. JESSETT. 1981. Influence of temperature on the growth in cell culture of malignant catarrhal fever virus. Res. Vet. Sci. **31:** 164-168.

22. HAMDY, F. M., A. H. DARDIRI, C. MEBUS, R. E. PIERSON & D. JOHNSON. 1978. Aetiology of malignant catarrhal fever outbreak in Minnesota. Proc. 82nd Annu. Meet. U.S. Anim. Health Assoc.: 248-267.

23. ROIZMAN, B. 1992. The family herpesviridae: an update. Arch. Virol. **123:** 425-449.

24. LAHIJANI, R. S., S. M. SUTTON, R. B. KLIEFORTH, M. F. MURPHY & W. P. HEUSCHELE. 1994. Application of polymerase chain reaction to detect animals latently infected with agents of malignant catarrhal fever. J. Vet. Diagn. Invest. **6:** 403-409.

25. ADAMS, S. W. & L. M. HUTT-FLETCHER. 1990. Characterization of envelope proteins of alcelaphine herpesvirus 1. J. Virol. **64:** 3382-3390.

26. HEUSCHELE, W. P. 1982. Malignant catarrhal fever in wild ruminants—a review. Proc. 86th Annu. Meet. U.S. Anim. Health Assoc.: 552-570.

27. MULUNEH, A. & H. LIEBERMANN. 1992. Studies concerning etiology of sheep-associated malignant catarrhal fever in Europe. Dtsch. Tierarztl. Wochenschr. **99:** 384-386.

28. WIYONO, A., S. I. F. BAXTER, M. SAEPULLOH, R. DAMAYANTI, P. W. DANIELS & H. W. REID. 1994. PCR detection of ovine herpesvirus-2 DNA in Indonesian ruminants—normal sheep and clinical cases of malignant catarrhal fever. Vet. Microbiol. **42:** 45-52.

Artificial Feeding Systems for Ixodid Ticks as a Tool for Study of Pathogen Transmission

A. S. YOUNG, S. M. WALADDE,[a] AND
S. P. MORZARIA[b]

International Livestock Research Institute (ILRI)
Post Office Box 30709
Nairobi, Kenya

[a]*International Center of Insect Physiology and Ecology (ICIPE)*
Post Office Box 30772
Nairobi, Kenya

INTRODUCTION

Among the blood-sucking arthropods, ixodid ticks have some of the most complicated feeding behavior requirements. Feeding is preceded by an attachment process, and each instar has to remain fully attached to the same feeding site for a period of 2-14 days. These biological demands have for a long time made it relatively difficult to develop viable methods were ixodid ticks could be fed without using their natural hosts. Literature survey shows that some of the *in vitro* feeding methods tried include capillary feeding,[1,2] feeding ticks through various membranes such as slices of bovine skin (dermatomes),[3] rabbit/bovine skins,[4] and silicone membrane.[5] The capillary method had severe limitations in that ticks had to be prefed on the natural host before fixing the diet filled capillaries to the mouthparts. Furthermore, the volumes of blood taken by the tick were very small because ticks had to feed in an unnatural manner. The use of host-skin tissue had certain problems in that the skin could rot easily. In cases where rotting could be prevented, it was observed that host skins were suitable for ticks with long mouthparts like *Amblyomma variegatum* but unsuitable for *Rhipicephalus appendiculatus* with short mouthparts. Advances in our knowledge on tick sensory biology and behavior[6] facilitated the development of artificial membranes and conditions conducive to the voluntary attachment and feeding to repletion of *R. appendiculatus,* a vector of *Theileria parva,* the causative agent of East Coast fever (ECF).[7,8]

Lack of efficient *in vitro* feeding systems for ixodid ticks has all along been a limiting factor to research efforts directed towards tick control and tick-borne disease transmission in particular. *In vitro* feeding systems can be used to study products injected by ticks into their hosts, the effects of substances ingested by ticks, and pathogen infection as well as transmission among ticks.[9] In this paper, we discuss

[b] Author to whom correspondence should be addressed.

211

observations on the *in vitro* feeding of *A. variegatum* and *R. appendiculatus* through host skin and artificial membranes respectively. The importance of the artificial membrane as a convenient tool applicable in tick pathogen transmission studies is also discussed. This is relevant in view of the fact that the artificial membrane can be tailored to suit the feeding requirements of various ixodid ticks.

MATERIALS AND METHODS

Host Skin Membranes

Feeding Apparatus/Responses of A. variegatum

A. variegatum was fed through either rabbit or bovine skin membranes. Pieces of rabbit skin 0.5–1.0 mm thick were obtained by skinning partly shaved slaughtered rabbits. The internal surface of the skin was sterilized with a mixture of gentamycin, penicillin/streptomycin, and fungizon. Bovine skin 0.8-1.5 mm thick was obtained from the lower abdomen of slaughtered calves and sterilized in the same manner as outlined above. Rabbit or bovine skin was used either fresh or after storage at −20°C for up to 3 months.

Making the feeding apparatus or unit began with holding in place with rubber bands a piece of host skin at one end of a 6-cm diameter plastic cylinder. The plastic cylinders used were originally cores of the zinc oxide plaster. A plastic cylinder fitted with a membrane constituted the blood chamber which was then installed into the tick chamber thus forming a feeding apparatus as shown in schematic diagram (FIG. 1). For details on the preparation of the feeding unit, refer to Voigt *et al.*[4] Each apparatus was placed in an incubator (Flow Laboratories, PLC, United Kingdom) with a 5-10% CO_2 atmosphere, temperature of 35-39°C, and relative humidity of 90-95%. Ticks were introduced into the tick chambers as follows: 20 males on day 0, 20 females on day 5, 100 nymphae on day 0, 200 larvae on day 0. Bovine blood was collected and handled under aseptic conditions, and coagulation was prevented by using each of the following methods: heparin at 50 IU/ml of blood, 0.5 M ethylene-diamine tetracetate sodium salt (EDTA) at a dilution of 100 µl/10 ml of blood or by defibrinating the blood with sterile glass beads and glass rods.

Tick Infection and Pathogen Transmission

In experiments designed to investigate tick infection and pathogen transmission, heparinized blood was collected from Boran calf experimentally infected with *Theileria mutans* Intona[10] by inoculation with blood stabilate[11] and from steers experimentally infected with *Cowdria ruminantium* Kiswani transmitted by *A. variegatum*. The cattle infected with *T. mutans* were bled when they had a parasitaemia of over 5% erythrocytes infected. But for the animal infected with *C. ruminantium*, bleeding was done when the animal showed febrile responses. Then the heparinized infected blood so collected was used to feed uninfected *A. variegatum* nymphae on membranes. Two months after moulting to adults, 45 ticks were applied to a heartwater-susceptible animal in an attempt to transmit the parasite.

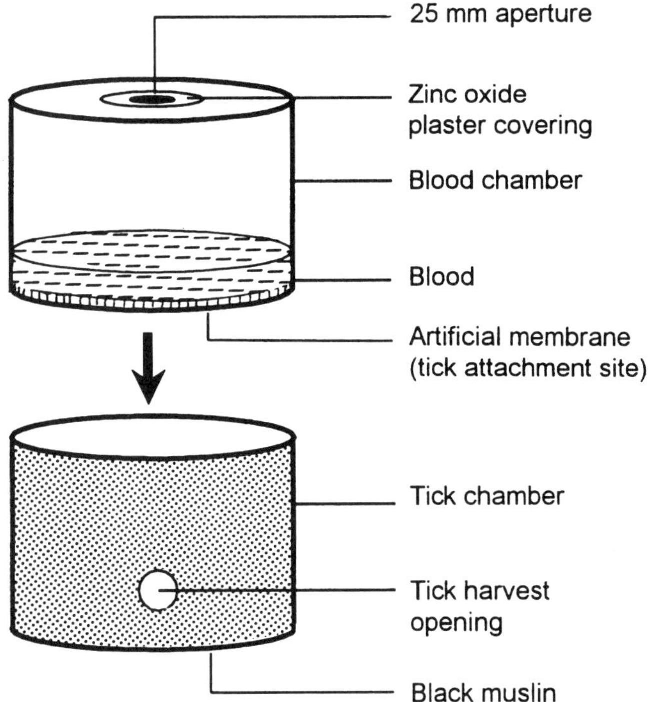

FIGURE 1. Drawing of the feeding apparatus.

Artificial Membrane

Feeding Apparatus/Responses of R. appendiculatus

The model of the feeding apparatus was similar to that used by Voigt *et al.*[4] (FIG. 1). However, in this case the blood chamber was made from the lower half of a 250-ml beaker with its bottom bearing a 25-mm aperture.[12] A modified Baudruch membrane[7,8] with its smooth surface facing inside the glass component was glued over the 7-cm diameter end of the cut beaker. The membrane then formed the base of the blood chamber, while the 25-mm aperture at the opposite end provided access into the chamber. For details of detoxifying the membrane and sterilizing the blood chamber refer to References 4 and 8. The external rough surface of the artificial membrane on the blood chamber was destined for tick attachment. It was made attractive for tick attachment by treating it with materials containing semiochemicals of host and tick origins. The materials bearing the semiochemicals were cattle/ tick washes and tick feces. In this work, emphasis was placed on feeding the *R. appendiculatus* nymphal instar. A cluster of 500–1000 nymphae was applied on the treated membrane surface before uniting the blood chamber onto the tick chamber. At the same time, 2000–3000 nymphae were applied on the shaved back of a Boran

TABLE 1. Infection of *R. appendiculatus* Nymphae with *T. parva* via Infected Cattle (*In Vivo*) and Feeding Apparatus (*In Vitro*) Donated with Blood from the Infected Cattle

Day	*T. parva* Infected Blood Donor Cattle	Feeding Apparatus Donated with Blood	Day
0	Sporozoite stabilate inoculated in healthy susceptible steer	–	–
13	Infested with clean nymphae	Infested with clean nymphae and supplied with clean blood	0
14	Nymphae attaching	Nymphae attaching	1
15	Nymphae attaching	Nymphae attaching	2
16	Few engorged nymphae start dropping	Supplied with infected blood from donor cattle	3
17	Dropping of engorged nymphae increased	Engorged nymphae start dropping	4
18	Dropping of engorged nymphae increased	Majority of engorged nymphae drop	5

steer from which heparinized blood supplied to the feeding apparatus or units was drawn. Blood in the tick feeding units was changed twice daily. The tick feeding units were kept in an incubator[4] maintained at 37°C, 75-80% RH, with an atmosphere of 2-3% CO_2. Nymphae that became replete and dropped off the membrane and the blood donor cattle were grouped according to the day of repletion and then assessed in terms of their engorged weights and moulting success.

Development of T. parva *Infections*

In feeding bioassay designed to examine development of *T. parva* infection in ticks fed *in vitro* versus those fed *in vivo,* uninfected heparinized blood was supplied to the blood chamber from day 0-2 after tick application to the membrane. From day 3 onwards, ticks were fed on *Theileria* piroplasm infected blood from infected blood donor cattle experimentally infected with *T parva* Muguga stabilate.[13] TABLE 1 shows the chronology of events on the infected blood donor cattle and the feeding apparatus with the artificial membrane.

Engorged nymphae from the *in vitro* and *in vivo* feeding systems were collected according to the day they became replete. They were allowed to moult in an incubator at 24°C and 85% RH. Six weeks after repletion, a portion of the resultant adults were prefed on rabbits from 4 days. The salivary glands of 30 males and 30 females were examined to assess their infection with *T. parva*[14] in terms of parasite prevalence and abundance. Prevalence was defined as the percentage of ticks infected as judged by assessment of the salivary gland infection with *T. parva,* and abundance as the mean number of infected salivary gland acini per tick examined.

RESULTS AND DISCUSSION

Feeding Responses of A. variegatum

Rabbit skin membrane of 0.5-0.6 mm thickness was the most suitable attachment/feeding substrate for *A. variegatum* larvae. Nymphae and adults attached and fed well on both cattle and rabbit skin membranes, but on many occasions, adults tended to perforate the rabbit skin thus causing extensive leakages. An important finding in this study was that whole blood treated with heparin as anticoagulant was superior to defibrinate blood in the *in vitro* feeding system. Defibrinated blood took second place, while EDTA was not a suitable anticoagulant. Similar observations were reported[8] on *R. appendiculatus* fed through artificial membranes. The weight achieved by membrane-fed ticks was usually lower than that achieved by ticks fed on cattle. The moulting success and egg-laying of membrane-fed ticks were comparable to those fed on cattle.[4]

Aggregation behavior of nymphal and adult ticks was seen during feeding on the membrane, but the larvae did not aggregate to the same extent. Investigations have already shown that in *Amblyomma* ticks, the attraction/aggregation attachment pheromone (AAAP)[15,16] produced by the dermal glands[17] is responsible for inducing the aggregating behavior.

Infection and Transmission of T. mutans/C. ruminantium

Adult ticks fed as nymphae on *T. mutans* infected blood through skin membranes were applied to a bovine susceptible to *T. mutans.* The ticks fed to repletion, and *T. mutans* infections developed in this animal with a prepatent period to piroplasms of 30 days after application of ticks. Schizonts were detected once, and that was on day 34 after application of the infected ticks, and maximum parasitemia of 20% erythrocytes infected was observed on day 46 after application of the infective ticks. Similarly, adults from nymphae fed on *C. ruminantium* infected blood were applied on another susceptible animal. The ticks fed to repletion, and the animal developed fever with body temperature of more than 39.5°C on day 15 after application of ticks and it died on day 25. Postmortem examination showed typical heartwater characteristics with colonies of *C. ruminantium* in brain smears stained with Giemsa's stain.

Feeding, Moulting Success, T. parva *Infections, and Potential Artificial Membranes*

The performance of membrane-fed nymphae in terms of engorged weights and moulting success was comparable to that of cattle-fed nymphae (TABLE 2). A similar

TABLE 2. Mean Feeding Values for Moulting Success and *Theileria parva* Infection Rates of *Rhipicephalus appendiculatus* Nymphae Fed through an Artificial Membrane Compared to Those Fed on Cattle (Means ± SE)[a]

Days after Tick Application and Parameters Assessed	Membrane-Fed Ticks	Cattle-Fed Ticks
Day 4		
Mean engorged weight (mg)	8.4 ± 0.12	8.5 ± 0.28
Percentage moulting success	93.0 ± 2.1	96.4 ± 1.50
Prevalence and Abundance	72.8 & 25.5	77.6 & 56.7
Day 5		
Mean engorged weight (mg)	8.5 ± 0.23	8.9 ± 0.27
Percentage moulting success	92.7 ± 2.1	97.1 ± 0.98
Prevalence and Abundance	68.6 & 25.2	76.0 & 56.9
Day 6		
Mean engorged weight (mg)	8.0 ± 0.27	8.9 ± 0.30
Percentage moulting success	92.3 ± 3.0	97.1 ± 1.27
Prevalence and Abundance	67.2 & 29.8	77.0 & 73.8

[a] Prevalence = percentage of ticks infected. Abundance = number of acini infected per tick examined.

pattern was observed for the infection parameters, namely, prevalence and abundance (TABLE 2). Ticks with that kind of *T. parva* infection level had a very good potential of transmitting the pathogen to healthy cattle as confirmed by earlier studies.[8]

In these studies, it was observed that when 189 adult ticks that had fed to repletion on day 4 as nymphae through membranes with heparinized *Theileria* piroplasm infected blood were applied to an animal, seventy-two females engorged successfully and the animal developed clinical theileriosis. Schizonts were detected on day 8 and fever on day 16. The animal recovered by day 26. Piroplasms were first detected on day 18 and reached a maximum parasitemia of 10.8% erythrocytes infected on day 24. The animal developed antibodies to *T. parva*. In another experiment, 270 adult ticks fed and infected in the same way as described above were also applied on a healthy cattle. That animal developed a *T. parva* infection with a prepatent period to schizonts of 14 days, and fever occurred on day 16, piroplasms were detected on day 17, and reached a maximum parasitemia of 23% erythrocytes infected on day 26. The animal had a severe *Theileria* reaction and was euthanized on day 30.

Observations on the host skin and the artificial membrane showed clearly that intraerythrocytic pathogens remain viable in heparinized blood, and the transmission of pathogens using heparinized blood appears to be efficient and should have many applications for studies on the transmission of these organisms. Use of skin membranes has several limitations, but an artificial membrane has an advantage as it provides a system independent of experimental animals and artificial membranes can be tailored to suit the requirements of a wide range of ticks as already reported.[18,19]

Previous attempts to infect *R. appendiculatus* nymphae with *T. pava* consisted of the rabbit perfusion method,[20] inoculation of infected cattle blood into splenectomized or irradiated rabbits on which nymphae *R. appendiculatus* were feeding,[21-23] capillary feeding of semiengorged nymphal ticks on infected cattle blood,[24] and inoculation of rabbit-fed replete nymphae with infected cattle blood.[25] None of these methods was as effective, convenient, or reproducible as the artificial membrane feed method discussed here. This is indicative of the potential of the method as a research tool that ought to be perfected.

REFERENCES

1. CHABAUD, A. G. 1950. Sur la nutrition artificielle des tiques. Ann. Parasitol. Hum. Comp. **25:** 142-147.
2. PURNELL, R. E. & L. P. JOYNER. 1967. Artificial feeding technique for *Rhipicephalus appendiculatus* and transmission of *Theileria parva* from the salivary secretions. Nature London **216:** 484-485.
3. KEMP, D. H., D. KOUDSTAAL, J. A. ROBERTS & J. D. KERR. 1975. Feeding of *Boophilus microplus* larvae on partially defined medium through thin slices of cattle skin. Parasitology **70:** 243-254.
4. VOIGT, W. P., A. S. YOUNG, S. N. MWAURA, S. G. NYAGA, G. M. NJIHIA, F. N. MWAKIMA & S. P. MORZARIA. 1993. *In vitro* feeding of instars of the ixodid ticks *Amblyomma variegatum* through skin membranes and is application to the transmission of *Theileria mutans* and *Cowdria ruminantium*. Parasitology **107:** 257-263.
5. STONE, B. F., M. A. COMMINS & D. H. KEMP. 1983. Artificial feeding of the Australian paralysis tick *Ixodes holocyclus* and collection of paralysing toxin. Int. J. Parasitol. **13:** 447-454.
6. WALADDE, S. M. & M. J. RICE. 1982. The sensory basis of tick feeding behaviour. *In* Physiology of Ticks. F. D. Obenchain & R. Galun, Eds.: 77-118. Pergamon Press. Oxford, England.
7. WALADDE, S. M., S. A. OCHIENG & P. M. GICHUHI. 1991. Artificial membrane feeding of the ixodid tick *Rhipicephalus appendiculatus* to repletion. Exp. Appl. Acarol. **11:** 297-306.
8. WALADDE, S. M., A. S.YOUNG, S. A. OCHIENG, S. N. MWAURA & F. N. MWAKIMA. 1993. Transmission of *Theileria parva* to cattle by *Rhipicephalus appendiculatus* adults fed as nymphae *in vitro* on infected blood through an artificial membrane. Parasitology **107:** 249-256.
9. BUTLER, J. F., W. R. HESS, R. G. ENDRIS & K. H. HOLSCHER. 1982. *In vitro* feeding of *Ornithodoros* ticks for rearing and assessment of disease transmission. Proc. 6th Congr. Acarology: 1075-1081.
10. MUTUGI, J. J. 1986. Transmission studies of a pathogenic strain of *Theileria mutans*. *In* Ticks and Tick-Borne Disease. R. W. Sutherst, Ed.: 95-96. Australia Center for International Agricultural Research. Canberra, Australia.
11. KATENDE, J. M., B. M. GODDEERIS, S. P. MORZARIA, C. G. NKONGE & A. J. MUSOKE. 1990. Identification of a *Theileria mutans*-specific antigen for use in an antibody and antigen detection ELISA. Parasite Immunol. **12:** 243-254.
12. WALADDE, S. M., A. S. YOUNG, S. N. MWAURA, G. M. NJIHIA & F. N. MWAKIMA. Optimization of the *in vitro* feeding of *Rhipicephalus appendiculatus* nymphae for the transmission of *Theileria parva*. Parasitology. (In press.)
13. MUSOKE, A. J., S. P. MORZARIA, C. NKONGE, E. JONES & V. NENE. 1992. A recombinant sporozoite surface antigen of *Theileria parva* induces protection in cattle. Proc. Natl. Acad. Sci. USA **89:** 514-518.
14. BÜSCHER, G. & B. OTIM. 1986. Quantitative studies of *Theileria parva* in salivary glands of *Rhipicephalus appendiculatus* adults. Int. J. Parasit. **16:** 93-100.

15. NORVAL, R. A. I., D. E. SONENSHINE & T. PETER. 1992. Attraction/aggregation/attachment/ pheromone (AAAP) of the bont leg ticks, *Amblyomma hebraeum* and *A. variegatum. In* Proceedings of the First International Conference on Tick-Borne Pathogens at the Host-Vector Interface. U. G. Munderloh & T. J. Kurtii, Eds.: 324–327. St. Paul University of Minnesota.

16. APPS, P. J., H. W. VILJOEN & V. PRESTORIUS. 1988. Aggregation pheromone of the bont tick *Amblyomma hebraeum:* identification of candidates for bioassay. Onderstepoort J. Vet. Res. **55:** 135–137.

17. SCHOENI, R., E. HESS, W. BLUM & K. RAMSTEIN. 1984. The aggregation-attachment pheromone of the tropical bont tick *Amblyomma variegatum.* Fabricius (Acari:Ixodidae): isolation, identification and action of its components. J. Insect. Physiol. **30:** 613–618.

18. COAN, M. E. & D. STILLER. 1993. *In vitro* feeding by ixodid ticks through Waladde's modified Baudruche membrane: unfed adult *Dermacentor andersoni* (Stiles) and *Boophilus microplus* (Canestrini) fed to repletion on defibrinated bovine blood. Proceeding of the Ninth International Veterinary Hemoparasite Disease Conference. p. 19. Meridan, Mexico, 6–9 Oct. 1993.

19. KUHNERT, F., P. A. DIEHL & P. M. GUERIN. 1995. The life-cycle of the bont tick *Amblyomma hebraeum in vitro.* Int. J. Parasitol. **25:** 887–896.

20. IRVIN, A. D., R. E. PURNELL & M. A. PEIRCE. 1970. Infection of *Rhipicephalus appendiculatus* ticks with *Theileria parva,* using a rabbits head perfusion technique. Res. Vet. Sci. **11:** 493–495.

21. BROCKLESBY, D. W. & B. O. VIDLER. 1986. Further attempts to infect laboratory animals with *Theileria parva.* Vet. Rec. **75:** 1277.

22. IRVIN, A. D., R. E. PURNELL, C. G. D. BROWN, M. P. CUNNINGHAM, M. A. LEDGER & R. C. PAYNE. 1974. Application of an indirect method of infecting ticks with piroplasms for use in the isolation of field infections. Br. Vet. J. **130:** 280–288.

23. PURNELL, R. E., A. D. IRVIN, C. D. KIMBER, P. L. OMWOYO & R. C. PAYNE. 1974. East Coast fever: further laboratory investigations on the use of rabbits as vehicles for infecting ticks with theilerial piroplasms. Trop. Anim. Health Prod. **6:** 145–151.

24. PURNELL, R. E. 1970. Infection of the tick *Rhipicephalus appendiculatus* with *Theileria parva* using an artificial feeding technique. Res. Vet. Sci. **11:** 403–405.

25. SCHREUDER, B. E. C. & G. UILENBERG. 1976. Studies on Theileriidae (Sporozoa) in Tanzania. V. Preliminary experiments on a new method for infecting ticks with *Theileria parva* and *Theileria mutans.* Tropenmedizin und Parasitologie **27:** 422–426.

Culicoides variipennis and Bluetongue Disease

Research on Arthropod-Borne Animal Diseases for Control and Prevention in the Year 2000[a]

W. J. TABACHNICK, M. A. ROBERTSON, AND
K. E. MURPHY[b]

Arthropod-Borne Animal Diseases Research Laboratory
USDA-ARS
Laramie, Wyoming 82071

[b]*Department of Biology*
The Citadel
Charleston, South Carolina 29409

Arthropod-borne diseases cause economic losses to the livestock industries of the United States and many other countries around the world. The number of diseases is large; the number of potential arthropod vectors involved is also large. US arthropods vector vesicular stomatitis viruses, bluetongue viruses, epizootic hemorrhagic disease viruses, babesia, and anaplasma among other pathogens that are transmitted by ticks, mosquitoes, or midges. A primary goal of US research on arthropod vectors of animal pathogens is to provide strategies to reduce the consequent agricultural losses due to the diseases they cause. Of equal importance is the goal to protect US animal populations from other vector-borne diseases that have heretofore been exotic to the US. Among these are rift valley fever, akabane, African horse sickness, ibaraki, bovine ephemeral fever, heartwater, East Coast fever, as well as exotic forms of the bluetongue and epizootic hemorrhagic disease viruses.

Reducing the impact of these diseases requires knowledge of the vector, the pathogen, and the host. Protecting a country from the danger of the entry of foreign pathogens requires a detailed understanding of the complex interactions of the components of these disease cycles. In order to predict the risk or danger of entry of a pathogen, detailed knowledge of the underlying mechanisms of vector-pathogen, vector-host, as well as pathogen-host interactions is needed.

Current knowledge and research directions involving bluetongue disease in the US provide an appropriate model with which to explore the major issues confronting arthropod-borne disease research elsewhere in the world. The issues related to providing for the health of US animal populations are instructive for other countries. A

[a] The authors were supported by USDA, Agricultural Research Service, and by USDA, CSRS, NRICGP 93-37204-9156 to WJT and KEM, and 92-37204-7960 to W. C. Black IV, WJT and MAR.

more general discussion of bluetongue disease issues throughout the world may be found elsewhere.[1,2]

BLUETONGUE DISEASE IN THE UNITED STATES

Bluetongue disease is primarily a disease of domestic ruminants and is caused by the bluetongue viruses in the genus *Orbivirus,* family Reoviridae. Clinical disease is most commonly observed in sheep, where symptoms include inflammation and congestion, leading to cyanosis, hemorrhages, and ulceration of the mucous membranes. In the US, infected sheep have morbidity rates usually less than 30% and generally mortality rates of 5% have been reported.[3] Cattle are susceptible to infection with bluetongue viruses, and although cattle have a prolonged viremia of several weeks, usually less than 1% of infected cattle exhibit clinical symptoms.

There are several attenuated vaccines available.[4] Bluetongue attenuated vaccines protect only against the specific serotype of bluetongue for which they were developed. The attenuated vaccines against US bluetongue serotypes 10, 11, 13, and 17 have not been available for general use outside the state of California due to the absence of required data that these vaccines will not revert to virulence in the arthropod vector. A new generation of recombinant bluetongue vaccines are under development that use only specific proteins as immunogens.[5] There should be no danger of reversion to virulence since these vaccines use only a small portion of the viral genome. Safe, efficacious, and cost-effective vaccines would be potent tools to protect livestock and reduce the economic significance of bluetongue disease. However, even should such vaccines become available for all of the bluetongue virus serotypes, there is still a need for understanding the mechanisms of bluetongue virus transmission by the insect vectors. The expense of vaccination could be reduced by using vaccinations only in regions where there was risk of disease due to the presence of competent vectors. Strategies to reduce the impact of the vector would provide a powerful tool, which when coupled with targeted vaccination would reduce disease in animal populations.

Bluetongue disease is important to the US because of its direct effect on animal health, and also because the presence of bluetongue viruses in the animals of a country results in restrictions on the movement of livestock and livestock germplasm to bluetongue-free countries. The Office of International Epizooties (OIE) places bluetongue on the List A of animal diseases as "those diseases which spread rapidly, the scope of which extends beyond national border. These diseases have particularly serious socio-economic or public health consequences, and are of major importance in the international trade of animals and animal products."[6] As a consequence, US livestock must go through expensive and time-consuming certification to determine bluetongue status prior to movement to bluetongue-free countries in the European Union. The resulting losses in trade to the US livestock industries have been estimated at $120 million annually. Losses worldwide have been estimated at $3 billion annually.[7]

The primary arthropod vector of the bluetongue viruses in North America is a biting midge, *Culicoides variipennis* Coquillet (Diptera: Ceratopogonidae). Genetic and morphologic studies of *C. variipennis* found there are at least three subspecies present in the US.[8,9] *Culicoides v. sonorensis* in the south, southwest, and west, *C. v. occidentalis* in the far west, and *C. v. variipennis* in the east, northeast, and central states. Genetic analyses of US *C variipennis* populations have found regional population genetic variation and diversity consistent with the hypothesis that the three subspecies are different species.[10,11]

Culicoides v. sonorensis is believed to be the primary vector of the bluetongue viruses in North America.[9,12] The evidence in support of this is virus isolations during bluetongue outbreaks from only this subspecies, research with *C. v. sonorensis* colonies, and a close relationship between the distribution of bluetongue antibody in cattle and the range of *sonorensis*.[8,13–16] Tabachnick and Holbrook[8] review epidemiologic information showing that the seroprevalence of bluetongue antibody in cattle has consistently been less than 2% within the north-northeastern tier of states (ME, NH, VT, MA, CT, RI, NY, PA, NJ, DE, MD, OH, WV, MI, WI, ND).[17] This low seroprevalence has been observed for nearly 20 years[18] in spite of the movement of bluetongue seropositive and probably viremic animals into this area from the areas of the US that are clearly endemic for bluetongue. Lopez *et al.*[19] found only 13 seropositive animals among more than 30,000 cows tested in New York State, and all 13 were traced to a point of origin from the US bluetongue endemic region. In addition, *Culicoides v. sonorensis* populations were found to be susceptible to oral infection with the bluetongue viruses, while *C. v. variipennis* from New York, and Maryland were found to be significantly more resistant (Jones unpublished, Tabachnick unpublished).[14,15,19] The preponderance of the current evidence collectively suggested that the northeastern US has not had bluetongue transmission among cattle. This region therefore should be considered a bluetongue-free area for the international trade of ruminant livestock due to the absence of a competent vector.[9,12,15,16] Acceptance of this finding by the international community would ease trade restrictions on the movement of US livestock and germplasm, considerably improving trade with bluetongue-free countries.

CURRENT ISSUES AND RESEARCH NEEDS FOR *CULICOIDES VARIIPENNIS* AND BLUETONGUE

Government regulatory officials, responsible for protecting the health of their country's animal populations, must be conservative in devising regulations. If a convincing case is to be made for the status of bluetongue in the US, or for the disease status of any region, then information must be obtained that explains the mechanism causing bluetongue disease epidemiology. The General Agreement on Tariff and Trade (GATT) is specific on the need for such information when regulating trade to protect a country's animal populations. Information of this nature will be critical for all countries and regions seeking to convince the international community of the disease status and/or risk of disease due to trade in livestock and livestock germplasm with that country.

The current information on bluetongue disease throughout the world demonstrates that the presence or absence of bluetongue in a region is based primarily on the presence or absence of suitable competent vectors in the genus *Culicoides*.[15,20] The situation within the US is consistent with this observation. However, in order to provide conclusive evidence, it will be necessary to understand the *C. v. sonorensis* mechanisms controlling vector ability for bluetongue viruses in comparison to *C. v. variipennis,* to show specifically why there is no bluetongue transmission in the northeastern US.

A variety of complex traits comprise an arthropod's ability or capacity to be a vector of a pathogen. Among these are traits such as host preference, feeding behavior, and vector competence which includes susceptibility to infection with the pathogen and the ability of the arthropod to transmit the pathogen to a host animal. The variation in these traits from individual to individual within any vector species is probably under the control of different arthropod genes, as well as unknown environmental factors.[21] As a first stage to understand the controlling mechanisms, it will be necessary to characterize the genes controlling these important traits. Once the controlling genes have been identified, then the responses of these genes under different environmental conditions will have to be determined. The range of phenotypes of a genotype in a variety of environments is called the "norm of reaction." For example, if the genotypes controlling *C. v. variipennis* vector competence for bluetongue viruses and the norms of reactions were known, we would have the possibility of understanding why certain individuals within the species are not vectors of the pathogen. Field populations could be characterized for vector competence gene frequencies, and it could be shown, for example, that the absence of bluetongue virus transmission in any region was due to the low frequency of competent vectors. Identifying genes and norms of reaction provides the ability to characterize the danger posed by arthropod populations.

We are using two different strategies to identify *C. variipennis* vector competence genes. One strategy, which we call the top-down strategy, is based on first identifying a specific function that may be directly involved in vector competence, e.g., *C. variipennis* virus gut receptor proteins, vector viral tissue tropisms, proteins correlated with susceptibility to infection differences between families. The specific trait is then studied to determine the genetic basis. For example, we have identified a single locus that controls susceptibility to infection in two *C. v. sonorensis* laboratory colonies.[22] We have made isogenic pools of *C. v. sonorensis* starting with bluetongue virus susceptible and resistant families from these colonies, and we have identified a ca. 80 kilodalton protein that is a candidate protein causing *C. v. sonorensis* resistance to infection with the bluetongue viruses (Murphy and Tabachnick, unpublished). A *C. variipennis* cDNA clone has been identified that produces the protein (Murphy, unpublished). These findings provide opportunities to determine the function of the protein from its nucleotide sequence, and to determine the effect of variation in this protein on vector competence in field collected *C. variipennis*. The disadvantage of this strategy is that the identified function may have little to do with controlling vector competence variation in the field.

An alternative approach is the bottom-up strategy. This strategy uses recombinational genetic mapping to identify controlling genes of the vector competence trait itself. Therefore one identifies the presence of phenotypic variation, e.g., *C. variipennis* susceptibility to infection with bluetongue virus, and uses molecular genetic markers to identify the specific genes controlling this variation. The strategy does not depend on an assumption about the significant traits. It is a powerful strategy since it will identify the locations of the genes controlling the phenotypic variation in the particular vector strains used in the mapping program. The disadvantage of this strategy is that only those loci controlling variation in the specific insect mapping crosses will be identified. The choice of strains used in mapping is critical, since failure to use appropriate strains may result in missing important loci controlling variation in field populations. Tabachnick[21] provided a review of the methodology to be used to map vector competence loci in an arthropod vector. A genetic mapping program is under way for *C. variipennis* using a variety of DNA molecular markers. The DNA markers are identified using the polymerase chain reaction in the form of simple sequence length polymorphisms (SSLPs), single-stranded conformation polymorphisms (SSCPs) and randomly amplified polymorphic DNAs (RAPDs).[23-27] Sixty of these DNA markers are being used to locate major loci controlling *C. variipennis* susceptibility to infection with the bluetongue viruses to specific chromosomal locations flanked by molecular markers (Robertson and Tabachnick, unpublished). Once this is accomplished, identification and cloning of the controlling genes will become high priority. Molecular cloning strategies will benefit from physical mapping of genes by such techniques as YAC (yeast artificial chromosome) libraries, and FISH (fluorescent *in situ* hybridization)[28] and may also require inbred and isogenic lines, chromosome balancer stocks, and deletion stocks. These tools are not yet available for *C. variipennis,* but it is anticipated that developments with *C. variipennis* and other insects, i.e., *Drosophila, Aedes aegypti,* and *Anopheles gambiae,* will provide new technologies and methods to allow work of this nature with *C. variipennis.*

Once genes and norms of reaction are identified, individuals and populations of *C. variipennis* could be characterized with assurance to determine vector competence. The relation of *C. v. variipennis* to the absence of bluetongue disease in the northeast US could be determined with assurance. The identified genes and environmental factors would offer new opportunities to develop novel control strategies by changing competent populations to incompetent populations through genetic strategies[29] or the use of environmental manipulations.[21]

RESEARCH NEEDS FOR ARTHROPOD VECTORS OF DISEASES OF LIVESTOCK

Genetic mapping programs for arthropod vectors of disease provide exciting opportunities for developing information on the underlying mechanisms controlling important vector traits. Mapping and identification of genes controlling a variety of other important arthropod traits will be possible using the same strategies. The immediate need is to apply the molecular markers for identifying vector competence, vector capacity, pesticide resistance, host preference, mating behavior, and other critical traits. Then appropriate genetic tools for isolating and cloning actual arthropod genes will need to be developed, i.e., cosmid and YAC libraries, transposon tagging, chromosomal deletion, and balancer stocks. These undertakings will be laborious, expensive, and require the efforts of several well-funded laboratories.

The benefits for preventing and controlling arthropod-borne animal diseases are enormous. In addition to the benefits of being able to define regions at risk for disease and for devising appropriate animal health regulations, the understanding of mechanisms of pathogen-vector interactions will provide information that is critical for predicting vector-pathogen interactions causing pathogen evolution, changing pathogen virulence, and for predicting the entry or emergence of new pathogens in a region. The emergence of new pathogens is a critical danger of our times.[30] There is precious little information on the potential for exotic pathogens to change in new environments, including exposure to different arthropod vectors. For example, the danger posed by the entry of exotic bluetongue serotypes is unknown. Will *C. variipennis* transmit exotic bluetongue serotypes? How will exotic serotypes behave in a new vector environment? Will pathogenesis of the bluetongue viruses be effected? What is the danger of the bluetongue viruses becoming more pathogenic? What about the danger of other heretofore exotic viruses, e.g., African horse sickness, Rift Valley fever? The answers to these questions require the information on vector-pathogen interactions that we have described.

Research on vector-pathogen interactions must be supported. The resulting information will be the basis for developing new disease-control methods, approaches to reduce the impact of arthropod pests of animals, and to protect animal and human populations from the emergence of new diseases.

ACKNOWLEDGMENTS

We thank W. C. Black IV, F. R. Holbrook, G. J. Hunt, and B. R. Miller for reviewing earlier drafts of this manuscript.

REFERENCES

1. GIBBS, E. P. J. & E. C. GREINER. 1988. Bluetongue and epizootic hemorrhagic disease. *In* The Arboviruses: Epidemiology and Ecology. T. P. Monath, Ed.: 39–70. CRC Press. Boca Raton, FL.

2. WALTON, T. E. & B. I. OSBURN, Eds. 1992. Bluetongue, African Horse Sickness and Related Orbiviruses. Proceedings of the Second International Symposium. CRC Press. Boca Raton, FL.

3. PARSONSON, I. M. 1990. Pathology and pathogenesis of bluetongue infections. *In* Bluetongue Viruses. P. Roy & B. M. Gorman, Eds.: 119–141. Springer-Verlag. Berlin, Germany.

4. WALTON, T. E. 1992. Attenuated and inactivated orbiviral vaccines. *In* Bluetongue, African Horse Sickness and Related Orbiviruses. Proceedings of the Second International Symposium. T. E. Walton & B. I. Osburn, Eds.: 851–855. CRC Press. Boca Raton, FL.

5. ROY, P. & B. J. ERASMUS. 1992. Second generation candidate vaccine for bluetongue disease. *In* Bluetongue, African Horse Sickness and Related Orbiviruses. Proceedings of the Second International Symposium. T. E. Walton & B. I. Osburn, Eds.: 856–867. CRC Press. Boca Raton, FL.

6. OFFICE OF INTERNATIONAL DES EPIZOOTIES. 1986. OIE International Zoo-Sanitary Code. 5th Edition. OIE. Paris, France.

7. BATH, G. O. 1989. Bluetongue. *In* 2nd International Congress for Sheep Veterinarians: 349–357. Massey University. New Zealand.

8. TABACHNICK, W. J. 1992. Genetic differentiation among populations of *Culicoides variipennis* (Diptera: Ceratopogonidae), the North American vector of bluetongue virus. Ann. Entomol. Soc. Am. **85:** 140–147.

9. TABACHNICK, W. J. 1992. Genetics, population genetics and evolution of *Culicoides variipennis:* implications for bluetongue virus transmission in the USA and its interna-

tional impact. _In_ Bluetongue, African Horse Sickness and Related Orbiviruses. Proceedings of the Second International Symposium. T. E. Walton & B. I. Osburn, Eds.: 262–270. CRC Press. Boca Raton, FL.

10. HOLBROOK, F. R. & W. J. TABACHNICK. 1995. The _Culicoides variipennis_ (Diptera: Ceratopogonidae) complex in California. J. Med. Entomol. **32:** 413–419.

11. HOLBROOK, F. R., W. J. TABACHNICK & R. C. BRADY. The _Culicoides variipennis_ (Diptera: Ceratopogonidae) complex in New England. Med. Vet Entomol. (In press.)

12. TABACHNICK, W. J. & F. R. HOLBROOK. 1992. The _Culicoides variipennis_ complex and the distribution of bluetongue viruses in the United States. Proc. US Anim. Hlth. Assoc. **96:** 207–212.

13. JONES, R. H. & N. M. FOSTER. 1978. Relevance of laboratory colonies of the vector in arbovirus research—_Culicoides variipennis_ and bluetongue. Am. J. Trop. Med. Hyg. **27:** 168 177.

14. JONES, R. H. & N. M. FOSTER. 1978. Heterogeneity of _Culicoides variipennis_ field populations to oral infection with bluetongue virus. Am. J. Trop. Med. Hyg. **27:** 178–183.

15. TABACHNICK, W. J. 1995. _Culicoides variipennis_ and bluetongue virus epidemiology in the United States. Ann. Rev. Entomol. **41:** 23–43.

16. WALTON, T. E., W. J. TABACHNICK, L. H. THOMPSON & F. R. HOLBROOK. 1992. An entomologic and epidemiologic perspective for bluetongue regulatory changes for livestock movement from the USA and observations on bluetongue in the Caribbean Basin. _In_ Bluetongue, African Horse Sickness and Related Orbiviruses. Proceedings of the Second International Symposium. T. E. Walton & B. I. Osburn, Eds.: 952–960. CRC Press. Boca Raton, FL.

17. PEARSON, J. E., G. A. GUSTAFSON, A. L. SHAFER & A. D. ALSTAD. 1992. Distribution of bluetongue in the United States. _In_ Bluetongue, African Horse Sickness and Related Orbiviruses. Proceedings of the Second International Symposium. T. E. Walton & B. I. Osburn, Eds.: 128–139. CRC Press. Boca Raton, FL.

18. METCALF, H. E., J. E. PEARSON & A. L. KLINGSPORN. 1981. Bluetongue in cattle: a serologic survey of slaughter cattle in the United States. Am. J. Vet. Res. **42:** 1057–1061.

19. LOPEZ, J. W., E. J. DUBOVI, E. W. CUPP & H. LEIN. 1992. An examination of the bluetongue status of New York State. _In_ Bluetongue, African Horse Sickness and Related Orbiviruses. Proceedings of the Second International Symposium. T. E. Walton & B. I. Osburn, Eds.: 140–146. CRC Press. Boca Raton, FL.

20. TABACHNICK, W. J., P. S. MELLOR & H. A. STANDFAST. 1992. Working team report on vectors: recommendations for research on _Culicoides_ vector biology. _In_ Bluetongue, African Horse Sickness and Related Orbiviruses. Proceedings of the Second International Symposium. T. E. Walton & B. I. Osburn, Eds. CRC Press. Boca Raton, FL.

21. TABACHNICK, W. J. 1994. The role of genetics in understanding insect vector competence for arboviruses. Adv. Dis. Vect. Res. **10:** 93–108.

22. TABACHNICK, W. J. 1991. Genetic control of oral susceptibility to infection of _Culicoides variipennis_ for bluetongue virus. Am. J. Trop. Med. Hyg. **45:** 666–671.

23. ORITA, M., Y. SUZUKI, T. SEKIYA & K. HAYASHI. 1989. A rapid and sensitive detection of point mutations and genetic polymorphism using polymerase chain reaction. Genomics **5:** 874–879.

24. WILLIAMS, J. G. K., A. R. KUBELIK, K. J. LIVAK, J. A. RAFALSKI & S. V. TINGEY. 1990. DNA polymorphisms amplified by arbitrary primers are useful as genetic markers. Nucleic Acids Res. **18:** 6531–6535.

25. CORNALL, R. J., T. J. AITMAN, C. M. HEARNE & J. A. TODD. 1991. The generation of a library of PCR-analyzed microsatellite variants for genetic mapping of the mouse genome. Genomics **10:** 874–881.

26. JACOB, H. J., K. LINDPAINTNER, S. E. LINCOLN, K. KUSUMI, R. K. BUNKER, Y-P MAO, D. GANTEN, V. J. DZAU & E. S. LANDER. 1991. Genetic mapping of a gene causing hypertension in the stroke-prone spontaneously hypertensive rat. Cell **67:** 213–224.

 ANNALS NEW YORK ACADEMY OF SCIENCES

27. RAICH, T., J. L. ARCHER, M. A. ROBERTSON, W. J. TABACHNICK & B. J. BEATY. 1993. Polymerase chain reaction approaches to *Culicoides* (Diptera: Ceratopogonidae) identification. J. Med. Entomol. **30:** 228-232.
28. LAWRENCE, J. B. 1990. A fluorescence in situ hybridization approach for gene mapping and the study of nuclear organization. *In* Genetic and Physical Mapping. K. E. Davies & S. M. Tilghman, Eds. **1:** 1-38. Cold Spring Harbor Press. Cold Spring Harbor, NY.
29. CRAMPTON, J., A. MORRIS, G. LYCETT, A. WARREN & P. EGGLESTON. 1990. Transgenic mosquitoes: A future vector control strategy? Parasitol. Today **6:** 31-36.
30. MORSE, S. S. 1993. Examining the origins of emerging viruses. *In* Emerging Viruses. S. S. Morse, Ed.: 10-28. Oxford University Press. Oxford, England.

Age and Weight Relationship in *Boophilus microplus* (Ixodoidea: Ixodidae) Larvae

R. DE LA VEGA[a] AND GRACIELLA DIAZ[b]

[a]*Laboratorios Biologicos Farmaceuticos*
Ciudad Habana, Cuba

[b]*Facultad de Biologia*
Universidad de la Habana
Cuba

INTRODUCTION

The study of the age of free-living stages of ticks is not a frequent subject in acarology research. Besides the baseline knowledge obtained in such studies, some applications in experiments about tick survival as well as in practical tick control programs are to be considered.

In the present paper, the relationship between age and weight of *Boophilus microplus* larvae is discussed.

MATERIALS AND METHODS

Ticks

Sixty engorged ticks from artificially infested cattle were used.

Incubation Conditions

Ticks, eggs, and larvae were maintained at $28 \pm 1°C$ and 100% relative humidity (RH).

Procedure

Eggs were observed daily, and the beginning of eclosion noted. Those vials where eclosion started the same day were grouped. Age zero of larvae was taken to be 10 days after the first day of eclosion. At similar incubation conditions,[1] once eclosion started, it took 7 days to be completed. Furthermore, larvae showed negative geotropism 3 days after their emergence from eggs.[2] This estimation of age zero has been employed by the authors[3] in larva survival forecast by means of a mathematical model; this model was successfully validated in natural simulated conditions.[3] Larvae

227

were weighed in 3 × 3 cm fine paper envelopes in groups that varied from 150 to 400 larvae. Weighing was performed on days 0, 8, 13, 21, 28, 36, 43, and 50 of age.

Method of Weighing

Ten envelopes were marked and weighed in an analytical balance (precision = 0.0001g) each day. Larvae were introduced into the envelopes; these were closed and weighed. Then envelopes with larvae were placed inside a stove at 50°C for 48 hours and weighed again. Larvae were counted under the stereoscopic microscope in each envelope and put in the same conditions of temperature and for the same time in the stove. Finally, dried envelopes were weighed.

Expression of Age as Effective Temperature Summings (ETS)

As in all poikilothermal beings, the age of ticks cannot be expressed in time units but in °C-time (°C-day in this case) if age is going to be compared at different incubation conditions. Life span is not only dependent on time but on a combination of time and temperature. If two groups of larvae are maintained at two different incubation temperatures for 10 days, it is not correct to express that both groups have the same age. Those larvae held at the highest temperature will live less long than the other group. This is based on an old publication by Sanderson and Peairs.[4] De la Vega and Diaz[3] employed minimum thermic threshold (MTT) of 14°C when environmental temperatures were ≥24°C. The effective temperature (ET) was 14°C (28°C in the incubator minus MTT). In the above-mentioned paper,[3] it was proved that for maximum larval survival, the ETS is 760°C-day, approximately.

Statistical Analysis

A one-way analysis of variance was done for both the live and dry weight with respect to larval age. Simple linear regressions were employed, the live and dry mean weights of larvae being the dependent variables and their ages, the independent one.

RESULTS

At age zero, all larvae had defecated. Due to technical reasons, weights of 21 days were discarded. At 57 days of age, only a few larvae almost motionless to breath stimulation were observed; thereafter no living larva remained. The maximum ETS observed was 798°C-day comparable with that obtained by De la Vega and

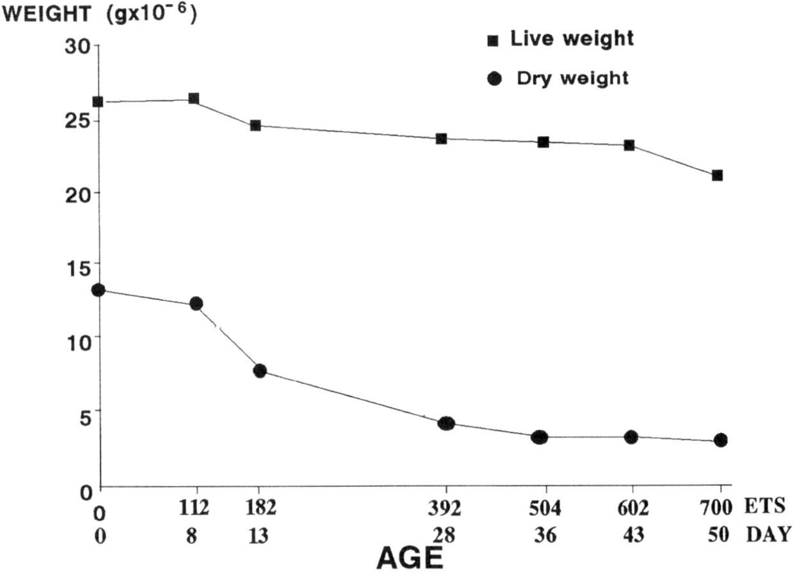

FIGURE 1. Larval weight and age. The weight is expressed in grams (g × 10⁻⁶), and the age in days and in effective temperature summings (ETS).

Diaz.[3] FIGURE 1 shows live weight means (LWM) and dry weight means (DWM) of larvae as a function of age. LWM slightly declined from 0 to 700°C-day (0 to 50 days in the present work). The regression was significant with a negative slope of 0.005. DWM sharply decreased from 0 to 504°C-day (0 to 36 days). The DWM of 504°C-day was not significantly different from those of 602 and 700°C (43 and 50 days, respectively). The coefficient of determination (r^2) was 0.95. The straight line equation was: $Y = 16.04 - 0.02X$, Y being the larval weight expressed in g × 10⁻⁶ and X the age expressed in ETS. In FIGURE 2 it can be appreciated that the absolute water content of larvae increases as they age from age 0 to 504°C-day, after that there is a decline in absolute water content. In the same figure, the dry weight loss of larvae in relation to age is shown. In the first 13 days (182°C-day), there is a marked loss of dry weight; after that the dry weight loss is less marked, and at 504°C-day (36 days) all dry material susceptible to be converted into energy was consumed. It has to be stressed that at the same time, water content begins to decline.

DISCUSSION

As has already been mentioned, LWM declines slightly from 0 to 700°C-day. The same phenomenon was noticed by Schunter and Tatchell[5] in *B. microplus* larvae held at 30°C and 90% RH from 8 to 36 days of age. These authors considered the first day of eclosion as age zero. They observed a great decline of living larval weight from 0 to 8 days. This was interpreted as losses due to defecation and other metabolic

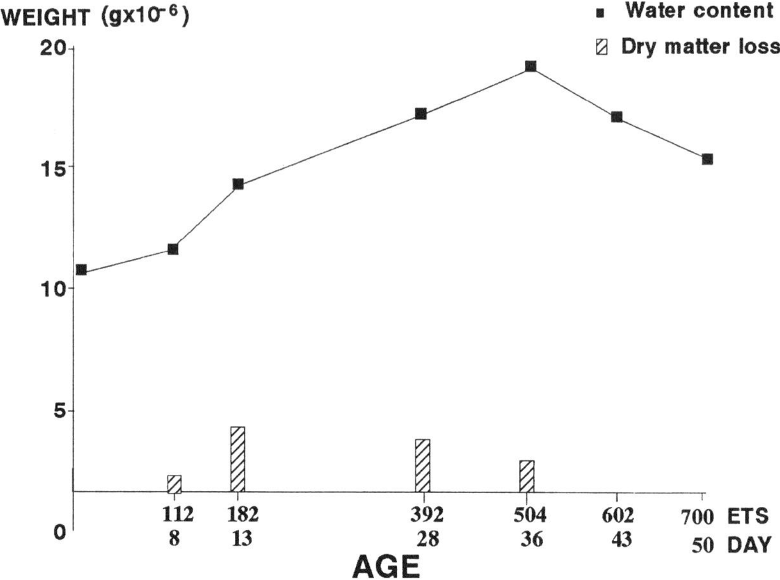

FIGURE 2. Water content and loss of dry matter in relation to larval age. The weight is expressed in grams (g × 10⁻⁶), and the age in days and in effective temperature summings (ETS).

wastes. In the present experiment, age zero was assumed to be 10 days after the first eclosion, and this phenomenon[5] (curve with two slopes) was not observed.

In FIGURE 2, it can be observed that the LWM lost as dry material is almost regained by the increase of water content. This fact explains by itself why LWM (FIG. 1) declines only slightly with aging. It seems that larvae need to maintain their body volume in order to continue to be viable. Maybe it can be explained because the more engorged the tick is, the less surface/volume ratio it has and thence evaporation is less marked. Another explanation, which does not exclude the first hypothesis, is that the larva needs to dilute the metabolic wastes produced in the process of getting energy from body reserves.

The relative humidity is above the critical equilibrium humidity.[6,7] This term was recently named critical equilibrium activity.[8] Uptake of water is accomplished by both passive and active mechanisms explained by the above-cited authors.[7,8] Larval drinking can be discarded because larvae have no possibility of acquiring water droplets.[5,9]

As shown (FIG. 2), when dry weight losses increase, the absolute amount of water increases. This was suggested by Hefnawy *et al.*[10] in adults of *Argas (Persicargas) arboreus* held at 28 ± 1°C and 96% RH for 105 days. This fact, which is to a great extent explained by passive uptake of water, seems to be complemented by active uptake. In effect, in FIGURE 2 it can be appreciated that as soon as dry weight ceases to decrease, water content stops increasing and begins to diminish. It might be interpreted that when reserve materials are completely depleted, the larva has no more energy for the active uptake of water. The remaining dry weight should correspond to structural elements of the larva. Another alternative hypothesis can be defended. In the degradation of organic matter, the final products are CO_2, water, and adenosine triphosphate. By means of this, all the facts that are discussed here can be explained. Nevertheless, it is more logical to think that both hypotheses may occur independently or simultaneously. Moreover, Knulle and Rudolph[7] have expressed that they can largely exclude changes in the water status brought about by the production of metabolic water. From the above results, two facts of possible practical importance can be inferred. Firstly, the larva can show vital signs and, in spite of this phenomenon, it should not be viable at all. When its metabolic reserves are no longer enough to assure active water uptake even at a high RH, this larva should not be able to climb up to a host, perforate its skin, and take the blood meal needed to finish its development. In the present experiment, larvae reach this condition at the age of 504°C-day. The last living larvae were seen at 798°C-day which means that practically in the final third of their lives, larvae are not viable. Based on this finding, cattle could begin grazing in deparasitation paddocks[11] sooner than different authors have estimated.[11,3] Secondly, the DMW is related to the age from 0 to 504°C-day. A high coefficient of determination was found ($r^2 = 0.95$) which indicates that there is a great predictive potential in the regression equation.

The most interesting relationship for practical purposes is age in dependence of larval weight, in order to estimate field larva age as a function of field collected larval weight. From a statistical point of view, to find out this relation by changing the variables with data obtained in this work (age in the ordinates and weight in the abscissa) would not be correct. Since the age of larvae was not taken in a random way, the relation between larval weight and age is not a correlation but a regression. Although a lot of work is still needed to achieve this prediction, the obtained results are encouraging.

SUMMARY

The study of the age of free-living stages of ticks is not a frequent subject in acarology research. Baseline knowledge and some possible applications both in research and in tick control are to be considered. Sixty engorged females were incubated at 28 ± 1°C and 100% relative humidity. Age zero of larvae was established at 10 days after the beginning of eclosion. Larvae were weighed in paper envelopes in groups ranging from 150 to 400 each. Live weight means (LWM) and dry weight means (DWM) were obtained. All larvae have defecated at age zero.

Maximum survival was 57 days. Larval age was also expressed as effective temperature summings (ETS). LWM declines slightly with age. DWM has a linear relationship to age from 0 to 504°C-day with a determination coefficient of 0.95. Absolute water content increases from 0 to 504°C-day; further water content diminishes. Dry matter weight declines just to 504°C-day of age. It might be theoretically possible to estimate the age of larvae in pasture by weighing groups of them.

REFERENCES

1. DE LA VEGA, R., G. DÍAZ & J. LUGO. 1984. Dinámica de la preoviposición y la eclosión en *Boophilus microplus* (Acarina: Ixodoidea). Reporte de Investigación del Instituto de Zoología. No. 18, Academia de Ciencias de Cuba: 1-12.
2. DE LA VEGA, R. 1976. Contribución al estudio de la biología de *Boophilus microplus* (Canestrini) en Cuba. Serie Biológica No. 64. Academia de Ciencias de Cuba: 1-18.
3. DE LA VEGA, R., F. FARRADÁ & G. DÍAZ. 1993. Aplicación de las constantes térmicas en el control de la garrapata del ganado vacuno VII. Despoblamiento. Rev. Salud Anim. **15**(2): 163-171.
4. SANDERSON, E. D. & L. M. PEAIRS. 1913. The relation of temperature to insect life. New Hampshire College Agricultural Experiment Station. Technical Bulletin No. 7: 1-125.
5. SCHUNTER, C. A. & R. J. TATCHELL. 1970. Drinking by larval cattle ticks, *Boophilus microplus* (Acarina: Ixodidae) J. Parasitol. **56**(6): 1239-1247.
6. HITCHCOCK, L. F. 1955. Studies of the non-parasitic stages on the cattle tick *Boophilus microplus* (Canestrini) (Acarina: Ixodidae). Aust. J. Zool. **3**: 295-311.
7. KNULLE, W. & D. RUDOLPH. 1983. Humidity relationships and water balance of ticks. *In* Physiology of Ticks. F. D. Obenchain & R. Galun. Eds.: 43-70. Pergamon Press. Oxford & New York.
8. SONENSHINE, D. 1991. Water balance in non-feeding ticks. *In* Biology of Ticks **1**: 398-438. Oxford University Press. Oxford, England.
9. WILKINSON, P. R. 1953. Observations on the sensory physiology and behaviour of larvae of the cattle tick *Boophilus microplus* (Can.) (Ixodidae). Aust. J. Zool. **1**: 345-356.
10. HEFNAWY, T., S. I. BISHARA & T. T. M. BASSAL. 1975. Biochemical and physiological studies of certain ticks (Ixodoidea): effects of relative humidity and starvation on the water balance and behaviour of adult *Argas (Persicargas) arboreus* (Argasidae). Exp. Parasitol. **38**: 14-19.
11. HARLEY, K. L. S. & P. R. WILKINSON. 1971. A modification of pasture spelling to reduce acaricide treatments for cattle tick control. Aust. Vet. J. **47**: 108-111.

Screwworm Eradication Program in Central America

THOMAS J. GALVIN AND JOHN H. WYSS

United States Department of Agriculture
Animal and Plant Health Inspection Service
International Services
Screwworm Eradication Program
Mexico City, D.F., Mexico

INTRODUCTION

The sterile male technique for the eradication of the screwworm *Cochliomyia hominivoraz* (Coquerel) was made possible by the work of Knipling and Bushland.[1] Their findings culminated in a large-scale field test on the island of Curacao in 1954. The success of this test resulted in a screwworm eradication program in the southeastern United States starting in 1957.[2] That program was extended to the southwestern United States in 1962,[3] and resulted in eradication of screwworms from the United States in 1966.

A biological barrier, sustained by continued dispersal of sterile flies over much of the Mexico-United States border area, and augmented by inspection and dipping of animals crossing into the United States from Mexico, was established. Following requests by Mexican livestock producers to extend the eradication program into Mexico, observations that the barrier would be cheaper (a 190 km width at the Isthmus of Tehuantepec compared to 2400 km at the United States-Mexico border) to maintain if the barrier zone were not so wide, and realization that a barrier farther from the border with Mexico would afford more protection for the United States, an agreement was signed on August 28, 1972, to form the Mexico-United States Commission for the Eradication of Screwworms (Commission). The plan was to eradicate screwworms in Mexico from the border with the United States to the Isthmus of Tehuantepec and establish a permanent barrier at that point. This objective was achieved in 1984. The barrier, in addition to continued releases of sterile screwworm flies, consisted of inspecting all livestock crossing from the south to the north of the barrier, with treatment and quarantine of screwworm infested animals and dipping or spraying of all livestock at three permanent and four temporary inspection stations covering all major roads in the area. Over the next few years, it was found that a barrier at the Isthmus of Tehuantepec tended to divide the country, with livestock producers to the south claiming the government was showing favoritism to producers north of the barrier. Data collected at the inspection stations showed that from nearly 400,000 to over 700,000 cattle crossed the barrier each year, from screwworm infested areas in the south to eradicated areas in the north, to reach livestock markets. The large numbers of livestock crossing the barrier created concern that some infested livestock might escape detection at the inspection stations and cause outbreaks of screwworm infestations in eradicated areas.

Further studies showed that there was a much better site for a permanent biological barrier.[4] A barrier in Panama, extending from near the Panama Canal to the border with Colombia, would require only 40 million sterile flies per week, compared to 150 million per week at the Isthmus of Tehuantepec. A barrier in Panama would provide considerably more protection to the United States than one in Mexico. A barrier in Panama would not divide the country into eradicated and noneradicated parts. A barrier in Panama would be more secure since there are no roads traversing the Darien Gap from Colombia to Panama. Following indications of interest in screwworm eradication from all Central American countries and Panama, a plan was developed in 1985[5] to extend the Screwworm Eradication Program through Central America and establish a permanent biological barrier in the eastern half of Panama.

The last screwworm infestation in Mexico, prior to an official declaration of freedom from screwworms on February 25, 1991, occurred July 10, 1990. However, an outbreak occurred starting in the southern part of the country on January 22, 1992. Although not proven, it is likely that the outbreak was initiated through importation of screwworm infested cattle from Central America through Guatemala. It is known that large numbers of cattle were being imported, both legally and illegally, at that time. This outbreak, with a total of 59 screwworm infestations, eventually involved 5 states (Campeche, Chiapas, Veracruz, Tabasco, and Tamaulipas). One focus in the state of Tamaulipas was only about 200 km from the Texas border. The last case found in this outbreak was discovered on September 30, 1992. Aerial dispersal of sterile screwworm flies commenced in each outbreak area in two days or less after the case was reported and continued for an additional 6 months after the last case in the area. On May 21, 1993, the first of five additional cases occurred in the states of Tamaulipas and Veracruz. It was assumed that these cases were the result of unreported cases cycling in low numbers during the winter from the 1992 outbreak rather than a new introduction from another country. The last of the five infestations was detected June 17, 1993. Again, sterile flies were dispersed over the infested areas starting within days of the report and continuing for 6 months after the last case in the area was found.

ACTIVITIES IN CENTRAL AMERICA

Guatemala

The Commission signed an agreement on December 10, 1986, with the Guatemala Ministry of Agriculture, Livestock, and Food (MAGA) to eradicate screwworms from Guatemala. A program office was established in Guatemala City in 1988. Aerial dispersal of sterile flies over the large northern state of Peten started in September 1988; the area dispersed was extended until the entire country was covered by January 1991. At the peak of dispersal, approximately 115 million flies per week were being dispersed over Guatemala. Case numbers collected each year in Guatemala from 1988 to 1995 are illustrated in FIGURE 1. The numbers are somewhat biased because field surveillance activities did not commence in the southeastern part of the country until well after the number of cases in the north had already started dropping because of sterile fly dispersals. The last autochthonous screwworm infestation in Guatemala

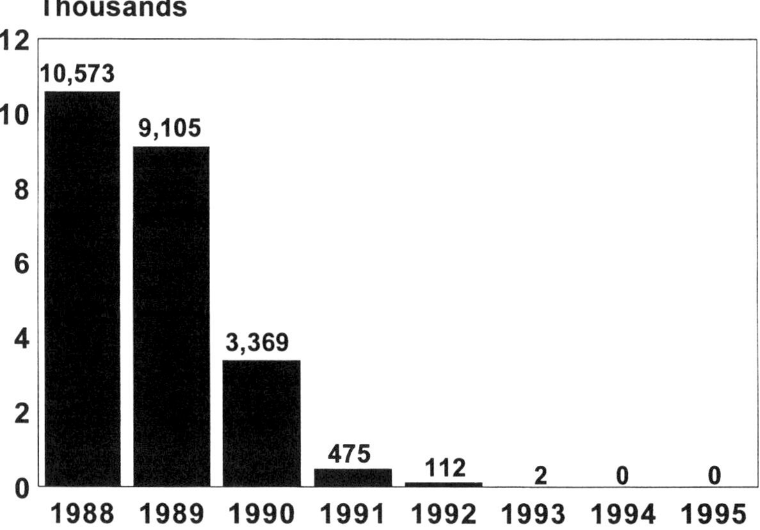

FIGURE 1. Screwworm infestations in Guatemala January 1988-April 1995.

was found May 10, 1992. Eight introduced infestations were found later in 1992 at the Playitas inspection station, one of a number of inspection stations situated near the borders with Honduras and El Salvador on main livestock transportation routes.

Two additional introduced cases, in February and April 1993, were found at the same inspection station. The number of flies dispersed was decreased starting with the northern state of Peten until by January through June 1993, an average of only 41 million sterile flies per week was dispersed. The number of flies dispersed continued to decrease until all dispersals ceased at the end of December 1993. An official declaration of freedom from screwworms was held on May 20, 1994. The program office was closed on June 30, 1994.

Belize

The Commission signed an agreement on August 2, 1988, with the Belize Ministry of Agriculture and Fisheries (MAF) to eradicate screwworms from Belize. The program office in Belmopan was dedicated on August 25, 1989. Dispersal of sterile flies commenced over the northern 40% of the country on August 26, 1989; 100% coverage of the country with sterile flies was achieved on April 10, 1990. Screwworm case numbers by year from 1989 to 1995 are given in FIGURE 2. This number is somewhat biased, because sterile fly dispersal started in the north, where the majority of cattle are found, nearly 3 months before all field inspectors were hired, trained, and provided with transportation. Interviews with livestock producers indicated that much higher numbers of screwworm infestations had occurred prior to the start of dispersal. A monthly high, after the program began, of 88 cases occurred in November

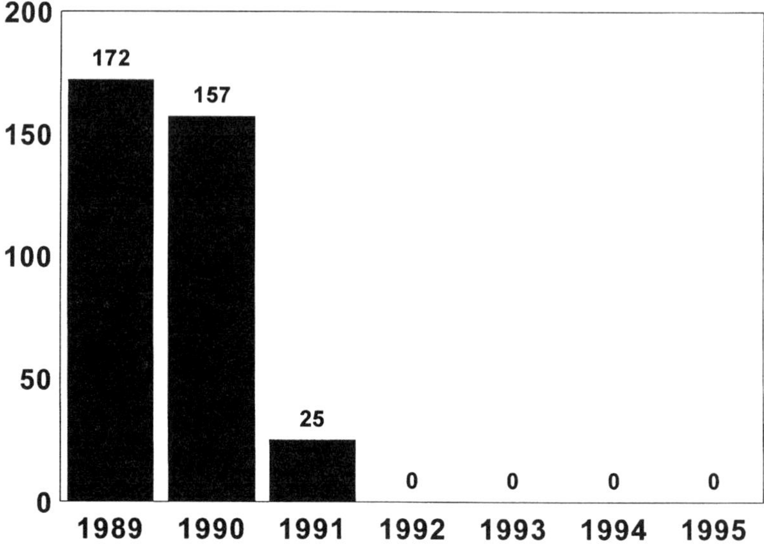

FIGURE 2. Screwworm infestations in Belize October 1989–April 1995.

1989. At the peak of dispersal, an average of 24 million sterile flies per week was dispersed. The last autochthonous screwworm infestation was discovered on July 1, 1991. No cases have been found since then. Dispersal of sterile flies over the northern half of the country ceased on June 1, 1992, and all dispersals ceased on December 31, 1992. The country was officially declared screwworm free on May 22, 1994. The program office was closed June 30, 1994.

El Salvador

The United States Department of Agriculture (USDA) signed an agreement on July 24, 1991, with the El Salvador Ministry of Agriculture and Livestock (MAG) to eradicate screwworms. A program office had already been opened and field surveillance activities initiated, resulting in an early start on obtaining data on the number of screwworm infestations. The number of infestations reported each year starting in October 1991 is given in FIGURE 3. Aerial dispersal of sterile flies over the entire country started in October 1991. At the peak of dispersal, an average of 24 million flies per week was being dispersed. A maximum of 800 cases in 1 month occurred in December 1991. Only small numbers of infestations were found after March 1993. Only two infestations occurred in 1994. One, in January, was autochthonous and another, in May, was an introduced case. No cases have occurred in 1995. All dispersal of sterile flies ceased in July 1994. An official declaration of freedom from screwworms is scheduled for June 19, 1995. The program office is scheduled to be closed on June 30, 1995.

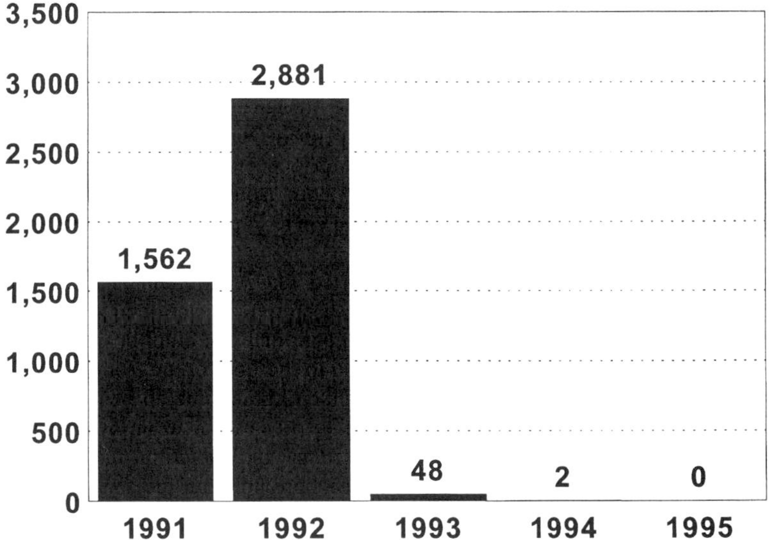

FIGURE 3. Screwworm infestations in El Salvador October 1991–April 1995.

Honduras

The USDA signed, on July 26, 1991, an agreement with the Honduras Secretariat of Natural Resources (SRN) to eradicate screwworms. A program office was placed in operation in September 1991, and field surveillance activities were initiated in the southwestern half of the country shortly afterwards. Sterile fly dispersal over the southwestern half of the country commenced in November 1991. The area of dispersal was increased until 100% of the country was being dispersed by June 1993. At the peak of dispersal, 120 million flies per week were scheduled to be dispersed. The area of dispersal began decreasing from the southwestern part of the country in June 1993 at about the same time that dispersal started in the easternmost part of the country. Only about 33% of the country is being dispersed in the first half of 1995, and plans are to decrease this to about 16%, dispersing only in areas adjacent to the border with Nicaragua, beginning July 1995. If the program in Nicaragua continues to progress satisfactorily, plans are to eliminate all dispersal of sterile flies in Honduras on September 30, 1995.

The number of screwworm infestations found since October 1991 is given in FIGURE 4. The number is somewhat biased, because surveillance country wide did not commence until well after the number of infestations in the southwestern half of the country had markedly decreased because of program activities.

Nicaragua

The USDA signed an agreement, on November 26, 1991, with the Nicaragua Ministry of Agriculture and Livestock (MAG) to eradicate screwworms. A program

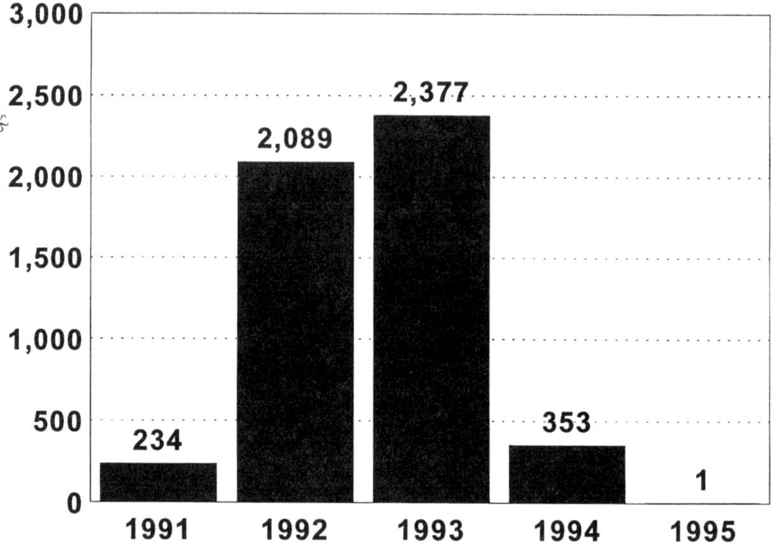

FIGURE 4. Screwworm infestations in Honduras October 1991–April 1995.

office was in operation in May 1992, and field surveillance country wide was initiated in July 1992. The number of screwworm infestations reported by year is given in FIGURE 5. A peak in one month of 3595 cases, in June 1993, was found. Dispersal of sterile flies over the northern half of the country commenced in July 1993, and 100% coverage was achieved in July 1994. Approximately 120 million flies per week are required to disperse programmed numbers over the entire country. The number of infestations country wide has dropped markedly. Only 137 infestations were found in March 1995. Plans are to continue dispersal over 100% of the country until the program in Costa Rica is well under way.

Costa Rica

The USDA signed, on November 29, 1993, an agreement with the Ministry of Agriculture and Livestock (MAG) to eradicate screwworms. The Screwworm Program office is being set up and will be in operation in June 1995, with field surveillance activities scheduled to start by September 1995. Dispersal of sterile flies over 100% of the country is scheduled to commence in January 1996. Approximately 60 million sterile flies per week will be needed to cover the entire country at the programmed rate.

Panama

The USDA signed an agreement on February 11, 1994, with the Ministry of Agriculture and Livestock Development (MIDA) to eradicate screwworms, construct

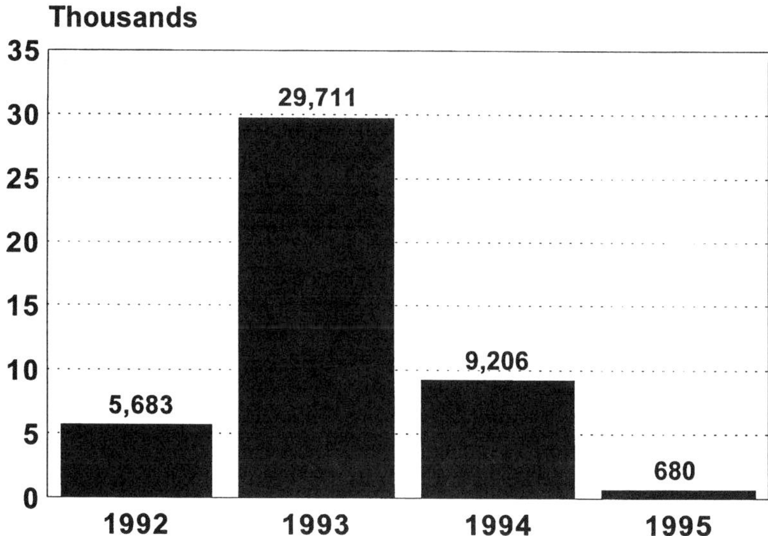

FIGURE 5. Screwworm infestations in Nicaragua July 1992-March 1995.

a sterile fly production facility, and establish a permanent biological barrier in Panama. The Regional Screwworm Program office is scheduled to relocate in Panama in July 1995, and field surveillance activities are planned to commence in 1996. Dispersal of sterile flies over the entire country is scheduled to commence as early as January 1997. Approximately 80 million sterile flies per week will be needed to cover the entire country at the programmed rate.

Plans call for the construction of a new sterile fly production plant east of Panama City near the Pacora River as soon as funds become available. A new plant would permit the current plant near Tuxtla Gutierrez, Chiapas, Mexico, to be closed. That plant has been in continuous operation 24 hours per day, 365 days per year since 1976. The costs of operating and maintaining that plant are steadily increasing. The cost of aerial transportation of pupae from Tuxtla Gutierrez, Mexico, to dispersal centers in countries where the program is operating is increasing as the distance from the plant to dispersal centers is increasing. The danger of fertile flies escaping so far inside eradicated zones is an ever-increasing concern.

SUMMARY

The Screwworm Eradication Program has been extremely successful in its efforts to achieve its goal of eradication of screwworms through Central America and establishment of a permanent biological barrier in the eastern half of Panama. Following eradication of screwworms from Mexico in 1991, eradication was achieved in Belize in 1992, in Guatemala in 1993, and in El Salvador in 1994. Honduras has been free of screwworms since January 1995, and the number of cases in Nicaragua has dropped, as of April 1995, to about 4% of the average number of cases found during the period June-August 1993.

REFERENCES

1. KNIPLING, E. F. 1985. Sterile insect technique as a screwworm control measure: the concept and its development. Misc. Publ. Entomol. Soc. Am. No. **62:** 4-7.
2. MEADOWS, M. E. 1985. Eradication program in the southeastern United States. Misc. Publ. Entomol. Soc. Am. No. **62:** 8-11.
3. BUSHLAND, R. C. 1985. Eradication program in the southwestern United States. Misc. Publ. Entomol. Soc. Am. No. **62:** 12-15.
4. BRAM, R. A. 1985. Future screwworm eradication programs. Misc. Publ. Entomol. Soc. Am. No. **62:** 37-40.
5. SNOW, J. W., C. J. WHITTEN, A. SALINAS, J. FERRER & W. H. SUDLOW. 1985. The screwworm (*Cochliomyia hominivorax*) (Diptera: Calliphoridae) in Central America and proposed plans for its eradication south to the Darien Gap in Panama. J. Med. Entomol. **22:** 353-360.

Central America Regional Screwworm Eradication Program (Benefit/Cost Study)

JOHN H. WYSS AND THOMAS J. GALVIN

United States Department of Agriculture
Animal and Plant Health Inspection Service
International Services
Screwworm Eradication Program
Mexico City, D.F., Mexico

INTRODUCTION

The screwworm fly, *Cochliomyia hominivorax* (Coquerel), is a parasite that attacks all warm-blooded animals including humans. Infested animals, unless detected and treated, will eventually die from the infestation or from secondary bacterial infection. This parasite has caused significant losses to the livestock industries of the Americas through death and weight loss, and increased veterinary, labor, medicine, and insecticide costs.

Screwworm eradication is a technically feasible, effective, and proven program.[1] A screwworm eradication program was initiated in the southeastern United States in 1957 and was successful in 1959.[2] Due to the successful program in the southeastern United States, a program was begun in the southwestern United States in 1962 and progressed until the United States was declared screwworm free in 1966.[3] To keep the United States free of screwworms, it was necessary to maintain a 325-km-wide barrier consisting of weekly aerial dispersal of sterile screwworm flies along the Texas-Mexico border during the winter months and the entire United States-Mexico border during the spring, summer, and fall months. Even with these precautions, the United States routinely became reinfested, with large screwworm outbreaks occurring in 1968 and 1972-1976. Beginning in 1972, to reduce the cost of maintaining the barrier, to increase protection for the United States, and to expand the benefits of screwworm eradication to Mexico, an agreement was signed between Mexico and the United States establishing the Mexico-United States Commission for the Eradication of Screwworms (Commission). The Commission progressively moved the barrier from the United States-Mexico border to the Isthmus of Tehuantepec in southern Mexico by 1984. This measure was successful in protecting the United States from reinfestation; the last screwworm outbreak in the United States occurred on August 31, 1982. Since that time, the annual producer benefits to the US livestock producers have been equivalent to $500 million in 1994 US dollars per year.[4] The Commission expanded the screwworm eradication program to cover all of Mexico in 1985, and Mexico was declared screwworm free on February 25, 1991. Since that time, annual producer benefits to the Mexican livestock producers have been equivalent to $275 million per year in 1994 US dollars.[5] The Commission entered into agreements with

Guatemala and Belize in1986 and 1988, respectively. Both Guatemala and Belize were declared screwworm free in 1994. Currently, Cooperative Screwworm Eradication Programs between the United States and other Central American countries, with the sterile flies being provided by the Commission, are in various stages in El Salvador, Honduras, and Nicaragua. Cooperative program activities are expected to begin in Costa Rica and Panama in 1995. The plan is to eradicate screwworms throughout Central America and establish a permanent sterile screwworm fly barrier in the eastern half of Panama. Panama is the ideal location for a permanent sterile fly barrier of this type and would be the most cost effective, because of its geographical location with reduced land mass and no animal movements from South America.

Although screwworm eradication is an expensive undertaking, the benefits of eradication justify the costs. The economics of living with screwworms and the benefit-to-cost ratio of eradication are important factors in selling an eradication program to foreign government officials. Foreign officials are often presented with information from other countries that they find hard to interpret and relate to their own country. This paper describes how the results of studies done in the United States and Mexico were extrapolated and updated to estimate the benefits that could be expected in Central America from a screwworm eradication program and how this information was used to personalize the estimates for each country to gain support for a regional screwworm eradication program.

MATERIALS AND METHODS

The information from a study titled "Evaluation of the Mexican-American Screwworm Eradication Program in Mexico"[5] carried out by the Texas A&M Extension Service, of the Texas A&M University System, under a US Department of Agriculture contract, was reviewed and the economic information updated to 1994 US dollars using the consumer price index published in the 1995 edition of the Economic Report of the President.

Livestock population data were obtained from the 1993 Production Year Book, published by the Food and Agriculture Organization (FAO) Statistics Division.[6] This animal population information was gathered and compiled for each Central American country and the region as a whole.

The total cattle populations of Central America and each country individually were allocated, using the criteria of the Texas A&M study, to the following groups: dairy cows, beef cows, calf crop, replacement heifers, bulls, stocker cattle, and feeder cattle. The only exception was with respect to dairy cows where actual available information was used. For the purpose of the Texas A&M study, stocker cattle were considered cattle on native pasture and feeder cattle were considered cattle on improved pasture. The animal census information was converted to expansion animal units, with an animal expansion unit being defined as female breeding animals, stocker cattle, feeder cattle, adult horses, and work animals. Expansion animal units for cattle were calculated based on the actual dairy cow numbers and beef cows estimated at 36% of total herd. Stocker and feeder cattle were calculated at 50% of the remaining herd after subtracting dairy and beef cows and an additional 35% to account for calf crop, replacement heifers, and bulls. The resulting number was

TABLE 1. Central America Animal Populations (Thousands)[a]

Country	Total Cattle	Dairy Cows	Swine	Sheep	Goats	Horses	Mules and Donkeys
Belize	51	6	26	4	1	5	4
Guatemala	2077	382	769	438	77	114	47
El Salvador	1243	290	308	5	15	95	26
Honduras	2388	386	587	8	27	172	91
Nicaragua	1600	252	570	4	6	250	53
Costa Rica	2175	318	244	3	2	114	12
Panama	1399	112	256	0	5	156	4
Total	10,933	1,746	2,760	462	133	906	237

[a] 1993 Production Yearbook, FAO Statistics Division.

distributed equally between the stocker and feeder cattle. Other livestock species were converted to expansion animal units using the same criteria as used by the Texas A&M study. Sixty percent of the entire population of sheep and goats, 10% of swine, 12.25% of horses, and 69.3% of the remaining horses and all other work animals were considered in determining expansion animal units.

Producer benefits were calculated by multiplying the various categories of expansion animal units by the updated benefit per expansion animal unit.

TABLE 2. Cattle Population Breakdown (Estimate) 1994 Central America

Categories		Animals
Dairy cows	16.0% of total herd	1,746,000
Beef cows	36.0% of total herd	3,935,880
Calf crop	15.0% of total herd	1,639,950
Replacement heifers	15.0% of total cows	852,282
Bulls	5.0% of total cows	284,094
Stockers	50.0% of remaining animals	1,237,397
Feeders	50.0% of remaining animals	1,237,397
Total cattle population		10,933,000

TABLE 3. Other Livestock Populations and Expansion Animal Units 1994 Central America

Categories	Total Population	Percent of Total Herd	Expansion Animal Units
Swine	2,760,000	10.00%	276,000
Sheep	462,000	60.00%	277,200
Goats	133,000	60.00%	79,800
Horses	906,000	12.25%	110,985
Work animals	237,000	Adults	787,945

To reflect the forward and backward linkages in the economy, the annual producer benefits were multiplied by 3.5, a factor used in previous studies. This is considered a conservative factor. This number was doubled to include the effect on the general economy of producer benefits and consumer benefits.

Producer benefits to perpetuity were calculated using 3%, 6%, and 8% discount rates. The three figures were calculated by dividing the discount rate into 100 and multiplying the quotient by the annual producer benefits. Using actual and estimated program costs for each country, the benefit-to-cost ratios (B/C) were calculated for both the 3% discount rate and the 8% discount rate.

TABLE 4. Total Annual Producer Benefits Expected by Livestock Category 1994 Central America

	Expansion Animal Units	Benefit $/Head	Total Benefits
Cattle			
Dairy operations	1,746,000	$15.23	$26,591,580
Cow-calf operations	3,935,880	$8.48	$33,376,262
Stocker cattle	1,237,397	$3.58	$4,429,881
Feeder cattle	1,237,397	$1.72	$2,128,323
Annual Producer Benefits Cattle			$66,526,047
Other livestock			
Swine	276,000	$18.53	$5,114,280
Sheep	277,200	$1.67	$462,924
Goats	79,800	$3.25	$259,350
Horses	110,985	$2.98	$330,735
Work animals	787,945	$0.95	$748,548
Annual producer benefits other livestock			$6,915,837
Total annual producer benefits all livestock			$73,441,884

TABLE 5. Screwworm Eradication Annual Producer Benefits

Country	Producer Benefits	Percent of Total
Belize	$347,135	0.5%
Guatemala	$15,410,565	21.0%
El Salvador	$9,385,248	12.8%
Honduras	$15,983,554	21.8%
Nicaragua	$11,051,356	15.0%
Costa Rica	$13,458,899	18.3%
Panama	$7,805,127	10.6%
Total producer benefits	$73,441,884	100%

RESULTS

The animal populations of the Central American countries are shown in TABLE 1.

The breakdown by categories of the total Central American cattle herd is shown in TABLE 2.

The dairy cows, beef cows, stocker cattle, and feeder cattle represent the expansion animal units for cattle. The populations and the expansion animal units for other livestock categories are shown in TABLE 3.

The annual producer benefits by livestock category and the total for Central America are shown in TABLE 4.

The share of each of the Central American countries of the annual producer benefits is shown in TABLE 5.

The livestock producer benefits have an additional positive effect on the general economy, due to the forward and backward linkages in the economy of $257 million. Consumer benefits were estimated to be equal in amount. Adding annual producer benefits, producer benefit effects on the general economy, and consumer benefits gives a grand total of $588 million overall effect on the economy of Central America. The benefits to perpetuity were determined using three different discount rates for all of Central America. The producer benefits to perpetuity ranged from $918 million to $2.4 billion depending on the discount rate used (TABLE 6).

TABLE 6. Total Producer Benefits to Perpetuity 1994 Central America

Discount Rate	Years	Total Benefits Annually	Total Benefits to Perpetuity
3%	33.3	$73,441,884	$2,448,062,800
6%	16.7	$73,441,884	$1,224,031,400
8%	12.5	$73,441,884	$ 918,023,550

TABLE 7. Screwworm Eradication Programs Benefit/Cost Ratios 1994 Central America

| Country | Annual Producers Benefits | Producer Benefits to Perpetuity | | Program Cost | B/C Ratio 8% DR | B/C Ratio 3% DR |
		8% Discount Rate	3% Discount Rate			
Belize	$357,135	$4,339,188	$11,571,167	$2,000,000	2.2	5.8
Guatemala	$15,410,565	$192,632,063	$513,685,500	$28,000,000	6.9	18.3
El Salvador	$9,385,248	$117,315,600	$312,841,600	$13,917,454	8.4	22.5
Honduras	$15,983,554	$199,794,425	$532,785,133	$42,395,342	4.7	12.6
Nicaragua	$11,051,356	$138,141,950	$368,378,533	$50,566,186	2.7	7.3
Costa Rica	$13,458,899	$168,236,238	$448,629,967	$35,000,000	4.8	12.8
Panama	$7,805,127	$97,564,088	$260,170,900	$28,368,833	3.4	9.2
Total	$73,441,884	$918,023,550	$2,448,062,800	$200,247,815	4.6	12.2

The projected benefit-to-cost ratios for the regional screwworm eradication program, as well as for individual countries, were calculated. This information is shown in TABLE 7.

DISCUSSION OF RESULTS

The annual producer benefits of $73,441,884 resulted from decreases in death loss, veterinary expenses, veterinary medicines, insecticides, inspection and handling costs, and increases in meat and milk production. These producer benefits have an additional positive effect on the general economy of each country and the region in general due to the forward and backward linkages in the economy. This multiplier effect was considered, on the basis of previous studies, to be 3.5. The additional effect on the general economy of the producer benefits was therefore estimated to be $257 million. In addition, it is estimated that consumer benefits are of equal magnitude. Therefore, this number can be doubled to $514 million to include both the general effect on the economy of the producer benefits and consumer benefits. Adding to this the producer benefits results in an overall positive effect of $588 million annually on the general economy of the Central American countries.

The benefits to perpetuity were calculated at 3%, 6%, and 8% discount rates. While the 8% discount rate represents the true cost of money, it is customary to use a 3% discount rate for programs for the public good.

The benefit-to-cost ratio for the screwworm eradication programs averaged from 4.6 to 12.2, depending on the discount rate used. Even the smaller number is very supportive of an eradication program.

In addition, screwworm eradication has a significant human and wildlife health component which has not been included in this analysis. Also, once screwworm is eradicated, the elimination of the use of insecticides to control this pest is an additional benefit.

The exercise clearly demonstrated that the economic benefit to the livestock producers in Central America would be significant, and strongly supported the idea of a regional screwworm eradication program. This information proved to be very helpful in selling the eradication program in Central America and obtaining significant levels of support from the participating countries.

REFERENCES

1. KNIPLING, E. F. 1985. Sterile insect technique as a screwworm control measure: the concept and its development, Entomological Society of America. Miscellaneous Publications No. 62: 4–7.
2. MEADOWS, M. E. 1985. Eradication program in the southeastern United States. Entomological Society of America, Miscellaneous Publications No. 62: 8–11.
3. BUSHLAND, R. C. 1985. Eradication program in the southwestern United States. Entomological Society of America, Miscellaneous Publications No. 62: 12–15.
4. SCRUGGS, C. G. 1975. The Peaceful Atom and the Deadly Fly. Jenkins Publishing Co. Pemberton Press. Austin, Texas.
5. DAVIS, E. E. & C. E. HOELSCHER. 1985. Evaluation of the Mexican-American Screwworm Eradication Program in Mexico. Texas Agricultural Extension Service, The Texas A&M University System. Contract No. 12-16-93-142, Vol. 1, II, and III.
6. 1993. Animal Health Year Book. FAO, OIE, and WHO, No. 33: 122–123.

Isolation and Identification of Bacteria Associated with the Screwworm Fly *Cochliomyia hominivorax*, Coquerel and Its Myiasis[a]

M. CABALLERO,[b] G. HERNÁNDEZ,[b]
F. POUDEVIGNE,[c] AND I. RUIZ-MARTÍNEZ [d]

[b]*Bacteriology Laboratory*
PIET
Post Office Box 304
Escuela de Medicina Veterinaria
UNA
Heredia, Costa Rica

[c]*ARS Cooperator*
United States Department of Agriculture
Agriculture Research Service
Screwworm Research Program
Panama

[d]*Department of Animal Biology and Ecology*
Area of Parasitology
Faculty of Experimental Sciences
University of Jaén
E-23071-Jaén, Spain

INTRODUCTION

The relationship between flies and disease has been of great interest for a long time. But since the beginning of the 20th century and the development of determinative bacteriology, the importance of these insects as vectors of disease has been established. It is a well-known fact that flies could be mechanical vectors for the transmission of disease, but it is necessary to demonstrate that these insects carry the germs that could cause disease.[1] There are several studies on bacteria associated with insects, but few studies have been done with the normal microflora.

The first studies done concerned the house fly (*Musca domestica*). There are other anthropozoogenic flies that have been studied.[2] These studies dealt with the

[a]This investigation was funded in part by a grant given to F. Poudevigne by the United States Department of Agriculture, Agriculture Research Service, Screwworm Research Program, Central America.

flora associated with the adult stages of these insects, and the 1960s is when studies had been conducted with the flora associated with larval stages of these flies.[3] Few studies dealt with flies that caused myiasis, such as *Lucilla cuprina* Wied,[4] *Cochliomyia hominivorax*,[5] and *Wohlfahrtia magnifica* (Schn.).[3]

The eradication of the screwworm has been successful from the United States of America and Mexico by means of the sterile male technique (SMT), and it is being used in the Central American countries.[6] The evaluation of wild populations and the effects of SMT has been carried out through several trapping and spontaneous oviposition techniques, such as the Animal Sentinel (SA). We developed an artificial wound (AW) as an alternative method to obtain spontaneous oviposition of wild flies.[7] In this way, we studied several factors about myiasis-producing flies (pH, temperature, and especially, associated microflora). The concept is to consider the myiasis as a whole microhabitat. In this study we have attempted to isolate and characterize the microflora on different life cycle stages of the screwworm *Cochliomyia hominivorax* Coquerel (Diptera: Calliphoridae) and the associated microflora in the wound myiasis.

MATERIALS AND METHODS

This study was conducted from September to December 1993 and from July to November 1994. The samples were taken from a Research Farm in Alto de Ochomogo, Cartago, Costa Rica, and the animals were kept in the farm of the School of Veterinary Medicine, Universidad Nacional, Heredia, Costa Rica.

The flies were bred under laboratory conditions; the original flies were captured from rotten beef liver with entomological nets. The female flies were gravid and ready to lay their eggs. They were placed in entomological cages and they were fed a mixture of water and honey (1:1) for three days. Black filter papers were placed so that the eggs should be laid there; these papers were moistened with larvae food (mixture of blood power, powder milk, powder chicken egg, gelatin, and formol). They were kept on a Petri dish for 12 to 24 hours. After the larvae instar 1 were hatched, they were transferred to a plastic container and kept in a water bath (35-40°C). The filter paper was eliminated after the instar 2 emerged. After 24 hours, the instar 3 emerged and the larvae were placed under a metallic container with wood dust over the water bath. Three days later the pupae were placed in entomological cages for the emerging of the adult flies.

We used three female sheep for each experiment, two sheep for the myiasis and one for control. The wound was made with the following size: 10 cm/diameter and 2 cm/deep.

SAMPLES

Six and 96 hours after the wound was made, the first swabs were taken from the edge and the bottom of the wound. The gravid flies were put on the wound. Twenty four hours after the ovoposition, 10 larvae instar 1 were taken under sterile conditions and swabs from the edge, the wall, and the bottom of wound. The same procedure was done for larvae instar 2, instar 3, crawl off, young pupae, and mature pupae. The adult flies were washed in sterile saline solution, the solution was discharged, and the flies were macerated for analysis of the internal flora.

We analyzed 731 samples, 443 swabs from the different stages of the wound, and larvae from the different instars, 26 pupae, and 62 adults.

Bacteriological Studies

The swabs were placed on three bacteriological media, blood agar, Mc. Conkey agar, and salt manitol agar (Difco Laboratories, Detroit, MI). The plates were incubated at 37° (± 1°C) for 24 hours ($\pm$ 2h). Bacteria were isolated and further identified by the Api Systems (Analytical Profile Index API 20E, API RAPID STREP, API NE, API STAPH).

The samples from the larvae were washed in 5 ml of sterile saline solution 0.85% and shaken for five minutes (Vortex-genie and k-550-g). The resulting fluid was analyzed for external microflora, following the procedure done with the swabs. The washed larvae, pupae, and adults were placed in sterile saline solution 0.85%, homogenized in a mortar, and 1 g was analyzed for the internal flora, following the procedure done with the swabs.

All data were recorded and processed on the DBASE Plus III Archives.

RESULTS AND DISCUSSION

Forty different bacteria were isolated and identified from the swabs, as shown in TABLE 1. We isolated *Escherichia coli* 65 times, the bacterium most frequently isolated in our study, *Proteus* sp. was isolated 43 times, *Proteus mirabilis* 28 times, *Staphylococcus* sp. was isolated 49 times, and *Streptococcus* sp. was isolated 45 times. Probably the two later species are related to the flora of the wound. We found that the gram-negative flora was more predominant than the gram-positive flora. As shown in TABLE 1, the amount of different bacterial species increased with the development of the myiasis. The adults carried many different bacteria that are transmitted to the host.

The external flora of larvae is shown in TABLE 2. Few bacterial species were isolated. Thirty-three gram-negative bacteria were isolated and 46 gram-positive bacteria; isolated far more frequently was *Staphylococcus aureus*.

The internal flora of the larvae, pupae, and flies is shown in TABLE 3. We found *Escherichia coli* 26 times, *Proteus* sp. 29 times, *Proteus mirabilis* 36 times, *Proteus vulgaris* 5 times, *Providencia rettgeri* 10 times, *Providencia stuartii* 3 times. Among the gram-positive bacteria, *Staphylococcus* sp. was identified 25 times, *Streptococcus* sp. 2 times only in larvae instar 3 when they are more active, and probably this means that these bacteria were fed from the wound. Among all the isolates, gram-negative bacteria were by far more abundant than gram-positive bacteria (158 isolates vs. 58 isolates). We found more *Staphylococcus* and *Streptococcus* species than other authors.[5] We feel that more research has to be done to determine this discrepancy.

There are some bacterial species like *Moraxella phenylpyruvica*, *Pasteurella pneumotrophyca*, and *Enterococcus faecalis* that were isolated on the insects but

TABLE 1. Bacteria Isolated from the Swabs of Myiasis Caused by *Cochliomyia hominivorax*, Coqueral[a]

Bacteria	S1	S2	S3	S4	S5	S6	S7	S8	S9
Agrobacterium radiobacter	1	—	—	—	—	—	1	2	—
Aetromaonas hydrophila	—	1	—	—	—	—	—	—	—
Aeromonas salmonicida	—	1	—	—	—	—	—	—	—
Alcaligenes xiloxidans	—	—	—	2	—	—	—	—	—
Aerococcus viridans	—	—	1	—	—	1	—	—	—
Bacillus sp.	1	1	—	—	1	—	—	—	—
Citrobacter sp.	—	1	—	—	1	—	—	—	—
Citrobacter freundii	—	—	—	—	—	—	1	—	—
Cryseomonas luteola	—	—	—	1	—	—	—	—	—
Escherichia coli	6	8	15	5	15	6	4	6	—
Enterobacter sp.	1	—	—	—	—	—	—	—	—
Enterobacter cloacae	2	2	2	—	1	—	—	—	—
Enterobacter sakasakii	—	2	1	1	—	—	—	—	—
Enterococcus faecium	4	—	—	1	5	—	—	—	—
Klebsiella sp.	1	1	1	1	—	—	1	—	—
Klebsiella pneumoniae	—	1	—	—	—	—	—	—	—
Klebsiella oxytoca	—	1	1	1	—	—	—	—	—
Moraxella phenylpyruvica	—	—	—	—	—	—	2	—	—
Oligella ureolytica	—	—	—	—	—	1	—	—	—
Proteus sp.	5	1	11	6	6	4	10	—	—
Proteus mirabilis	—	1	1	11	4	4	7	—	—
Providencia stuartii	—	—	—	—	—	—	—	2	—
Providencia rettgeri	1	—	—	—	1	—	—	—	—
Pseudomonas sp.	—	—	—	—	—	1	—	—	—
Pseudomonas aeruginosa	—	—	2	2	—	—	—	—	—
Pseudomonas stutzeri	—	1	—	—	—	—	—	—	—
Salmonella sp.	—	—	—	—	2	1	—	—	—
Streptococcus sp.	2	—	—	—	—	—	15	26	4
Staphylococcus sp.	18	6	3	10	2	3	5	1	1
Staphylococcus aureus	3	—	—	—	—	—	6	11	4
Staphylococcus hycus	1	—	—	—	—	—	1	—	—
Staphylococcus sciuri	—	—	—	—	—	—	—	2	—
Staphylococcus simulans	—	—	—	—	—	—	3	2	—
Staphylococcus epidermidis	1	—	—	—	—	—	—	—	—
Staphylococcus lentus	1	—	—	—	—	—	—	—	—
Staphylococcus warneri	1	—	—	—	—	—	—	—	—
Staphylococcus xiloxus	4	—	—	—	—	—	2	—	—
Xantomonas maltophila	—	—	—	—	1	—	—	—	—

[a] S1: swab at 6 hours after excision, S2: during ovoposition, S3: 24 hours after ovoposition (larvae 1), S4: 48 hours after ovoposition (larvae 2), S5: 72 hours after ovoposition (young larvae 3), S6: 96 hours after ovoposition (mature larvae 3), S7: 120 hours after ovoposition (crawl off), S8: 144 hours after ovoposition (young pupae), S9: 168 hours after ovoposition (mature pupae).

TABLE 2. External Bacteria Isolated from Larvae of *Cochliomyia hominivorax,* Coquerel[a]

Bacteria	L1	L2	JL3	ML3
Aeromonas salmonicida	—	—	1	—
Bacillus sp.	1	—	—	—
Citrobacter sp.	—	—	2	—
Enterobacter sp.	—	—	1	—
Enterococcus faecium	—	—	—	1
Escherichia coli	1	2	4	—
Klebsiella sp.	—	1	—	—
Moraxella phenypyruvica	—	—	1	—
Proteus mirabilis	—	3	3	1
Proteus rettgeri	—	—	1	—
Pasteurella pneumotrophyca	—	—	1	—
Proteus sp.	—	1	1	1
Providencia stuartii	—	—	1	—
Pseudomonas aeruginosa	—	1	—	—
Staphylococcus aureus	6	5	8	8
Staphylococcus epidermides	—	—	—	1
Staphylococcus sciuri	—	1	—	—
Staphylococcus xilosus	2	1	—	2
Staphylococcus sp.	3	1	1	1
Streptococcus sp.	1	—	1	3

[a] L1 = larvae 1. L2 = larvae 2. JL = juvenile larvae 3. ML3 = mature larvae 3.

never were isolated in the wounds. This can suggest that these bacteria are truly microflora of this insect.

This is just a preliminary study about the microflora of the screwworm and the myiasis that it causes. More studies have to be done to elucidate several aspects, specially the role of bacteria like *Staphylococcus* and *Streptococcus*. The importance of this study is that we have isolated many bacteria that can be used for artificial wounds and other purposes.

SUMMARY

We studied eight different myiasis of sheep caused by screwworm flies under laboratory conditions. Swabs were taken from the wound, before, during, and after the myiasis. Seven hundred and thirty-one samples were streaked on different bacteriological media. All samples were identified by Api System (bioMerieuex). We found thirty-eight different bacterial species in the exudates from the myiasis (before, during, and after the parasitic cycle). The analysis of bacterial flora of the screwworm showed, on larvae stage 1, 10 different bacterial species, on larvae 2, 12 bacterial species; larvae 3, 15 bacterial species; crawl off, 15 bacteria species, pupa, 9 bacterial species; and adults, 2 bacterial species and on the pioneer fly, 14 different bacterial species were isolated and identified.

TABLE 3. Internal Bacteria of the Fly *Cochliomyia hominivorax* Coquerel[a]

Bacteria	PF	L1	L2	JL3	ML3	CO	JF	MP	A
Aeromonas hydrophyla	—	—	—	1	—	2	—	—	—
Aerococcus viridans	1	—	—	—	—	—	—	—	—
Alcaligenes xiloxidans	2	—	—	—	—	—	—	—	—
Bacilbus sp.	1	—	—	—	—	—	—	—	—
Citrobacter sp.	4	—	—	2	—	—	—	—	—
Citrobacter freundii	—	—	2	—	—	—	—	—	—
Enterobacter sp.	—	1	—	1	—	1	—	—	—
Enterobacter cloacae	1	—	2	1	—	1	2	—	—
Enterobacter sakasakii	—	2	—	—	—	—	—	—	—
Escherichia coli	8	5	3	1	—	4	4	—	1
Klebsiella sp:	4	—	1	—	—	2	1	—	—
Klebsiella oxytoca	6	2	1	—	—	—	—	—	—
Klebsiella pneumoniae	—	—	—	—	—	—	1	—	—
Moraxella phenylphyruvica	—	—	—	—	—	2	—	—	—
Proteus sp.	6	1	2	3	3	9	1	4	—
Proteus mirabilis	11	—	4	5	9	1	1	5	—
Proteus vulgaris	4	—	—	—	—	—	1	—	—
Providencia rettgeri	6	—	—	—	—	4	—	—	—
Providencia stuartii	—	—	—	2	—	1	—	—	—
Pseudomonas vesicularis	—	—	—	—	—	1	—	—	—
Pseudomonas chlorophis	—	—	—	—	—	1	—	—	—
Salmonella sp.	—	—	—	—	1	—	—	—	—
Streptococcus sp.	—	—	—	—	2	—	—	—	—
Staphylococcus sp.	3	3	1	1	4	6	1	3	2
Staphylococcus aureus	6	1	4	3	5	5	5	—	—
Staphylococcus xiloxus	—	1	—	—	—	1	—	—	—
Staphylococcus warnerii	—	—	—	—	1	—	—	—	—

[a] PF: pioneer fly, L1: larvae 1, L2: larvae 2, JL3: juvenile larvae 3, ML3: mature larvae 3, CO: crawl off, JF: juvenile pupae, MP: mature pupae, A: adult

ACKNOWLEDGMENTS

The authors appreciate the technical assistance of Mr. Javier Gonzalez, Mr. Gerardo Cervantes on field and laboratory studies, and Ing. Marco J. Murillo on the statistical analysis.

REFERENCES

1. RAMOS, C. J., R. PORTUONDO & C. DURANZA. 1985. Presencia de Bacterias Patógenas en las Moscas. Rev. Cub. Hig. y Epide. **23:** 147-151.
2. GRAHAM-SMITH, G. S. 1912. An investigation into the possibility of pathogenic microorganism being taken up by the larva and subsequently distributed by the fly. 41 st. Ann. Rep. & Supp. Med. Off. Col., 1911-1912. B: 330-335.

3. Ruiz-Martínez, I. 1990. Contribución al conocimiento de varios aspectos del díptero productor *Wohlfahrtia magnifica* (Schiner, 1962). Tesis doctoral. Universidad de Granada. Servicio de Publicaciones UGRA.

4. Emmens, R. L. & M. D. Murray. 1982. The role of bacterial odours in oviposition by *Lucilla cuprina* (Wiedemann) (Diptera: Calliphoridae), the Australian Sheep Blowfly. Bull. Ent. Res. **72:** 367–375.

5. Bromel, H., F. M. Duh, G. R. Erdmann, L. Hammack & G. Gassner. 1983. Bacteria associated with the screwworm fly (*Cochliomyia hominivorax,* Coquerel) and their metabolites. Endocytobiology **2:** 791–800.

6. Graham, O. H. 1985. Symposium on the eradication of the New World Screwworm in the United States and Mexico. ESA Miscellaneous Publications, no. 62.

7. Poudevigne, F. A., I. R. Martínez & J. B. Welch. 1995. pH and oviposition of the screwworm fly (*Cochliomyia hominivorax* Coquerel) (Diptera: Calliphoridae): application to an artificial wound. Laboratory and field results. 3rd Biennial Meeting Society for Tropical Veterinary Medicine. May, 1995. San José, Costa Rica.

The Application of Probabilistic Scenario Analysis for Risk Assessment of Animal Health in International Trade

A. S. AHL[a]

*USDA
Animal and Plant Health Inspection Service
Policy and Program Development
Riverdale, Maryland 20737*

INTRODUCTION

Dealing with risk is not new. People frequently alter their behavior to avoid the occurrence of hazards or unwanted events in their personal lives. Likewise, veterinary professionals, dealing with animal health issues, are required to evaluate risk on a daily basis. However, until recently, these professional evaluations of risk were made more informally, perhaps intuitively, without reference to any formal analytical system.

International travel and trade have always been popular. Ancient peoples traveled in camel or horse caravans, by ship, by foot. The travel was slow, and animals that were ill or diseased infrequently survived the rigors of travel to carry disease agents into new territory. Animal diseases therefore traveled slowly. However, international travel and trade are now most likely to occur by plane, so that an animal on the range today in one country can be on the range half-way around the world tomorrow.

In addition to the speed of international travel and transportation, the cost of international travel has decreased, making it accessible to many more citizens of the world. This has led to a very rapid increase in the number of travelers and the amount of trade in the past 10 years. All evidence suggests that these trends will continue.

International trade has also grown in popularity. In the past three years, passage of two international trade agreements, General Agreement on Tariff and Trade (GATT) and North American Free Trade Agreement (NAFTA), has spurred interest in developing systematic and scientific ways to evaluate risks to animal health. The trade agreements require that countries freely trade animals and animal products with each other, using regionalization and risk assessment to evaluate them. Refusal to import can only be based on a demonstration of threat to animal health grounded in systematic and scientific risk assessment.

---

[a] Current position; Director, USDA/Office of Risk Assessment and Cost-Benefit Analysis, Washington, DC 20250-3800.

To understand the significance of these developments, a brief review of general trade practices may be helpful. For most of the 20th century, imports from a given country have been based on the animal health status of that country. In some cases, the country may be divided into regions recognizing areas of that country that have varying status with respect to animal health. This is the concept of *regionalization.*

Regionalization for animal health has long been recognized domestically within the US. For example, infection of cattle with *Brucella abortus,* i.e., brucellosis, occurs only in a few states inside the US. States or areas without brucellosis are careful about accepting cattle into their borders that may carry the agent. US trading partners may also recognize this fact. In many situations, however, countries are not regionalized and the trading status for all parts of the country are based on a single designation of the country as "free" or "not free" of a given animal disease agent.

Veterinary professionals must use systematic and science-based evaluation of animal health in considering imports. Risk analysis, especially its analytic component, risk assessment, is the field that provides the structure for this activity. One method that has proved particularly useful is probabilistic scenario analysis (PSA). This paper lays out the principles of risk analysis and the general approach to PSA; the paper that follows provides an example of the use of this method as applied to animal import activities. Regionalization and risk assessment will support active and free international trade in animals and their products while providing the safety needed to maintain good health in our national herds.

NOMENCLATURE

As in all new fields, the nomenclature is often confusing and ambiguously used. Lack of standarization and consistency in communicating concepts and ideas hinders working relations. As a relatively new field, risk analysis is no different. In animal agriculture, our resources are few and our challenges are many. If agreement can be reached on the meaning of a few basic terms, it will facilitate basic communication with each other. In that way the agenda of promoting international trade can more easily move forward. To that end, the definitions of several words and phrases are suggested here.[1]

Risk analysis: the process that includes risk assessment, risk management, and risk communication.

Risk assessment: the process of identifying a hazard and evaluating the risk of a specific hazard, either in absolute or relative terms. It may be qualitative or quantitative.

Risk management: the pragmatic decision-making process concerned with regulating the risk.

Risk communication: open two-way exchange of information and opinion about risk leading to better understanding and better risk management decisions.

Safety: the degree to which risks are judged acceptable; a subjective measure of the acceptability of risk.

To the extent that the international animal health community can agree to careful use of these terms, we can increase our efficiency and effectiveness. By carefully avoiding the confusion of babble, we can preserve our scarce agricultural resources and boost productivity at the same time.

RISK ASSESSMENT

Risk assessment is a formal analytical system that provides estimates of the probability of occurrence of unwanted future events. This information is provided to a decision maker who must decide on the best course of action. Risk assessment has been used in engineering,[2] financial forecasting,[3] and environmental assessment[4] since the middle of this century, though in animal health, it is in its infancy.

Risk assessment basically answers three questions. The first, "what can go wrong?," provides the identification of the hazard of interest. The second question, "what is the probability it will go wrong?," evaluates how likely it is that the hazard will occur. The third question asks "what are the consequences if the hazard does occur?." The answer to the latter two questions is defined as "risk."[5] Though formally defined as probability and magnitude of the occurrence of the hazard, the term "risk" is used informally in several other ways. It may refer to the hazard itself, the probability of an event's occurrence, or to behavior or activity that could bring harm to an individual. When hearing the word risk, it is necessary to think about the context in which the word is used.

Risk assessment is the domain of the sciences, making use of concepts from many fields: epidemiology, pathology, virology and bacteriology, physiology, biochemistry, nutrition, probability, systems analysis, decision theory, and others. Risk assessment also utilizes information such as anecdotal evidence, expert opinion, and other items from less structured sources. However, the evidence used in a risk assessment and the sources of that evidence must be displayed so that there is no ambiguity about the kind of information used in the analysis. The pertinent information is structured to assist the risk manager in making a decision about mitigating the likelihood or consequences of the adverse events.

Risk management is the domain of the decision maker who chooses among courses of action, the acceptability of the risk posed by each, and the use of mitigation measures to decrease the risk. Larger spheres of policy, diplomacy, politics, economics, and legal issues are also a part of risk management. It is important to remember that the risk assessment process is initiated by risk managers and the output is utilized by them in making decisions.

At minimum, the requirements are that risk assessments be transparent, flexible, well documented, and used consistently. To be well documented, the risk assessment must show the model, the evidence for the model, and the source of the evidence for the model in writing in a careful and organized fashion. Transparency means that the risk assessment is well organized so that an educated layperson can read and understand the document. To be flexible, a risk assessment must allow for rapid reevaluation should new information become available. Consistency refers to the way in which risk assessment is applied to trading partners. Methodologically, it means that if one country is treated one particular way for a given import, other countries must also be so treated in evaluating them for importation into your country. It does not mean that if a given country accepts an import product "x" from one country (or region), the importing country must also accept a similar import from another country (or region). The evaluation and risk management decision must be made by region and be based on the risk assessment.

As new data become available, risk assessments must be updated. A decision made today based on current information can be perfectly defensible. However, a similar import reviewed in the future may result in a different decision based on evaluation of new information reflecting changes in animal health status of the region from which the export is proposed. Risk assessments become powerful analytical tools that can support good decisions and promote animal health on the farm, in transportation, and in markets, and all the way to the presentation of the animal or product for final destination in the importing country. Since data needs and models used for risk assessment are closely related, a discussion of useful models is important to considering data needs for good animal health.

Models and Risk Assessment

Risk assessment models represent tools that provide support to the decision maker. Just as there are many different tools for building a house, there are a variety of risk assessment methods for evaluating animal health. The importance of any risk assessment model is that it can help standardize the approach and define data needs. Qualitative and quantitative models are both useful, each having its special role in evaluation. The most powerful allow the quantitation of both risk and uncertainty. With increasing sophistication of the field of risk assessment, the biological variability inherent in both animal populations, testing procedures, and the humans who carry them out may also be modeled.[6]

Models and the risk assessments developed from them can also provide a focus for discussion, with input from all interested communities: scientists, businesses, regulators, and consumers. This interaction initiates the process of risk communication. Properly conducted risk assessment assures that the decision-making process is open to scrutiny by all who are affected by the decisions.

One of the most familiar models is that presented in the 1983 "Red Book"[4] consisting of four steps: (1) hazard identification, (2) dose-response assessment, (3) exposure assessment, and (4) risk characterization. The process was designed to characterize the potential adverse health effects of human exposure to environmental hazards and to be used as a tool in environmental and human health management. The model has been widely used in many parts of government and in many regulatory activities.

This model has been useful for assessing risk of toxins and residues in animals, since dose-response and exposure assessment can help evaluate outcomes of agents that do not multiply. However, for bacteria, viruses, and other agents that multiply, the four-step model may not be as useful as others.[7] One reason is that the "dose" of pathogen, for example, in an animal on the farm may be very small, existing at a level undetectable by current testing methods. Husbandry conditions during transport to slaughter may favor growth or shedding of a particular pathogen in an animal's gastrointestinal tract so that when the animal arrives in the new country, the agent has magnified in number and can infect other animals. Thus, an agent that replicates requires a different approach from those materials for which the "Red Book" model was developed. Much of current discussion in animal health centers around agents (e.g., viruses and bacteria) that reproduce in their living host.

In this paper, the use of a powerful tool, probabilistic scenario analysis, is discussed for its utility in risk assessment for international movement of animals and their products.

PROBABILISTIC SCENARIO ANALYSIS

The first use of probabilistic modeling as a research tool, in its modern form, stems from work on the atomic bomb with techniques developed by Ulam, von Neumann, and Fermi.[8] The method has been well tried and proved useful in many fields, including plant health,[9] animal health.[10,11] It is an excellent tool for estimating the probability or frequency of an unwanted event occurring.

Scenario Analysis

Probabilistic scenario analysis follows a series of activities to assure systematic consideration of the problem at hand. First, it requires a careful statement of the question or problem. At this point, the hazard or hazards are identified.

The next step is to list all the events that can initiate the occurrence of the hazard(s) identified. These initiating events (IE) may lead to a pathway of success or failure. The desire is for every IE to lead to success, defined as the absence of hazard occurrence. That is, everything goes well, the IE leads to an end point in which the hazard does not occur. This is the "success scenario." However, at each step along the way, there is a possibility for something to go wrong, for an event to occur that leads to the hazard(s).

In developing this part of the model, then, the next procedure is to outline this series of events which leads from each IE through all the events that must occur to reach the end point (EP). This is called an event tree (ET). Each segment along the ET length is called a branch; each branch begins with a node. A given ET may have several branching pathways, each providing separate EP. Thus, PSA provides an ET that describes graphically and in writing the sequence of events from the IE to the EP that leads to the occurrence of the hazard. FIGURE 1a shows a simple tree with a single branch and end point and FIGURE 1b shows a branching tree. An example of how an ET can help evaluate a hypothetical importation illustrates the PSA process (FIG. 2). Consider that the decision maker must evaluate, systematically and scientifically, the request to import an untreated animal product from country A. Thirty percent of livestock used as a source of this product are infected with agent X. Six months after harvest, 50% of the product no longer harbors agent X. Transit time from harvest to arrival at the portals of the importing country is 6 months.

Ideally, in this import, the animals yielding the product for import are not infected, the product is not infected by contact with the agent during transit, the product arrives in the importing country and does not infect any livestock. This is the desired "success scenario." On the other hand, livestock transactions, like many other things in life, are subject to chance events, or events that are unplanned and unexpected. The

Fig. 1a. Simple Event Tree

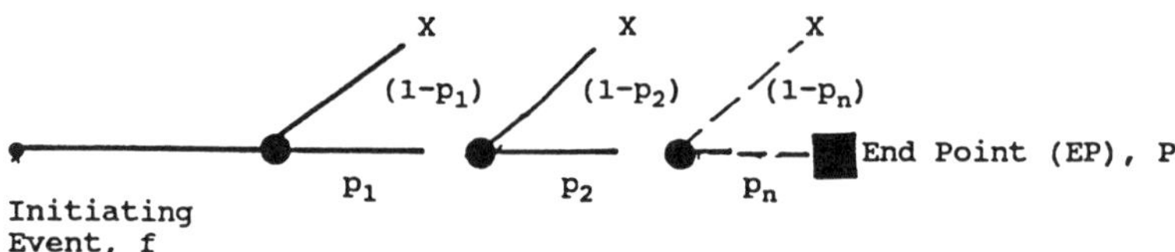

P = probability of End Point event
  = (p_1) (p_2) (p_n)

F = frequency of End Point event
  = (f) (P)

Fig. 1b. Branching Event Tree

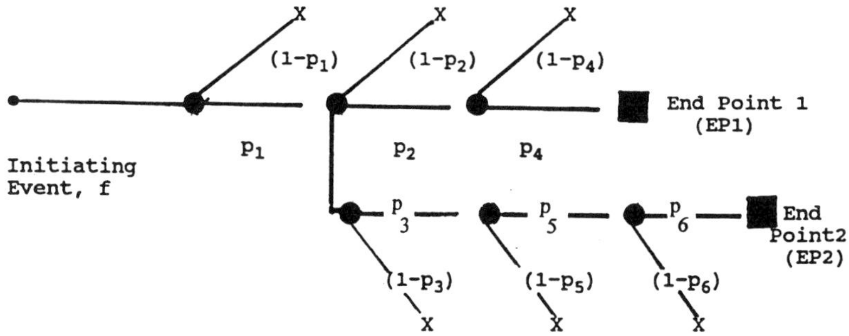

P 1 = probability of End Point 1 event
    = (p_1) (p_2) (p_4)

P 2 = probability of End Point 2 event
    = (p_1) (p_3) (p_5) (p_6)

P_{1+2} = [(p_1) (p_2) (p_4)] + [(p_1) (p_3) (p_5) (p_6)]

F = frequency of the End Point Events for EP 1 and EP 2

  = [(f) (P 1)] + [(f) (P 2)]

F = frequency of End Point event
  = (f) (P)

FIGURE 1. Not all initiating events that lead to hazard occurrence can be modeled as a simple event tree (FIG. 1a); some require more complex modeling (as in FIG. 1b). Notice that the total risk incurred in FIGURE 1b is a summation of the risk in each separate branch of the event tree.

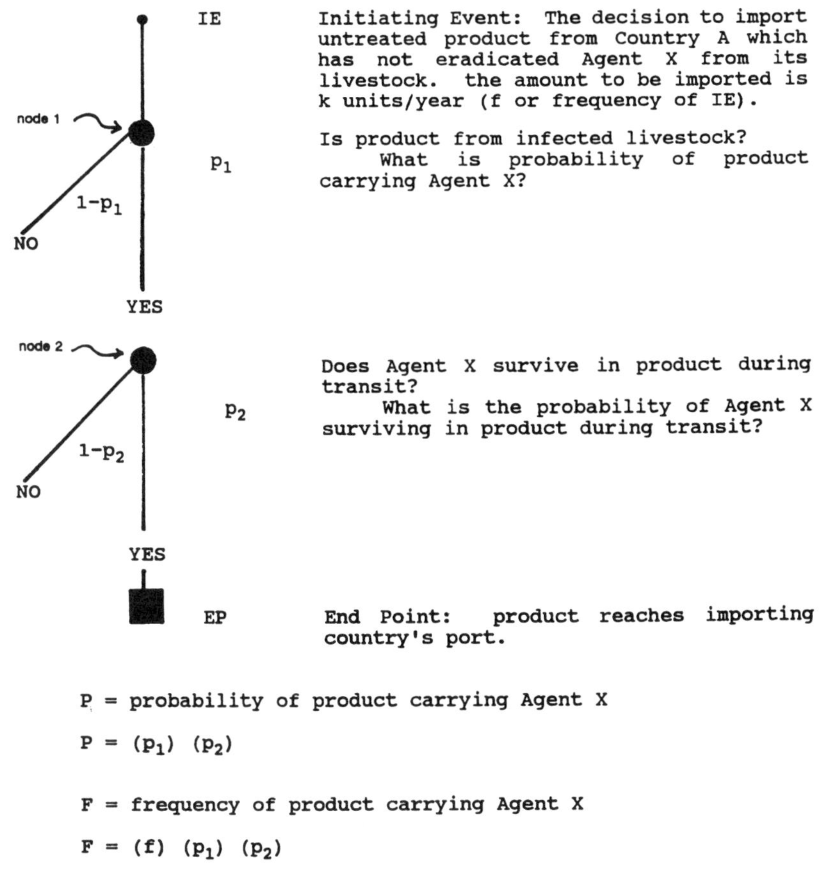

FIGURE 2. A simple event tree to illustrate the process of probabilistic scenario analysis. A product is being imported from a country in which 30% of the livestock carry agent X; agent X does not occur in the importing country. Six months after harvest, agent X has disappeared from half of the product. From harvest to arrival in the importing country is 6 months.

decision maker for the country potentially receiving this product must evaluate all events carefully.

The as-planned success scenario is for the untreated products to be taken from livestock free of agent X. The end point is that untreated product is delivered to the importing country where it is not the source of any disease or problem for the importing country. This is the success scenario.

FIGURE 2 shows these events in a format that is useful in the analysis. Notice that the IE is defined as a frequency, f. At node 1 the first question is: "Is product from livestock infected with agent X?" Note that a "no" response leads to the product not posing a risk, thus leading to the "no risk" branch at this node. If the answer is "yes," then the branch toward hazard occurrence is activated.

However, the actual answer to such questions cannot always be known with absolute certainty. If we knew with 100% certainty that no animal yielding exportable product was infected with agent *X,* then risk assessment would be unnecessary. The decision could be made on the certain knowledge. Since decisions in animal health are rarely so simple, analysis proceeds with the second question asked at node 1: "What is the probability (p_1) of product carrying agent *X?.*"

Node 2 is concerned with survival of agent *X* in the product during transit. The question becomes "what is the probability (p_2) that agent *X* the "no" answer leads to the no risk pathway.

To constitute a hazard, agent *X* in or on the untreated product must enter the importing country, and then come into contact with susceptible livestock, infect those animals, and spread to others. These events could be modeled at subsequent nodes. However, in animal health risk assessment, it is often assumed that the probability of each of these events is 1.0. This may especially be true when dealing with agents on the OIE List A of diseases[12] since these agents are known for their rapid spread often by a variety of means.

Evidence for the Event Tree

Once the ET is completed, evidence must be gathered to evaluate each node in the tree. Each piece of evidence should be recorded and associated with the proper node; one piece of evidence may be associated with one or more nodes, as appropriate. Each piece of evidence should be referenced in a bibliography so the source of each piece of evidence is documented. Labels on the tree support easy reference to the evidence and bibliographic information associated with each node.

Evaluating the Evidence Qualitatively

The next task is to describe and evaluate the evidence for each node.[13] The summary of evidence evaluation may be described qualitatively, such as "high, medium, or low." This allows the ET to be used for a kind of qualitative estimate at each node, which when considered all together, can give the assessor a general feel for the urgency of the problem. Recent work from Centers for Disease Control and Prevention[14] and Centers for Epidemiology and Animal Health[15] uses a qualitative evaluation of an event tree to estimate probability.

Another approach is to use an ordinal ranking system for evaluating the evidence. For example, the ranking might be 1 for the lowest level of risk up to 5 for the highest. The risk level for each node on a branch of the ET tree can be averaged (total points divided by the number of nodes) for an ordinal estimate of the total risk of that branch.

Qualitative or semiquantitative evidence evaluation for the event tree is an excellent way to rapidly evaluate a particular requested import. This approach is also useful to compare several problems that need attention, thus screening them to set priorities for further analysis (screening risk assessment). Such an approach is also useful in ranking risks for comparison purposes. When time and resources are short and a risk assessment must be made to provide guidance to a risk manager (decision maker) very quickly, this use of probabilistic scenario analysis can furnish support in a timely way. The same event tree that is used in a qualitative or semiquantitative manner can also be quantified to provide more precise risk estimates.

Quantifying the Event Tree

Perhaps the qualitative analysis has not provided clear direction for the decision maker. Or perhaps, the decision to be made is of great importance. For example, if the risk assessment is to support a decision developing a major regulation with respect to trade. In other situations, precision in the estimate of risk from the risk assessment is necessary. In all these instances, it is desirable to quantify the event tree for more precise analysis of the hazard in question. The evidence available for each node must be weighed very carefully. In many instances, more data or more precise evidence may be sought for this quantitation.

The event tree can be populated with a variety of kinds of data. For example, a sophisticated epidemiological survey may provide a mean and standard deviation for a given event. These data points, both mean and standard deviation, can be used directly at one node in the model. At another node along a branch there may be only sketchy data from anecdotal sources or a casually conducted survey. This less rigorously analyzed information may also be used in the model. The information is captured as a (minimum) three-point curve, or triangular probability distribution, the three points being the most likely estimate, and the highest and lowest values consistent with the evidence available. These three points incorporate both what is known and what is unknown in this format known as the probability distribution function (PDF). Highest and lowest values close to each other indicate a higher degree of certainty about an estimate than do those further apart, just as small standard deviations indicate more certainty than do very large ones (FIG. 3).

The total probability for the event tree (FIG. 2), that is, the probability of the hazard occurring, is the product of the probabilities in each node of the branch:

P = probability of product carrying agent X entering country
$P = (p_1) (p_2)$

The expected frequency of infected product is expressed:

F = frequency of entry of product carrying agent X
$F = (f) (p_1) (p_2)$

If the data are provided as point estimates, it is simple enough to multiply the values. If the data are provided as ranges, with a most likely expected value (i.e., as a PDF), then computation can be accomplished by employing a software package such as @Risk to complete a Monte Carlo simulation. If the ET has more than one branch, the probability associated with each branch must be calculated and summed to provide the total probability estimate. The PDF can be transformed into a cumulative distribution function (CDF) which allows the risk manager to read the frequency on the x-axis and the cumulative probability of frequency on the y-axis. FIGURE 4 shows the PDF and CDF for the agent X example. This quantitative output is a statement

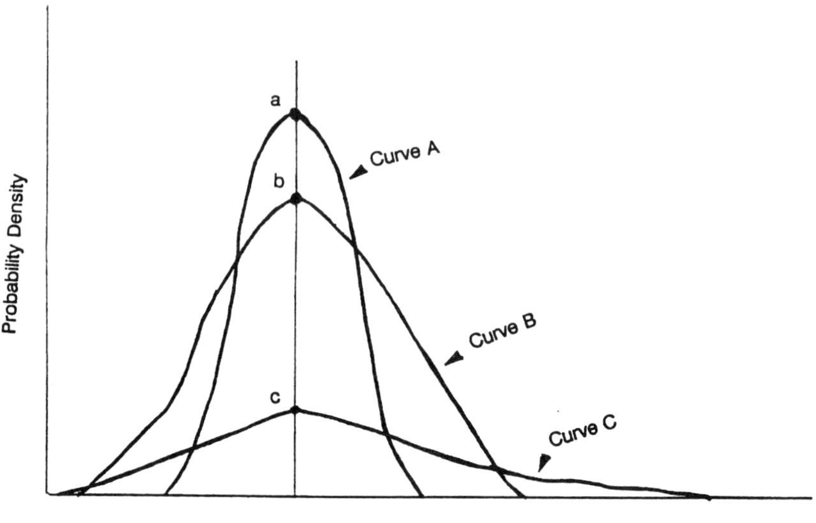

FIGURE 3. Probability density functions. A probability density function (PDF) captures the entire range of the state of knowledge and displays uncertainty. Points a, b, and c show an identical frequency. If only a point estimated were reported, the information contained in curves A, B, and C would be missing. Note that curve B illustrates the most confidence (least uncertainty) while curve C illustrates the least confidence (most uncertainty). This kind of information can be very important to decision makers. (Adapted from Roberts, Ahl, and McDowell.)[13]

of the probability of the hazard occurring. It is used by the risk manager, along with other information, to make decisions about the acceptability of an import.

DEFINING THE MAGNITUDE

The second part of the definition of risk is a concern for the magnitude of the outcome should the hazard occur. Many countries have economic studies and cost-benefit analyses of incursions of foreign animal diseases into their territory in the past. In some cases, these analyses are updated to consider the magnitude of the effects in current fiscal terms. In the original definition of risk, these economic studies were precisely what was called for. In fact, by multiplying probability and magnitude, the outcome, expected value, was a number that could be used to provide decision makers with a single, simple value on which to manage the risk.

However, further consideration makes this seem less useful. First, consider that a highly probable event with very low magnitude of effects and a low-probability event with high magnitude of effects could have the same expected value. However, the management of two such sets of events might best be regulated in different ways.

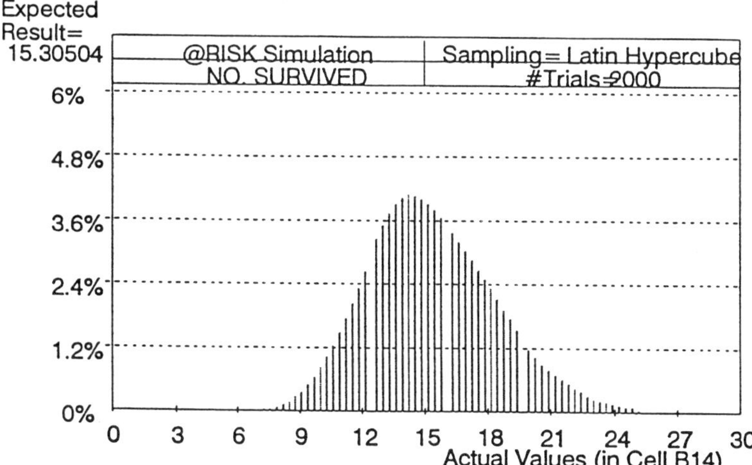

PDF--number of products contaminated with Agent X entering the importing country annually, per 100 shipped.

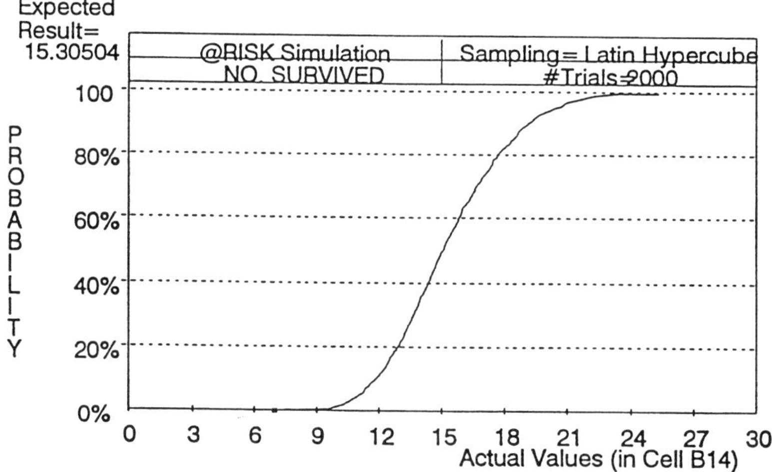

CDF--number of products contaminated with Agent X entering the importing country annually, per 100 units shipped. Notice the probability of up to 14 units is approximately 0.4.

FIGURE 4. The probability density function (PDF) curve can be transformed to a cumulative density function (CDF) that allows one to read probability directly from the *y*-axis.

Therefore, the tradition has developed to report both the probability and the magnitude, but to keep the two analyses separate.

A second trend has further changed the way risk assessment is evaluated. Magnitude often meant economic and social effects. Concomitantly, there has been emphasis on keeping the risk assessment as the biological analysis, placing economic, social, legal, and other considerations into some other category such as risk management or risk characterization. In this schema then, risk becomes the probability of the hazard occurring and the magnitude becomes a description of the biological effects if the hazard should occur. Cost-benefit analysis is then used as a mechanism to discuss the "magnitude" concept that was originally incorporated into the definition of risk.[16]

There are a whole set of other associated effects unanalyzed and undescribed. These include economic (other than cost-benefit considerations), legal, political, social, and technological effects. In addition, the alternatives for action and a comparison with similar risks may also be important to the decision maker. These are often very important factors that must be considered before a decision is made about a particular hazard. For now, the tradition is developing to place these other important factors in the category known as *risk characterization.*

This developing definition of risk assessment clarifies the classic definition provided earlier in this paper. By placing all the accessory nonbiological information important to the decision maker in the category risk characterization, the entire risk analysis becomes more explicit, and hopefully more useful. The risk assessment becomes a biological statement, and all nonbiological information is separately available.

THE SPECIAL CASE FOR ANIMAL HEALTH

Animal health risk assessment is guided by principles established by the international governing body, Office of International Epizootics, or OIE. Housed in Paris, France, OIE oversees protection of animal health worldwide. Countries engaged in international trade in animals or animal products provide information to OIE, which collects and makes available to all members information on the status of animal health of its members. Each member country has delegates to OIE who meet together to set policy for animal health information collection and dispersal. In addition, OIE provides a list of animal diseases and categorizes them by their severity, both economic and biological.[12] This list, A and B, serves the international animal health community as a hazard identification. In addition, changes in the classification of disease under consideration will include information about the magnitude of disease effects, both biological and economic, in this categorization. Therefore with both hazard identification, magnitude, and part of the risk characterization accounted for by OIE classification, the most important consideration remaining is to estimate the probability of incursion of an unwanted disease/agent along with a desired importation.

INTERPRETING THE RISK ASSESSMENT

As discussed earlier, the output of the risk assessment is a PDF or a CDF (FIG. 4). This provides the decision maker with information about the probability of occurrence of the hazard, but is often insufficient to permit a reasonable decision. One item lacking is a consideration of the use to which the import will be put.

The use of the imported animal or animal product is an important consideration. For example, if the imported animals are half a dozen pigs, what is the expected use of these animals. If they are to be taken to a research facility on Manhattan island in New York City there may be reason to tolerate a higher risk than if the pigs are headed to a breeding herd in central Iowa. The number of animals in the import, the frequency with which the animals are imported into the US, the tools available to deal with a disease should the agent occur and escape. There are a whole host of other questions that may be considered when making the management decision about the level of risk to accept. Management schemes have been proposed,[17,18] but our experience with them is still meager.

SUMMARY

Abstract: Probabilistic scenario analysis (PSA) is a method of risk assessment that has had wide usage in many fields, including engineering, nuclear safety, and financial analysis. It is fast becoming the ''gold standard'' for risk assessment in many fields, including human health and the environment. It has been successfully applied in animal and plant health issues as well. PSA begins with the identification of a hazard and the development of a step-by-step scenario from some initiating event to the end point at which the hazard occurs. In PSA, the pathway leading to the end point is outlined by a model, called an event tree. Each step leading to the occurrence of the hazard is carefully outlined, called the node on the event tree. At each of these nodes, the probability of the event leading to the end point (hazard occurrence) is evaluated. The evaluation of the probability at each node may be qualitative or quantitative. The evidence used may come from standard epidemiological studies. The model can also accommodate expert opinion, anecdotal evidence, or any other information that can be verified that is pertinent to the event leading to hazard occurrence. The PSA is also quite flexible for it can be quickly revised when new data become available. With careful statement of the evidence and linkage back to bibliographic or other sources, it can provide a transparent, flexible, well-documented approach to risk assessment for animal health. Risk assessment along with regionalization is the key to healthy national herds and free international trade.

ACKNOWLEDGMENT

The author appreciates the help and support of Mr. Robert McDowell, USDA/APHIS/PPD, in devising the example used to illustrate PSA and in developing the data shown in FIGURE 4.

REFERENCES

1. AHL, A. S., J. A. ACREE, P. S. GIPSON, R. M. MCDOWELL, L. MILLER & M. D. MCELVAINE. 1993. Standardization of nomenclature for animal health risk analysis. Review of Science and Technology of the Office of International Epizootics **12**(4): 1045-1055.

2. HOLLOWAY, C. A. 1979. Decision Making Under Uncertainty: Models and Choices. Prentice-Hall. Englewood Cliffs, NJ.
3. HERTZ, D. B. 1964. Risk analysis in capital investment. Harvard Business Rev. **42:** 95-108.
4. NATIONAL RESEARCH COUNCIL. 1983. Risk Assessment in the Federal Government: Managing the process ("The Red Book"). National Academy Press. Washington, DC.
5. KAPLAN, S. & B. J. GARRICK. 1981. On the quantitative definition of risk. Risk Analysis **1**(1): 11-27.
6. FREY, H. C. 1992. Quantitative analysis of uncertainty and variability in environmental policy making. Report from Center for Energy and Environmental Studies, Department of Engineering and Public Policy, Carnegie Mellon University. Pittsburgh, PA.
7. AHL, A. S. 1994. From farm to table: applications of risk analysis to the safety of foods of animal origin. Technology: J. Franklin Inst. **331A:** 77-82.
8. HAMMERSLEY, J. M. & D. C. HANDSCOMB. 1964. Monte Carlo Methods. John Wiley & Sons. New York, NY.
9. KAPLAN, S. 1993. The general theory of quantitative risk assessment—its role in the regulation of agricultural pests. B. North Am. Plant Pest Org. **11:** 123-146.
10. MILLER, L., M. D. McELVAINE, R. M. McDOWELL & A. S. AHL. 1993. Developing a quantitative risk assessment process. Review of Science and Technology, Office of International Epizootics **2**(4): 1153-1163.
11. McELVAINE, M. D., R. M. McDOWELL, R. W. FITE & L. MILLER. 1993. An assessment of the risk of foreign animal disease introduction into the United States of America through garbage from Alaskan cruise ships. Review Science and Technology, Office of International Epizootics **2**(4): 1165-1173.
12. KELLAR, J. A. 1993. The application of risk analysis to international trade in animals and animal products. Review of Science and Technology, Office of International Epizootics. **2**(4): 1023-1044.
13. ROBERTS, T., A. S. AHL & R. McDOWELL. 1995. Risk assessment for microbial hazards. *In* Proceedings of the USDA/ERS Workshop on Tracking Foodborne Pathogens from Farm to Table, January 9-10, Washington, DC.
14. CENTERS FOR DISEASE CONTROL AND PREVENTION (CDC). 1994. Guidelines for preventing the transmission of *Mycobacterium tuberculosis* in health-care facilities. Morbid. Mortal. Weekly Rep. October 28 **43**(#RR-13).
15. CENTERS FOR EPIDEMIOLOGY AND ANIMAL HEALTH (CEAH), USDA/APHIS/VS. 1994. Food and mouth disease: sources of outbreaks and hazard categorization of modes of virus transmission. December, 1994.
16. P.L. 103-354. The US Department of Agriculture Reorganization Act of 1994. Signed by President Clinton on October 13, 1994.
17. METCALF, H. 1994. Summary for general criteria for animal disease regionalization. USDA/APHIS/VS Report.
18. McNAB, W. B. 1994. Risk Assessment Models of the Animal and Plant Health Risk Assessment Network. Report from Agriculture and Agri-Food Canada, July 7, 1994.

Importation of Bovine Genetics

A Quantitative Risk Assessment of Disease Transmission by Bovine Embryo Transfer[a]

P. SUTMOLLER [b]

[a]*Pan American Foot and Mouth Disease Center (PAHO/WHO)*
Caixa Postal 589
20001-970 Rio de Janeiro, Brazil

[b]*2430 Crowncrest Drive*
Richmond, Virginia 23233

Many countries have imposed major obstacles against the international trade in bovine genetic materials by their prohibitive import regulations. For instance, most countries that are free of major epidemic animal diseases prohibit the importation of semen and embryos from countries with foot and mouth disease (FMD), even though research and field experience have shown that, with appropriate restrictions, these materials can be moved internationally with little risk of disease transmission. Reluctance on the part of veterinary authorities to relax the existing regulations to permit legal international movement will probably be best overcome by providing satisfactory estimations of the real risk involved. A major drawback of the "zero risk" import concept is that farmers cannot legally obtain desirable genetic material, and this policy stimulates illegal movement, which in itself is a risk. Also, "zero-risk" policies will be untenable in the near future in view of the NAFTA and GATT trade agreements, under which health requirements cannot be used as nontariff trade barriers. Instead, international trade restrictions must be based on risk assessment and risk management.

The problem is to define the risk of a specific import activity and acceptable levels of risk relative to the consequences of disease introduction by such an import. Bovine embryo transfer is used as the basis for the present discussion because there is a well-established protocol for the international movement of bovine embryos.[1] FMD is one of the main disease risks, so it is assumed that the embryos will originate from an FMD-infected region, but one with good epidemiological information.

Based on published scientific data, the embryo transfer industry maintains that embryos washed according to protocols given in the Manual of the International Embryo Transfer Society[2] (IETS Manual) are safe for international movement. But the regulatory veterinarian objects; saying "How do I know that the washing was done properly?" The weakness of each approach is that both parties in the argument are looking at only *one link in the chain of events*. It should be emphasized that each link in the chain provides a certain level of protection against disease transmission and that it is the *total* chain that determines the safety of the importation.

Presently, embryo transfer in livestock is an important business in many parts of the world. At least 350,000 embryo transfers are done annually, and some 35,000 embryos are moved between countries.[3]

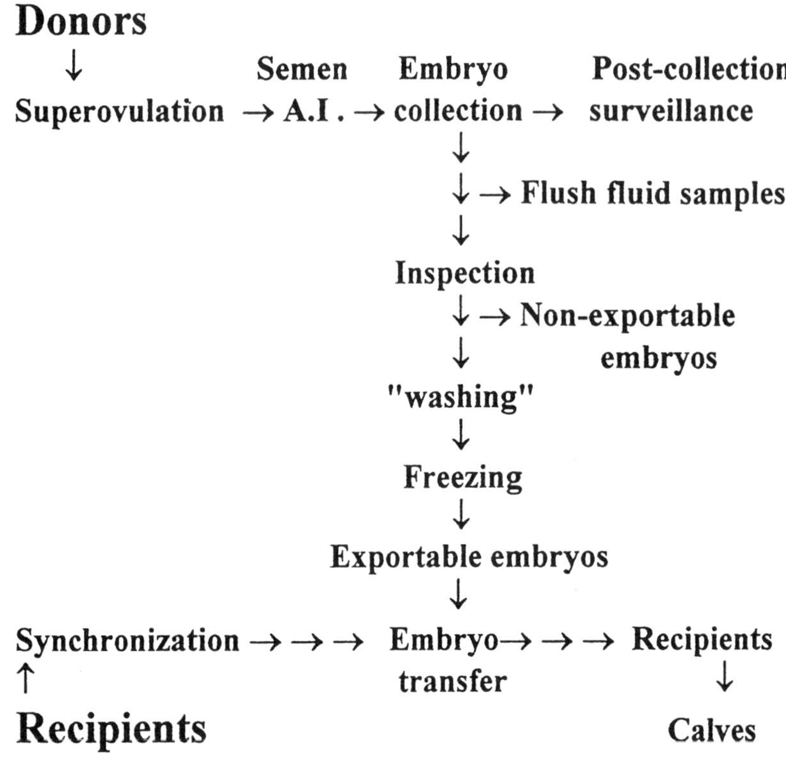

FIGURE 1. Schematic presentation of bovine embryo transfer procedures.

FIGURE 1 summarizes the most important component of this technique, particularly with regard to the safety of embryos for disease prevention. Embryo donor cows are superovulated by a series of hormone injections which are usually administered by the owner's own veterinarian or by farm personnel under the direction of the embryo collection team (ECT).[2] The donors are observed closely for signs of estrus and then inseminated. Embryos are usually recovered six to eight days after insemination by flushing the uterus with collection fluid. The embryos and uterine debris are separated and transferred into an embryo recovery dish, and the contents of the dish are systematically examined at low magnification to locate embryos. The embryos are then transferred to smaller dishes with fresh medium and evaluated.

Since the membrane surrounding the embryo, the so-called zona pellucida (ZP), is an important barrier for pathogens, it s checked for intactness and defects. Embryos destined for international movement must be washed an additional 10 times, and this "washing" involves transferring the embryos through 10 changes of medium in disposable plastic multiwell cell-culture plates, during which they are gently agitated. Each wash must constitute at least a one-hundred-fold dilution of the medium of the previous wash. The embryos are observed under low power during the washings,

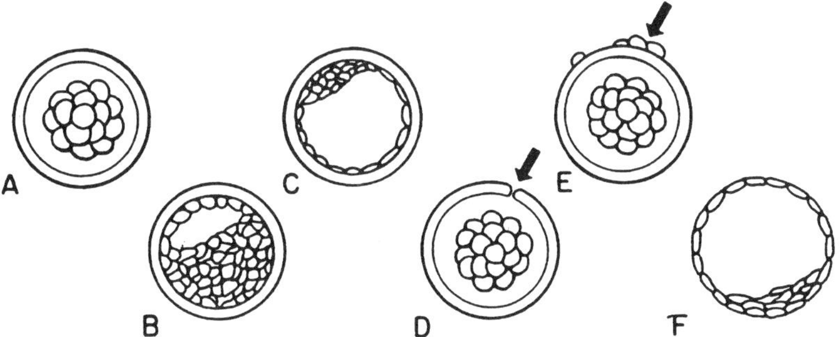

FIGURE 2. A. ZP-intact morula, B. ZP-intact blastocyst, C. ZP-intact expanding blastocyst, D. Morula with a defect in the ZP, E. Morula with extraneous material adhering to the ZP, F. Hatched blastocyst without ZP. A, B, and C are acceptable embryos in terms of disease control, D, E and F are unacceptable. Based on IETS Manual.[2]

but they are finally checked under higher magnification to ensure that the ZP is still intact and free from extraneous cellular debris.

Embryos for international trade are stored frozen from the time of their collection until they are used. Ideally, the recipient should be at the same stage of the estrus cycle as the donor when the embryo was collected. After thawing, the embryos are placed in the uterus of the recipients, and, hopefully result in genetically valuable calves nine months later.

There are three lines of defense against introduction of disease by embryos:

The *first level* consists of a favorable disease situation in the country and in the area or on the farm of origin, the health of the embryo donor, as well as the pathogenesis of the disease agent. Of primary concern to regulatory officials are measures taken in exporting countries to ensure a very low probability that the embryo donors will have been exposed to disease agents.

The next questions to be considered are: How is the disease transmitted? and Does the agent reach the embryo? For a pathogen to be transmitted by an embryo, it must be able to associate itself with the embryo or the surrounding tissues and fluid. The probability of this happening depends very largely on the pathogenesis of the disease, the characteristics of the disease agent, and the immune status of the donor. Therefore, detailed information about the pathogenesis of diseases of concern is of great importance for risk assessment.

The *second level of defense* against the disease includes measures for the sanitary handling and processing of the embryos by the ECT. Washing bovine embryos, provided their ZPs are intact, has been shown to be highly effective for removing many pathogens. However, the effectiveness of embryo washing also depends on whether the disease agent adheres to the ZP. Some pathogens tend to stick rather firmly and may be trapped in crevices or submicroscopic defects in the ZP.

FIGURE 2 shows some different categories of bovine embryos: A, B, and C are embryos that are acceptable in terms of disease control; D, E, and F are unacceptable.

TABLE 1. Scenario Pathway for Disease Transmission by Bovine Embryo Transfer

P_1	Disease in the Region?	No	No risk
		Yes	$\rightarrow P_2$
P_2	Disease on donor farm?	No	No risk
		Yes	$\rightarrow P_3$
P_3	Infection on donor farm observed by Animal Health Surveillance System?	Yes	No export
		No	$\rightarrow P_4$
P_4	Infection on donor farm observed by Embryo Collection Team?	Yes	No export
		No	$\rightarrow P_5$
P_5	Disease agent reaches embryonic environment of infected donor?	No	No risk
		Yes	*Contaminated* embryos$\rightarrow P_6$
P_6	Nonexportable (F_1) embryos detected?	Yes	Individual embryo discarded
		No	*Infectious* embryos$\rightarrow P_9$
P_7	Exportable embryos (1-F1) properly washed?	No	*Infectious* embryos $\rightarrow P_9$
		Yes	$\rightarrow P_8$
P_8	Disease agent adheres to *zona pellucida?*	No	Washing effective$\rightarrow P_9$
		Yes	*Infectious* embryos$\rightarrow P_9$
P_9	Total risk of *infectious* embryos (P6, P7, and P8)	Total risk *infectious* embryos$\rightarrow P_{10}$	
P_{10}	Clinical disease on infection on donor farms during postcollection quarantine?	Yes	No export
		No	$\rightarrow P_{11}$
P_{11}	Disease agent detected in infective collection fluid?	Yes	No export
		No	$\rightarrow$*Export*

The *third line of defense* includes observations on the health status of the herd of origin, during the time the embryos are in postcollection storage, as well as possible postcollection tests on flushing fluids and any unfertilized embryos.

Based on information of the three lines of defense a scenario pathway can be developed that follows the flow of risk-associated events from the time activity starts on the donor farms to the time of exporting the embryos (TABLE 1). Next, all events are considered that may possibly influence the chance that pathogens are carried from an infected donor cow to the recipient cow by the embryo. The questions asked for each event in the scenario pathway are: What can go wrong?, How likely is that to happen?, and What are the consequences? For instance, if FMD is detected on the donor farm, the export of embryos should be canceled. Similarly, if the ECT finds a nonintact ZP embryo, that embryo should be discarded for export. However, failure to detect either case may mean that high-risk embryos will be processed and exported.

The number of farms required depends on the number of embryos to be exported, the number of transferable embryos produced per donor animal, and the number of donors per farm. These numbers vary according to the livestock production system.

P_1 is the presence of the disease in the region during the time that the embryos for export are collected and it in turn determines the chance that the disease is on the donor farm (P_2). The next events are the detection of the disease in the herd by the animal health surveillance system (P_3), or by the ECT (P_4). If the disease in question is detected, there should be no risk, because the whole operation should be canceled.

If the disease escapes detection and the disease agent contaminates the embryos in the genital tract of the donor, they are referred to as *contaminated* embryos (P_5)

P_6 is the failure by the ECT to detect and discard nonexportable embryos (FIG. 2). Any undetected nonexportable *contaminated* embryos are now referred to as *infectious* embryos. The next question (P_7) to be answered is: Does the ECT fail to properly wash embryos? If so, then a proportion of the embryos also may remain *infectious.* If the ECT washes the embryos properly, but the disease agent adheres to the ZP (P_8), the pathogen may not be completely removed and they too will remain *infectious.* Thus, *infectious* embryos (P_9) are a function of the three events: nonexportable embryos not removed, embryos not properly washed, and disease agent not removed by the washing.

P_{10} is the occurrence of the disease concerned on the donor farm while the embryos are in storage. The exportation of the embryos must be canceled if the disease is detected.

P_{11} is the failure of laboratory tests to detect the virus. For FMD, the scenario pathway stops here, but for some less catastrophic diseases, risk factors in the receiving country may also be considered, such as the minimal infective dose of the pathogen or the existence of vectors when the disease is insect borne.

In TABLE 2 an attempt is made to quantify the probabilities of the occurrences in the scenario pathway for FMD by establishing a minimum/maximum range, as well as the most likely value. The minimum, maximum, and most likely values form what is known as a "three-point estimate," which not only attempts to express the best knowledge about the event, but also provides a guide to the degree of uncertainty of information related to the event. It would take too much space to go over all the evidence that was collected to quantify these values, but basically, the information was obtained from own experience, interviews with embryo transfer practitioners, and from the literature. The epidemiological data originate from the Continental Vesicular Disease Surveillance and Information System, coordinated by the Pan American FMD Center in Rio de Janeiro,[4,5] and from the State Animal Disease Surveillance System of São Paulo.[6] While the values used for the present estimations are based on currently available information, the assessment can, if required, be easily recalculated when new information becomes available.

For the risk assessment calculation the first thing to be considered is the probability that an infected farm is included among those selected for the collection of the embryos. This probability can be calculated from the number of donor farms and the disease incidence in the region.

In summary the basis for the estimation in TABLE 2 are as follows:

TABLE 2. Quantification of Risk for Foot and Mouth Disease Transmission by Embryo Transfer

		Probability		
		Minimum	Most Likely	Maximum
N_1	Number of embryos to be imported.		200	
N_2	Number of donor farms required for N1.	10	15	30
P_1	Herd prevalence (infected herds/herds).	0.0001	0.001	0.005
P_2	Probability of including at least one infected farm equals $1-(1-P_1)^{N2}$.			
P_3	Probability that disease surveillance fails to detect FMD on the donor farm.	0.01	0.05	0.1
P_4	Probability that ECT fails to detect FMD on donor farm.	0.001	0.01	0.05
P_5	Probability of FMD virus reaching the embryo.	0.001	0.01	0.1
F_1	Fraction of nonexportable embryos.	0.005	0.02	0.1
P_6	Probability that the ECT fails to remove nonexportable embryos (F1).	0.01	0.05	0.5
P_7	Probability of exportable embryos (1-F1) being washed inadequately.	0.0001	0.001	0.01
P_8	Probability of FMD virus adhering to exportable washed embryos[a]	0		0.01
P_9	Equals $1-\{[1-(P6 \times F1)] \times [1-(P7 \times (1-F1))] \times [1-(P8 \times (1-F1)):]\}$.			
P_{10}	Probability of failure to detect FMD on the donor farm.	0.001	0.01	0.05
P_{11}	Probability that laboratory tests fail to detect FMD virus in the collection fluid.	0.01	0.05	0.1

[a] Uniform distribution.

N_1 number of donor farms required. A batch of 200 embryos for export would probably require the flushing of some 50 donor cows. Based on these numbers and local estimates, the 200 embryos are likely to require some 15 farms, but not less than 10 and not more than 30 farms.

P_1 disease prevalence of the region. For the purpose of illustrating the risk assessment, a region was selected in the northwestern part of the State of São Paulo, Brazil. The incidence for the years 1992 and 1993 in the region were 20 and 15 herds with FMD per 10,000 farms, respectively. Since in the near future a slight improvement of the epidemiological situation can be expected, the most likely prevalence is estimated to be 10 infected herds per 10,000 herds. It is unlikely that the prevalence of FMD will surpass 50 per 10,000, because that would suggest the presence of an epizootic situation and all embryo export activities would probably then be suspended.

P₂ probability of including at least one infected donor farm. P_2 can be calculated from the number of donor farms required and the incidence of FMD in the region. The probability that at least one infected donor farm is included equals $1 - (1 - P_1)^{N2}$. However, in the range of values used, $N_2 \times P_1$ gives a similar numerical result.

P₃ probability that the animal health surveillance system fails to detect FMD on the donor farm. In this particular region of Brazil, it is unlikely that a significant number of outbreaks would go undetected. Moreover, embryos for export will probably be collected on well-managed farms with high animal health standards. The range of the probability that the surveillance system fails to detect FMD on the donor farms is estimated to be 1-10%, with a most likely probability of 5%.

P₄ probability that the ECT fails to detect FMD on the donor farm. The planning and execution of an embryo transfer program are complex and require frequent contacts between the farm management personnel and the ECT, sometimes over a period of several weeks. For a successful embryo transfer program, both farm personnel and the ECT must pay close attention to herd health aspects during this time. Thus it is unlikely that the ECT will collect embryos on a farm with clinical FMD, but of course this cannot be fully excluded. The values for the probability of failure of the ECT to detect FMD on the donor farm is estimated to be somewhat lower than for P_3.

P₅ probability that FMD reaches the genital tract and contaminates the embryo. In the case of FMD, the frequency with which the virus reaches the embryonic environment depends on the likelihood of viremia, which in turn depends on the immune status of the embryo donor. Since donor farms are likely to maintain a good FMD vaccination program, a conservative estimate of the likelihood of viremia and of contamination of the embryos is below 10% and most probably in region of 1%.

F₁ probability of nonexportable embryos. As shown in FIGURE 2 certain types of embryos must be excluded from being transferred for reasons of potential disease transmission. The range for the most likely to maximum occurrence of 2-10% was obtained from embryo transfer practitioners.

P₆ and P₇ probability of failure of the ECT. Risk units P_6 and P_7 relate to the collection and handling of the embryos by the ECT. Any existing risk will be greatly reduced if handling procedures are done correctly. The most important sources of risk are: ''nonexportable embryos are not removed'' (P_6) and ''washing of exportable embryos is not done correctly'' (P_7). Human error is an important source of risk. Therefore, estimates for this section must reflect the level of confidence that officials in the importing country have in the competence and integrity of the ECT. The estimated maximum probability of 50% that the ECT may fail to remove nonexportable embryos may seem rather high, but this is to take account of the fact that ECTs in Brazil are not officially regulated. However, this does not necessarily reflect adversely on the quality of work by any individual ECT.

P₈ probability of FMD virus adhering to exportable washed embryos. For estimations of ''virus adheres to the ZP,'' published research data were considered. Singh et al.[7] found that all FMD virus was removed from 169 *in vitro* contaminated bovine embryos by washing. McVicar et al.[8] collected 48 embryos and oocytes from heifers at the peak of FMD viremia, and, after washing, these also were to be found free of virus. Based on these data, we can state with 95-99% confidence that the risk of FMD virus not being removed by the washing will not exceed 1%.[9] If the washing procedure introduces a dilution factor of 10^{-20} and even if the starting titer of virus contamination is of the order of 10^5, then the probability of not removing virus will, in practical terms, be zero. Thus, the risk range for the failure of the washing procedures for FMD is 0-1%.

P_9 *probability of infectious embryos.* This probability is a function of 3 probabilities: nonexportable embryos not removed, exportable embryos not properly washed, and FMD virus not removed by washing from exportable embryos.

$$P_9 \text{ equals } 1-\{[1\text{-}(P6 \times F1)] \times [1 - (P7 \times (1 - F1))] \times [1 - (P8 \times (1 - F1))]\}.$$

In the case of FMD, which has a high probability of being removed by the washing, the main factors that count are the "human error" factors P_6 and P_7.

P_{10} *probability that FMD is not detected on the donor farm while the embryos are in postcollection storage.* Ideally, farms from which embryos for export have been collected should afterwards remain under official supervision, which would make it unlikely that FMD on the farm or the area would go unreported. In the case of Brazil, there are no official requirements in this regard. The estimates for P_{10} are therefore based on the assumption that the importing country requests post collection surveillance as part of the export health certification process. The probability of failure of detection of FMD postcollection is estimated to be similar to P_4.

P_{11} probability that laboratory tests fail to detect FMD virus in the collection fluid from an infected donor. In the absence of actual statistics for the testing of embryo collection fluid for FMD virus, the most likely probability of false-negative results is to be 5%, but, this may be unduly pessimistic.

The above three-point estimates do not indicate the relative chance that the event will occur. For more precise estimations the Lotus 123/@RISK computer simulation program was used.[c] This software package permits the construction of a so-called probability density function (PDF) for each event (FIG. 3). A PDF shows the range of probabilities that the event will occur and also the relative chance that it will occur. For most simulations, a "triangular distribution" was utilized, based on a minimum, maximum, and most likely value. Only for P_8 was a "uniform distribution" used based on a minimum/maximum range with all probabilities within that range having the same chance of occurring.[10] The results of the calculation of an individual PDF for event P_5 (probability of contamination of embryos in the genital tract) is shown in FIGURE 3. Of particular interest is the *expected result,* which is the center of the distribution and indicates the most likely occurring outcomes of the simulation; and the *shape of the distribution,* indicating the chances of particular outcomes. The PDF of P_{11} (probability that laboratory tests will fail to detect FMD virus in the collection fluid) has a narrower range of frequencies on the X-axis, and this indicates a greater level of confidence in the estimates than those for P_5.

The final PDF for the importation risk is based on the product of the following probabilities: P_2 (disease on the donor farm) $\times$ P_3 (disease surveillance fails) $\times$ P_4

[c] Lotus and 1-2-3 are registered trademarks of Lotus Development Corporation. @Risk, Risk Analysis Add-in for Lotus 1-2-3, Version 2.01, 1992, Palisade Corporation, 31 Decker Rd., Newfield, NY 14867.

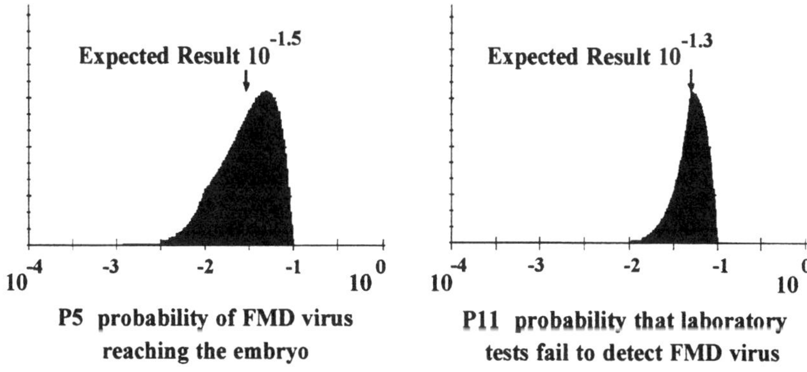

FIGURE 3. Probability density functions for P_5 and P_{11} of FMD transmission by bovine embryos. *x*-axis (horizontal): frequency of the event occurring. *y*-axis (vertical): probability of frequency.

(ECT fails to detect FMD) × P_5 (FMD contaminated embryos) × P_9 (ECT fails to do a good job) × P_{10} (postcollection surveillance fails) × P_{11} (laboratory tests fail).

The @Risk computer program elegantly solves this equation. The results, as presented in FIGURE 4, predict that the probability of including one or more infectious embryos in a batch of 200 embryos from the selected region in São Paulo is most likely to be in the order of 10^{-11}, with a 95% confidence level that this risk does not exceed 10^{-10}. In other words, for the estimate values used, the risk of importing a batch of 200 embryos containing one or more infectious embryos is one in 100 billion, but a small number of the outcomes may fall in the far right side of the distribution in the one in 10 billion range.

The first line of defense against the introduction of an exotic disease through embryo importation includes the disease situation in the region, the health status of the farms and donor cows from which the embryos are collected, and the pathogenesis of the disease agent. With the importation of a batch of 200 embryos from the region selected for this study, the predicted probability that this defense line will fail is 1 in a million, with a 5% confidence level that the risk does not exceed 1:200,000. Thus, it is obvious that the first line of defense is very important for preventing the introduction of FMD through bovine embryo transfer.

The second line of defense includes the proper handling and processing of embryos by the ECT. The present consensus among veterinary scientists is that for FMD, the washing protocol for bovine embryos basically eliminates the important risk. However, the QRA shows that the probability of failure by the ECT to remove nonexportable embryos and the probability of failure to adequately wash the embryos must also be considered. Because of these potential human errors, the risk reduction in this part of the scenario pathway is only about a 100-fold.

The third line of defense includes postcollection quarantine of donor farms and testing of collection fluids. The risk reduction of the third line of defense is about one in a thousand. For FMD we can question the need for testing collection fluid, since the remaining importation risk would still be in the order of one in 100 million.

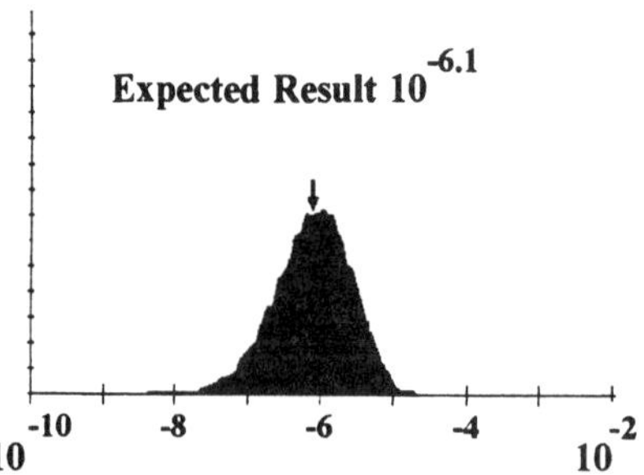

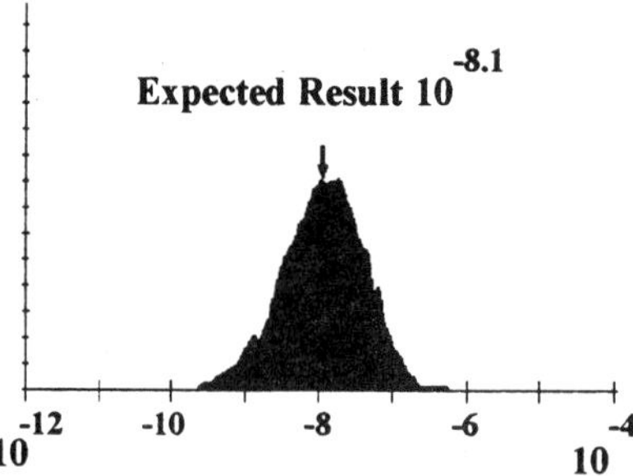

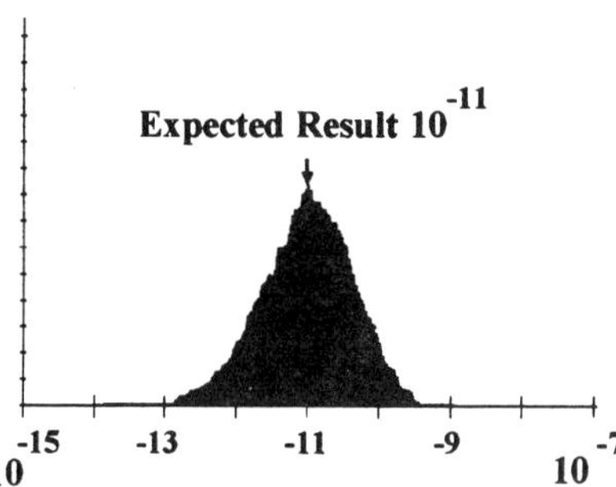

FIGURE 4. Probability density functions for the three lines of defense of FMD transmission by bovine embryos. X-axis (horizontal): frequency of the event occurring. y-axis (vertical): probability of frequency.

However, testing may be important for other diseases such as bluetongue which show less pronounced clinical signs than FMD.

An important part of risk analysis and a major challenge is the communication of QRA information to those making the political decisions. Potential users of the results of QRAs usually have very different backgrounds. For example, members of the embryo industry are mainly interested in the production of valuable, good quality embryos and in developing new markets, but some of them may not have a strong background in infectious disease control. Regulatory veterinary officials in the exporting and importing countries, on the other hand, may not be aware of all the details of embryo transfer procedures that determine their safety. Veterinary officials are often more comfortable with a zero-risk policy and reluctant to change regulations that, for a long time, seemed to have worked well. Somewhere in between is the livestock industry wishing to market or to obtain superior genetic material.

In order to address and convince these different groups effectively, QRAs must be transparent, consistent, and well documented. It is necessary to level the playing field with open discussions on the pathogenesis and epidemiology of the disease, and on procedures to which the animal product is submitted. Because of the importance of the first line of defense, it is essential to have detailed and reliable epidemiological information and to characterize the disease situation in the region from which the product originates. Risk analysis for international trade in animal products is intimately related to effective disease surveillance and information systems and regionalization.

REFERENCES

1. OFFICE INTERNATIONAL DES EPIZOOTIES. 1993. International Animal Health Code, Mammals, Birds and Bees. 1003 & 1994 Up-dates, 1993: 413–419.
2. STRINGFELLOW, D. A. & S. M. SEIDEL, Eds. 1990. Manual of the International Embryo Transfer Society. 2nd Edit. (1990). IETS. Champaign Il.
3. THIBIER, M. 1993. Statistics describing the international embryo transfer industry. IETS Embryo Transfer Newsletter **11:** 10–13.
4. ASTUDILLO, V. M. 1983. Information and surveillance system of vesicular diseases in the Americas. Use of grid maps for monitoring, data collection and reporting. Rev. Sci. Tech. Off. Int. Epiz. **2**(3): 739–749.
5. PAN AMERICAN FOOT AND MOUTH DISEASE CENTER (PAHO/WHO). Weekly and monthly epidemiological reports, Centro Panamericano de Febre Aftosa, Caixa Postal 589, CEP 20001-970, Rio de Janeiro (RJ), Brazil.
6. COORDENADORIA DE ASSISTENCIA TECNICA INTEGRAL. 1994. Combate a febre aftosa no Estado de São Paulo, 1994. CATI, Departamento de Defesa Agropecuaria, Caixa postal 960, CEP 13073-001, Campinas (SP), Brazil.
7. SINGH, E. L., J. W. MCVICAR, W. C. D. HARE & C. A. MEBUS. 1986. Embryo transfer as a means of controlling the transmission of viral infections. VII. The in vitro exposure of bovine and porcine embryos to foot-and-mouth disease virus. Theriogenology **26:** 587–593.
8. MCVICAR, J. W., E. L. SINGH, C. A. MEBUS & W. C. D. HARE. 1986. Embryo transfer as a means of controlling the transmission of viral infections. VIII. Failure to detect foot-and-mouth disease viral infection associated with embryos collected from infected donor cattle. Theriogenology **26:** 595–603.
9. SNEDECOR, G. W. 1956. Statistical Methods Applied to Experiments in Agriculture and Biology. 5th Edit. The Iowa State College Press. Ames, Iowa.
10. 1992. @RISK, Risk Analysis and Simulation Add-In for Lotus 1-2-3, Version 2.01, Users Guide, 1992. Palisade Corp. 31 Deckeer Rd., Newfield, NY. 14867.

Application of Risk Assessment to International Trade in Animals and Animal Products

H. E. METCALF,[a] J. H. BLACKWELL, AND
J. A. ACREE [b]

United States Department of Agriculture
Animal and Plant Health Inspection Service
Veterinary Services
National Center for Import and Export
Riverdale, Maryland 20737

[b]*Buena Vista, Colorado 81211*

INTRODUCTION

Ratification of the North American Free Trade Agreement (NAFTA)[1] and the General Agreement on Tariffs and Trade (GATT)[2] by the United States Congress has committed the United States to the goal of free trade while maintaining rules that protect countries and areas of the world that are free of certain disease agent(s) from becoming infected.[3] The concept of regionalization[4] and risk assessment[5] are the methods that must be used to achieve these goals.

Risk assessment is not new in dealing with animal trade. Intuitive and subjective risk assessments have been done for decades whenever a decision was made to import or deny entry of an animal or animal product. Application of objective quantitative risk assessments and analysis to the process is a change in approach.

The disease risk factors identified in trade in animals and animal products have been described as being in three categories as source risk factors, commodity risk factors, and destination risk factors.[6] Although often treated together in doing quantitative risk assessment (QRA), each of these broad categories can be treated separately.

Source area risk factors include those factors present at the origin of the animals or animal products involved in international trade and may include the following factors:

- Prevalence of a disease agent in the area exporting the animals or products.
- Geographic location and environment.
- The status of the adjoining or neighboring areas.
- Trading partners and practices.
- Infrastructure of exporting countries. The animal health infrastructure such as the number of veterinarians, etc., is an important factor, but the entire

---

[a] Address correspondence to 10347 Broom Lane, Seabrook, MD 20706.

structure of livestock industry, the legal structure, and the physical infrastructure need to also be evaluated. Evaluation of the legal infrastructure is one of the most nebulous parts of any evaluation. It is often easier to make laws in many countries of the world than to get compliance with or enforce the laws. Legal compliance is often impossible for the outside observer to evaluate.

- A surveillance system in an area is essential to be able to do an adequate risk assessment. The existence and type of surveillance system are dependent upon the infrastructure in the area. Knowledge about the occurrence and prevalence of disease agent(s) is dependent upon the existence of an adequate surveillance system in the exporting area.

Each of these factors needs to be evaluated in order to categorize the risk of the source country or region. The member countries of the International Office of Epizootics (OIE) are currently trying to develop standards for evaluating infrastructure and surveillance systems for import risk analysis procedures.

Commodity risk factors. Each type of commodity traded may alter the risk assessment for any disease agent(s). Calculation of commodity risks should be fairly forthright where there is sufficient scientific data. Commodity risk factors are often made more complex than necessary because there is a tendency to mix source and commodity factors together in evaluating the risk of the commodity. In order to determine the commodity risk factor, it is necessary to begin with the premise that the commodity is infected with the disease agent of concern and examine each step of the processing, handling, and storing of the commodity in order to determine how much the infection is reduced by each process. Commodities can be quite diverse and can include live animals, manufactured by-products from animals, or live cultures of pathogenic agents for investigative purposes. Each commodity carries a risk factor that, once determined, should be identical for every source. When commodity risks are determined, the assessment needs to be cataloged so that all future decisions and countries of the world use the same assessment parameters for determining the risk of a certain commodity. It would appear logical that OIE should keep and publish a catalogue of commodities and their respective risks.

Destination risk factors are those factors present in the importing country or region that may either enhance or inhibit the transmission of a disease agent into the new area or development of disease and include such factors as:

- Availability and susceptibility of *reservoirs,* and availability and competence of vectors for an imported disease agent.
- The intended use and distribution of the commodity and any limitations that may be placed on its use or distribution.
- *Political, environmental, and societal risks* that may occur if it is necessary to destroy all or part of a population of animals because of introduction of a disease agent.
- *Benefit/cost* analysis for the imports. It may be better to deny a particular high-risk import if the potential costs would far exceed any possible benefits that may accrue to the importer or to society.[7]

Each importing country or region must access these risks for itself and determine whether it is willing to assume a certain risk or whether it will require certain risk-mitigation procedures in order to reduce the risk to an acceptable level.

A more difficult area for risk evaluation is the risk of unknown hazards. Diseases such as bovine spongiform encephalopathy, porcine respiratory and reproductive syndrome, and human immunodefiency virus were not known to exist just a few years ago. How many unknown animal diseases are in the world that the United States has avoided by not permitting imports from countries affected with foot and mouth disease?

RISK CLASSIFICATION OF REGIONS

New Paradigms to Describe Risk

In order to begin talking about risk assessments in international livestock trade we must first discard the old paradigms regarding areas as either "free" or "not free". These must be replaced with the concept of "risk" whether it be a simple "low risk" vs. "high risk" or some more elaborate scheme. "Free" does not mean "risk free". "Not free" does not mean that all areas that fit into that category are necessarily of equal risk.

The primary purpose of a risk assessment procedure is to determine what risk management methods will be used to reduce an unacceptable unmitigated risk to an acceptable risk. It is not technically or administratively feasible to do elaborate quantitative risk assessments for each and every proposed import. What is possible is to arrive at some risk classifications for which a proscribed set of management tools can be applied. We have identified 6 such classifications which we believe can be codified into our regulations and provide our livestock industry maximum protection but still provide market access. The proposed categories range from "negligible risk" to "high risk". Risk management procedures for these categories would vary from a simple certification of origin for livestock and products from the lowest risk category to mandatory pre- and postembarkation and postentry testing and quarantine for the highest risk categories.

By developing this type of strategy, it will be much more feasible to recognize areas of a country as lower risk even though other parts of the country may be high risk.

Rationale for Risk Classification

"Freedom from disease" is a parameter that does exist and which we can estimate but can never be known for certain. It always comes back to the conundrum of proving a negative condition.

Quantitative risk assessment (QRA) procedures result in an abstract number (probability of failure) which varies from approaching zero to approaching 1 but never quite reaches either zero or one. It is difficult to arrive at an acceptable level of risk from the results of a QRA. What risk is acceptable versus unacceptable? Even the most unacceptable unmitigated risk may be made acceptable by using certain risk mitigation procedures to reduce the risk. There are a limited number of measures that can be taken to manage the risk of animal disease introduction. Ultimately the final determination of acceptable risk is the benefits vs. costs of success or failure in assuming the risk. The NAFTA refers to these by saying "each Party shall, in establishing its appropriate level of protection regarding the risk associated with the introduction, establishment or spread of an animal or plant pest or disease, and in accessing risk, also take into account the following economic factors, where relevant: (a) loss or production or sales that may result from the pest or disease; (b) costs of control or eradication of the pest or disease in its territory; and (c) the relative cost-effectiveness of alternative approaches to limiting risks."[8]

Among countries now classified as not known to be affected with certain restricted diseases there are different levels of risk involved in importing from those countries. An example would be the comparative risk of importing foot-and-mouth disease virus from Chile versus Mexico. Both these countries are currently considered "free," but Chile is bordered by known infected countries whereas Mexico has no neighbors known to be affected. Differences also exist between countries now classified as affected with restricted diseases as some countries have an effective eradication or control program whereas others may have neither surveillance or control.

For these reasons we have developed risk classifications that can be applied to all situations and to zones, regions, or to countries. These proposed risk classifications will include three classes for regions we would now consider currently not affected with a restricted disease agent and would qualitatively be described as *negligible, slight,* and *low-risk regions.* Areas that we now consider as affected with a restricted disease agent would be classified as a *moderate risk* for a low prevalence region, *high risk* for higher prevalence regions, and *unknown risk* for regions for which we simply do not have enough information to categorize. The unknown risk category may include some regions that could easily be in the negligible, slight, or low-risk categories if sufficient information were available. The following are the proposed codes for risk classes for areas

Risk class RN region (negligible risk) would be the nearest risk level to a zero risk. "Negligible" is defined as being "so small or unimportant or of so little consequence as to warrant little or no attention."[9] This is the goal of any region as far as disease control and would be the goal of any regional disease eradication program. The question is, how can we know that a region is a negligible risk as far as disease is concerned? In the past we have relied primarily on absence of reports of a disease condition in the country or region. This is probably fairly reliable for countries or regions where a large adequately trained animal health infrastructure is present, but in developing or undeveloped regions, the possibility of a disease condition being unreported is high. There are certain criteria that should be met before any region should be considered a "negligible risk" including: (1) physical isolation from any region affected with the disease agent, (2) existence of an adequately trained animal health infrastructure, and (3) absence of the disease condition during the lifetime of any living susceptible animals. This last criteria may be somewhat controversial, but the criteria for determining when a disease agent has actually been eradicated from a population of animals has been long debated, and it has been generally assumed that certain eradication has been achieved and active surveillance can be curtailed only after a complete generation of unexposed and unvaccinated animals has replaced any animals that may have been present when the disease was last known to exist. The need for surveillance in such a region would be limited to an adequate passive surveillance system. The requirements for an adequate passive surveillance system should include a mandatory requirement to report any putative occurrences of the restricted disease agent and an identified agency to receive and investigate any reports with adequate resources and training to properly investigate, diagnose, and take actions against any outbreak of the disease. Passive surveillance should also include continuing education for animal health professionals and a continuing public information program for the livestock industry and general public.

Risk class R1 (slight risk) regions are those where the risk is small but because of proximity to affected regions, trading practices with affected regions, or recent past history of the disease could not be considered as a negligible risk region. The criteria for an adequate animal health infrastructure are the same as for a negligible risk region, but there would be a need for a more active surveillance program in such a region because of the higher risk of introduction of disease. The greatest risk from these regions would be recent introductions of infected animals, agents, or infected products from affected areas. We would consider the likelihood of residual infection in these regions to be negligible.

Risk class R2 (low risk) regions are those where there are no recent outbreaks of disease but which have just recently reached such a point or where vaccination for the disease continues in the region. Such a region would also require an adequate animal health infrastructure and maintain an active surveillance program. There would be a concern about residual infection as well as recently introduced infection in these regions.

Risk class R3 (moderate risk) regions are those where the disease is known to exist at a low prevalence and that have an active program to eradicate the disease. An eradication program implies active continuous surveillance and adequate resources to investigate and eliminate any outbreaks of a disease that may occur in the region.

Risk class R4 (high risk) regions are those where the disease is known to exist and a regional control program exists, but it may not have resources or surveillance to be considered an eradication program, or the disease prevalence in the region is such that the region could not be considered a low prevalence region. There should be some active surveillance program in these regions, and there should be a minimal requirement of an animal health infrastructure to diagnose and report occurrences of disease.

Risk class RU (unknown or unclassified risk) regions are those where there is not adequate infrastructure, or an adequate surveillance program and any control programs are done only at the individual herd or premise level. Classification as an unknown risk may not always be due to inadequate infrastructure, resources, or surveillance but simply because the information has not been shared and cannot be evaluated for classification purposes. This would usually be because the country or region has not expressed any interest in exporting animals or animal products.

The ultimate goal of any commodity that moves in international trade is to be of negligible risk in introducing restricted disease agents into new environments. This can be accomplished by only importing from regions that are classified as a negligible risk or by taking necessary risk-mitigating steps to reduce the final risk to a negligible level even though the source of the commodity may be a higher risk region. Some of the mitigating factors include:

- Certification of origin of animals or products.
- Tests of imported animals or products.
- Tests of herds or premises of origin.
- Treatment of animals or products.
- Quarantine of imported animals.
- Restricted use or movement of imported animals or products at destination to reduce costs of failure.

CALCULATING THE UNRESTRICTED RISKS

Estimating the Magnitude of the Reservoir

The prevalence of reported disease in any area can be likened to the tip of the iceberg. Reported disease is generally of little concern to the official certifying an export unless there is outright fraud involved in the certification. Known infected herds or animals are of little risk since they can be easily avoided. More reliable estimates of prevalence should be based on periodic or continuous surveillance of the population. Usually in an eradication program, some type of surveillance program must be in place to adequately obtain accurate data about the disease prevalence. It is usually only under these conditions that adequate prevalence data can be obtained. If the reservoir of infection is in wild animals or vectors, then the prevalence data in livestock may not provide a reliable estimate of the magnitude of the reservoir in the region. In such a case, some type of active surveillance program to estimate the disease reservoir must be done.

Unreported disease is a problem in estimating many diseases and particularly affects prevalence reports. This category includes those animals that actually do become sick and may disseminate infection. There are a multitude of reasons a disease condition will go unreported including failure to recognize the disease, diseases that do not manifest readily apparent clinical signs (brucellosis, tuberculosis, bluetongue in cattle, etc.), lack of observation of the animals, considering that the disease is insignificant or unimportant, or actual fraudulent hiding of the disease. A general maxim is that the more commonplace a disease condition is in the animal population, the less likely the disease will be reported.

Inapparent or latent infection will vary for many disease agents and conditions. Knowing the natural pathogenesis and epidemiology of the disease is important in estimating the quantity of inapparent or latent infection in the population. Vaccination can often mask an infection by protecting animals that may otherwise become clinically apparent sentinels after coming in contact with a carrier.

Usually the only method of estimating the unreported disease and inapparent infection in a region is from the prevalence data for the disease. If there have not been any recent reports of the disease, then some type of projection analysis must be done from previous prevalence reports.

Introduced disease would be through animals, products, or vectors moved either legally or illegally into a region. Usually the only information available is for legal movements or importations. Evaluation of illegal movements can only be done subjectively by evaluation of the infrastructure controlling border stations or ports.

In developing a risk scenario tree for the unmitigated risk, we need to consider all of the above reservoirs of the disease agent that may exist in the region.

EXAMPLE OF A RISK ASSESSMENT

The example region has been endemically affected with foot-and-mouth disease (FMD) virus for over 100 years. In 1989, a major effort began to eradicate FMD from the region, and March 1994 was the first month in recent recorded history when there had been no reported FMD infection in the region. There were additional outbreaks in April 1994, which were quickly stamped out, and since April 1994 there have not been any further outbreaks. Semiannual vaccination of all cattle is done in all parts of the region using oil-adjuvant killed vaccine. The monthly prevalence history is shown for this region in TABLE 1.

Estimating the Source Risk

If we were going to do a risk assessment as of December 31, 1994 for this region for FMD and only used the current prevalence data as the first scenario parameter, then we would have a zero input and all subsequent scenarios would likewise be zero. We need to find a method to estimate residual unreported disease and latent infection in the population. To do this we need additional information as shown in TABLE 2.

Several dynamic factors occur in the animal population over time that must be considered when trying to calculate the residual infection in the population. With FMD virus, we are primarily concerned with animal reservoirs of infection rather than inanimate sources, since susceptible animals should provide adequate sentinels for infection from the environment.

The attrition rate of the population will affect the occurrence of potential residual infection. The attrition rate is the annual rate for which animals in the population die or are slaughtered. We generally use an attrition rate of 20% per year for cattle with a minimum of 15% and a maximum rate of 25%. This rate is conservative for all cattle in a population, and in fact we could probably use the proportion of cattle slaughtered each year plus a small additional proportion for cattle that die of natural causes if we have such data available.

The next factor we must consider is the rate of infection for exposed animals in the population. FMD virus is a highly infectious agent, and a large proportion of exposed susceptible cattle will probably become infected, but when dealing with a vaccinated population, few of these will show clinical signs or shed virus. We have used a rate of 30% infection rate for exposed animals in the population with a maximum of 50% and a minimum of 10% for infected animals in the population.

The third factor we must consider is the duration of infection in the infected cattle. It has been shown that FMD virus carriers may persist up to 15 months or longer after infection[10] but other studies have found that many experimentally infected cattle recover in 5 to 6 weeks.[11] Other studies have found that up to 3% of cattle may remain infected 180 days or more after exposure.[12] From these and other observations, it is difficult to arrive at an average duration of the carrier status in infected

TABLE 1. Monthly Reports of Outbreaks of Foot-and-Mouth Disease in Region—Example.

Year	Month											
	Jan.	Feb.	Mar.	Apr.	May	June	July	Aug.	Sep.	Oct.	Nov.	Dec.
1985	407	268	94	54	124	97	54	42	43	36	6	14
1986	42	41	10	21	109	129	67	97	142	128	14	13
1987	22	22	23	163	432	212	128	108	119	71	13	11
1988	11	99	5	9	37	62	41	67	38	54	17	9
1989	14	15	9	2	32	34	15	45	97	99	54	18
1990	13	11	104	133	120	105	65	40	122	79	25	24
1991	33	18	16	14	20	20	19	18	21	34	5	16
1992	8	20	34	18	47	62	27	26	34	42	25	7
1993	32	27	31	21	14	10	6	17	23	7	4	4
1994	15	1	0	2	0	0	0	0	0	0	0	0

TABLE 2. Annual Data for Infected Premises and Animals and Exposed Premises and Animals

Year	Premises Infected -a-	Animals on Affected Premises -b-	Premises Depopulated -c-	Animals Depopulated (est) -d-	Adjacent Exposed Premises -e-	Animals on Adjacent Premises -f-	Total Exposed Animals Retained[a] -g-
1985	1,240	688,412	0		4,927	4,948,387	5,636,799
1986	813	504,737	0		3,332	2,282,420	2,787,157
1987	1,324	967,854	0		4,303	3,119,675	4,087,529
1988	359	287,814	0		1,292	1,059,440	1,347,254
1989	434	238,137	0		1,692	1,248,696	1,486,833
1990	841	631,411	0		2,691	1,130,220	1,761,631
1991	234	166,366	0		959	602,529	768,895
1992	350	254,328	0		1,470	455,700	710,028
1993	196	208,641	4	4,256	1,108	644,139	848,518
1994	18	37,209	6	12,402	72	25,204	50,004

[a] Total animals exposed that are retained is the sum of columns b and f minus column d.

cattle. What is needed is the infective half-life for the virus. This is an area where some additional research would be helpful, but because of the potential cost of such a research project, it will probably never be done. Based on what information is available, we have used a 90-day expected half-life with a maximum of 120 days and a minimum of 60 days as a reasonable estimate of the half-life for FMD in vaccinated cattle.

Using these parameters and the data generated in TABLE 2, we calculated the projected number of infected carriers surviving in the population within the region in TABLE 3.

The total animal population in this region was 53,300,000 head. Therefore the probabilities are given for the expected, minimum, and maximum values of carrier cattle in the population in TABLE 4. This would be the prevalence of residual infection in the region and would be the likelihood that any single animal would be infected when it left the farm in the region. The likelihood of an infected animal being in a shipment would be contingent primarily on the number of animals in the shipment.[13]

Animals imported into the region from other regions with a higher prevalence or of a higher risk than the region being evaluated need to also be considered. This amounts to a separate risk assessment on each region from which the animals or animal products have been imported. The attrition rates and infection half-lives can be used on these animals in the same manner as for exposed animals in the assessed region. Usually these numbers are small, and imported animals may not be likely to be exported unless they were imported strictly for the purpose of laundering the animals for export. In our example 73,497 animals had been imported during the previous 5 years from other regions that were known to be affected with FMD virus. By a similar analysis as above, it was estimated that 53 of these were expected to

TABLE 3. Estimation of Infected Animals Retained and Numbers of Infective Animals Remaining in the Population on December 31, 1994

| Year | Total Exposed Animals Retained | Infected Animals Retained[a] | | | Number of Original Infected Animals Surviving[b] | | | Number Infected Carriers Surviving[c] | |
		Expected 30% Infection Rate	Minimum 10% Infection Rate	Maximum 50% Infection Rate	Expected 20% Annual Attrition Rate[d]	Minimum 25% Annual Attrition Rate	Maximum 15% Annual Attrition Rate	Expected 90 Day Half-life	Minimum 60 Day Half-life
1985	5,636,799	1,689,676	563,225	2,816,127	226,784	42,290	652,263	0	0
1986	2,787,157	836,147	278,716	1,393,579	140,282	27,903	379,737	0	0
1987	4,087,529	1,226,259	408,753	2,043,765	257,165	54,562	655,184	0	0
1988	1,347,254	505,502	168,501	842,503	132,514	29,989	317,750	0	0
1989	1,486,833	446,050	148,683	743,417	146,162	35,283	329,358	0	0
1990	1,761,631	528,489	176,163	880,816	216,469	55,739	459,791	1	0
1991	768,895	230,669	76,890	384,448	118,102	32,438	236,099	9	0
1992	710,028	213,008	71,003	355,014	135,325	39,939	256,498	173	2
1993	848,518	254,555	84,852	424,259	203,644	63,639	360,520	3,031	143
1994	50,004	15,001	5,000	25,002	15,001	5,000	25,002	1,831	136
Totals	19,484,648	5,945,356	1,981,785	9,908,927	1,592,480	386,782	3,672,301	4,544	281

[a] Infected animals retained = (exposed animals retained) (infection rate).

[b] Infected animals surviving = (infected animals retained) $\times$ (1-attrition rate)$^{\text{(years since infected)}}$.

[c] Animal surviving = (infected animals surviving) $\times$ 0.5 $^{\text{(days since infected + half-life)}}$.

[d] Attrition rates are the normal rates of culling, death, and slaughter of animals from the population.

TABLE 4. Estimation of Prevalence Rate from the Expected Number of Infective Animals in the Population and the Total Population in the Region

Risk of Residual Infected Animals in Population	Expected -a-	Minimum -b-	Maximum -c-
A. Estimated number of infective animals	4,544	281	20,446
B. Animal population in region	53,300,000	53,300,000	53,300,000
C. Prevalence rate of residual infection (A ÷ B)	8.5×10^{-5}	5.0×10^{-6}	3.8×10^{-4}

be infected at the time of import and 1 is still expected to be infected for a prevalence rate of 0.00000002 (2 per 10 million) with a maximum rate of 0.00000006 (6 per 10 million) and a minimum rate of 0.00000000002 (2 per 100 billion).

An outbreak of disease that is not reported is difficult to estimate. For the purposes of this assessment we have assumed that the unreported cases would be equal to the residual latent infection. This may be a high estimate in a largely unvaccinated population, but in a population where nearly all the susceptible animals are vaccinated, a single atypical mild case in a herd without any apparent spread to other immune animals in the herd we believe is a reasonable assumption since it has been reported that FMD in vaccinated animals often follows an atypical course. Using these data we arrive at the risk assessment scenarios in TABLE 5.

The combined risk SO–S2 from TABLE 5 becomes the unmitigated risk for any one animal being infected when exported from the region. This assumes that any exports are randomly selected from the entire population. Any attempts at selection

TABLE 5. The Unmitigated Assessment of an Animal Being Infected with FMD at the Time It Leaves a Premise of Origin[a]

Scenarios	Unmitigated Risk of FMD from Region	Expected[b]	Minimum	Maximum
S0	Prevalence of residual infection in region	8.5×10^{-5}	5.0×10^{-8}	3.8×10^{-4}
S1	Risk of introduction from affected regions	2.0×10^{-8}	2.0×10^{-11}	6.0×10^{-8}
S2	Outbreaks not detected on farms	8.5×10^{-5}	5.0×10^{-6}	3.6×10^{-4}
S0–S2	P(infected animals on farms)	1.7×10^{-4}	1.0×10^{-5}	7.7×10^{-4}

[a] This assumes no mitigating measures are taken such as inspection of examination of the animals before or after they leave the premise of origin.

[b] The expected, minimum, and maximum risks are the point estimates of the expected, best case, and worst case estimates respectively.

of segments of the population, etc., are risk mitigation procedures that further alter the final risk.

Estimating the Commodity Risk

Each and every live animal or product from an animal or in contact with an infected animal has its own commodity risk. Each commodity that is handled in exactly the same manner would have the same commodity risk regardless of the region where it is produced. For this purpose, we calculate the commodity risk with the presumption that the animal that produced the commodity is always infected. The final risk then would be the source risk as calculated above times the commodity risk. For our example we will look at the single commodity of fresh or frozen deboned beef that has been processed in the following manner:

1. All cattle are held overnight at the slaughter plant and inspected prior to entry into the slaughter plant. All sick or lame animals are removed and quarantined.
2. At slaughter, the head, oral cavity (including tonsils, lymph nodes), feet, nonedible and edible glands (including the rumen) are thoroughly inspected for lesions of FMD.
3. The carcass is allowed to hang in the cooler for at least 36 hours to allow pH changes associated with rigor mortis to occur in skeletal muscle.
4. All bone and lymphatic tissue is removed from the meat during the deboning process.
5. pH tests are routinely taken on carcasses prior to deboning and on individual cuts of meat after the deboning process. Any cut of meat that has a pH greater than 5.9 is rejected for export and can only be used in the local domestic market.

Using the above information, we can develop an estimate for the risk factor for fresh deboned beef as in TABLE 6.

Scenario S0 assumes that an FMD-infected animal arrives at a slaughter plant.

Scenario S1-disease not detected at antemortem inspection. All cattle are inspected just prior to entry into the auction site or the slaughter plant. Sick or lame animals are removed and quarantined. We estimate that 90% of the FMD-positive animals that get to this point would be missed due to lack of clinical signs since we are primarily dealing with vaccinated latent carriers. We estimate that this would range from 50% to 99% of the infected animals being missed.

Scenario S2-disease not detected on postmortem inspection. We estimate that 90% of the FMD positive carcasses would be missed due to lack of clinical signs with a range of 50% to 99%.

Scenario S3-infected material not removed during slaughter. Because FMD virus in latent carrier animals has been primarily shown to reside in the pharyngeal region and this tissue is removed, it is estimated that 90% of the infected products would be removed during the normal slaughter process with a range of 80% to 99%; therefore, 10% with a range of 1% to 20% would survive.

TABLE 6. Risk Assessment for Deboned Beef Assuming that an Infected Animal Arrives at a Slaughter Plant

Scenarios	Unmitigated Risk of FMD from Region	Expected[a]	Minimum	Maximum
S0	Infected animal arrives at slaughter plant	1.00	1.00	1.00
S1	Disease not detected at antemortem inspection	0.90	0.50	0.99
S2	Not detected on postmortem examination	0.90	0.50	0.99
S3	Infected tissue not removed during slaughter	0.10	0.01	0.20
S4	Virus survives maturation	0.002	0.0002	0.03
S5	Virus not eliminated during deboning	0.002	0.0002	0.02
S6	Infected meat not eliminated during deboning	0.02	0.002	0.05
S0–S1	p(Infection eliminated at antemortem)	9×10^{-1}	5×10^{-1}	9.9×10^{-1}
S0–S2	p(Infection not eliminated at postmortem)	8.1×10^{-1}	2.5×10^{-1}	9.801×10^{-1}
S0–S3	p(Infection not eliminated at slaughter)	8.1×10^{-2}	2.5×10^{-3}	1.9602×10^{-1}
S0–S4	p(Virus survives maturation)	1.62×10^{-4}	5×10^{-7}	5.8806×10^{-3}
S0–S5	p(Virus not eliminated during deboning)	3.24×10^{-7}	1×10^{-10}	1.7612×10^{-4}
S0–S6	p(Virus not eliminated by pH tests)	6.48×10^{-9}	less than 10^{-64}	6.8806×10^{-6}

[a] Expected, minimum, and maximum risk factors are the point estimates for expected, best case, and worse case scenarios respectively.

Scenario S4—virus survives maturation. The pH changes associated with rigor mortis occur in skeletal muscle within 24 to 48 hours after slaughter. A pH below 6.0 (5.8) is virucidal for most FMD virus.[14] However, there are occasions when there is not a uniform change in pH throughout the carcass, or strains of virus are present that may be pH resistant.[15]

Variable $S4_1$—survival of virus due to uneven pH changes. Experience with people who have tested the pH of meat indicate that the proportion of cuts rejected due to insufficient lowering of the pH is less than 2%. We have used a range of 0.01% to 2% with an expected value of 0.1% of the meat cuts will fail to reach a pH of 6.0 or less.

Variable $S4_2$—survival of virus due to pH resistant variants. The rates for pH-resistant viruses are not available, but since the workers who describe these variants seemed to not have a great deal of difficulty of finding them, we have assumed that 1 of every 1000 (0.1%) virus strains may have subpopulations of virus with this characteristic as a reasonable estimate with a minimum value of 0.01% and a maximum of 1%.

The combined risk for Scenario S4 would be $S4_1$ plus $S4_2$ thus the total expected risk for S4 would be 0.2% with a range of 0.02% to 3%.

Scenario S5—virus not eliminated during deboning and removal of lymph nodes. After the onset of FMD infection in a susceptible animal, progeny virus spreads via the circulatory system to various epithelial sites, including bone marrow and lymph nodes in the body. Thus the deboning process and removal of lymphatic tissue reduce the potential for the spread of FMD through bone-in meat.[14] We estimate that 0.2% of any residual FMD virus would not be eliminated by deboning with a range of 0.02% to 2%.

Scenario S6—virus not eliminated by pH tests of meat cuts. pH testing of carcasses and individual cuts of meat would largely eliminate cases where uneven pH changes would occur in the carcass. We have assumed that this process would eliminate all but 2% of any carcasses with virus that may survive due to uneven pH changes and those with virus that survive reduced pH. The expected survival would therefore be 2% with a range of 0.2% to 5%.

The combined risk scenarios for deboned beef handled in this manner from an infected animal would be 6.48×10^{-9} or that virus would survive in 1 of every 154 million *infected* carcasses slaughtered. In order to arrive at the risk of importing infected meat from the region in our example, one needs to take the product of the unmitigated risk and the commodity risk to arrive at the per-unit risk and calculate the number of units that will be expected to be shipped.

Where:

$P\{C\}$ = commodity risk factor
$P\{R\}$ = source region risk factor
n = number of animals represented
$P\{I\}$ = probability of one or more infected animals in shipment

Then:

$$P\{I\} = 1 - ([1 - (P\{C\} \times P\{R\})]^{(N)})$$

Using our example then

$$P\{R\} = 1.70 \times 10^{-4}$$
$$P\{C\} = 6.48 \times 10^{-9}$$
$$P\{C\} \times P\{R\} = 1.10 \times 10^{-12}$$

If we assume that meat from 2,000,000 head of cattle will be exported each year, then $n = 2,000,000$

$$P\{I\} = 1 - ([1 - (1.10 \times 10^{-12})^{2,000,000}]) = 0.000002 \text{ per year}$$

Expected frequency of introductions = 0.000002 = 500,000 years. Therefore the probability of exporting infected meat would be 0.000002 per year or once every 500,000 years infected meat would be expected to be exported from the region.

SUMMARY

NAFTA and GATT have specified that we must base import decisions on sound scientific information. This includes the use of risk assessment procedures to justify excluding or restricting the international movement of either live animals or products from animals.

Procedures and techniques have been described for calculating the risk of almost any hazard that one can imagine. Of particular concern to animal health professionals are the risks associated with introducing known or unknown disease agents into a population of animals that has been previously free of a specific disease.

Traditional risk assessment procedures calculate the probability that a disease agent transmission will occur and the cost of such an event. The responsible manager must decide on an acceptable threshold of risk. A set of criteria for determining what risk is acceptable has been developed that categorizes risks for importing animals or animal products from any region of the world based on classification of the region as negligible, slight, low, moderate, high, or unknown risk.

REFERENCES

1. 1993. North American Free Trade Agreement. Ratified by United States Congress December 1993, Effective January 1, 1994.
2. 1994. General Agreement on Tariffs and Trade. Ratified by United States Congress, December 1994, Effective January 1, 1995.
3. 1993. North American Free Trade Agreement. Basic Rights and Obligations, *Right to Take Sanitary and Phytosanitary Measures.* Chapter Seven, Section B, Article **712:** 30-31.
4. 1993. North American Free Trade Agreement. Adaptation to Regional Conditions, Chapter Seven, Section B, Article **716:** 34-36.
5. 1993. North American Free Trade Agreement. Risk Assessment and Appropriate Level of Protection, Chapter Seven, Section B, Article **715:** 33-34.
6. MORLEY, R. S. 1993. A model for the assessment of the animal disease risks associated with the importation of animals and animal products. Rev. Sci. Tech. Off. Int. Epiz. **12**(4): 1055-1092.
7. ACREE, J. A. & V. C. BEAL, JR. 1989. Animal health perspectives of international embryo exchange. Live Animal Trade Transport Magazine **1**(1): 33-38.

8. 1993. North American Free Trade Agreement. Risk Assessment and Appropriate Level of Protection. Chapter Seven, Section B, Article 715, Paragraph **2:** 33–34.

9. 1990. Webster's New Collegiate Dictionary. 9th Ed. Merriam-Webster Inc. Springfield, MA.

10. BURROWS, R. 1966. Studies on the carrier state of cattle exposed to foot-and-mouth disease virus. J. Hyg. (Camb.) **64:** 81–90.

11. McVICAR, J. W. & P. SUTMOLLER. 1974. Neutralizing activity in the serum and oesophageal-pharyngeal fluid of cattle after exposure to foot-and-mouth disease virus and subsequent re-exposure. Brief Rep. Arch. Gesamte Virusforschung **44**(2): 173–176.

12. BRUGI, M. 1928. Les metodes generales de la prophylaxie de la fievre aphteuse. Bull. OIE **I:** 564–569.

13. MACDIARMID, S. C. 1993. Risk analysis and the importation of animals and animal products. Rev. Sci. Tech. Off. Int. Epiz. **12**(4): 1093 1017.

14. COTTRAL, G. 1969. Persistence of foot-and-mouth disease virus in animals, their products and the environment. Bull. Off. Int. Epiz. **71:** 541–568.

15. MUSSGAY, M. 1959. Gewinnung einer pH-stabilin Variante des Virus der Maul-un Klauenseuche und ihre Vermehrung in Gewebekulturen mit sauerem Medium. Zentralbl. bact. **140:** 183.

Implications of International Trade Agreements for Global Animal Health

DAN J. SHEESLEY AND JOHN K. GREIFER

Trade Support Team
United States Department of Agriculture
Animal and Plant Health Inspection Service
International Service
14th and Independence Avenue SW
Room 1128 South Building
Washington, DC 20250

INTRODUCTION

The North American Free Trade Agreement (NAFTA), a trilateral trade pact between United States, Canada, and Mexico, went into effect in January 1994. The GATT Uruguay Round Agreement, a global trade pact involving 120 plus countries, went into effect in January 1995. The same month that GATT went into effect, the US Administration and 33 leaders from Central and South America and the Caribbean made a joint declaration of their commitment to building a hemispheric free area—known as the Free Trade Areas for the Americas (FTAA)—by the year 2005. These trade initiatives are all aimed at liberalizing and expanding regional and global trade.

In both the GATT and NAFTA negotiations, countries made a historic decision to reform agricultural trade. In addition to cutting tariffs and reducing other trade-distorting practices, such as export subsidies, NAFTA and GATT established a set of rules to control the use of technical health barriers. Most significantly, countries are required to document in a transparent fashion the scientific basis for any regulations that affect trade. Countries also agreed that import requirements should, to the greatest extent possible, be based on international standards.

What challenges lay ahead for regulatory officials charged with furthering these GATT and NAFTA trade objectives while ensuring the safe movement of animals and animal products? This paper reviews the basic sanitary and phytosanitary (SPS) rules established under GATT and NAFTA and discusses the contributions veterinary scientists and regulatory officials can make to advance free trade goals while ensuring that health risks associated with international commerce are effectively managed.

GATT AND NAFTA, A HISTORICAL PERSPECTIVE

GATT, established in 1947, has proven to be a highly effective treaty in reducing high tariffs on manufacturing commodities and stimulating international trade to new unprecedented levels. By the early 1980s, some countries, particularly the United

States, came to the view that this same success could be duplicated in the agricultural sector.

From the very beginning of the Uruguay Round, the United States took a strong position in advocating an aggressive negotiating agenda on agriculture to reduce and/ or eliminate tariffs, quotas, trade-distorting subsidies, and unjustified health barriers [also known as SPS measures (SPS)]. Countries eventually agreed that government interference in global agricultural markets had gotten out of hand. In the end, GATT signatory countries agreed to reduce some of this government interference by putting some controls on the use of export subsidies and reducing agricultural tariffs by an average of 40% by the end of a seven-year period.

Negotiators recognized that as efforts were made to eliminate tariffs and other forms of protection, the likelihood existed that countries would use health requirements as new protectionist barriers. For this reason, countries agreed to establish a code or set of disciplines to condition the use of SPS measures in trade. The end result was the GATT Uruguay Round Agreement on the Application of Sanitary and Phytosanitary Measures.

When US, Mexico, and Canada officials met to negotiate NAFTA in 1990, they addressed many of the same objectives that were being pursued in the GATT negotiations, including the goal of reforming agricultural trade and the use of technical health standards. In the end, NAFTA replicated the same SPS rules and principles established under GATT.

SANITARY AND PHYTOSANITARY PROVISIONS

The GATT[a] and NAFTA[b] agricultural texts describe a number of important rules and principles. However, the most basic rule is that health-related import restrictions have a transparent, scientific rationale. These SPS rules include the following:

Basic Rights (NAFTA: Article 712 and GATT: Article 2)

Both GATT and NAFTA maintain every country's right to adopt any measure it believes is necessary to protect the health of its consumers, plants, and animals, including measures that may be more stringent than existing international standards.

Countries are also encouraged to take into account the objective of minimizing negative trade effects and to avoid arbitrary or unjustifiable distinctions in the level of protection in different circumstances if such distinctions result in unjustifiable discrimination against goods from another trade partner. In other words, countries should (1) use the least trade distorting option when it considers imposing some regulatory measure and (2) seek consistency in its risk-management practices.

[a]North American Free Trade Agreement, Chapter 7, Section B, Sanitary and Phytosanitary Measures.

[b]World Trade Organization Agreement on the Application of Sanitary and Phytosanitary Measures.

Risk Assessment (NAFTA: Article 715 and GATT: Article 5)

The requirement that measures be based on science means that import decisions must be based on a risk assessment that uses scientific data and methodologies. GATT and NAFTA describe a number of factors that would be considered in a risk assessment, including:

- Risk assessment methodologies and techniques developed by the OIE, IPPC, or regional standards organizations.

- Relevant scientific evidence.

- Relevant processes and production methods.

- Relevant inspection, sampling, and testing methods.

- Prevalence of diseases of concern, including the existence of disease-free or low-prevalence areas.

- Relevant ecological and environmental conditions.

- Relevant treatments, including quarantines.

- Relevant economic factors, such as production or sales losses and control costs if a particular disease were to be introduced.

Clearly, this particular GATT and NAFTA commitment will require countries to make some investments in its regulatory institutions in order to carryout these risk-assessment functions. APHIS, for the most part, has already developed its risk-assessment capabilities. However, many countries lack the institutional wherewithal to perform these functions. Recognizing this situation in many less developed countries, GATT Article 14 gives less developed countries a grace period of two or five years, depending on its economic status, to adapt their regulatory systems and institutions to fulfill this and other SPS obligations.

Equivalence (NAFTA: Article 714 and GATT: Article 4)

Countries are committed to recognizing another country's SPS measure(s) as equivalent to their own when the exporting country can objectively demonstrate that its approach provides the importing country's desired level of protection.

Equivalence encourages countries to recognize that different procedures could and can be used to achieve the level of protection demanded by the importing country. Equivalence focuses on procedures (such as inspection, certification, testing, surveying, trapping, fumigation, and other treatments or practices) that countries use to ensure or certify the safety of a particular commodity.

The burden is on the exporting country to objectively demonstrate that its system or practices produce equivalent results or meet the sanitary or phytosanitary concerns of the importing country.

Role of International Standards Organizations (NAFTA: Article 713 and GATT: Article 3)

The reason for promoting the development and use of international standards is to harmonize, to the greatest extent possible, countries' SPS measures and practices. Trade negotiators believed that by reducing unnecessary variances between countries' technical health standards, we would, thereby, reduce the source of many technical trade differences and disputes.

With this end in view, GATT and NAFTA officially recognized the Office of International Epizootics (OIE) as the forum for global standards in animal health, Codex Alimentarius (Codex) for food safety standards, and the International Plant Protection Convention (IPPC) for plant health standards.

Regionalization (NAFTA: Article 716 and GATT: Article 6)

GATT and NAFTA commit countries to recognizing disease and pest-free areas as well as areas of low pest and disease prevalence. This concept is perhaps the most significant policy and regulatory issue facing APHIS and our trade partners. It is expected to create new opportunities for the United States, as well as other countries, to export from areas that are demonstrated to be free of particular diseases even though the diseases or pests of concern may exist elsewhere within the national territory.

APHIS' position is that a standardized and transparent risk assessment process must include criteria for evaluating an exporting country's veterinary or plant health infrastructure, including its disease-control capabilities, surveillance and monitoring systems, and animal and plant import policies.

Furthermore, an evaluation of a free zone or area will need to consider the following risk factors: (1) proximity of the area to infected areas, (2) historical disease incidence within the area, (3) vaccination policies and practices within areas, (4) type of physical separation from higher risk areas, (5) type of importation of animal and animal products from higher risk areas, (6) disease surveillance within the area, (7) disease control policies and resources within area, and (8) demography and infrastructure within the area.

It should be noted that GATT and NAFTA have provisions requiring countries to recognize not only disease-free areas, but areas of low disease or pest prevalence.

Transparency (NAFTA: Article 718–719 and GATT: Article 7)

NAFTA and GATT include provisions that encourage greater openness of the domestic rule-making process that affects trade. For example, a country must publish its requirements, provide advanced notification to other countries of any changes in its SPS requirements that may affect trade, make available the scientific basis for the proposed action, and allow countries an opportunity to comment on the proposed action. An exception to this advance notification rule exists for emergency disease or pest situations.

Developed countries, like the United States, for the most part, have the resources to publish their regulatory actions in an official national gazette and dedicated staff to review, analyze, and consider public comments and input. For example, as a result of long standing US law, the Administrative Procedures Act, APHIS has been publishing its import rules and inviting public input into the rule-making process since 1946. Hence, these transparency issues, or at least the notification and publication process, are not new to the United States. However, they are new to many countries in the world, and implementing these risk assessment and transparency requirements will require governments to find the resources and technical assistance. As indicated earlier, GATT provides a five-year grace period to the least developed countries to establish these processes.

Sanitary and Phytosanitary Committees (NAFTA: Article 722 and GATT: Article 12)

Both GATT and NAFTA call for the formation of Sanitary and Phytosanitary (S&P) Committees. The basic function of the committees is to oversee and monitor implementation of the agreements' S&P provisions. Furthermore, the GATT and NAFTA S&P Committees are expected to play a vital role in (1) facilitating consultations on specific S&P trade problems, (2) encouraging technical cooperation, and (3) addressing standard setting issues as they arise.

DISPUTE SETTLEMENT

In the event that countries depart from the SPS code, a complaining party has a right to challenge another country's health requirements through the dispute settlement system. Separate and distinct dispute settlement systems exist under NAFTA and GATT. A dispute arising between Mexico, Canada, or the United States would be first addressed through NAFTA. However, there is nothing in the agreements preventing these countries from using the GATT forum for resolving disputes.

GATT and NAFTA each established their own institutional structures to oversee implementation of their respective agreements and administer the dispute settlement procedures. For example, the GATT Uruguay Round agreement established the World Trade Organization (WTO) to facilitate the implementation of the Uruguay Round obligations, serve as a forum for further negotiations among members, administer the dispute settlement procedures, serve as a trade policy monitoring mechanism, and fulfill other functions regarding global economic policy issues.

Similarly, NAFTA established the Free Trade Commission to oversee implementation of our trilateral obligations and to address specific issues, including disputes, that may be raised by Mexico, the United States, or Canada.

While existing as separate systems, the GATT and NAFTA dispute settlement procedures share some common features. For example, in both systems, disputes begin with technical consultations at the bilateral level. If technical discussions at this level fail to resolve the issue, a complaining party may raise the issue to the appropriate NAFTA or GATT committee. If the complaint involves a health issue, the complaint would be raised at the SPS Committee level (committee functions were discussed earlier in this paper). If resolution is not achieved through consultations at the SPS Committee level, the complaining party has the right to invoke the formal NAFTA or GATT dispute settlement process. In short, before countries can invoke formal dispute settlement procedures at either the WTO or NAFTA commission level, the disputing parties must exhaust all efforts to resolve the issue at the bilateral and committee levels.

In both the GATT and NAFTA dispute settlement systems, panels may be formed to review complaints and make recommendations. Such panels may seek recommendations and advice from the relevant international standard setting organizations (i.e., OIE, Codex, IPPC, or their regional subsidiary organizations) or appoint a board of experts (from these or other appropriate source) to evaluate the technical aspects of a given issue.

Four points about the process should be underscored:

1. *Emphasis will be placed on resolving technical disputes and complaints at the technical level (i.e., SPS Committee level) in order to prevent setting off the more formal and complex dispute settlement process.*

2. *The challenging party has the burden of proving that a particular SPS measure or standard is in violation of GATT or NAFTA.*

3. *Panels will determine whether the measure subject to dispute is based on a scientific risk assessment and whether data and the process used in that assessment were:*

   - *collected in a scientific fashion.*
   - *based on international standards if such standards exist.*
   - *not unnecessarily and unfairly discriminatory in nature.*
   - *transparent.*

4. *If a panel issues an opinion that a standard is in violation, that country has the option of either changing the offending measure or keeping the GATT or NAFTA inconsistent measure and compensating the complaining party for the value of impaired trade. If compensation is not provided, the complaining party would be permitted to suspend some trade concessions of equivalent value to lost trade.*

CONTRIBUTION OF VETERINARY SCIENCE

Central to the dispute settlement process is the use of panels, which may be convened to review disputes and provide findings and recommendations to the WTO. These panels will, in turn, rely on international sources of expertise for advice on technical trade matters, including, but not limited to, veterinary expertise available at the OIE.

A key question panels are likely to ask when reviewing an issue or complaint is what is the relevant international standard or guideline and what is the expert advice of the relevant international standard setting body? With this in mind, it becomes apparent that these international standard setting bodies will be vitally important forums for addressing international animal health and trade issues and developing future international standards. OIE standards, for example, are likely to prevail when the WTO is called upon to review technically based trade disputes.

Clearly, the OIE, like the IPPC and Codex, will increasingly focus its efforts in developing internationally accepted standards in their respective health areas and veterinary and regulatory officials from around the world will play a crucial role in these efforts. The contributions that veterinary scientists and officials can make in the international trade policy area include the following:

- Providing technical advice and analysis to the WTO and NAFTA SPS Committees as well as panels that may be formed to examine disputes involving a sanitary or technical health issue.

- Accelerating the development of technically and operationally feasible standards for conducting risk assessments, particularly at the regional and international levels (i.e., OIE).

- Developing risk-management strategies to reduce disease risks to an acceptable level and thereby allow the safe movement of animal products. New risk management tools and technologies may reduce a risk to an acceptable level to allow for trade.

- Widening and deepening the level of discourse between regulators and scientists around the world to accelerate the diffusion and transfer of regulatory ideas, approaches, and technologies (i.e., quarantine practices, data, surveillance, and monitoring systems and techniques).

APHIS for its part has made it a top priority to provide leadership on these issues and intends to do this by engaging both its foreign regulatory counterparts and domestic industries in discussions on risk assessment, regionalization, and other key standard-setting issues.

Assessment of the Risk of Introducing Foot-and-Mouth Disease into Panama via a Ferry Operating between Cartagena, Colombia and Colon, Panama

W. R. WHITE,[a] R. L. CROM,[b] AND K. D. WALKER[b]

[a]United States Department of Agriculture
Animal and Plant Health Inspection Service
International Services
4700 River Road, Unit 67
Riverdale, Maryland 20737-1233

[b]United States Department of Agriculture
Animal and Plant Health Inspection Service
Veterinary Services
Centers for Epidemiology and Animal Health
555 South Howes Street
Fort Collins, Colorado 80521

INTRODUCTION

Although foot-and-mouth disease (FMD) is enzootic in much of South America, and occurred earlier this century in the United States, Mexico, and Canada, the disease has never been reported in Panama and Central America. Direct spread from South America has been prevented by the Darien Gap, approximately 100 kilometers of impassable territory shared by Colombia and Panama, which makes direct contact between cattle populations difficult.

FMD entered South America in 1870 by a cattle importation from Europe, with disease occurring simultaneously in Argentina, Brazil, Chile, and Uruguay.[1] Throughout the 1950s, numerous outbreaks occurred in most South American countries, becoming enzootic in the major cattle-raising areas by the 1960s, including Colombia. Colombia experienced its first FMD outbreak in 1950.

A comprehensive list of the modes of transmission of FMD and their categorization into low, medium, or high hazard has been developed.[2] Ninety-nine animals and 97 animal products or fomites were identified as possible sources of FMD transmission. Several of these animals, including domestic livestock, camelids, capybaras, and peccaries, and most of these food and nonfood animal products, are present in South America. Live animals, bovine semen, fresh meat, salted hides, garbage, and contaminated shoes were ranked as high hazard; cheese and pasteurized milk as low to medium hazard; and vehicles as medium hazard.

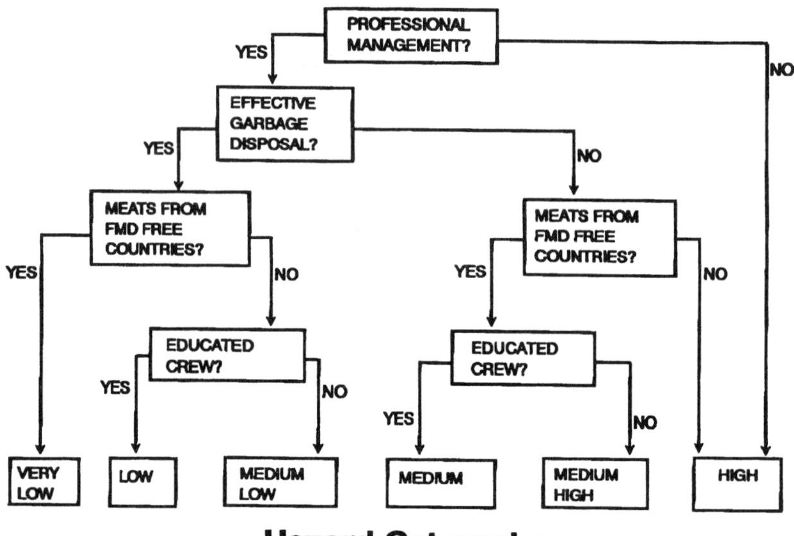

FIGURE 1. Methodology for hazard categorization of ferry operations.

The Crucero Express ferry is a cruise ship that advertizes itself as the connecting link to the Pan American Highway, as a vacation cruise, and as a business cruise for Colombian merchants to shop in the duty-free zone in Colon, Panama. The ferry began operation December 4, 1994, and carries passengers, cargo, and vehicles, but Panamanian policies now prevent vehicles from traveling to Panama. The ferry makes three round trips per week, and has a local Cartagena cruise on Saturdays for relaxation and on-board duty-free shopping.

In August 1994, the Executive Director of the Organismo Internacional Regional de Sanidad Agropecuaria (OIRSA) requested that the US Department of Agriculture (USDA) perform a risk assessment for the introduction of FMD into Panama by the ferry. An on-site visit of the ferry and ports of Colon and Cartagena was performed in March 1995 in preparation for the risk assessment.

The purpose of this paper is to describe the methodology and results of the risk assessment. Because of the nature of the project, a qualitative method using hazard categorization methodology was selected.[2]

METHODS

Hazard Categorization of Crucero Express Ferry Components

Four ferry components were identified as FMD hazards: management of the ferry and transport of passengers, cargo, and vehicles. The methodology for categorizing the hazards is in FIGURE 1 (ferry operations) and FIGURE 2 (passengers, cargo, and

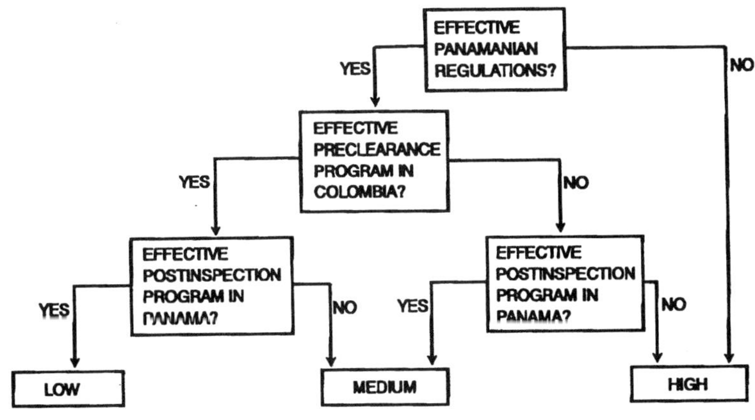

FIGURE 2. Methodology for hazard categorization of passengers, cargo, and vehicles.

vehicles). Each of the four components was assigned a hazard category based on how the questions in the diagrams were answered.

Ferry Operations: FIGURE 1

Information on the owners and management of the ferry was obtained through interviews with the General Director of the Crucero Express, employees at Crucero Express office in Cartagena, and the ferry and crew members. An on-site inspection was also made of the ferry, the Cartagena and Colon ferry terminals, and the Crucero Express office in Cartagena.

Four questions were asked with regard to the hazard presented by the operation of the ferry. Firstly, is the operation professional, experienced, and financially stable? Secondly, does the ferry have an effective garbage disposal system that would protect Panama's shores from the threat of FMD virus contaminated animal products? Thirdly, does the ferry purchase its meat and dairy products for passenger and crew use only from FMD-free countries? Fourthly, have the boat crew, managers, and owners been educated regarding modes of transmission, clinical signs, and potential economic consequences of an FMD outbreak to Panama, Central America, and North America, and have the crew been informed not to bring meat and dairy products from FMD-affected countries for personal consumption?

Passengers, Cargo, and Vehicles: FIGURE 2

Information on passengers, cargo, and vehicles was gathered through interviews and data requests from agricultural officials, and by an on-site visit to the ferry and ferry terminals in Colon and Cartagena.

Three questions were asked regarding the hazards of passengers, cargo, and vehicles. Firstly, are Panamanian regulations effective: do they restrict the entry of FMD-susceptible animals, animal products, and animal by-products from FMD-affected countries; are regulations enforced and are there penalties for breaking such regulations; and are import permits, health certificates, and cargo manifests required for entry of animals and animal products from any country into Panama?

Secondly, is there a preclearance program in Colombia for passengers, cargo, and vehicles? For passengers, is there a professional and thorough baggage inspection, are confiscations monitored by the Colombian Agricultural Institute (ICA) at the port, and are passengers warned not to bring plant and animal products on their voyage to Panama? For cargo, does the Port Authority at Cartagena maintain control over movement of the Crucero Express ferry and the cargo it carries? For vehicles, are they closely inspected for smuggled animals and animal products, and is there an effective cleaning and disinfection program?

Thirdly, is there a postinspection program in Panama for passengers, cargo, and vehicles? Elements for postinspection are similar to those of preclearance, e.g. is baggage thoroughly inspected; do Panamanian authorities closely examine cargo manifests and declarations, and are dirty containers cleaned and disinfected when leaving the ferry; and are vehicles closely inspected, and cleaned and disinfected?

Passenger Profile

A short questionnaire was administered to help define demographically the type of passenger using the Crucero Express, and to determine if passengers within two weeks of departure had contact with FMD-susceptible animals. The questionnaire was administered once, on a Sunday departure from Cartagena, Colombia, and was given when all the passengers were assembled during the safety presentation soon after departure. Interviews with Crucero Express management were also performed to identify the type of passenger targeted by marketing campaigns for the ferry.

RESULTS

Hazard Categorization of the Crucero Express Ferry

Ferry Operations

Question 1: Professional Management of the Ferry: Yes. The Crucero Express ferry was built in France in 1973 and remodeled in Norway in 1994 (1994 Crucero Express informational brochure). It has a gross weight of 11,676 tons, and is 141.46 meters long by 21.93 meters wide. The ferry is owned by a Norwegian company, Colorline, and is leased for three years by the Crucero Express Company (Mr. Rodrigo Gomez, General Director, Crucero Express, Colon, Panama, personal communication). Effjohn International, ferry transportation specialists from Sweden and Finland who own the Silja and Commodore cruise lines of Scandinavia and Miami, Florida, respectively, were consultants to the ferry from the initial startup until March 1995. The Crucero Express Company is a Panamanian subsidiary of the Flota Mercante

Grancolombiana S.A. (FMG), which has had 40 years experience in international sea transportation in the region. The principal financial backers of the FMG are the Colombian National Federation of Coffee Growers (CNFCG) and the National Developmental Bank (Banco Nacional de Fomento) of Ecuador, contributing 80% and 20% of the budget, respectively.

Question 2: Effective Garbage Disposal: Yes. Mixed garbage from the Crucero Express is collected in plastic bags and placed in 3-m³ dumpsters stored in a room on the vehicle deck. Garbage is collected by a private contractor who drives on board at Cartagena and then delivers the bags to the municipal dump for burial. Disinfection of the ferry deck is performed after the truck leaves, according to Crucero Express officials. Panamanian port authorities allow no off-loading of ferry waste onto Panamanian shores. There is no incinerator on the ferry, at the Colon or Cartagena ports, or at the municipal dumps. The ferry presently averages 9 m³ of garbage per round trip.

Question 3: Buy Animal Products from FMD-Free Countries: No. Panamanian policies prevent the purchase of meat or dairy products from FMD-affected countries by the Crucero Express. When the food lockers were examined during the on-site visit, a few fresh legs of lamb from Cartagena and both fresh and processed sausages from Colombia were observed. In addition, during an earlier inspection in March 1995, a USDA employee observed frozen sausages from Colombia, and fresh and frozen pork purchased that day in Colombia. Crucero Express officials report that the majority of food items are purchased in the Free Zone of Colon where meat products from FMD-affected countries are not allowed.

When docking in Colon, the Crucero Express presents to agricultural quarantine inspectors of the Ministry of Agricultural Development (MIDA) a general declaration, cargo manifest, passenger list, and computer printout of all food stock items. The food stock printout lists the food item, price, quantity, and date purchased, but does not list where the food item was produced or purchased. Also, MIDA port inspectors do not presently look closely at the labels on the meat and dairy items to determine country of origin.

Question 4: Education of the Crew Regarding FMD: No. In conversations with management of the Crucero Express, as well as with the ferry captain and a few crew members, there was a general awareness of the importance of FMD. We were unable to interview any manual laborers, most of whom were Colombian or Filipino. The crew list presented on the day of our inspection had 166 names. There have been no presentations on FMD by ICA or MIDA to the ferry management and crew. However, an information pamphlet that discusses the dangers of FMD and other exotic diseases has been made available by MIDA to the ferry management. Crew members that board at Cartagena are not allowed to bring food products onto the ferry; however the ferry is docked at Cartagena several hours before Customs and ICA agents arrive for duty, and crew members can move freely from shore to the ferry during this period.

Final Rating: Medium Low. The Crucero Express has professional and experienced management with financial backing for the long term. There is an effective garbage-disposal system that would prevent FMD viral contaminated meat from reaching Panamanian shores. Animal products from FMD-affected countries are being purchased in small amounts for consumption on the ferry. And no formal education of crew members regarding the risk and transmission of FMD has occurred. The hazard category for the ferry operations is medium low.

Passengers

Question 1: Effective Panamanian Regulations: Yes. Decree Number 57 of February 7, 1956, prohibits the importation of FMD-susceptible animals and products. Decree Number 168 of August 18, 1965, requires import permits, official health certificates endorsed by the local Panamanian consulate, and manifests by exporters. Decree Number 15 of May 18, 1967, prohibits the importation of animals and animal products without a permit and authorizes fines of US $50 to $5000, and prison, for violators. Fines have been levied against the Crucero Express on two occasions to date relating to a shipment of Canadian-origin pigs feet.

Question 2: Effective Preclearance Program in Colombia: Yes. When passengers enter the ferry terminal at Cartagena, they see a prominent red sign warning passengers not to take animal and plant products on their voyage to Colon due to risks of FMD and other exotic diseases. When receiving their boarding passes, a 2 × 1 inch sticker is attached with a message in red print advising passengers not to take animal and plant materials. A thorough inspection of baggage is performed by the Colombian military for narcotics (two soldiers per table with two tables total) and ICA port officials for agricultural products (one per table with a third inspector boarding and inspecting the ferry at arrival). After inspection, a 3.75 × 3.25 inch ICA inspection sticker is placed on the side of the bag.

A log of confiscations has been kept at the Cartagena ferry terminal by ICA since January 1995 for monitoring purposes. During 28 round trips from January 24 through March 28, 1995, there were 13 confiscations from passengers going to Colon and 10 confiscations from passengers returning from Colon. During this period there were 3307 passengers going to Colon and 3306 passengers returning from Colon. All confiscations were fruits or vegetables. From January 24 through April 21, the only confiscated animal products were a sausage and one half pound of cheese on April 2 from passengers going to Colon.

Question 3: Effective Postinspection Program in Panama: Yes. Five MIDA port inspectors attend to each arrival of the ferry: 1 or 2 board the ferry, 1 inspects the disinfection of cargo containers leaving the ferry, and 2 inspect the baggage. There are two tables with 1 MIDA inspector and 2 customs agents per table. Inspectors going on board examine the general declaration, cargo manifest, passenger list, and food stock list, and then perform a general inspection of the ferry, including the kitchen, food-storage rooms, restaurants, cabins, garbage dumpsters, and cargo containers.

Passengers disembarking the ferry must walk across a "sanitary carpet" to disinfect the bottoms of their shoes before entering customs. The sanitary carpet consists of a 3 × 5 foot piece of outdoor carpeting sitting in a shallow metal reservoir and soaked with an iodophor disinfectant (Vanodine, Pfizer Veterinary Division, Costa Rica). Baggage is thoroughly inspected by Customs and MIDA officials. Any confiscated meat or plant materials are placed back on the ferry for return to Colombia. Confiscations have dropped over time, probably due to preclearance activities.

Final Rating: Low. Effective Panamanian regulations exist to prevent the entry of animals and animal products by passengers, and thorough and professional preclearance and postinspection programs exist in Colombia and Panama, respectively. The hazard category for passengers is low.

Cargo.

Question 1: Effective Panamanian Regulations: Yes. Import regulations for animals and animal products are defined under Decrees 57, 168 and 15 (see Passengers, Question 1).

Question 2: Effective Preclearance Program in Colombia: Yes. According to the Port Authority Captain in Cartagena, the Crucero Express has a permit to carry containers, and it must obey the import regulations of Panama. Before each trip, the ferry provides a cargo manifest and trip itinerary to the Cartagena Port Authority, as verified by a computer printout provided by the Port Authority Captain. The Crucero Express also provided copies of cargo manifests from January 31, when transport of cargo from Cartagena to Colon was first permitted, through March 28, 1995. Sixty nine containers were sent from Cartagena to Colon, with mainly chemical and industrial products, but no animal and plant products were listed on the manifests provided.

A failure in the preclearance program occurred when the Crucero Express attempted to transport to Colon on two subsequent occasions a container of frozen pigs feet originating in Canada that had transitted in Colombia. The container had been transported from Canada to Colombia on a cargo boat owned by the FMG, and then transferred to the Crucero Express for shipment to Panama. The manifest stated the contents of the container. Although it is unclear whether this was a deliberate attempt to bypass Panamanian regulations or simply a misunderstanding of the regulations, it underscores the necessity for better communication between the Crucero Express management and ICA officials. Panamanian regulations do not allow transit of animal products through FMD-affected countries without special permit.

On balance, the preclearance program in Colombia for cargo was given a qualified yes for effectiveness.

Question 3: Effective Postinspection Program in Panama: Yes. MIDA port inspectors examine the cargo and cargo manifest provided by the Crucero Express, and inspect the off-loading and disinfection of containers. Both 40- and 20-foot containers are shipped on trailer beds with wheels, and are driven off the ferry by trucks. The exteriors of the containers are sprayed with a disinfectant-insecticide mixture (Vanodine and malathion) while the wheels of the trailer bed are driven through a trough containing 550 gallons of Vanodine (12 ml per gallon of water) or 4% sodium carbonate. In addition, MIDA inspectors detected the prohibited shipments of pigs feet mentioned above, indicating they are monitoring cargo closely.

Final Rating: Low. Effective Panamanian regulations exist to prevent the entry of animals and animal products in cargo, a moderately effective preclearance program exists in Colombia, and an effective postinspection program exists in Panama. Cargo transported by the Crucero Express contains no animal or plant material as indicated on cargo manifests. The hazard category for cargo is low.

Vehicles

Question 1: Effective Panamanian Import Regulations: Yes. Import regulations for animals and animal products are defined under Decrees 57, 168, and 15 (see Passengers, Question 1). Present Panamanian policy allows movement of vehicles on the Crucero Express from Panama to Colombia only, but not from Colombia to Panama due to FMD concerns. From December 4, 1995, through March 28, 1995, 101 cars were shipped from Panama to Colombia.

Question 2: Effective Preclearance Program in Colombia. A preclearance program does not presently exist for vehicles in Colombia since vehicles do not travel from Colombia to Panama. However, an effective program would consist of a thorough visual inspection of vehicles by ICA port inspectors for live animals and animal products, and thorough cleaning and disinfection. Cleaning could be done using any method to remove organic debris, with special attention given to the tires and undercarriage, e.g. steam cleaning. Disinfection could be done using a noncorrosive disinfectant effective against FMD, e.g., 2% chlorine dioxide (Oxine, Bio-Cide International, Norman, OK) or 4% sodium carbonate with 0.1% sodium silicate. Oxine was very effective against FMD during laboratory trials (Dr. Carol House, USDA-Foreign Animal Disease Diagnostic Laboratory, Plum Island, New York, 1994, personal communication). Sodium carbonate is a controlled substance in Colombia, which will affect its availability as a disinfectant. Vanodine is effective but corrosive.

Question 3: Effective Postinspection Program in Panama. A postinspection program does not presently exist for vehicles in Panama since vehicles do not travel from Colombia to Panama. MIDA presently uses a system based on OIRSA standards for disinfecting cargo containers leaving the ferry, and this system could be adapted for vehicles. In addition, the cleaning and disinfection methods described in Question 2 above would be very effective against FMD and noncorrosive to vehicles.

Final Rating: No risk at present; Low hazard is possible. Since current Panamanian policies prevent transport of vehicles from Colombia to Panama on the Crucero Express, there is no risk. However, if thorough car inspections and thorough cleaning and disinfection procedures described above were implemented in both Colombia and Panama, the hazard category for vehicles would be low.

Passenger Profile

The results of the questionnaire are in TABLE 1. Eighty five of 142 passengers (60%) responded to the questionnaire. Only one passenger from 83 responses (1.2%) identified having had contact with FMD-susceptible livestock within the last two weeks, a housewife with pig contact. Of 83 responses, 41 occupations were listed, but none were agricultural related. Of the responders, 93.3% had middle or upper middle incomes. Finally, passengers on the Crucero Express are probably demographically similar to airplane travelers, since 84% of Crucero Express passengers responding to the questionnaire said they would travel to Panama by airplane if the ferry was not available.

From customs declaration forms, managers of the Crucero Express report that 60% and 40% of the passengers were engaged in business and tourism, respectively.

TABLE 1. Results of "Crucero Express" Questionnaire, April 2, 1995

Variable	Frequency	Percentage
Country of Residence		
Panama	46	55.4%
Colombia	31	37.3%
Canada	2	2.4%
France	2	2.4%
US	2	2.4%
Total	83	100.0%
Days in Colombia		
One	7	15.6%
Two to four	31	68.9%
Five or more	7	15.6%
Total	45	100.0%
Areas visited in last two weeks		
Cartagena only	25	31.6%
Colombia elsewhere	26	32.9%
South America	3	3.8%
Other countries	25	31.6%
Total	79	100.0%
Contact with livestock in previous two weeks		
Yes	1	1.2%
No	82	98.8%
Total	83	100.0%
Age		
Less than 18 years	0	0%
18 to 34 years	31	36.9%
35 to 64 years	49	58.3%
65 years or Greater	4	4.8%
Total	84	100.0%
Sex		
Female	45	53.6%
Male	39	46.4%
Total	84	100.0%
Occupation		
Professional/managerial	47	56.6%
Merchant	14	16.9%
Housewife	7	8.4%
Student	7	8.4%
Laborer services	5	6.0%
Retiree	3	3.6%
Total	83	100.0%
Average weekly income		
Less than US $100	6	7.7%
US $100–300	22	28.2%
More than US $300	50	64.1%
Other modes of travel to Panama you would use		
Airplane	68	84.0%
Would not travel	7	8.6%
Other	6	7.4%
Total	81	100.0%

During the first two months of 1995, 62% of sales were in Colombia and 38% in Panama. The majority of passengers had middle to upper middle incomes. From December 4, 1994, when ferry operations began, through March 28, 1995, a total of 11,945 passengers traveled between the countries (5917 from Cartagena to Colon and 6028 from Colon to Cartagena).

The ferry service is marketed to tourists in both countries as a minicruise with shopping and sightseeing, to Colombian merchants wishing to shop in the Free Zone at Colon, and to corporate groups for conferences on the ferry. Passengers can travel by bus to hotels in Panama City as part of a travel package. Approximately five million potential customers with middle to upper middle incomes are targeted in Colombia (Mr. Rodrigo Gomez, personal communication).

One of the original objectives of the ferry was to serve as the final link for vehicles on the Pan American Highway. However, Panamanian policies have prevented vehicles from entering Panama. This has forced the Crucero Express to solicit cargo containers to replace the expected vehicles. They ship 40- and 20-foot containers faster and more cheaply than regular cargo boats, charging US $1200 and $1600, respectively (Rodrigo Gomez, personal communication), compared to the average rate of US $1400 and $2200 (Cartagena Port Authority Captain, personal communication).

DISCUSSION

When compared to airplane and cargo boat operations, the risk from the Crucero Express is no doubt lower. Airplanes transport many moore passengers and cargo containers to Colombia than the Crucero Express, they serve Colombian origin meats to passengers with the waste incinerated at the Tocumen airport in Panama City, there are no signs in airport terminals warning passengers against bringing animal and plant materials to Panama, and the knowledge of FMD by airplane management and crew is probably no greater than that of the Crucero Express.

Cargo boats ship many more containers to Panama than the Crucero Express, they probably bring meats of Colombian origin for consumption by crew members, and managements certainly had no informational training on FMD.

The greatest risk for the introduction of FMD into Panama from Colombia may not be from the Crucero Express ferry, cargo boats, or airplanes, whose movement and operations are regulated by authorities in Panama and Colombia, but from the large, uncontrolled movement of small boats (1 to 50 tons) from Colombia to Panama. These boats are not controlled in Colombia, and they leave with meat of Colombian origin, and rarely live goats and pigs, for consumption during the trip. These boats trade along most of the Atlantic coast of Panama, including ports in the San Blas District (trade for coconuts), the Cocosolo dock in Colon (buy electronics), and the Upper Coast north of Colon which all have small pig populations. There is also some trade with the lower Pacific coast of Panama. In 1994, 11 confiscations of animals or animal products were made from small boats at Puerto Obaldia on the Atlantic coast, including 4 live pigs during one confiscation.

SUMMARY

It should be emphasized that the proposed ferry hazard categorizations do not represent absolute risks for introducing FMD into Panama, but instead provide a systematic method for comparing and estimating risks in the absence of quantitative data. A hazard rating of high may not necessarily represent a high quantitative risk for the introduction of FMD, but is high when compared to other scenarios. A low hazard rating may estimate a low quantitative risk of importing FMD, but economic consequences of a potential outbreak should also be considered.

When further data become available, a more complete assessment of the risks of the Crucero Express compared to airplanes, cargo boats, and small boats can be performed. At present, the risk of the Crucero Express is at least as low as the other transport modes described above. Since vehicles are not presently allowed transport from Colombia to Panama, they present no risk to Panama, but with proper cleaning and disinfection procedures, vehicles can be permitted with low risk. However, the Crucero Express can carry 125 vehicles, and thorough cleaning and disinfection of this many cars will require modern and efficient facilities not yet present at either port.

Addendum: When this paper was written, vehicles were not allowed passage on the Crucero Express Ferry from Colombia to Panama. In January 1996, Panama changed its policy to allow entry of vehicles from Colombia and South America on the Crucero Express Ferry. Vehicles must be washed with a hot water vapor spray at a local car wash in Cartagena supervised by an official from ICA. The undercarriage must then be sprayed with Vanodine solution, and the vehicle driven through a disinfectant trough of Vanodine solution just prior to boarding the ship. Panamanian officials from MIDA periodically monitor the vehicle disinfection process.

REFERENCES

1. WOODHAM, C. B. 1991. Prospects for the Exportation of South American Meat and Livestock Products to the European Community after the Advent of New Regulations in 1992. Vet. Epi. and Economics Research Unit, Univ. Reading. Chp. 4: 4.1.1.
2. USDA-APHIS-VS. CENTERS FOR EPIDEMIOLOGY AND ANIMAL HEALTH. 1994. Foot-and-Mouth Disease: Sources of Outbreaks and Hazard Categorization of Modes of Virus Transmission. Fort Collins, CO.

Preharvest Food Safety

A. B. CHILDERS AND B. WALSH

Department of Veterinary Anatomy and Public Health
Texas A&M University
College Station, Texas 77843-4458

Preharvest food safety is essential to the protection of our food supply. The production and transport of livestock and poultry play an integral part in the safety and wholesomeness of food products derived from such animals and birds. The goals of such safety assurance include provisions that these animals and birds are as free as possible from pathogenic microorganisms, disease, and parasites; they must also be free from potentially harmful residues and physical hazards.

The functions of this safety assurance should be based on hazard analysis and critical control points (HACCP) from producer to slaughter plant. The emphasis should be on prevention of identifiable hazards rather than on removal of contaminated products. This requires a partnership involvement between producers and marketers. The production goal is to minimize infection and insure freedom from potentially harmful residues and physical hazards. The marketing goal is control of exposure to pathogens and stress.[1]

The criterion for safety assurance is that producers and marketers should have functional HACCP management programs. The essential operations should be covered in a program similar to the National Milk Producers-Milk and Dairy Beef Residue Prevention Protocol.[2] Other important elements are certification of producers using "integrated quality assurance" and personnel training. Essential HACCP programs would include production procedures, agricultural chemical usage, production of feedstuffs, feed manufacturing, drug usage, immunization, implantation, and treatment practices, assembly and transportation, animal identification, and complete record management.

Production HACCP programs should exclude animals (from slaughter) with acute infectious conditions or parasitisms or those with chronic, persistent, or latent infections that might be transmitted by food.[3] Those with harmful residues from medicated feeds, biologicals, drugs, or agricultural chemicals should also be excluded. Animal identification procedures should be used that would permit trace-back of livestock to the farm of origin. Marketing HACCP programs are also important to prevent infectious agent transmission or contamination from other animals. Livestock may be exposed to infectious agents as a result of the unsanitary condition of transport vehicles and pens. Repeated use of unclean vehicles to transport livestock should be avoided as well as housing in unclean pens. Livestock-handling personnel should wear clean clothing, and footwear should be disinfected daily. Stress, extended time in movement, and improper handling may weaken animals and enhance the spread of infectious agents and outbreaks of disease.[4]

HACCP plans must be based on risk assessment principles, and the procedures must be defined. Other elements of such a program would include preslaughter certification (vendor accountability), environmental protection, control of chemical hazards, use of live animal drug testing procedures, identification of physical hazards, and an analysis of the prevalence of risks.[5]

Elements of such a program would concentrate on breeding and parturition practices, feed manufacturing, and production practices. A successful HACCP program would assess breeding and birth dates, treatment of a dam during gestation and after parturition, drug use at parturition, and treatment of a newborn. Feed manufacturing may include use of legal and/or illegal drugs, feed additives, and the withdrawal periods associated with them. It also assures quality feed ingredients that are clean and hazard free, suing approved drugs and following extralabeling precautions. The producer must follow indications for use in appropriate species, proper route of administration, adhere to withdrawal periods, and be aware of public health risks.[6] The producer must have a valid veterinarian/client/patient relationship.[1] Drug compounding should be done with legal drugs while maintaining complete records and avoiding incompatibilities and violative residues. Over the counter drugs and off-label drug use are considered illegal drug usage and should be avoided. Parasite and vermin control should be maintained. Control of biologicals and pharmaceuticals is also necessary. Identification and accurate accompanying records are vital.

Animals must also pass preslaughter certification under a successful HACCP program. Important components of certification are vendor accountability (i.e. liability), transgenic biotechnology, and record maintenance.[5]

Environmental protection should encompass continual sanitation and cleaning of premises and regular visual inspection for hazards. It should also include proper storage and labeling of hazardous material and complete record management.[1]

Preharvest control of chemical hazards includes the following: antimicrobial, growth promotants/anabolic steroids, tranquilizers, parasiticides, rodenticides, feed additives, micronutrients (i.e., selenium), and pesticides/fungicides/herbicides. Also to be considered are mycotoxins including aflatoxins and environmental pollutants.[1,6] The two major classes of environmental pollutants to be considered are the heavy metals including at least lead, zinc, cadmium, and arsenic and chemicals that include at least PCBs, PBBs, and dioxins.

In order to accurately identify preharvest hazards at slaughter, a herd history is essential. Veterinarians and food safety inspectors must observe animals and birds at slaughter and then report their findings to the producers and their veterinarians. If necessary, laboratory confirmation may be required.[7]

Conditions that warrant closer observation of livestock are moribund or disabled/injured animals, cull cows with mastitis, bob veal, skin conditions, and injection sites. Other areas of concern include epithelioma of the bovine eye, central nervous system disorders, and infectious conditions including arthritis, upper respiratory tract problems, and pneumonia, septicemia, and enteritis.[3]

A chemical testing procedure should be implemented and test for growth promotants or anabolic steroids, pesticides, parasiticides, fungicides and herbicides, heavy metals and environmental pollutants, and antimicrobial agents.[8] Specific tests for antimicrobial agents include live animal swab test (LAST) which should be producer initiated, swab test on premises (STOP), sulfa on site (SOS), and calf antibiotic and sulfa test (CAST).

There are several physical hazards that would have to be identified in an effective HACCP program. These physical hazards include traumatic radiculitis ("hardware disease"), hypodermic needles, surgical materials, shot pellets, and fitting violations of show livestock.[1]

The National Residue Program has identified the presence of numerous residues that represent a risk to human and animal health in all classes of meat and poultry slaughtered in the United States.[9] The following is the most current list of the percentage of positive samples for each class of compounds:

Antimicrobials:
Beef cow and bulls—1.8%
Calves—1.8%
Steers and heifers—0%
Swine—0.35%
Lambs—0.35%
Poultry—0.2%
Growth Promotants or anabolic steroids—0%
Pesticides—0.12%
Parasiticides/fungicides/herbicides—0.14%
Heavy metals/environmental pollutants—0.34%
Mycotoxins—0%

SUMMARY

Preharvest food safety is essential for the protection of our food supply. The production and transport of livestock and poultry play an integral part in the safety of these food products. The goals of this safety assurance include freedom from pathogenic microorganisms, disease, and parasites, and from potentially harmful residues and physical hazards. Its functions should be based on hazard analysis and critical control points from producer to slaughter plant with emphasis on prevention of identifiable hazards rather than on removal of contaminated products. The production goal is to minimize infection and insure freedom from potentially harmful residues and physical hazards. The marketing goal is control of exposure to pathogens and stress. Both groups should have functional hazard analysis and critical control points management programs which include personnel training and certification of producers. These programs must cover production procedures, chemical usage, feeding, treatment practices, drug usage, assembly and transportation, and animal identification. Plans must use risk assessment principles, and the procedures must be defined. Other elements would include preslaughter certification, environmental protection, control of chemical hazards, live-animal drug-testing procedures, and identification of physical hazards.

REFERENCES

1. CHILDERS, A. B. & J. Y. HASLAM. 1995. *In* The Report of the Technical Analysis Group: Slaughter. G. W. Beran, Ed. USDA Food Safety and Inspection Service. Washington, DC.

2. AMERICAN VETERINARY MEDICAL ASSOCIATION & NATIONAL MILK PRODUCERS FEDERATION. 1994. Milk and Dairy Beef Residue Prevention Protocol. Schaumburg, IL.
3. WORLD HEALTH ORGANIZATION. 1976. Microbiological Aspects of Food Hygiene. Report of a WHO expert committee with participation of FAO. Technical Report Series No. 598. World Health Organization. Geneva, Switzerland.
4. SNIJDERS, J. M. A. 1994. Modernization of meat inspection: why and how? *In* Proceedings: 40th International Congress of Meat Science and Technology, The Hague, the Netherlands.
5. HASLAM, J. Y. & A. B. CHILDERS. 1995. *In* The Report of the Technical Analysis Group: Slaughter. G. W. Beran, Ed. USDA Food Safety and Inspection Service. Washington, DC.
6. FOOD AND AGRICULTURE ORGANIZATION/WORLD HEALTH ORGANIZATION. 1988. Evaluation of Certain Veterinary Drug Residues in Food. Technical Report Series No. 763. Food and Agriculture Organization. Rome, Italy.
7. WEISS, M., T. ROBERTS & H. LINDSTROM. 1993. Food safety issues: modernizing meat inspection. Agri. Outlook **1979:** 32.
8. ROBERTS, T. & D. SMALLWOOD. 1991. Data needs to address economic issues in food safety. Am. J. Agri. Eco. **73:** 933.
9. UNITED STATES DEPARTMENT OF AGRICULTURE. 1993. National Data Residue Program, USDA FSIS, Science and Technology. Washington, DC.

Multiple Gene Expression in Baculovirus System

Third Generation Vaccines for Bluetongue Disease and African Horsesickness Disease

POLLY ROY [a]

Laboratory of Molecular Biophysics
University of Oxford
United Kingdom
and
University of Alabama at Birmingham

INTRODUCTION

The complexity of the mechanisms of protective immunity against infections is appreciated fully only recently due to modern immunology. With the advent of newer technologies and greater understanding of the molecular biology of pathogens, the conventional empirical approaches to vaccine development have given way to attempts to rationally design vaccines. Attention has been focused on defining the relevant proteins and sequences involved in eliciting protective immune responses to viral and other pathogens and on systems to produce and present proteins and protein sequences to elicit the required protection. My laboratory over the last few years has been involved in defining the biochemical and molecular basis of particular viruses, arthropod-borne orbiviruses, and to rationally design suitable vaccines. Orbiviruses have similar morphological and biochemical characteristics, with distinctive pathobiological properties and host ranges. Three different orbiviruses are under study, bluetongue virus (BTV) of sheep, epizootic haemorrhagic disease (EHD) virus of deer, and African horsesickness (AHS) virus. These viruses consist of seven structural proteins organized in four concentric layers.[1,2] The innermost components of the virus particle are composed of three minor proteins (VP1, VP4, and VP6) together with the 10 segments of double-stranded (ds) RNA genome. For BTV and presumably for AHSV and EHDV, these are surrounded by a layer of a single major protein, VP3, 120 copies of which are arranged in 12 pentameric configurations to form the subcore.[3] This subcore is encapsidated by 260 trimers of another core protein, VP7, to form the core particle. The core has a knobby surface appearance in an icosahedral symmetry.[2,3] The outer capsid also has an icosahedral configuration and consists of two other major proteins, VP2 and VP5, each with distinctive shapes, one globular and the other sail-shaped.[4] The globular proteins, 120 in number, sit neatly in the

[a] Address correspondence to NERC Institute of Virology and Environmental Microbiology, Mansfield Road, Oxford, OX1 3SR, United Kingdom.

channels formed by each of the six-membered rings of VP7 trimers. The sail-shaped spikes, which project 4nm beyond the globular proteins, are located above 180 of the VP7 trimers and form 60 triskelion-type motifs that cover nearly all the VP7 molecules. It is likely that these spikes are the viral hemagglutinating protein VP2, which also contains the virus-neutralizing epitope, and that the globular proteins are VP5.

BACULOVIRUS EXPRESSION VECTORS

Recent advances in gene manipulation have made it possible to express foreign genes in heterologous systems. The productivity and flexibility of insect baculovirus expression vectors and the ability of the baculovirus genome to incorporate (and express) large amounts of foreign DNA have permitted this system to be used for the expression of not only a single gene, but also for the simultaneous expression of dual and multiple genes.[5] To accomplish this, several expression vectors have been developed based on the resident promoters of the nuclear polyhedrosis virus of *Autographa californica* (AcNPV). One vector that has found particular favor is the strong promoter of a so-called very late polyhedrin gene (ph). The promoter of the ph gene has been exploited to express one, two, or multiple foreign gene products either by replacing the ph gene or by duplicating the promoter together with appropriate transcription termination signals.[5] A second very late gene, p10, which is responsible for the synthesis of a nonstructural protein, has been successfully used to express foreign proteins either by replacing the resident p10 gene or by duplicating the p10 promoter and inserting it elsewhere in the AcNPV genome, e.g., in proximity to the ph gene site.[6]

Using these various expression vectors, we have expressed all 10 genes of BTV and synthesized BTV proteins and analyzed structure-function relationship of each gene and gene product. In addition, relevant genes of ASHV and EHDV are also expressed similarly.[7,8] Subsequently we have expressed cassettes of BTV gene simultaneously and prepared core and viruslike particles (CLPs and VLPs, respectively) and used them for answering some fundamental questions regarding viral assembly as well as developing candidate vaccines, as described below.

PRODUCTION OF BTV PROTEINS AND STRUCTURES

The viral dsRNA segments were isolated from infected cell extracts, individual RNA segments were separated, and full length DNA clones representing the 10 RNA segments were obtained.

The cDNA clones of the 10 RNA segments of BTV-10 were subsequently manipulated to derive 10 recombinant baculoviruses, each synthesizing BTV-10 gene product. When each recombinant protein was characterized biochemically and immunologically, and compared with authentic BTV-10 proteins, all appeared to be identical to the corresponding BTV proteins.[7] The functions of some of these gene products in virus replication and assembly processes have been subsequently investigated. The preliminary data obtained from these studies are summarized in TABLE 1.

To determine whether the expressed proteins could indeed interact with each other and form complex structures within the insect cells, an attempt was made to synthesize BTV corelike particles (CLPs) by simultaneous expression of two major core proteins VP3 and VP7 by one dual recombinant baculovirus. To achieve this, the cDNA copies of RNA segment 3 (which encodes VP3) and RNA segment 7 (which encodes VP7) were inserted into a dual transfer vector.[9] They were under the control of two polyhedrin promoters, organized in opposite orientations, and from this, a recombinant baculovirus expressing both VP3 and VP7 proteins was successfully derived.[9]

Electron microscopy of *Spodoptera frugiperda* insect cells infected with the recombinant baculovirus showed large aggregates of foreign material in the cytoplasm which, under high magnification, appeared to consist of spherical particles. These particles were isolated by purifying the cell extracts on discontinuous sucrose gradient. When examined under the electron microscope, the products were found to consist of empty CLPs whose size and appearance were similar to authentic cores prepared form BTV (see FIG. 1). The stoichiometry of VP3 and VP7 in the expressed particles was estimated to be similar to that of these components derived from infectious BTV.

To confirm that the CLPs are indeed identical to the BTV core structures, three-dimensional reconstructions were obtained using image analysis of cryo-electron micrographs and compared with that of authentic core structures.[3] These images have revealed a similar, if not identical, icosahedral structure to that of authentic cores[2] but have also yielded information not available from previous analyses of the structure of the core particle including the two domains of VP7 trimers. Recent analyses of the x-ray crystallographic atomic structure of baculovirus synthesized VP7 confirmed that indeed VP7 exists as a trimer, thus confirming the reconstruction studies.[10]

SYNTHESIS OF DOUBLE-CAPSID VIRUSLIKE PARTICLES

To determine whether double-capsid particles can be synthesized by the baculovirus expression system, a second dual recombinant baculovirus was constructed to express the two outer capsid proteins, VP2 and VP5. Coexpression of VP2, VP3, VP5, and VP7 in insect cells using both dual expression vectors have led to the assembly of double-capsid VLPs containing all four major structural proteins.[11] These VLPs were of various sizes and contained different amounts of VP2 and VP5 proteins of the outer capsid. In addition, some CLPs were also present in the preparations. However, the VLPs were similar in diameter to virions (81 nm) and could be purified to homogeneity by a one-step process involving sucrose gradient centrifugation of infected cell lysates (see FIG. 1).

For optimum synthesis of VLPs, we have recently developed a quadruple (pAcAB4) expression vector using p10 and polyhedrin promoters.[12] When all four proteins were expressed by a quadruple recombinant baculovirus, the VLPs that were synthesized were virtually homogeneous, in contrast to the results from coinfection with two viruses.

TABLE 1. BTV Genes and Gene Products

Genome Segment	No. of bp	Encoded Protein(s)	Predicted Mass (Da)	Location	Function
L1	3954	VP1	149 588	Inner core	RNA polymerase
L2	2926	VP2	111 112	Outer shell	Type-specific structural protein
L3	2772	VP3	103 344	Core	Structural protein, forms scaffold for VP7 trimers
M4	2011	VP4	76 433	Inner core	Capping enzyme, guanylyl-transferase
M5	1639	VP5	59 163	Outer shell	Structural protein
M6	1769	NS1 tubules	64 445	Nonstructural	Unknown, may be involved in morphogenesis
S7	1156	VP7	38 548	Core surface	Group-specific structural protein
S8	1123	NS2 VIBs	40 999	Nonstructural	Binds mRNA
S9	1046	VP6	35 730	Inner core	Binds ssRNA, dsRNA
S10	822	NS3	25 572	Nonstructural	Glycoproteins, aid virus release
		NS3A	24 020	Nonstructural	Glycoproteins, aid virus release

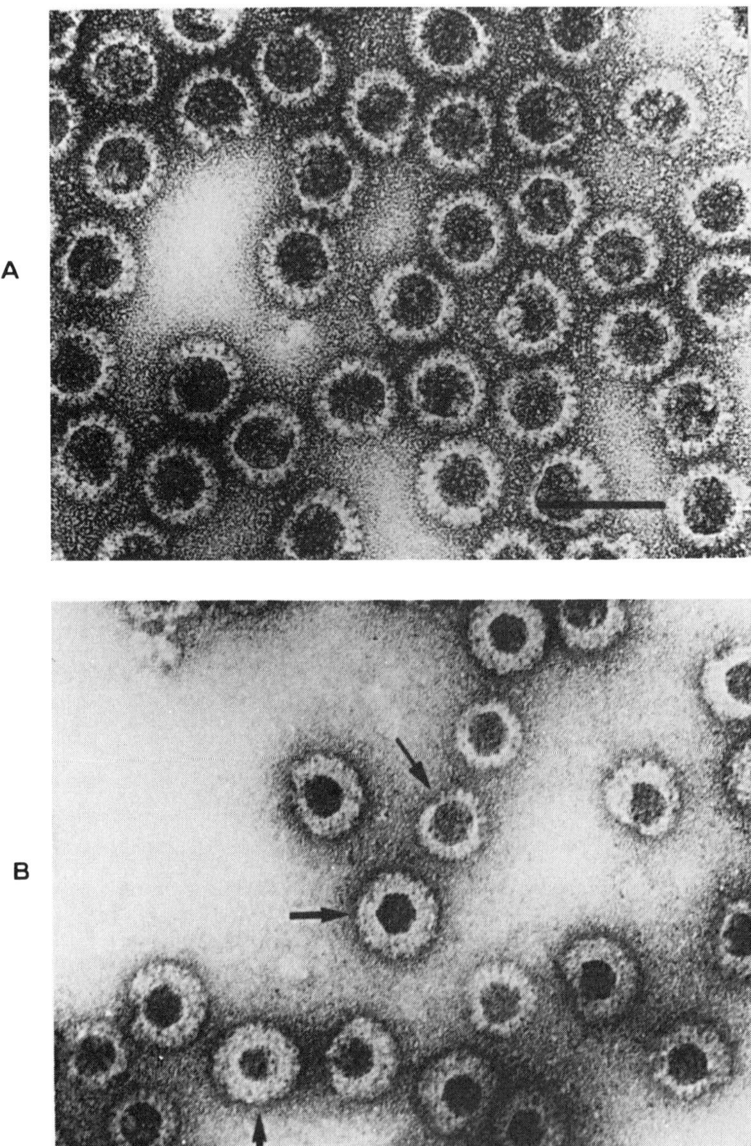

FIGURE 1. Recombinant baculovirus expressed corelike particles (A) and viruslike particles (B) purified on self-forming CsCl gradients.

IMMUNIZATION OF SHEEP WITH VLPS AFFORDS LONG-LASTING PROTECTIVE IMMUNITY AGAINST BTV CHALLENGE

VLPs synthesized by recombinant baculoviruses are biologically and immunologically similar to the virion. VLPs exhibit high levels of hemagglutination activity similar to those of authentic BTV. Antibodies raised to the expressed particles contain high titers of neutralizing activity against the homologous BTV serotype.[11] To analyze the protective effects and duration of VLP vaccination, a number of successful vaccination trials have been performed using BTV-susceptible, 1-year-old Merino sheep (BTV free) in insect-proof isolation status.[13,14] Here, we describe one such trial in which VLPs (10 μg or 50 μg per sheep) representing BTV-10 and BTV-17 were employed.[14,15]

Each group of eight sheep was immunized subcutaneously with VLPs in saline containing the indicated amount of protein (TABLE 2) and suspended in 50% Montanide Incomplete Seppic Adjuvant (ISA-50, Seppic, Paris). Each animal received 2 ml of the mixture. For control experiments, one group of sheep received only saline. The vaccinated animal was boosted with the same amounts of protein on day 21.

The neutralizing antibody titers of the vaccinated sheep were determined at weekly intervals and over a 60-week period after the booster. Both types of VLP elicited (to various levels) antibodies that neutralized the homologous virus. In almost all cases, these neutralizing titers remained high throughout the 60-week period. The neutralizing antibody titers for the animals that received 50-μg doses of VLPs were not significantly higher than those that received the 10-μg doses. Sheep vaccinated with the mixture of the two types of VLPs induced antibodies that neutralized both types of virus as well as some related heterologous viruses (e.g., BTV-4) when tested by plaque reduction assays. As expected, the six control sheep that were inoculated with saline remained seronegative.

All the sheep were challenged 14 months after the booster vaccination by the subcutaneous injection of virulent BTV. The animals that were challenged with the homologous viruses (BTV-10, BTV-17) were completely protected and showed no clinical reactions, even those that received 10-μg doses of VLP. Also, no viremias were detected in these animals after challenge. In addition, some animals with 50 μg VLPs were also protected in low level when challenged with heterologous virus. By comparison, the control animals developed high or moderate signs of disease [BTV-10, clinical reaction index (CRI): 7.1-8.0; BTV-17, CRI: 1.6-2.7] and produced viremias (data not shown).

In summary, the data showed that long-lasting protection against homologous BTV challenge was provided by vaccination with VLPs. Some preliminary evidence was obtained for cross-protection, depending on the challenge virus and the amounts of antigen used for vaccination.[15]

The data clearly demonstrate that VLPs are highly immunogenic even at low doses. The sheep given only 10 μg of VLPs actually received 1-2 μg of VPs (10-20% of the VLP mass). This compares well with the much higher amounts required for VP2 alone, or VP2 and VP5 mixtures in protection studies.[16] There are several possible explanations. First, the conformational presentations of the relevant epitopes on VP2 probably mimic those on the authentic virus. Second, both VP2 and VP5

TABLE 2. Neutralizing Antibodies of Sheep at Various Intervals after Vaccination with Bluetongue Virus–like Particles and Their Protective Responses following Challenge (after 14 Months) with Homologous and Heterologous Virulent Bluetongue Viruses[a]

Sheep Number	Antibody to:	Week 3	Week 5	Week 11	Week 15	Week 19	Week 23	Week 27	Week 31	Week 35	Week 39	Week 43	Week 47	Week 51	Week 56	Challenge Virus	CRI[b]
1	BTV-10	24	32	24	32	32	32	32	32	32	24	16	16	16	16	BTV-10	0
2	BTV-17	64	128	128	128	64	64	64	64	64	32	32	32	32	32	BTV-17	0
3	BTV-10	96	256	64	32	32	32	32	32	32	32	32	32	32	32	BTV-10	0
4	BTV-10	48	128	64	64	64	64	64	64	64	64	48	48	48	48	BTV-10	0
5	BTV-17	32	64	64	64	64	64	64	64	64	64	64	64	64	64	BTV-17	0
6	BTV-17	4	4	4	4	4	4	4	4	<4	<4	<4	<4	<4	<4	BTV-17	0
7	BTV-10	16	16	16	16	16	16	16	8	8	8	8	8	8	8	BTV-4	0
	BTV-17	16	16	12	8	4	4	<4	<4	<4	<4	<4	<4	<4	<4		
8	BTV-10	4	4	16	ND	16	16	16	16	16	16	16	16	16	16	BTV-11	3.1
	BTV-17	<4	<4	8	8	8	8	12	8	8	8	8	8	8	8		
9	BTV-10	48	256	64	64	64	64	64	64	64	64	32	32	32	32		
	BTV-17	128	256	256	128	128	128	128	128	128	64	64	64	64	54	BTV-11	3.0
	BTV-4	8	12	4	<4	<4	<4	<4	<4	<4	<4	<4	<4	<4	<4		
10	BTV-10	32	32	16	16	16	12	16	12	12	12	12	12	12	12		
	BTV-17	64	64	32	32	32	32	32	32	32	32	32	32	32	32	BTV-4	0.5
	BTV-4	4	8	4	<4	<4	<4	<4	<4	<4	<4	<4	<4	<4	<4		

11	BTV-10	32	256	64	64	32	32	32	32	32	32	32	32	32	16		
	BTV-17	128	256	64	64	32	32	32	32	32	32	32	32	16	16	BTV-11	2.3
	BTV-4	16	32	12	<4	8	<4	<4	<4	<4	<4	<4	<4	<4	<4		
12	BTV-10	64	64	32	32	32	32	32	32	32	32	32	32	32	32	BTV-4	0
	BTV-17	64	128	128	128	128	128	128	64	64	64	64	64	64	32		
13	BTV-10	32	32	32	32	24	32	32	16	16	16	16	16	16	16	BTV-11	0
	BTV-17	96	192	128	128	64	64	64	64	64	32	32	32	32	32		
14	BTV-10	24	32	24	16	16	16	16	16	12	16	16	8	8	8		
	BTV-17	24	32	32	32	32	32	32	32	16	8	8	8	8	8	BTV-11	2.3
	BTV-4	16	24	16	<4	6	<4	<4	<4	<4	<4	<4	<4	<4	<4		
15	BTV-10	<4	<4	<4	<4	<4	<4	<4	<4	<4	<4	<4	<4	<4	<4	BTV-17	1.6
	BTV-17	<4	<4	<4	<4	<4	<4	<4	<4	<4	<4	<4	<4	<4	<4		
16	BTV-10	<4	<4	<4	<4	<4	<4	<4	<4	<4	<4	<4	<4	<4	<4	BTV-17	2.7
	BTV-17	<4	<4	<4	<4	<4	<4	<4	<4	<4	<4	<4	<4	<4	<4		

[a] NT: Not tested due to loss of identification; sheep 1, 2 received 2 doses of 10 μg BTV-10 VLPs; sheep 3, 4 received 2 doses of 10 μg BBTV-17 VLPS; sheep 5, 6 received 2 doses of 50 μg BTV-10 VLPs; sheep 7, 8 received 2 doses of 50 μg BTV-17 VLPs; sheep 9-13 received 2 doses of mixtures BTV-10 (10 μg) and BTV-17 VLPs (10 μg); sheep 14-18 received 2 doses of mixtures of BTV-10 (50 μg) and BTV-17 VLPs (50 μg); sheep 19-24 received no VLPs. The indicated weeks represent the weeks after the second VLP inoculation. Challenge virus was administered following the week 56 serum sample.

[b] Clinical Reaction Index.

TABLE 3. [a]

Sheep Number	Neutralizing Antibody	CRI	Viremia
	CLP		
1	320	1.5	+
2	160	5.6	+
3	320	3.8	+
4	160	2.7	+
5	160	4.0	+
	Controls		
1	*320*	11.4	+
2	*320*	9.5	+
3	*320*	12.3	+
4	*320*	6.8	+
5	*160*	5.1	+

[a] The immunized and control sheep developed viremias at high neutralizing antibody titers. However, clinical signs of bluetongue were reduced for the vaccinated animals (CRI 1.5-5.6, average 3.5) compared to the controls (CRI 5.1-12.3, average 9.0). Only the controls showed the characteristic mouth and foot lesions.

are present. Third, the VP3 and VP7 may provide a necessary scaffold for VP2 and VP5 antigen presentation. Fourth, any of the four BTV proteins might have a direct role in eliciting cell-mediated immunity induced by the BTV VLPs.

It can also be anticipated from the results obtained that this technology has much to offer for development of vaccines for both veterinary disease and human diseases. Moreover, there is every reason to believe that it should be possible to make vaccine chimeras representing different BTV serotypes (e.g., involving the expression of several BTV VP2 genes).

CORELIKE PARTICLES WITHOUT THE OUTERCAPSID CAN MITIGATE THE VIRAL INFECTION

The question of whether CLPs containing the two conserved proteins VP3 and VP7 would provide a measure of homologous and heterologous BTV protection by cell mediated mechanism was investigated. For initial studies two groups of five sheep each were used. One group of five sheep was inoculated with 50 µg BTV-10 CLP in ISA50; and a second group of five sheep was boosted on day 21, and challenged with BTV-10 two weeks later.

All postchallenged sheep developed viremias and neutralizing antibodies (TABLE 3).[15] However, with the exception of fever, the vaccinated sheep developed only slight clinical reactions, whereas controls showed characteristic mouth and feet lesions in addition to fever. The average CRI of the vaccinated sheep was 3.5 whereas that of the control sheep was 9.0.

In summary, partial protection against BTV challenge was afforded by CLP vaccination. Further experiments are planned to study this phenomenon further.

PROTECTION AFFORDED IN HORSES BY RECOMBINANT BACULOVIRUSES AGAINST VIRULENT VIRUS CHALLENGE

For AHSV vaccine, recent data have been obtained with recombinant viruses synthesizing AHSV-derived antigens, not particulate structures. Six horses were inoculated with 50 or 300 μg of VP2 proteins in different preparations. To prepare the materials, a dual recombinant virus expressing VP2 and VP5 of AHSV-4 that has been reported recently[17] has been utilized. Since the level of VP5 synthesis in insect cells by this recombinant virus is low, the amount of VP5 was not possible to estimate. The infected cells were initially lysed by freezing and thawing and subsequently by 25 mM NaHCO3. Clarified supernatants obtained after centrifugation of the lysate as well as the precipitated materials were used for vaccination. In addition, purified VP7[18] was also added to one of the preparations, as indicated in Table 4. Since the primary sequence of the core protein VP3 of AHSV-4 and BTV-10 is highly conserved,[19] we have also added VP3 of BTV-10 to one of these preparations, as AHSV-4 VP3 is not available yet.

In addition to unpurified VP2 and VP5, both proteins were also purified using affinity columns as described previously.[17] Although purified VP2 and VP5 could not be solubilized, nevertheless, the insoluble materials were also resuspended in the adjuvant and were inoculated. However, the level of neutralizing antibodies rose rapidly after the booster. Surprisingly, the two horses that received the purified outercapsid protein, as well as the core protein (No. 5 and no. 6) had less antibodies compared to horse no. 4, which received the same amount of VP2 and VP5, but no core protein. The least antibodies were detected in the serum of horse no. 6, which had all four proteins. Whether BTV-VP3 protein had interfered with the immune response is not clear. As expected, the control animal was completely negative of neutralizing antibodies.

All horses were challenged by subcutaneous injection with infective horse blood containing virulent South African serotype 4 at day 110 (i.e., 4 months) after the primary immunization. As shown in Table 4, all horses showed anamnestic response, eliciting increasing levels of neutralizing antibodies. None of these showed any clinical reaction against the virulent virus challenge. The control horse, on the other hand, had a fever from day 9 of postchallenge and died at day 12. The animal had typical mixed lesions of horse sickness disease. The postchallenge blood samples were examined for viremia using embryonated chicken eggs.[16] As expected, none of the vaccinated samples had any indication of viremia, confirming that indeed complete protection was afforded by the recombinant proteins.

CLPs AND VLPs AS VEHICLES FOR FOREIGN IMMUNOGENS

Synthesis of morphological entities in eukaryotic cells by an expression system opens up a wealth of possibilities to define mechanisms of protein-protein assembly under natural intracellular conditions. For example, by manipulating the structures

TABLE 4. Vaccination of Horses with Recombinant AHSV Proteins[a]

Horse No.	Antigen	0	21	42	49	66	86	93	100	110	Serum-Neutralization Titers against AHSV-4			Clinical Rx	Viremia
											7 p.c	14 p.c	21 p.c		
1	~300 μg VP2 + VP5* [=CCL]	Neg	>640	>640	>640	640	640	640	640	320	640	>640	1280	Neg	Neg
2	~300 μg VP2 + VP5* [=CCL] = 1 mg VP7 [purified]	Neg	>640	>640	>640	320	320	240	240	160	160	>640	1280	Neg	Neg
3	~50 μg VP2 + trace VP5* [=CLS]	Neg	>640	>640	>640	320	320	240	160	160	160	>640	20480	Neg	Neg
4	~300 μg VP2 + 30 μg VP5 [purified]	Neg	<10	640	640	320	320	320	320	320	640	>640	5120	Neg	Neg
5	~300 μg VP2 + 30 μg VP5 + 100 μg VP7 [purified]	Neg	<10	160	80	80	80	80	80	80	640	>640	10240	Neg	Neg
6	~300 μg VP2 + 30 μg VP5 + 100 μg VP7 + 100 μg BTV VP3 [purified]	Neg	<10	80	30	20	20	<20	<20	<20	30	160	1280	Neg	Neg
7	Control (saline)	Neg	0	0	0	0	0	0	0	0	†	†	†	†	†

[a] All proteins are AHSV-4 unless otherwise stated. CCL + Crude Cell Lysate = baculovirus expressed AHSV-4 VP2/5 infected Sf9 cells, freeze-thaw lysed; CLS = Cell Lysate Sup. = above sample further lysed in 20 mM NaHCO$_3$; clarified supernatant; *VP5 difficult to quantitiate. Purified VP2 and VP5 are insoluble—collaboration with Baltus Erasmus, Onderstepoort, South Africa. † Severe African horsesickness disease, clinical reaction from day 9 10 12, died at day 13, spleen and lung positive.

of various proteins, it will be possible to develop particulate antigen delivery systems for the presentation of immunogenic epitopes against other infectious diseases.

Towards this end, the major protein, of the core, VP7 was selected for such manipulation. To insert the four epitopes without perturbing the overall structure of the protein, it was necessary to analyze the atomic structure of the molecule. The protein was therefore purified, crystallized and the three-dimensional structure of crystallized VP7 (expressed by baculovirus protein) analyzed recently. The x-ray crystallographic analysis revealed that the VP7 molecule exists as trimers and each molecule has two domains. The three VP7 molecules interact end to end to form the trimer.[10] The upper domain of the molecule mainly contains β-sheet and α-strands with a number of loops. The lower domain, on the other hand, is mainly composed of tight helices which cannot be disturbed. However, there are several sites within the various loops of upper domain that can be manipulated for insertion of foreign epitopes as indicated in the ribbon diagram of the trimer (FIG. 2).

As indicated in the VP7 diagram, a number of regions appeared to be accessible to the surface of the molecule including an Arg-Gly-Asp (aa. 168-170) motif. Therefore, this site as well as two other sites (Ala aa. 145 and Gly aa. 238), was chosen for insertion of foreign epitopes. All three sites have been manipulated, and restriction sites were created for receiving foreign epitopes. The RGD site was used for insertion of the synthetic double-stranded DNA fragment coding for the Hepatitis B virus surface antigen (HBsAg). Chimeric HBsAg-VP7 protein was expressed and tested for its ability to form CLPs. Indeed the chimeric protein could form CLPs that were similar to CLPs produced with the normal VP7. They had a complete or nearly complete set of VP7 spikes and appeared to be as stable as nonchimeric CLPs. The other two sites were used to insert immunogenic region of surface glycoprotein gp51 of bovine leukemia virus (BLV). Both sites appeared to be suitable for insertion of extra amino acid residues without disturbing CLP formation (unpublished observation). The immunogenic and protective properties of these chimeric particles are now under investigation.

CONCLUSION

Recent developments in biotechnology have made it possible to synthesize BTV double-shelled VLPs that mimic authentic virions but lack the genetic material and viral replicating machinery. Both biochemically and immunologically, these particles are similar to the authentic virus. In sheep, doses as low as 10 μg of VLPs elicited immune responses that were sufficient to protect sheep against virulent virus challenge. Moreover, protective immunity was also afforded against certain heterologous virus challenges. Whether sheep are protected for periods longer than 14 months has not yet been tested.

Generally, virus-antibody interactions that result in the neutralization of virus infectivity are the basis for the development of many viral vaccines, including BTV. The serotype-specific BTV VP2 protein is considered to be the main antigen that is responsible for inducing virus-neutralizing responses in vaccinated animals. The roles of other antigens in procuring a protective immune response, such as that involving cell-mediated responses, are not known. To investigate this issue we initiated a study

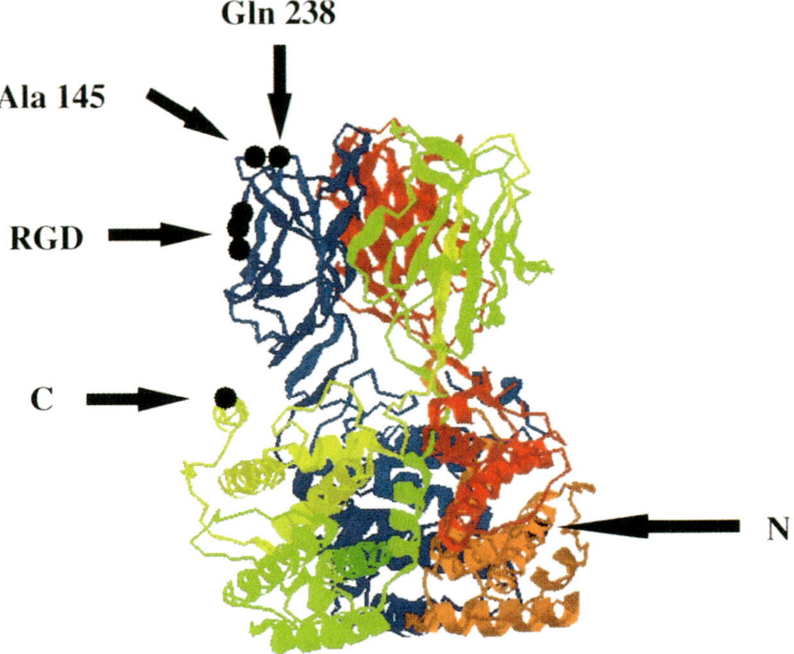

FIGURE 2. The ribbon diagram of the three dimensional structure of a BTV VP7 monomer and the residues that are manipulated for insertion of foreign epitopes. The amino (N) and carboxy (C) termini are indicated.

in sheep using CLPs in order to determine whether they elicit protection, specifically cross-protection since the BTV VP3 and VP7 antigens are highly conserved. Preliminary data indicate that CLPs stimulate partial protection in sheep, suggesting that cellular immunity may play a role in the overall protection process.

In addition, it is anticipated that the availability of large quantities of viral components, and of CLPs and VLPs, will not only facilitate an understanding of their three-dimensional structure at the atomic level, but due to the authenticity in their biological and immunological properties, the subunit vaccine (VP2 for AHSV) and the assembled VLPs are ideal as viral vaccines (for BTV), as they lack any genetic material and cannot replicate.

Since CLPs and VLPs are highly immunogenic and elicit protective immunity in vertebrate hosts, BTV CLPs and VLPs are now being developed to deliver multiple peptide components representing viral epitopes in order to elicit protective immunity. Since BTV CLPs and VLPs are large multiprotein structures, they can incorporate alternative protein forms of the structural proteins (e.g., chimeric VP7), including alternative forms of these proteins (e.g., VP7a, VP7b, etc.). In addition, there is a possibility of using more than one of the protein types to deliver antigens (e.g., VP2, VP5, VP7). This feature has important implications, since to decrease the chance of nonresponsiveness, vaccines designed to elicit immunity should not be based on a single viral sequence or antigen but should include as many immunogenic sequences as feasible. Moreover, VLPs based on alternative BTV serotypes can be used for successive immunizations in order to evade the anti-BTV response elicited in a primary vaccination. This is another novel feature of the BTV system, not currently available in other antigen delivery systems (e.g., vaccinia, polio, Ty particles, etc.).

BTV CLPs and VLPs offer several particular advantages over other systems. First, as noted above, large quantities can be produced due to the expression capabilities of baculovirus vectors (ca. 20-30 mg per liter culture, produced in serum-free medium, stable to freeze-drying, etc.). Second, CLPs and VLPs can be purified using a one-step generic protocol based on the physical properties of the particle (gradient centrifugation of cell lysates). Third, they are devoid of any detectable amount of insect, or baculovirus proteins, RNAs, or DNAs. Fourth, the purification procedure is gentle enough to maintain the morphological structure of the particles in their native conformations. Fifth, the particles cannot replicate, although multiple epitopes could be accommodated. Lastly, and most importantly, VLPs have inherent properties of inducing both B cell and T cell responses in vertebrate hosts.

ACKNOWLEDGMENT

All vaccine trials have been performed in collaboration with Dr. B. J. Erasmus and his associates at the Onderstepoort Vaccine Factory, Department of Agriculture, Onderstepoort, South Africa.

REFERENCES

1. ROY, P. & B. M. GORMAN. 1990. Bluetongue viruses. *In* Current Topics in Microbiology and Immunology. P. Roy & B. M. Gorman, Eds.: 1-200. Springer-Verlag. Heidelberg, Germany.
2. PRASAD, B. V. V., S. YAMAGUCHI & P. ROY. 1992. Three-dimensional structure of single-shelled BTV. J. Virol. **66:** 2135-2142.
3. HEWAT, E. A., T. F. BOOTH, P. T. LOUDON & P. ROY. 1992. 3-D reconstruction of bluetongue virus core-like particles by cryo-electron microscopy. Virology **189:** 10-20.
4. HEWAT, E. A., T. F. BOOTH & P. ROY. 1992. Structure of bluetongue virus particle by cryo-electron microscopy. J. Struct. Biol. **109:** 61-69.

5. BISHOP, D. H. L. 1992. Baculovirus expression vectors. Semin. in Virol. **3:** 253–264.
6. O'REILLY, D. R., L. K. MILLER & V. A. LUCKOW. 1992. *In* Baculovirus Expression Vectors: A Laboratory Manual. W. H. Freeman and Company.
7. ROY, P. 1992. Bluetongue virus proteins. (Review article.) J. Gen. Virol. **73:** 3051–3064.
8. ROY, P., P. P. C. MERTENS & I. CASAL. 1994. African horsesickness virus structure. (Review article.) *In* Comparative Immunology Microbiology and Infectious Diseases. **17** (3/4): 243–273. Elsevier Science Ltd. Cambridge, United Kingdom.
9. FRENCH, T. J. & P. ROY. 1990. Synthesis of bluetongue virus (BTV) core-like particles by a recombinant baculovirus expressing the two major structural core proteins of BTV. J. Virol. **64:** 1530–1536.
10. GRIMES, J., A. BASAK, P. ROY & D. I. STUART. 1995. The crystal structure of bluetongue virus VP7; implications for virus assembly. Nature **373:** 167–170.
11. FRENCH, T. J., J. J. A. MARSHALL & P. ROY. 1990. Assembly of double-shelled, virus-like particles of bluetongue virus by the simultaneous expression of four structural proteins. J. Virol. **64:** 5695–5700.
12. BELYAEV, A. S. & P. ROY. 1993. Development of baculovirus triple and quadruple expression vectors: co-expression of three or four bluetongue virus proteins and the synthesis of bluetongue virus-like particles in insect cells. Nucleic Acids Res. **21:** 1219–1223.
13. ROY, P., T. J. FRENCH & B. J. ERASMUS. 1992. Protective efficacy of virus-like particles for bluetongue disease. Vaccine **10:** 28–32.
14. ROY, P., D. H. L. BISHOP, H. LEBLOIS, S. OLDFIELD & B. J. ERASMUS. 1994. Long-lasting protection of sheep against bluetongue challenge after vaccination with virus-like particles: evidence for homologous and partial heterologous protection. Vaccine **12:** 805–811.
15. ROY, P., J. CALLIS & B. J. ERASMUS. 1994. Protection of sheep against bluetongue disease after vaccination with core-like and virus-like particles: Evidence for homologous and partial heterologous protection. *In* Proceedings for 97th Annual Meeting, USA Health Association: 88–97.
16. ROY, P., T. URAKAWA, A. A. VAN DIJK & B. J. ERASMUS. 1990. Recombinant virus vaccine for bluetongue disease in sheep. J. Virol. **64:** 1998–2003.
17. MARTINEZ-TORRECUADRADA, J., H. IWATA, A. VENTEO, I. CASAL & P. ROY. 1994. Expression and characterization of the two outer capsid proteins of African horse sickness virus. The role of VP2 in virus neutralization. Virology **202:** 348–359.
18. CHUMA, T. H. LE BLOIS, J. M. SANCHEZ-VIZCAINO, M. DIAZ-LAVIADA & P. ROY. 1992. Expression of the major core antigen VP7 of African horse sickness virus by a recombinant baculovirus and its use as a group-specific diagnostic reagent. J. Gen. Virol. **73:** 925–931.
19. IWATA, H., T. CHUMA & P. ROY. 1992. Characterization of genes encoding two of the major capsid proteins of epizootic haemorrhagic disease virus indicates a close genetic relationship to bluetongue virus. J. Gen. Virol. **73:** 915–924.

A Blocking ELISA for Detection of Antibody to a Subgroup-Reactive Epitope of African Horsesickness Viral Protein 7 (VP7) Using a Novel Gamma-Irradiated Antigen[a]

J. A. HOUSE, J. L. STOTT,[b] M. T. BLANCHARD,[b]
M. LaROCCO, AND M. E. LLEWELLYN

USDA/APHIS/VS/NVSL Foreign Animal Disease
Diagnostic Laboratory
Greenport (FADDL), New York 11944

[b]*University of California*
School of Veterinary Medicine
Davis, California 95616

INTRODUCTION

African horsesickness (AHS) is a highly lethal viral disease of equidae caused by 9 serotypes of orbiviruses in the subgroup African horsesickness viruses.[1] The disease occurs enzootically in sub-Saharan Africa, but on numerous occasions, it has extended outside of its endemic area and caused significant loss of equidae.[2]

Serological diagnostic techniques for AHS include agar gel immunodiffusion, complement fixation, indirect fluorescent antibody, each of which detects antibodies against subgroup-reactive epitopes, and virus neutralization (VN) that is serotype specific.[3] Various ELISA procedures have been reported.[3–7] This paper describes the evaluation of monoclonal antibody–based blocking ELISA using a gamma-irradiated inactivated cell culture derived antigen.

MATERIALS AND METHODS

Horse Serums

Serums ($n = 301$) from normal equidae in the United States were kindly provided by Dr. James Pearson (Chief, Diagnostic Virology Laboratory, USDA, APHIS, VS, National Veterinary Services Laboratories, Ames, IA). Immune serums (69) were produced, at the FADDL, from animals that were vaccinated, vaccinated and chal-

[a]This work was supported in part by a United States Department of Agriculture (Animal and Plant Health Inspection Service) cooperative agreement No. 12-34-94-1089-CA.

lenge-inoculated, or challenge inoculated then recovered from AHS serotypes 1, 2, 3, 4, 6, 7, 8, and 9.

Virus Neutralization Test

The standard VN test procedure was used.[2] Neutralization of 75% of the cytopathogenic effect (CPE) at a final serum dilution of 1:10 or greater was established as positive.

Antigen

A cell culture antigen was prepared by infecting a freshly confluent monolayer of Vero cells with AHS virus (AHSV), usually AHSV serotype 4 (AHSV-4). The virus was adsorbed onto the monolayer for 2 hours at 37°C, then 40 ml of Eagles minimum essential medium (EMEM) (Biowhittaker, Walkersville, MD) was added to a 150-cm^2 tissue culture flask (Corning Inc., Corning, NY). The culture was incubated at 37°C until 85-95% CPE occurred (30 to 48 hours). The cells were scraped from flasks, and the suspension was centrifuged at 900 × g for 20 min. The supernatant fluid was decanted and discarded. The cells were suspended in a 10% lactose-phosphate buffer (5% 0.1 M NaH$_2$PO$_4$ and 95% 0.1 M Na$_2$HPO$_4$) at the rate of 15 ml per culture flask and sonicated for 4 min in an ice bath using a cup horn (Model W-225, Heating Systems–Ultrasonics Inc., Farmingdale, NY) adjusting the power output to obtain maximum sonication as indicated by turbulence (usually 70-90% maximum power). The suspension was centrifuged for 10 min. at 900 × g. The supernatant fluid was saved, and the pellet resuspended in 5 ml of lactose-phosphate buffer per culture flask. The pellet suspension was then sonicated and centrifuged as above. The supernatant fluids were pooled and the pellet discarded. The antigen was aliquoted into cryopreservation vials (Cryovials^{tm} Vanguard International Inc., Neptune, NJ) and stored at −70°C. When removed from storage, it was exposed to gamma radiation from a ^{60}Cobalt source (Gammacell 220, Nordion International Inc., Kanata, Ontario, Canada) while held at −70°C. Temperature was controlled in the chamber using gas from evaporating liquid nitrogen. The dose of gamma radiation was calculated by standard procedures and a standard curve for radiation dose decay.

The virus present in selected antigen batches was titrated in Vero cells prior to irradiation using standard procedures.

Monoclonal Antibody (MAB)

The characteristics of MAB F9H have been described.[8] This IgM antibody reacts with a subgroup-reactive epitope, present on viral protein 7 (VP7) of all strains of 9 serotypes of African horsesickness viruses tested. This reaction is blocked by AHS immune serum.

ELISA Procedure

Microtiter plates (Immunol 2^{tm}, Dynatech Laboratories, Chantilly, VA) were coated with 100 μl per well of antigen in carbonate/bicarbonate coating buffer pH 9.6 (Sigma Chemical Co.; St. Louis, MO) and adsorbed for 60 min at 37°C. All incubations were done on a plate shaker. The plates were decanted and washed 3 times with phosphate-buffered saline containing 0.05% Tween 20 (PBST) (Aldrich Chemical Co. Inc., Milwaukee, WI) by flooding the wells and soaking for 1 min between washes, and tapping the plates dry. Fifty (50) μl of each test serum diluted 1:10 in PBST with 4% bovine calf serum (BS) (Hyclone Laboratories, Inc., Logan, UT) and 4% normal horse serum (HS) (ICN Biomedicals Inc., Costa Mesa, CA) was added to 2 wells and incubated for 45 min at 37°C. Leaving the test serum in the well, 50 μl of mouse ascitic fluid origin MAB F9H (in serum diluent) was added to each well of test serum and 4 wells that contained no test serum (MAB reference well); 4 wells contained only 100 μl of diluent and were used as blanks. After incubation for 45 min at 37°C, the plates were washed. One hundred (100) μl of goat anti-mouse IgM horseradish peroxidase conjugate (Sigma Chemical Co.) in serum diluent were added to all wells and incubated for 45 min at 37°C. The plates were decanted and washed. Substrate, 3, 3′, 5, 5′ tetramethylbenzidine dihydrochloride (TMB) (Sigma Chemical Co.) in pH 5.0 phosphate (0.0514 M) citrate (0.0243 M) buffer with 33 μl of 3% hydrogen peroxide per 10 ml of substrate, was added (100 μl per well) and incubated at room temperature for 20 min. The reaction was stopped by adding 50 μl of 2 M H_2SO_4 per well. Optical density (OD) was read at 450 nm (MR 700 Microplate Reader, Dynatech Laboratories, Chantilly, VA) using 630 nm filter as a nonspecific absorbance reference.

The following formula was used to calculate the percent inhibition (%INH) of blocking of the MAB binding by specific antibody:

$$\%\text{INH} = [1 - \left[\frac{t - b}{m - b}\right]] \times 100$$

where: t = average OD of 2 test wells, b = average OD of 4 blank wells, and m = average OD of 4 MAB wells.

Conjugate and Antigen Titration

A block titration was done to determine the optimal dilutions of goat anti-mouse IgM horseradish peroxidase conjugate and AHSV-4 antigen. On each of 3 respective microtiter plates, 12 twofold dilutions of antigen (1:16 through 1:32, 768) in coating buffer were adsorbed onto each well in the 12 columns of 3 respective microtiter plates for 60 min at 37°C. After washing, 50 μl of diluent was added to all wells and incubated for 45 min at 37°C. Without decanting or washing, 50 μl of MAB F9H, diluted 1:2000, was added to all wells and incubated for 45 min at 37°C. Plates were decanted and washed, and 100 μl of 8 twofold dilutions of conjugate (1:250 through 1:32,000) was added to each of the 8 respective rows and incubated for 45 min at 37°C. Plates were decanted, then washed. Substrate was added and incubated. Color development was stopped and read as above. The average readings of the dilutions were used to generate a 3-dimensional mesh graph (Sigma Plot^{tm}, Jandel Scientific, San Rafael, CA 94901) that was used to determine optimal dilutions of the reagents.

Effect of Gamma Radiation on Antigen

Ampules of antigen were exposed to doses of 0, 3, 6, 9, 12, and 15 megarads (Mrads) of gamma radiation. ELISA was performed on gamma-irradiated antigens by coating plates with 50 μl of antigen and reacting with a 1:2000 dilution of MAB F9H, as described above.

An aliquot of antigen removed prior to irradiation was titrated for virus. Irradiated antigen (0.5 ml) was inoculated onto each of two 150-cm^2 cultures of Vero cells. After incubation for 9 days at 37°C, the cultures were observed, frozen, and thawed and passed again into Vero cells. The passage and incubation were repeated a third time to detect any remaining live virus.

Conjugate Reactivity with Test Components

Three commercial conjugates (designated 1, 2, and 3) were studied to determine their reactivity with components in the ELISA. Components used were AHSV-4 antigen, Vero cell soluble antigen prepared with uninfected Vero cells, HS, BS, MAB F9H, and blank wells. Each of the components was adsorbed onto microtiter wells, then washed. Conjugates were added, plates were washed, substrate was added, and ODs were determined after stopping the reaction with H$_2$SO$_4$.

Blocking Agents

Commercial HS and BS were tested for their ability to diminish nonspecific conjugate reactions. Horse serum at 5% and HS (5%) plus BS (5%) were used in the diluent to determine their ability to minimize nonspecific reactions.

Relationship of %INH to Virus Neutralization (VN) Titer

Virus neutralizing antibody titers of 59 AHS immune serums were used in a scatter plot against %INH for these serums. Regression analysis was performed to determine the relationship of the reactions.

RESULTS

Block titration of antigen and conjugate yielded a pattern of decreasing ODs with dilution (FIG. 1). By adjusting the Z-axis to the OD desired in the test (OD = 1.0), the combinations of antigen and conjugate dilutions needed to give that OD were displayed at the intersection of the XY plane in a 3-dimensional mesh plot (FIG. 2).

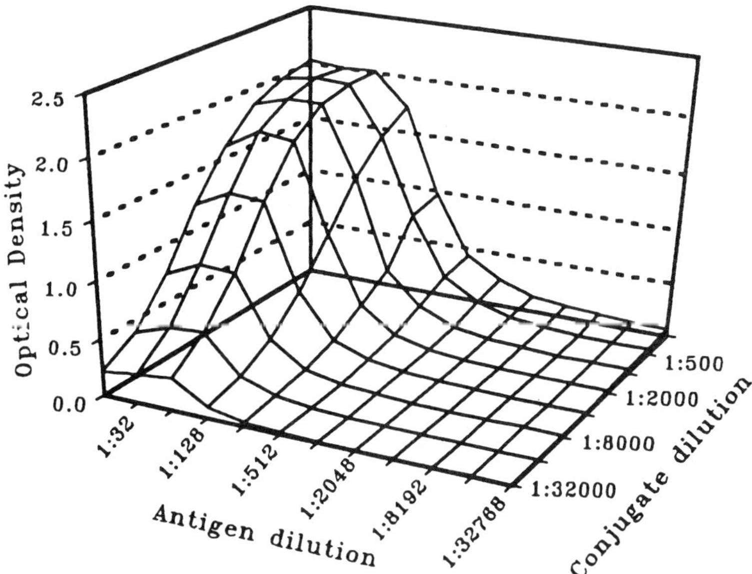

FIGURE 1. A 3-dimensional mesh graph of a block titration of AHSV antigen irradiated with 9 Mrads of gamma radiation and a commercial anti-mouse IgM HRPO conjugate.

The exposure of AHSV-4 antigen to 3 and 6 Mrads of gamma radiation resulted in a slight increase in antigen potency. Reactivity of the antigen was greatly enhanced at exposure to 9 Mrads, then decreased as radiation was increased to 12 and 15 Mrads (FIG. 3). By block titration, the working dilution of nonirradiated antigen was determined to be 1:8 with an MAB F9H dilution of 1:1000. After irradiation with 9 Mrads of gamma radiation, the working dilution was 1:128 with an MAB F9H dilution of 1:1000.

Four different serials of AHSV ELISA antigen contained $10^{7.7}$, $10^{7.7}$, $10^{7.9}$, and $10^{7.9}$ TCID$_{50}$ of virus per ml for an average of $10^{7.8}$ TCID$_{50}$ per ml of antigen. Antigen irradiated with 9 Mrads of gamma radiation had no sign of viral CPE after three passages on Vero cells.

AHSV-4 antigen, normal Vero cell antigen, HS, and BS reacted nonspecifically with commercial conjugates to varying degrees. Conjugates 1 and 2 both reacted strongly with AHSV-4 antigen and HS. Conjugate 1 reacted strongly with normal Vero cells. Conjugate 2 reacted strongly with BS. Conjugate 3 did not have any significant reaction to any test component except the specific reaction to the MAB (TABLE 1).

Blocking agents (HS and BS) used in the test diluent enhanced the sensitivity of the test by increasing the %INH of an immune serum tested at 2 dilutions. The greatest increase in sensitivity was observed when HS and BS were used together (TABLE 2).

When normal horse serums were tested in the blocking ELISA (B-ELISA) using nonirradiated antigen, there was a relatively normal distribution with a slight tailing

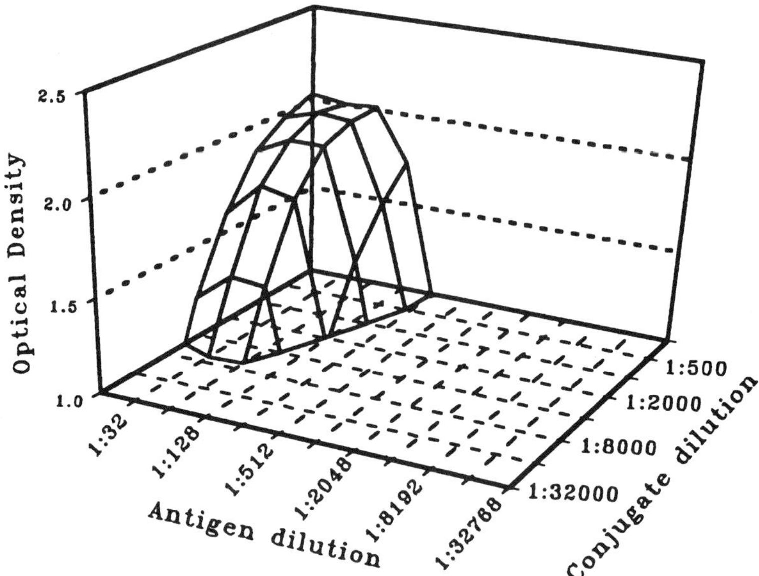

FIGURE 2. A 3-dimensional mesh graph of a block titration of AHSV antigen irradiated with 9 Mrads of gamma radiation and a commercial anti-mouse IgM HRPO conjugate showing the interface of reactions of the reagents at OD = 1, the desired OD of the MAB F9H indicator system.

toward negative readings. Using antigen irradiated with 9 Mrads, the readings were skewed toward the positive and then almost truncated above 10 %INH; there was a long tailing toward negative %INH (FIG. 4). Immune serums had a similar distribution with the nonirradiated and irradiated antigens (FIG. 5).

The normal and immune serums showed two clear population distributions in the B-ELISA when using 9 Mrad irradiated antigen (FIG. 6 and TABLE 3). The results

TABLE 1. AHS B-ELISA: Reactivity of Conjugates with Test Components

Component	CONJ[a] 1	CONJ 2	CONJ 3
AHSV 4 Antigen	1.00[b]	1.10	0.03
Vero cells	0.93	0.37	0.02
Horse serum	1.60	1.78	0.11
Bovine serum	0.14	1.71	0.00
MAB F9H[c]	1.69	1.87	1.70
Blank	0.05	0.03	0.04

[a] CONJ = conjugate.

[b] Data are optical density.

[c] MAB = monoclonal antibody F9H.

TABLE 2. AHS B-ELISA: the Effect of Blocking Agents

Serum	PBST[a]	5% HS[b]	5% HS + 5% BS[c]
Pony 7 1:10	96[d]	98	98
Pony 7 1:1000	25	36	42
Normal pony 1:10	9	3	2

[a] PBST = Phosphate-buffered saline with 0.05% Tween 20.

[b] HS = horse serum.

[c] BS = bovine serum.

[d] Data are % inhibition.

for the populations of normal and immune samples interfaced at 19 %INH where there was 1 normal serum and 1 immune serum. At 20 %INH there was 1 immune serum.

The VN test results on immune serums ranged from 1:10 to 1:960. The majority of VN titers were in the range of 1:10 to 1:150; and in that range, the %INH ranged from 20 to 96%. There was not a significant relationship between the VN titer and %INH ($r^2 = 0.36$; FIG. 7).

DISCUSSION

A B-ELISA was evaluated for detecting specific antibody to a subgroup reactive epitope on AHSV VP7 that reacted with MAB F9H. A novel gamma-irradiated

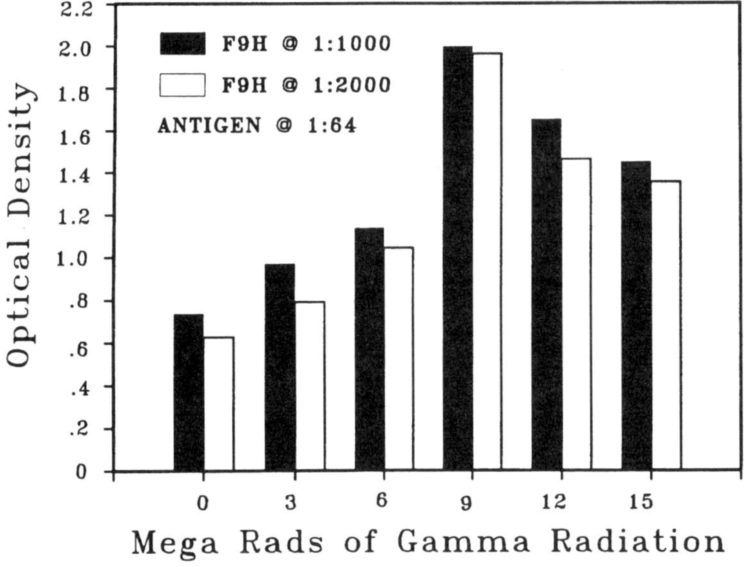

FIGURE 3. The effect of various doses of gamma radiation on AHSV antigen.

TABLE 3. Population Statistics for Reactions in the AHS B-ELISA

	Normal Serums		Immune Serums	
Population	Nonirradiated Antigen	Irradiated Antigen	Nonirradiated Antigen	Irradiated Antigen
Number	301	301	69	69
Average[a]	3.1	1.1	61	60.6
Maximum[a]	20	19	96	98
Minimum[a]	−22	−40.5	20	19
Standard deviation[a]	6.5	10.3	20.4	20.4

[a] Data expressed in % inhibition.

antigen was used in the test. Antigen exposed to 9 Mrads of gamma radiation had approximately 16 times the titer of the nonirradiated antigen. It is likely that the gamma radiation disrupted the viral particles and exposed the viral inner core that consists of VP3 and VP7 of the AHSV. The specific reactions accounting for the increased reactivity of the antigen could be the subject of a future investigation.

The irradiation of the AHSV antigen provides an advantage in addition to increasing potency, namely, safety. The D_{10} (dose of gamma radiation needed to reduce viral infectivity by 1 $\log_{10}$) for bluetongue, another orbivirus, is 0.83 Mrads.[9] A dosage of 9 Mrads is calculated to kill $10^{10.8}$ TCID$_{50}$ of orbivirus. Thus, there is a safety factor of 10^3 TCID$_{50}$ considering that the average titer of virus in the 4 serials of

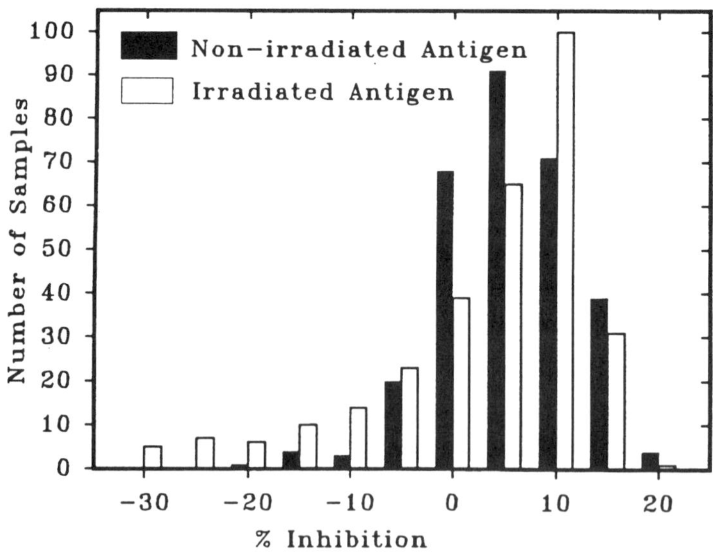

FIGURE 4. The distribution of %INH of serums from normal equidae in the US.

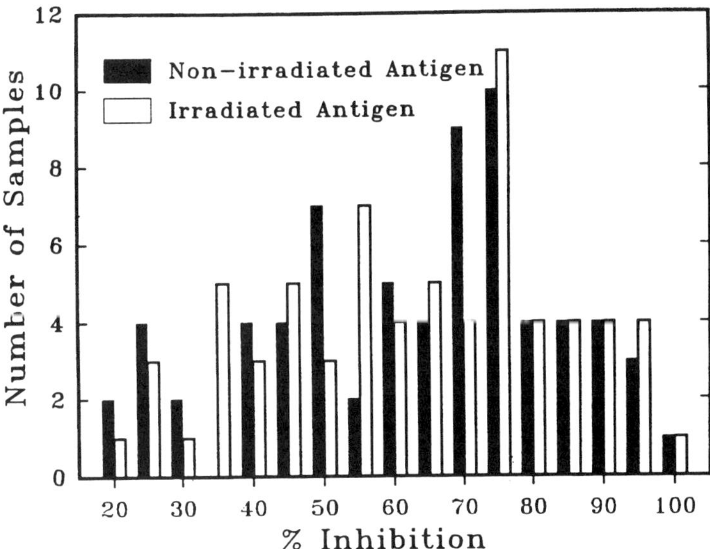

FIGURE 5. The distribution of %INH of serums from equidae with antibody to AHSV antigens.

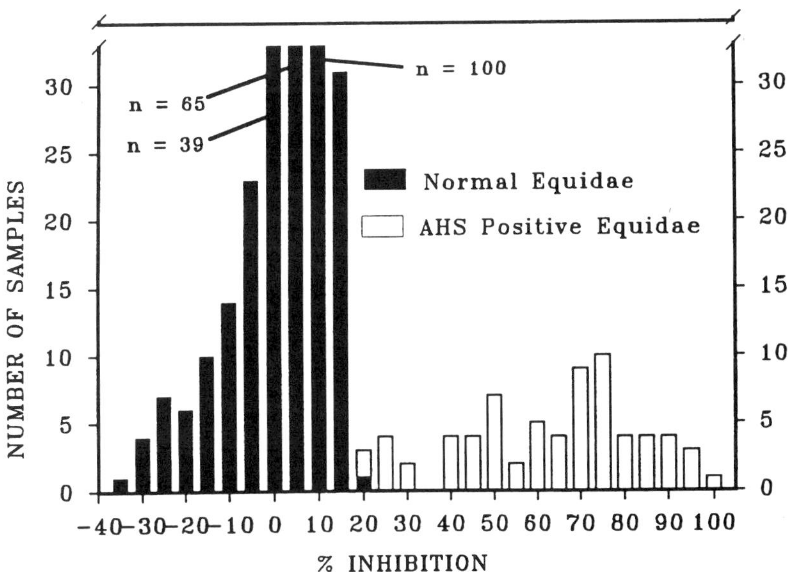

FIGURE 6. The distribution of AHS B-ELISA reactions of serums from normal equidae and immune equidae.

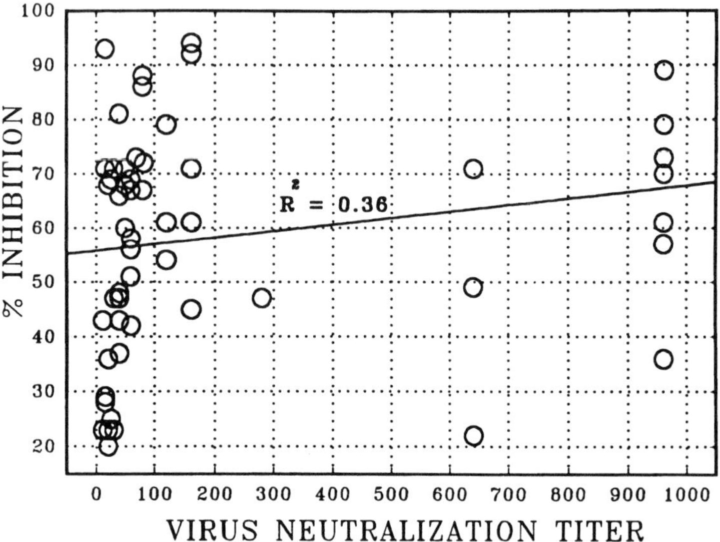

FIGURE 7. A scatter diagram showing the distribution of AHS *B*-ELISA reactions and virus neutralizing titers or serums from equidae with serum antibody to AHSV antigens.

AHSV antigen was $10^{7.8}$. No virus was detected in a typical serial of AHSV antigen irradiated with 9 Mrads of gamma radiation after 3 blind passages in Vero cells, confirming the inactivation. None of the previously reported ELISAs[3–5, 7] that used cell-culture-derived antigen described procedures for inactivation of their antigen. Considering that AHS is a foreign animal disease in the greater majority of the world, such antigens may represent a risk. The VP7 antigen expressed in *Escherichia coli*[6] does have this safety advantage.

Commercial conjugates reacted nonspecifically with various test components. In a blocking ELISA, this can obscure the blocking by an immune test serum and result in a false-negative reaction with immune serum. Even with the conjugate that did not react appreciably with test components, incorporating the HS and BS blockers notably enhanced the sensitivity of the test by reducing low levels of interfering reactions. Thus, the careful evaluation of conjugates is essential to the accuracy of an ELISA.

When normal serums were tested using irradiated antigen, there was a notable skewing and almost truncation in the reactions toward the positive, but there was a tailing toward the negative compared with reactions observed using nonirradiated antigen. The reactions of serums in the immune population were very similar regardless of the antigen used. We cannot explain why the reactions of the normal population changed their distribution when tested using gamma-irradiated antigen.

Regarding the interpretation of the test, the interface between B-ELISA reactions of samples from the normal and immune populations was at 19 %INH. There is no set rule for selecting a cutoff point for considering a test reaction as positive. Based upon the population statistics and the reactions observed, we designated reactions above 19 %INH as positive. For confirmation, virus neutralization may be used.

It was not surprising that there was not a significant correlation between the %INH and VN antibody ($r^2 = 0.36$). At least one neutralizing epitope is on VP2,[10] while MAB F9H reacts with an epitope on VP7. It was apparent that animals responded independently to the immunological stimulation by the epitopes on the two viral proteins.

SUMMARY

A novel gamma irradiated inactivated cell culture derived African horsesickness viral (AHSV) antigen was used in a blocking ELISA (B-ELISA) for detecting antibody to a subgroup-reactive epitope of AHSV. A monoclonal antibody (MAB), class IgM, against an epitope on African horsesickness (AHS) viral protein 7 (VP7) was developed in BALBc mice and used in the B-ELISA. The MAB, designated F9H, was blocked by 69 serums from equidae with antibody to AHS, but its binding activity was not appreciably affected by 301 serums that did not contain antibodies to AHS virus. An ELISA protocol using a blocking format is described.

ACKNOWLEDGMENTS

We wish to thank Drs. C. A. Mebus and B. I. Osburn for their support of the work and helpful suggestions and Dr. C. House for performing the virus neutralizing antibody titers on the serums.

REFERENCES

1. 1991. Classification and nomenclature of viruses. Arch. Virol. Suppl. **2:** 188-192.
2. HOUSE, J. A. 1993. African horse sickness. Vet. Clin. North Am. **9:** 355-364.
3. HOUSE, C., P. E. MIKICIUK & M. L. BERNINGER. 1990. Laboratory diagnosis of African horse sickness: comparison of serological techniques and evaluation of storage methods of samples for virus isolation. J. Vet. Diagn. Invest. **2:** 44-50.
4. WILLIAMS, R. 1987. A single dilution enzyme-linked immunosorbent assay for the quantitative detection of antibodies to African horsesickness virus. Onderstepoort J. Vet. Res. **54:** 67-70.
5. HAMBLIN, C., S. D. GRAHAM, E. C. ANDERSON & J. R. CROWTHER. 1990. A competitive ELISA for the detection of group-specific antibodies to African horse sickness virus. Epidemiol. Infect. **104:** 303-312.
6. WADE-EVANS, A. M., T. WOOLHOUSE, R. O'HARA & C. HAMBLIN. 1993. The use of African horse sickness virus VP7 antigen, synthesized in bacteria, and anti-VP7 monoclonal antibodies in a competitive ELISA. J. Virol. Methods **45:** 179-188.
7. WILLIAMS, R., D. H. DU PLESSIS & W. VAN WYNGAARDT. 1993. Group-reactive ELISAs for detecting antibodies to African horsesickness and equine encephalosis viruses in horse, donkey and zebra sera. J. Vet. Diagn. Invest. **5:** 3-7.

8. STOTT, J., J. A. HOUSE, M. T. BLANCHARD, R. E. SHOPE, C. A. MEBUS & B. I. OSBURN. Development and characterization of monoclonal antibodies specific for African horse sickness viral proteins. J. Virol. (Submitted.)
9. HOUSE, C., J. A. HOUSE & R. J. YEDLOUTSCHNIG. 1990. Inactivation of viral agents in bovine serum by gamma irradiation. Can J. Microbiol. **36:** 737–740.
10. BURRAGE, T. G., R. TREVEJO, M. STONE-MARCHAT & W. W. LAGREID. 1993. Neutralizing epitopes of African horsesickness virus serotype 4 are located on VP2. Virology **196:** 799–803.

Serum Antibody to Rift Valley Fever Virus in African Carnivores

CAROL HOUSE,[a] KATHLEEN A. ALEXANDER,[b]
PIETER W. KAT,[b] STEPHEN J. O'BRIEN,[c] AND
JOSEPH MANGIAFICO [d]

[a]*USDA
APHIS
NVSL
FADDL
Post Office Box 848
Greenport, New York 11944*

[b]*School of Veterinary Medicine
University of California
Davis, California 95616*

[c]*National Cancer Institute
Frederick, Maryland 21702*

[d]*USAMRIID
Fort Detrick
Frederick, Maryland 21702*

INTRODUCTION

Rift Valley fever (RVF) was first described in 1931 by Daubney *et al.*[1] as a highly fatal epizootic disease of sheep in Kenya. The causative agent, RVF virus (RVFV), is a Bunyavirus in the genus Phlebovirus[2] and is spread mainly by mosquitoes, particularly species of Aedes and Culex. While RVFV is known to be endemic in the Rift Valley, it has caused outbreaks in Egypt,[3] South Africa,[4] Mauritania,[5] and Madagascar.[6] Cattle, sheep, goats, dogs, cats, mice, hamsters, rats, camels, water buffaloes, horses, monkeys, and humans are known to be susceptible to RVFV, as recently reviewed by House.[7] This paper reports the results of an extensive RVF serological survey among a diversity of African carnivore species.

MATERIALS AND METHODS

Serum Samples

Sera used in this study were largely collected in conjunction with our carnivore disease and genetic surveys and were stored at −20 and −80°C until tested. Serum samples collected from lions (*Panthera leo*, Ngorongoro Crater, Tanzania, *n* = 16, 1991; Serengeti National Park, Tanzania, *n* = 60, 1986-1993; Kruger National Park,

South Africa, $n = 31$, 1991; various locations, Kenya, $n = 6$, 1987), cheetahs (*Acinonyx jubatus*, Serengeti National Park, Tanzania, $n = 60$, 1986-1992; various locations in Kenya, $n = 4$, 1987), domestic cats (*Felis catus*, Masai Mara, Kenya, $n = 43$, 1992), domestic dogs (*Canis familiaris*, Moremi, Botswana, $n = 35$, 1992; Masai Mara, Kenya, $n = 51$, 1989-1991), African wild dogs (*Lycaon pictus*, Moremi, Botswana, $n = 33$, 1991-1992; Masai Mara, Kenya, $n = 11$, 1987-1990; Kruger National Park, South Africa, $n = 4$, 1990-1991; Serengeti National Park, Tanzania, $n = 17$, 1986-1989), spotted hyenas (*Crocuta crocuta*, Masai Mara, Kenya, $n = 65$, 1990-1992), and jackals (*Canis* spp., various locations in Kenya, $n = 15$, 1988-1989; Chipangali Breeding Centre, Zimbabwe, $n = 7$, 1988).

Serological Tests

All samples were screened for antibody to RVFV at a 1:10 dilution in the hemagglutination inhibition test (HI).[8] Samples testing positive on the screening test were irradiated with 3 megarads of gamma radiation and forwarded to the United States Army Medical Research Institute of Infectious Diseases (USAMRIID) where they were tested using a serum dilution plaque reduction neutralization test (PRNT) described by Early *et al.*,[9] but modified for use in 12-well plastic plates. The virus used in the PRNT was the ZH501 strain of RVFV.[3] The neutralization was scored positive at the highest serum dilution that inhibited 80% or more of plaque formation compared to the virus control titration. The titer of the sample was the reciprocal of the last positive serum dilution.

RESULTS

The results of the HI and PRNT are given in TABLE 1. No hyenas, domestic cats, or domestic dogs were positive. Two of 64 cheetahs, 3 of 22 jackals, 7 of 65 African wild dogs and, 15 of 113 lions were positive on the screening test. There was sufficient quantity of the samples of the cheetahs, African wild dogs, and lions to be tested by the PRNT. No cheetahs or African wild dogs were PRNT positive. All lions testing positive by HI from the Ngorongoro Crater area in Tanzania and from South Africa were PRNT seropositive, with titers ranging from 1:40 to 1:640 (TABLE 2).

DISCUSSION

No neutralizing antibody to RVFV was found in the survey samples from cheetahs, domestic cats, hyenas, domestic dogs, or African wild dogs. The positive hemagglutination inhibition tests on 3 jackals could not be confirmed due to insufficient sample. The results in the lion survey quite conclusively show that antibody to RVFV exists in high titers in lions. This is the first report of RVFV antibody in lions. It is possible that the animals, sampled in 1991, were exposed to RVFV during the 1989 rainy season, which resulted in the latest reported outbreak of RVF.[10-12]

Transmission of RVF to lions is probably mosquito vectored, as the animals of the Tanzanian Ngorongoro Crater area, characterized by high rainfall and permanent

TABLE 1. Hemagglutination Inhibition end Plaque Reduction Neutralization Test Results

Sample	Date	Location (Number)	Percent HI Pos.[a]	Percent PRNT Pos.[b]
Lions	1987	Kenya (6)	0	
	1991	Kruger, S. Africa (31)	9.7	9.7
		Tanzania		
	1986–93	Serengeti (60)	5.0	0
	1991	Ngorongoro Crater (16)	56.3	56.3
Cheetahs	1986–92	Serengeti (60)	3.3	0
	1987	Kenya (4)	0	0
Domestic cats	1992			
		Kenya (43)	0	
Domestic dogs	1992	Botswana (35)	0	
	1989–91	Kenya (51)	0	
African wild dogs	1991–92	Botswana (33)	9.1	0
	1987–90	Kenya (11)	27.3	0
	1990–91	Kruger, S. Africa (4)	25	0
	1986–89	Serengeti (17)	0	
Hyenas	1990–92	Kenya (65)	0	
Jackals, black-backed	1988–89	Kenya (15)	20	NT[c]
	1988	Zimbabwe (1)	0	
Jackals, side-striped	Zimbabwe (6)	0		
1988				

[a] Percent samples positive on hemagglutination inhibition test.
[b] Percent samples positive on plaque reduction neutralization test.
[c] Not tested.

TABLE 2. Plague Reduction (80%) Neutralization Test Results of Lion Serum Samples

Location	Number of Animals	Titer
Ngorongoro Crater	2	640
	1	320
	2	160
	2	80
	2	40
South Africa	1	320
	1	160
	1	80

swamps on the crater floor, experienced a 56.3% incidence (9 of 16 lions tested) while all animals of the drier Tanzanian Serengeti Plain area ($n = 60$) were negative for neutralizing antibody to RVFV. The Ngorongoro Crater is on the southeast edge of the Serengeti Plain. In the relatively arid Kruger National Park of South Africa, 3 of 31 lions had neutralizing antibody to RVFV, an incidence of 9.7%. The association of rainfall, green vegetation index, mosquitoes and incidence of RVF is well established; indeed, satellite remote sensing imagery has been used to calculate the green vegetation index and to then predict RVF activity.[13]

The pathology of RVF in the lion, the impact of RVF on lion populations in areas experiencing RVF epizootics, and the role of the lion in maintaining a RVFV reservoir are new and important questions.

ACKNOWLEDGMENTS

We would like to thank the following individuals for contributing samples to this study: K. Holekamp, L. Smale, J. W. McNutt, M. K. Laurenson, V. Wilson, and M. G. L. Mills.

REFERENCES

1. DAUBNEY, R., J. R. HUDSON & P. C. GARNHAM. 1931. Enzootic hepatitis or Rift Valley fever: an undescribed virus disease of sheep, cattle and man from East Africa. J. Pathol. Bacterol. **34:** 545–579.
2. BISHOP, D. H. L., C. H. CALISHER, J. CASALS, M. P. CHUMAKOV, S. YA. GAIDOMAVICH, C. HANNOUN, D. K. LVOV, I. D. MARSHALL, N. OKER-BLOM, R. F. PETTERSON, J. S. PROTERFIELD, P. K. RUSSELL, R. E. SHOPE & E. G. WESTAWAY. 1980. Bunyaviridae. Interviorlogy **14:** 125–143.
3. MEEGAN, J. M. 1979. The Rift Valley fever epizootic in Egypt 1977–78. 1. Description of the epizootic and virological studies. Trans. Soc. Trop. Med. Hyg. **73:** 618–623.
4. VAN VELDEN, D. J. J., J. D. MEYER, J. OLIVIER, J. H. S. GEAR & B. MCINTOSH. 1977. Rift Valley fever affecting humans in South Africa. A clinicopathological study. South African Med. J. **51:** 867–871.
5. JOUAN, A., I. COULIBALY, P. LENA, J. L. SARTHOU, B. LEGUENNO, J. MEEGAN & T. KSIAZEK. 1989. Filter paper confetti in a serological Rift Valley fever survey. Res. Virol. **140:** 169–173.
6. MORVAN, J., D. FONTENILLE, J. F. SALUZZO & P. COULANGES. 1991. Possible Rift Valley fever outbreak in man and cattle in Madagascar. Trans. R. Soc. Trop. Med. Hyg. **85:** 108.
7. HOUSE, J. A., M. J. TURELL & C. A. MEBUS. 1992. Rift Valley fever: present status and risk to the Western hemisphere. *In* Tropical Veterinary Medicine: Current Issues and Perspectives. J. C. Williams, K. M. Kocan & E. P. J. Gibbs, Eds.: 233–242. Ann. N.Y. Acad. Sci. **653.**
8. SHOPE, R. E., C. J. PETERS & J. S. WALKER. 1980. Serological relation between Rift Valley fever virus and viruses of the phlebotomus fever serogroup. Lancet **I:** 886–887.
9. EARLY, E., P. H. PERALTA & M. JOHNSON. 1967. A plaque reduction method for arboviruses. Proc. Soc. Exp. Biol. (NY) **125:** 741–747.
10. LOGAN, T. M., K. J. LINTHICUM, F. G. DAVIES, Y. S. BINEPAL & C. R. ROBERTS. 1991. Isolation of Rift Valley fever virus from mosquitoes (Diptera: Culicidae) collected during an outbreak in domestic animals in Kenya. J. Med. Enterol. **28:** 293–295.
11. DAVIES, F. G., T. M. LOGAN, Y. BINEPAL & P. JESSEN. 1992. Rift Valley fever virus activity in East Africa in 1989. Vet. Rec. **130:** 247–248.

12. LOGAN, T. M., F. G. DAVIES, K. J. LINTHICUM & T. G. KSIAZEK. 1992. Rift Valley fever antibody in human sera collected after an outbreak in domestic animals in Kenya. Trans. R. Soc. Trop. Med. Hyg. **86:** 202-203.
13. LINTHICUM, K. J., C. L. BAILEY, F. G. DAVIES & C. J. TUCKER. 1987. Detection of Rift Valley Fever viral activity in Kenya by satellite remote sensing imagery. Science **235:** 1656-1659.

The Use of Polymerase Chain Reaction to Identify *Pasteurella granulomatis* from Cattle

H. P. VEIT, G. R. CARTER, F. RIET-CORREA,[a] AND
S. S. BROWN

Department of Biomedical Sciences and Pathobiology
Virginia-Maryland Regional College of Veterinary Medicine
Virginia Tech
Blacksburg, Virginia 24061-0442

[a]Faculty of Veterinary Medicine
Federal University of Pelotas
96010-900 Pelotas RS, Brazil

INTRODUCTION

A bacterium first described as *Pasteurella haemolytica*-like was isolated from subcutaneous granulomas in cattle from southern Brazil, and was named *Pasteurella granulomatis* (Pg).[1,2] This organism (Pg) is associated with a disease called "lechiguana," also known as "bovine focal proliferative fibrogranulomatous panniculitis."[3,4] Typically, there are numerous microabscesses within proliferating subcutaneous connective tissue, with eosinophilic lymphangitis, and less-frequent regional lymphadenitis.[4] Pg has been isolated from 14 of 18 spontaneous bovine lechiguana cases; however, experimental reproduction of the disease has not been fully accomplished.[4] Untreated field cases often result in death 3 to 8 months after subcutaneous swelling is noted.[4] It is suspected that the human warble fly (*Dermatobia hominis*) participates in transmission of Pg.[4] The present study was initiated to use polymerase chain reaction (PCR) amplification in order to recognize Pg DNA from cultures and possibly from bacteria within bovine or insect tissue. Such identification of Pg would have epidemiological uses and enable detection of Pg in clinical materials where cultural examination fails. Successful use of the PCR procedure also was expected to show the relatedness of Pg strains to each other and to other *Pasteurella* and *Actinobacillus spp.*

METHODS

Isolate Acquisition and Storage

The Pg strains were isolated originally in the Regional Diagnostic Laboratory of Dr. Franklin Riet-Correa and provided by Dr. G. R. Carter and Ms. S. S. Brown. The *Pasteurella multocida* and *P. haemolytica* strains were provided by Dr. G. R. Carter; the *Actinobacillus lignieresii* strain was obtained from Dr. M. M. Chengappa, Kansas State University, Manhattan, Kansas. All of the strains were stored in trypticase soy broth with 0.3% (w/v) yeast extract and 10% (v/v) glycerol (TSB) in liquid nitrogen.

Polymerase Chain Reaction Procedures for Bacterial Isolates

Cultures of all four species were grown overnight (18-24 h) at 37°C on Colombia Blood Agar Base (Difco) supplemented with 5% sterile defibrinated ovine blood (CBA). Colonies (1-6) were picked into 25 µl distilled water in a 500-µl microcentrifuge tube (Treff Lab, Tekmar Co., Cincinnati, OH) using a 200-µl tip on a Gilson Pipetman P-200 (Rainin) set on 25 µl. Colonies and water were sucked up and down multiple times in the pipet tip to evenly disperse the bacteria. A 2-µl aliquot was withdrawn and streaked onto CBA for storage in TSB in liquid nitrogen for future reference.

The remainder of the suspension was microwaved in tubes mounted in a styrofoam collar to keep them upright and prevent the water from boiling off. Four or fewer tubes were heated in 20 ml distilled water in a 50-ml beaker at full power (Sears Kenmore Auto Defrost household microwave oven, model 566.8724980) for 5.5 min. Five or more tubes were heated in 50 ml distilled water in a 600-ml beaker at full power for 6.0 min. These times allowed for five full minutes of boiling. The tubes were cooled, then centrifuged (Eppendorf microcentrifuge, model 5413) for 15 min. The supernatants were drawn off and stored in 500-µl microcentrifuge tubes at 4°C for use as the DNA template for amplification.

The PCR procedures were a slight modification of that previously reported for *Actinobacillus pleuropneumoniae*.[5] Primers were synthesized by Integrated DNA Technologies, Inc., Coralville, IA. Universal M13 primer (TTA-TGT-AAA-ACG-ACG-GCC-AGT) with M13 reverse sequencing primer (GGA-AAC-AGC-TAT-GAC-CAT-G) (M13R), and T3 primer (GCA-ATT-AAC-CCT-CAC-TAA-AG) with T7 primer (GTA-ATA-CGA-CGC-ACT-ATA-G)[5] were used in pairs for the amplification. Primers were received lyophilized; each was resuspended in distilled water to produce 0.1 nmol per 10 µl, sufficient for one amplification reaction.

A master mix was made by combining 100 µl GeneAmp 10× PCR Buffer II, 20 µl each dATP, dCTP, dGTP, and dTTP, 5 µl *Taq* DNA polymerase (5 U/µl), and 160 µl 25 mM $MgCl_2$. All reagents were from Perkin Elmer (Norwalk, CT). Each 100 µl reaction mixture was made by combining, in a 500-µl microcentrifuge tube, 34.5 µl master mix, 40.5 µl distilled water, 10 µl of each primer in the desired pair, and 5 µl template DNA (the supernatant prepared from the colonies). Negative controls contained no template DNA and 5 µl additional distilled water. The reaction mixtures were combined thoroughly, quick-spun in the microcentrifuge, then overlaid with 100 µl mineral oil (Sigma molecular biology grade).

Amplification was performed in a thermocycler (Coy TempCycler, model no. 60), programmed for 2 cycles of 5 min at 94°C, 5 min at 40°C, and 5 min at 72°C, followed by 35 cycles of 1 min at 94°C, 1 min at 60°C, and 1 min at 72°C, with the fastest available ramp time between temperatures, holding at 4°C at the completion of amplification.

Electrophoresis was performed using a 1% agarose gel. Minigels were produced by combining 0.4 agarose (electrophoresis grade, Gibco BRL) with 40 ml TBE buffer

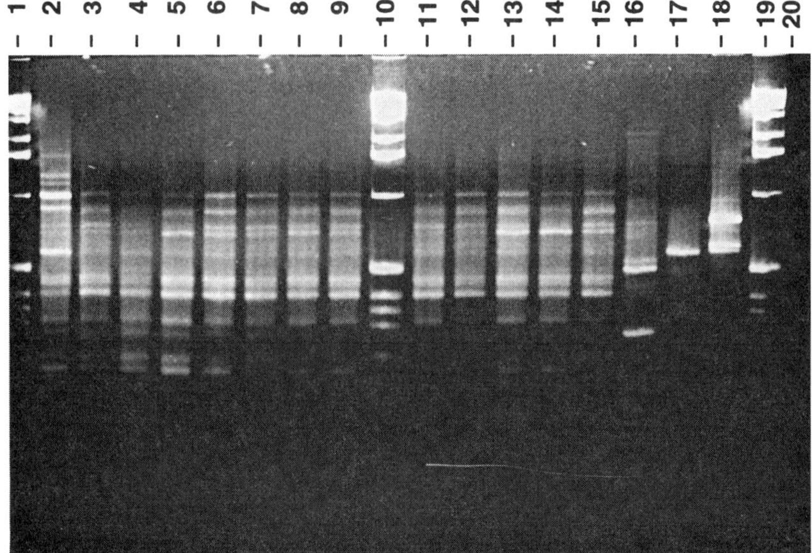

FIGURE 1. Electrophoresis gel of amplified DNA bands of *Pasteurella granulomatis* (Pg), *P. multocida* (Pm), *P. haemolytica* (Ph), and *Actinobacillus lignieresii* (AI) using T3 and T7 primers. Lanes 1, 10, 19 = 1 kb DNA standard, 2 = Pg no. 294, 3 = Pg no. 13B, 4 = Pg no. 10A, 5 = Pg no. 26, 6 = Pg no. 448, 7 = Pg no. 212/92, 8 = Pg no. 15B, 9 = Pg no. 115, 11 = Pg no. 145/91, 12 = Pg no. 162/92, 13 = Pg no. 148, 14 = Pg no. 147/95, 15 = Pg no. 352/ 94, 16 = Pm no. 2225, 17 = Ph no. P1148, 18 = AI no. KSU-94, 20 = negative control.

(10.8 g Tris base, 5.5 g boric acid, and 4.0 ml 0.5 M EDTA, pH 8.0, per liter of solution) in a 125-ml erlenmeyer flask. This was covered and heated in the microwave for 3 min at 80% power to melt the agarose. It was cooled somewhat, 4 µl ethidium bromide stock solution (10 mg/ml) was added, and the solution was swirled and poured with a comb in place. Wide minigels were poured with twice this volume, and heated for 6 min at 80% power.

A 3.0-µl aliquot of loading buffer (Maniatis type III: 0.25% bromphenol blue, 0.25% xylene cyanol, and 30% glycerol in distilled water, heated at 90°C for 15 min) was combined with either 8.5 µl reaction product or 1 µl 1 kb DNA ladder (Gibco BRL). The maximum volume possible was loaded in each well of the gel. The gel was electrophoresed using TBE buffer in a BioRad MiniSub DNA Cell or Wide Mini Sub Cell at 65 V for 1.5–1.75 h using a power supply (E-C Apparatus Corporation, Model EC-103) set for constant voltage.

PCR Procedure with Bovine Tissue

PCR was carried out on bovine tissue collected aseptically from a freshly dead adult Holstein cow submitted for a routine postmortem examination. Approximately 0.5 g of popliteal lymph node and 0.5 g of nearby subcutaneous connective tissue

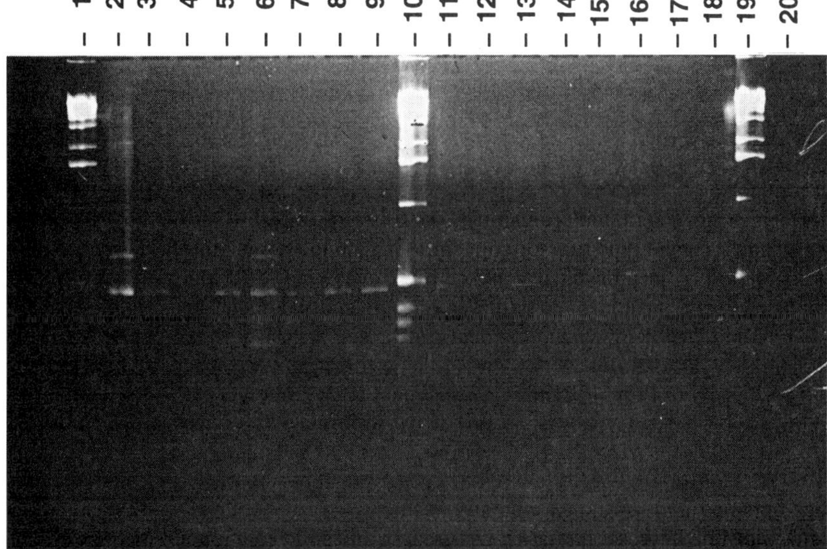

FIGURE 2. Electrophoresis gel of amplified DNA bands of *Pasteurella granulomatis* (Pg), *P. multocida,* (Pm) *P. haemolytica* (Ph), and *Actinobacillus lignieresii* (AI) using M13 and M13R primers. Lanes 1, 10, 19 = 1 Kb DNA standard, 2 = Pg no. 294, 3 = Pg no. 13B, 4 = Pg no. 10A, 5 = Pg no. 26, 6 = Pg no. 448, 7 = Pg no. 212/92, 8 = Pg no. 15B, 9 = Pg no. 115, 11 = Pg no. 145-91, 12 = Pg no. 162/92, 13 = Pg no. 148, 14 = Pg no. 147-95, 15 = Pg no. 352/94, 16 = Pm no. 2225, 17 = Ph no. P1148, 18 = AI no. KSU-94, 20 = negative control.

were macerated in 1.0 ml sterile distilled water divided into ten equal aliquots in 500-μl microcentrifuge tubes, and boiled and centrifuged as were the bacterial isolates. The supernatant was amplified with T3/T7 primers, as previously described, either alone or combined with DNA from Pg strain 448, replacing equal volumes of distilled water in the reaction mixture.

RESULTS

Amplification of bacterial DNA was successful using both sets of primers (FIGS. 1 and 2), but it was easier to evaluate the 16 bacterial isolates using the T3 and T7 primers (FIG. 1). Examination of DNA bands of all *Pasteurella granulomatis* strains indicated a high degree of similarity, with 11 common bands using the T3/T7 primers and two common bands using the M13/M13R primers. One isolate, strain 294, had three additional bands (150, 1315, and 1636 kb), compared to the other 12 Pg isolates using the T3/T7 primers, and one additional band (3100 kb) with the M13/M13R primers. Some Pg isolates were missing bands common to the other organisms (strain 10A, FIG. 1, TABLE 2, and strains 212-92, 15B, 145-91, 162-92, 148, 142-94, and 352-94, FIG. 2, TABLE 3).

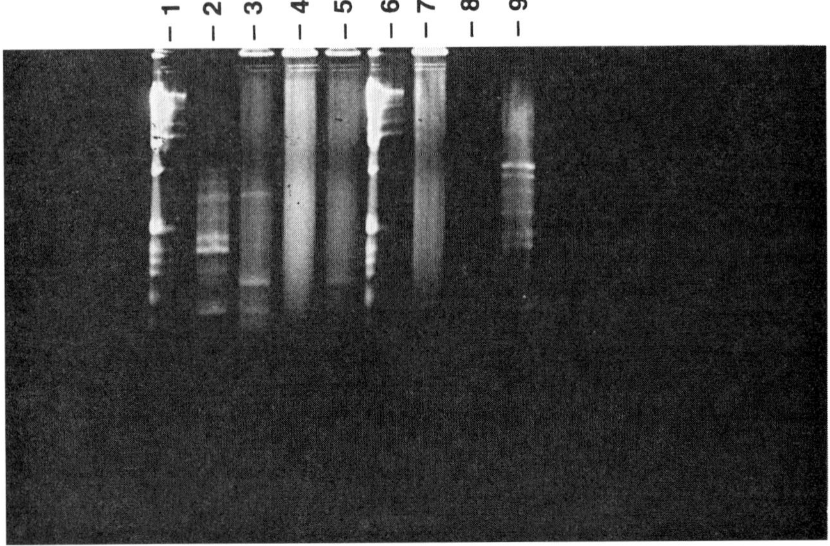

FIGURE 3. Electrophoresis gel of amplified *Pasteurella granulomatis* (Pg) DNA bands using T3 and T7 primers with or without bovine tissue supernatant. Lanes 1, 6 = 1 Kb DNA standard, 2 = Pg no. 448, 3 = Pg no. 448 and 10 µl of bovine tissue supernatant (BTS), 4 = Pg no. 448 and 30 µl of BTS, 5 = negative control with 10 µl of BTS, 7 = negative control with 30 µl of BTS, 8 = negative control, 9 = positive control (Pg no. 26).

There were strong and unique bands for each of the non-Pg strains run with the T3/T7 primer pair, which can be best appreciated in FIG. 1 and TABLE 2. The DNA bands of the non-Pg isolates were faint in the M13/M13R procedure (FIGURE 2, TABLE 3) and generally difficult to evaluate.

Amplification of Pg DNA (strain 448) with bovine tissue supernatant was only marginally successful (FIG. 3). The bovine tissue supernatant apparently reduced the band intensity and eliminated most of the Pg DNA bands. At the higher bovine tissue supernatant concentration, no Pg DNA bands were identifiable.

DISCUSSION

There was strong DNA homogeneity among the *Pasteurella granulomatis* isolates, based on the number and location of identifiable bands using both pairs of primers. Strain 294 (American Type Culture Collection, ATCC 439246) appeared to have three additional bands in the T3/T7 PCR and one additional band in the M13/M13R PCR runs. The remaining Pg isolates showed 11 similar bands (T3/T7 PCR), with some variability of band intensity (see FIGS. 1 and 2, TABLES 2 and 3). Two of these 12 isolates, ATCC 49244 (strain 26) and ATCC 49245 (strain 115), are apparently closely related, while strain number 294 (ATCC 49246) is less so, given the unique

TABLE 1. Bacterial Strains and Sources

Species	Strain	Source
Pasteurella granulomatis	10A	Isolated from *D. hominis* larva from Brazilian cow inoculated with strain 145/91.
P. granulomatis	13B	Isolated from *D. hominis* larvae from calf from Brazilian farm with no history of lechiguana.
P. granulomatis	15B	Isolated from *D. hominis* larva from calf from Brazilian farm with no history of lechiguana.
P. granulomatis	26	Isolated from field case of lechiguana in Brazil. Type strain = American type culture collection (ATCC) 49244.
P. granulomatis	115	Isolated from field case of lechiguana in Brazil = ATCC 49245.
P. granulomatis	148	Isolated from field case of lechiguana in Brazil.
P. granulomatis	294	Isolated from field case of lechiguana in Brazil = ATCC 49246.
P. granulomatis	448	Isolated from field case of lechiguana in Brazil.
P. granulomatis	145/91	Isolated from field case of lechiguana in Brazil.
P. granulomatis	162/92	Isolated from field case of lechiguana in Brazil.
P. granulomatis	212/92	Isolated from Brazilian cow with experimental lechiguana (inoculated with strain 145/91).
P. granulomatis	147/94	Isolated from field case lechiguana in Brazil.
P. granulomatis	352/94	Isolated from field case of lechiguana in Brazil.
P. multocida	P2225	Bovine isolate, serotype A:14.
P. haemolytica	P1148	Bovine isolate.
Actinobacillus lignieresii	KSU94	Bovine isolate.

bands previously mentioned. ATCC 49244 (strain 26) is designated as the type strain of the species.

Pasteurella multocida, P. haemolytica, and *Actinobacillus lignieresii* isolates were run with Pg isolates. This was done for taxonomic reasons and also to facilitate differentiation of Pg from the aforementioned species that may infect cattle. On the

TABLE 2. Amplified DNA Bands Using T3 and T7 Primers on 13 Isolates of *Pasteurella granulomatis*, and Single Isolates of *P. multocida, P. Haemolytica*, and *Actinobacillus lignieresii*[a]

	Genus, Species	Isolate Identification Strain	>200	200	220	280	350	370	400	450	506	516	530	540	750	780	820	900	920	1000	1020	1150	1315	1636	3100
1	*Pasteurella granulomatis* (Pg)	294		+	+	+		+	+	++			++		+	+		+		+	+	+	+	+	
2	Pg	13B		+	+	+		+	++	++			+		+	+		+			+				
3	Pg	10A		+	+	+		+	++	++			+		+	+		+			+				
4	Pg	26		++	++	+		++	++	++			+		+	+		+			+				
5	Pg	448		+	+	+		+	++	++			+		+	+		+			+				
6	Pg	212/92		+	+	+		+	++	++			+		+	+		+			+				
7	Pg	15B		+	+	+		+	++	++			+		+	+		+			+				
8	Pg	115		+	+	+		+	++	++			+		+	+		+			+				
9	Pg	145/91		+	+	+		+	++	++			+		+	+		+			+				
10	Pg	162/92		+	+	+		+	++	++			+		+	+		+			+				
11	Pg	148		+	+	+		+	++	++			++		+	+		+			+				
12	Pg	147/94		+	+	+		+	++	++			+		+	+		+			+				
13	Pg	352/94		+	+	+		+	++	++			+		+	+		+			+				
14	*Pasteurella multocida*	P2225				++					+	+			+	+		+	+			+			+
15	*Pasteurella haemolytica*	P1148	+				+				+		++												
16	*Actinobacillus lignieresii*	KSU-94											+	++			+			+			+		

[a] Degree of band density: blank = not detectable; + = faint; ++ = intense.

TABLE 3. Amplified DNA Bands Using M13 and M13R Primers on 13 Isolates of *Pasteurella granulomatis* and Single Isolates of *P. multocida, P. haemolytica,* and *A. lignieresii*[a]

Genus, Species	Strain	>200	200	230	285	290	340	490	520	710	730	800	900	3100
Pasteurella granulomatis (Pg)	294	+	+			+	+	++				+		+
Pg	13B	+	+			+	+	+				+		
Pg	10A	+	+			+	+	+				+		
Pg	26	+	++			+	+	+				+		
Pg	448	+	+			+	+	+				+		
Pg	212/92	+	+			+		+				+		
Pg	15B	+	+			+	+	+						
Pg	115	+	+			+	+	++				+		
Pg	145/91	+	+			+	+	+						
Pg	162/92	+						+						
Pg	148	+				+		++						
Pg	147/94	+						+						
Pg	352/94	+						+						
Pasteurella multocida	P2225			+		+					+		+	
Pasteurella haemolytica	P1148				+									
Actinobacillus lignieresii	KSU-94													

[a] Degree of band density: blank = not detectable; + = faint; ++ = intense.

basis of the most consistent and intense banding, Pg has characteristic bands at approximately 370, 400, 450, 750, 780, 900, and 1020 kb (FIG. 1, TABLE 2). Of these seven bands, four (370, 400, and 450, and 1020 kb) appear to be unique to Pg and not present in the other three species. With the M13/M13R procedure, only the 490-kb band was consistently present, easily identifiable, and unique to the Pg strains.

Based on our preliminary trials, bovine tissue supernatant (BTS) has significant inhibitory effects on the Pg DNA amplification, but this effect is not absolute (FIG. 3). Further work will be done to enhance Pg DNA amplification in the presence of BTS.

CONCLUSIONS

The PCR procedure used here, involving the T3/T7 primers, was successful in identifying the 13 Pg isolates as Pg and differentiating them from 3 other species (*Pasteurella multocida, P. haemolytica,* and *Actinobacillus lignieresii*). We conclude Pg is genetically distinct from the other 3 species.

One of the 13 Pg isolates, viz., strain 294, ATCC 49246, may be slightly genetically different from the other Pg isolates, but still shares much DNA, as denoted in 11 common bands. The other 12 Pg isolates appear genetically similar within the PCR procedures used in this study.

The presence of bovine tissue supernatant significantly inhibited Pg DNA amplification, but did not completely eliminate it. We conclude further refinements of our Pg DNA amplification procedures will be necessary for it to be sufficiently specific and sensitive to be used with bovine tissue samples.

REFERENCES

1. RIBEIRO, G. A., G. R. CARTER, W. FREDERICKSEN, & F. RIET-CORREA. 1989. *Pasteurella haemolytica*-like bacterium from a progressive granuloma of cattle in Brazil. J. Clin. Microbiol. **27**(6): 1401–1402.
2. 1990. Validation of the publication of new names and new combinations previously effectively published outside the IJSB. List No. 32. Int. J. System Bacteriol. **40**(1): 105–106.
3. RIET-CORREA, F., M. C. MÉNDEZ, A. L. SCHILD, M. C. A. MEIRELES & R. M. SCARSI. 1984. Doenças diagnosticadas no ano 1984. Laboratório Regional de Diagnóstico. Editora da Universidade, Pelotas, Brazil: 14–18.
4. RIET-CORREA, F., M. C. MÉNDEZ, A. L. SCHILD, G. A. RIBEIRO & S. M. ALMEIDA. 1992. Bovine focal proliferative fibrogranulomatous panniculitis (lechiguana) associated with *Pasteurella granulomatis*. Vet. Pathol. **29**: 93–103.
5. HENNESSY, K. J., J. J. IANDOLO & B. W. FENWICK. 1993. Serotype identification of *Actinobacillus pleuropneumoniae* by arbitrarily primed polymerase chain reaction. J. Clin. Microbiol. **31**(5): 1155–1159.

Role of *Pasteurella granulomatis* and *Dermatobia hominis* in the Etiology of Lechiguana in Cattle[a]

SÍLVIA L. LADEIRA, FRANKLIN RIET-CORREA,
DANIELA B. PEREIRA, AND GORDON R. CARTER [b]

Regional Diagnostic Laboratory
Faculty of Veterinary Medicine
University of Pelotas
96010-900, Pelotas RS Brazil

[b]*Virginia-Maryland College of Veterinary Medicine*
Virginia Tech
Blacksburg, Virginia 24061

INTRODUCTION

A new disease of cattle, defined as a focal proliferative fibrogranulomatous panniculitis, has been reported from the State of Rio Grande do Sul, southern Brazil.[1] The disease, known as lechiguana, is characterized by a large subcutaneous swelling that increases in size, and if untreated causes a progressive weakness resulting in death. Histologically, the lesion consists of focal proliferation of fibrous tissue, infiltrated by plasma cells, eosinophils, lymphocytes, and sometimes neutrophils. The primary lesion is an eosinophilic lymphangitis which results in eosinophilic abscesses which may become small granulomas with occasional rosettes in their centers.[1] More recently, lechiguana was also reported from other Brazilian states including Santa Catarina, Paraná, and Minas Gerais (unpublished data).

The name *Pasteurella granulomatis* has been given to the bacterium that has been consistently isolated from deep biopsies of lesions.[2] Because the area of anatomic distribution of lechiguana lesions is similar to that of infection by *Dermatobia hominis*, it was postulated that this insect may transmit the causative agent and produce lesions that lead to lechiguana. In attempts to reproduce the disease, a lesion similar to those observed in natural cases was reproduced in 1 of 12 cattle inoculated with *P. granulomatis*.[1]

The objectives of these experiments were (1) to determine if lechiguana can be reproduced by the inoculation of *P. granulomatis;* (2) to determine if the failure to reproduce the disease is due to loss of pathogenicity by *P. granulomatis;* (3) to determine if there is an association between *P. granulomatis* and *D. hominis* in the etiology of lechiguana; (4) to determine if the lechiguana lesion is due to the

[a] This work was funded by grants of the Conselho Nacional de Desenvolvimento Científico e Tecnológico (CNPq) and Fundação de Amparo à Pesquisa do Rio Grande do Sul (FAPERGS).

359

TABLE 1. Inoculation of *Pasteurella granulomatis* in Cattle in Experiments 1 to 12

	Inoculum		Cattle	
Experiment	Type	Amount and Route	(Number)	Age
1	Suspension of P.g[a] with oil adjuvant	2 ml SC[a]	3	>3 years
		4 ml SC	3	>3 years
2	Suspension of P.g with ground D.h[a]	2 ml SC	3	>3 years
		4 ml SC	3	>3 years
3	Suspension of P.g with gastric mucin	4 ml SC	6	>3 years
4	Suspension of P.g	2 ml SC	6	>3 years
5	BHI culture of P.g with saponin	4 ml SC	6	10 months
6	BHI culture of P.g with a previous passage in mice	5 ml SC	12[b]	>3 years
7	Suspension of P.g	3 ml SC	6[c]	>3 years
8	Suspension of P.g	4 ml IL[a]	3	10 months
9	Suspension of P.g with gastric mucin	4 ml IL	3	10 months
10	Suspension of P.g	Injection of 0.2–0.4 ml in lesions of D.h	11	>3 years
11	BHI culture of P.g	Lesions of D.h moistened with the P.g culture	6	>3 years
12	BHI culture of P.g	Lesions of cattle 9 (experimentally) infected with D.h moistened with P.g culture	9	>3 years

[a] SC = subcutaneous; IL = intralymphatic; P.g = *P. granulomatis;* D.h = *D. hominis.*
[b] These cattle had been inoculated previously in experiments 3 and 4.
[c] These cattle were immunodepressed.

multiplication of *P. granulomatis* in a concurrent inflammatory process, or to an immunologic inflammation; and (5) to determine the possibility of reproducing the disease in immunodepressed cattle.

MATERIAL AND METHODS

Inoculation of Cattle with **P. granulomatis**

In experiments 1 to 12, *P. granulomatis* (strain 145/91) in different preparations was inoculated into 68 cattle. Details of these inoculations are presented in TABLE 1.

The bacteria inoculated in experiments 1, 2, 3, 4, 7, 8, 9, and 10 were from a 24-hour culture of *P. granulomatis* on blood agar, suspended in saline solution or nutrient broth to a concentration of approximately 2.3×10^8 CFU/ml. The inoculum used in experiments 5, 6, 11, and 12 was a 24-hour brain heart infusion (BHI) broth culture of *P. granulomatis* containing approximately 1.68×10^9 CFU/ml.

In experiment 1, the inoculum was a suspension of *P. granulomatis* in saline solution (0.85% NaCl) mixed (v/v) with complete Freund's adjuvant. In experiment 2, the inoculum was prepared with ground larvae of *D. hominis,* diluted in saline solution, centrifuged, and the supernatant fluid mixed (v/v) with the saline solution of *P. granulomatis* used previously. In experiments 3 and 9, the inoculum was the same suspension of *P. granulomatis* in saline solution mixed (v/v) with a 5% mucin (porcine stomach, Sigma Chemical Company, St. Louis, MO) solution. In experiment 5, the inoculum was a solution of 1 mg of saponin mixed (v/v) with a BHI broth culture of *P. granulomatis.* In experiment 6, BHI broth cultures of *P. granulomatis* with a previous passage in mice were inoculated into 12 cattle. These cattle had been inoculated 68 days before in experiments 3 and 4. The 6 cattle used in experiment 7 were immunodepressed by the daily inoculation of dexamethasone at 0.1 mg per kg of body weight, for 5 days. Three cattle were inoculated with *P. granulomatis* on the first day of treatment with dexamethasone and the other 3 on the last day. Thirty days after the initiation of the experiment, the 6 animals were treated again with 0.1 mg of dexamethasone per kg for 5 days.

In experiments 8 and 9, *P. granulomatis* was inoculated intralymphatically. For this purpose, methylene blue (5% aqueous solution) was injected in the interdigital space of the left hind limb. Five to ten minutes later, 2 incisions at right angles were made in the external lateral face of the metatarsus, and the skin was raised for the location of a lymphatic vessel. *P. granulomatis* was inoculated into the lymphatic vessel with an endovenous infusion set (Venescalp 23G, São Paulo, SP, Brazil).

In experiment 10, a suspension of *P. granulomatis* in nutrient broth was inoculated into lesions of *D. hominis* in 11 cattle. In each animal, 2 to 5 lesions of *D. hominis* were inoculated with 0.2 to 0.5 ml of inoculum. The cattle had been treated for *D. hominis* 4 days before inoculation, and thus the larvae were dead at the time of inoculation.

Experiment 11 involved 6 cattle with severe natural infection by *D. hominis.* In each animal, 50 to 100 lesions of *D. hominis* were moistened with swabs that had been dipped in a BHI broth culture of *P. granulomatis.*

In experiment 12, each of 9 cattle were infected experimentally in the right prescapular region with 30 larvae of *D. hominis.* Before inoculation, the skin of the region had been moistened with a BHI broth culture of *P. granulomatis.* After infection with the larvae at weekly intervals for 4 weeks, swabs saturated with similar BHI broth cultures of *P. granulomatis* were applied to the larval lesions.

Five to 25 days after inoculation, needle biopsies were taken from the abscesses induced by *P. granulomatis* in 1 animal in experiment 3, 2 in experiment 4, 2 in experiment 5, 4 in experiment 6, and 1 in experiment 7, the purulent exudate obtained was examined bacteriologically.

In experiment 11, 2 surgical biopsies were obtained 49 and 55 days after inoculation from the swelling produced in one animal. The material obtained was examined bacteriologically and histologically. The biopsies were fixed, sectioned, and stained by standard procedures.[3]

TABLE 2. Isolation of *P. granulomatis* from Lesions Caused by *D. hominis* and from Larvae

Farm	Cattle	No. of Lesions and Larvae Cultured		No. of Isolates of *P. granulomatis* Obtained from:	
		Larvae	Lesions	Larvae	Lesions
1	Cow	1	1	—	—
1	Calf	12	12	4	2
2	Steer	4	4	—	—
2	Steer	3	3	—	—
2	Cow[a]	1	1	—	—
3	Steer	1	1	—	—
3	Cow	1	1	—	—

[a] Cow with a spontaneous lesion of lechiguana.

In experiment 11, 35 days after inoculation, exudate from the lesions caused by *D. hominis* in each of the cattle was collected with swabs for bacteriological examination. The hemolymph from one larva of *D. hominis* was also examined bacteriologically. For this purpose, the larva was washed in saline solution and the hemolymph was obtained with sterile syringe and needle.

Bacteriologic Study of **D. hominis** *and Lesions Caused by the Larvae in Cattle*

Larvae of *D. hominis* and exudate from lesions caused by the fly in cattle were collected from 7 cattle of 3 farms (TABLE 2) and examined bacteriologically. On farms 1 and 3, lechiguana had never been observed. On farm 2, one cow had a lechiguana lesion at the time of collection. Collection and culturing of exudate from the lesions and hemolymph from the larvae were performed by the methods previously described.

RESULTS

Cattle Inoculation

In the 36 cattle inoculated with *P. granulomatis* in experiments 1 to 5 and 7, and in the 12 cattle reinoculated in experiment 6, small abscesses were produced at the injection site. The abscesses regressed spontaneously in 15 to 60 days. *P. granulomatis* was isolated from the purulent exudate in experiments 3 to 7. Only temporary subcutaneous edema was observed in the 6 calves inoculated by the intralymphatic route in experiments 8 and 9.

FIGURE 1. Heifer inoculated in experiment 11. Note the subcutaneous mass over the withers.

No significant changes were observed in the 20 cattle with lesions of *D. hominis* that had been inoculated with *P. granulomatis* in experiments 10 and 12.

Bacteriologic examination of the lesions caused by *D. hominis* in experiment 10 yielded various bacteria but not *P. granulomatis*. *P. granulomatis* was recovered from a *D. hominis* larva obtained from one animal at the end of experiment 12.

In experiment 11, no lesions were observed in the 6 cattle until day 31 postinoculation. On day 34, a hard, rounded subcutaneous swelling was observed on the withers in one animal. This region had many *D. hominis* larvae at the time of the inoculation. Ten days after the first observation, the lesion measured 22 cm craniocaudal, 8 cm dorsoventral, and 25 cm transversal (FIG. 1); 11 days later it measured 23 cm craniocaudal, 8 cm dorsoventral, and 27 cm transversal. At this time, the animal was treated, for 5 days, with 1.5 g of chloramphenicol daily. The swelling regressed rapidly, and regression was almost complete 8 days after the start of treatment. *P. granulomatis* was isolated in pure culture from the biopsies performed 10 and 21 days after the first observation of the swelling.

The histologic appearance of the first biopsy was characterized by proliferation of loose connective tissue infiltrated by numerous eosinophils and some neutrophils and lymphocytes. Many collagenous fibers were calcified (FIG. 2). Dilated lymphatic vessels containing eosinophils, neutrophils, and unidentified large mononuclear cells were observed within the connective tissue (FIG. 3 top). Deposits of eosinophilic granules were present along lymphatic walls (FIG. 3 bottom). Microabscesses consisted of mainly eosinophils with some neutrophils, mononuclear cells, and occasionally some eosinophilic acellular debris was also present (FIG. 3 top). Rosettelike structures were observed in some microabscesses, and colonies of gram-negative bacteria, either bacilli or coccobacilli, were present in some of the rosettes. Edema and some muscle

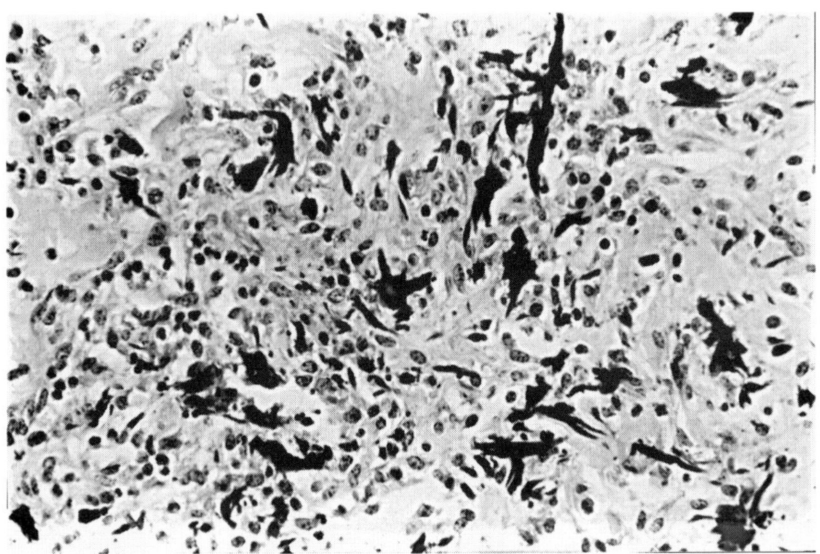

FIGURE 2. Subcutaneous mass. Heifer from experiment 11. Biopsy obtained 34 days after inoculation. The collagenous fibers are heavily calcified. HE × 200. (Reduced to 80%.)

fibers were observed within the connective tissue. The second biopsy, obtained 11 days after the first one, had a similar histologic appearance, but the connective tissue was more mature and the number of inflammatory cells within the collagenous fibers was reduced.

Bacteriologic Findings of D. hominis *Larvae and Lesions Caused by the Fly in Cattle*

P. granulomatis was isolated from one calf on farm 1, where lechiguana had never occurred (TABLE 2). Six isolates were obtained from the 12 larvae and 12 samples of exudate from this calf. On one occasion, the bacterium was isolated from both the larva and the sample of the exudate collected from the lesion caused by it. On 3 occasions, it was isolated only from larvae; and in 1, only from the exudate.

P. granulomatis was not isolated from 11 larvae and exudates obtained from other 6 cattle in the 3 farms.

DISCUSSION

Lechiguana is a disease of cattle characterized by a focal proliferative fibrogranulomatous panniculitis that has been associated with *P. granulomatis*. The role of *P. granulomatis* as the cause of lechiguana is supported by (1) its consistent isolation from deep biopsies of the lesions; (2) the recovery of all affected cattle treated with

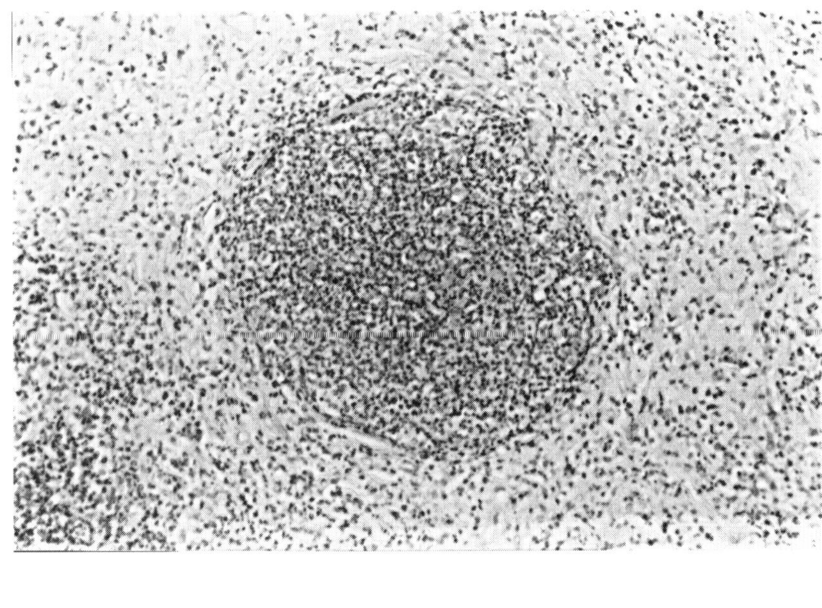

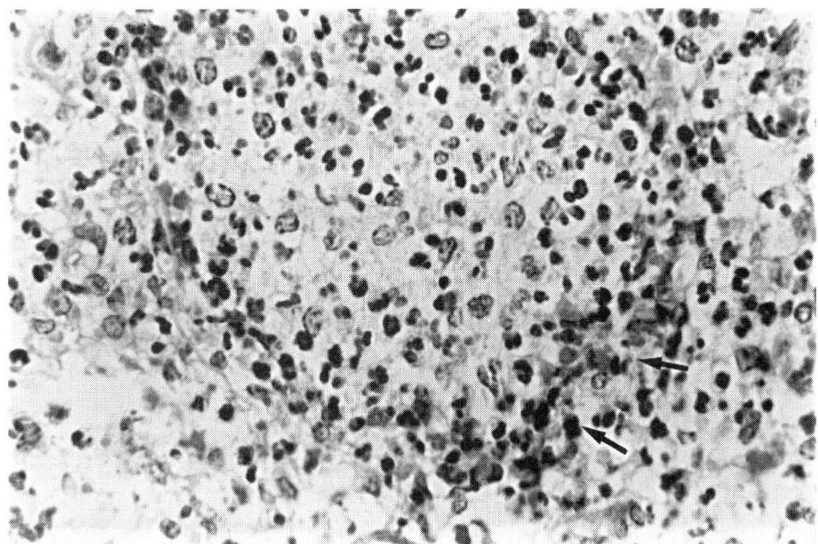

FIGURE 3. Subcutaneous mass. Heifer from experiment 11. Biopsy obtained on day 34 after inoculation. Top: Dilated lymphatic vessel containing many eosinophils and some neutrophils and unidentified large mononuclear cells. An abscess containing mainly eosinophils is formed in the lower left corner. HE × 50. Bottom: Lymphangitis with margination and degranulation of eosinophils (arrows). HE × 200. (Reduced to 80%.)

chloramphenicol and the death of untreated animals; (3) the histology of the lesions characterized by eosinophilic lymphangitis and microabscesses sometimes containing rosettes with gram-negative bacilli or coccobacilli in their centers; and (4) the experimental production of a swelling similar to lechiguana by the inoculation of *P. granulomatis* in a steer that had recovered from the disease following treatment with chloranphenicol.[1]

The attempts to reproduce the lesions characteristic of lechiguana in experiments 1 to 10 and 12 were unsuccessful. These attempts included the use of adjuvants to induce a concurrent inflammatory process (experiments 1 and 5); the repeated inoculation of the same cattle to determine if an immunologic mechanism, such as delayed hypersensitivity, was involved in the production of the lesion (experiment 6); the inoculation of cattle treated with dexamethasone to detect if an immune deficiency has some role in the pathogenesis of the disease (experiment 7). The possibility that the failure to reproduce the disease may have been due to a loss of pathogenicity of *P. granulomatis* was tested in experiments 7 and 12 by the use of mucin and in experiment 10 by the inoculation of *P. granulomatis* after the passage in mice. Except for the production of small abscesses, these efforts were negative. These results were similar to those obtained previously in 11 of 12 inoculated cattle.[1] Abscesses at the injection site were also produced by the inoculation of 2 cattle with biopsy material obtained from cases of lechiguana,[4] which seems to eliminate the possibility of a loss of pathogenicity of the bacterium or that another agent is responsible for the disease.

The intralymphatic inoculation was performed in experiments 8 and 9 because the primary lesion of lechiguana is a lymphangitis. However, this inoculation also failed to reproduce the disease. In previous experiments, inoculation by different routes including the intramuscular[1,4] and the intradermic[4] were unsuccessful. These results suggest that the route of inoculation may not be an important factor in the initiation of lechiguana.

That parasitism by *D. hominis* may have a role in the initiation of lechiguana and its subsequent transmission is suggested by (1) the anatomical distribution of the lesions observed in lechiguana, mainly in and around the scapular region, which is similar to the surface distribution of the lesions due to parasitism by *D. hominis;* (2) the geographical distribution mainly in hilly country with natural bush where *D. hominis* is frequent; and (3) the simultaneous decrease in the incidence of the parasitism by *D. hominis* and lechiguana since 1985. This reduction was apparently due to the use of pyrethroid compounds for tick control. These drugs control *D. hominis* by controlling the insects that are vectors of the eggs of the fly.

Four experiments (no. 2, 10, 11, and 12) were performed to demonstrate whether or not there was an association between *P. granulomatis* and *D. hominis* in the production of lechiguana. In experiment 2, the bacterium inoculated with ground *D. hominis* resulted in abscesses similar to those observed in experiments 1 and 3 to 7. In three experiments, *P. granulomatis* was inoculated into lesions caused by *D. hominis.* In experiments 10 and 12, the results were negative, but in experiment 11, a swelling identical to that observed in natural cases of lechiguana was produced in one heifer. Like the natural disease cases, the experimental lesion, from which *P. granulomatis* was isolated, was a hard, rapidly growing mass that displayed histologically a focal proliferative fibrogranulomatous panniculitis characteristic of lechiguana. The lesion regressed rapidly after treatment with chloramphenicol.

The experimental production of the disease in one animal with severe infection by *D. hominis* (50 to 100 larvae in 100 to 200 cm^2 of skin) and the fact that these lesions are almost always situated in the same anatomic region in lechiguana (mainly in and around the scapular region) suggest that in natural cases, the large lesions characteristic of lechiguana occur when *P. granulomatis* infects severe lesions caused by *D. hominis*. However, at least experimentally, such previous larval lesions are not strictly necessary for the development of lechiguana lesions, because in a previous experimental production of lechiguana in cattle, the bacterium was inoculated subcutaneously in an animal not infected by *D. hominis*.[1]

The isolation of *P. granulomatis* from larvae and lesions caused by *D. hominis* in one calf from a farm where lechiguana had never occurred suggests that *D. hominis* can carry and transmit *P. granulomatis*. This fact was also evidenced in experiment 12 with the isolation of *P. granulomatis* from one *D. hominis* larva.

The lechiguana lesion produced in experiment 11 in one animal was observed 34 days postinoculation. In a previous experimental production of the disease, the lesion also appeared 34 days postinoculation.[1] This apparent long incubation period, the negative results obtained in the other experiments, and also the infrequent occurrence of the natural disease suggest that lechiguana is a disease for which Koch's postulates are not easily fulfilled. Nevertheless, the experimental production of the disease and the recovery of *P. granulomatis* from *D. hominis* larvae suggest that lechiguana is caused by *P. granulomatis* and that *D. hominis* has a role in the initiation and transmission of the disease.

SUMMARY

Attempts were made to reproduce bovine lechiguana, a disease associated with *Dermatobia hominis* and *Pasteurella granulomatis* infections. Suspensions of *Pasteurella granulomatis* were mixed with each of the following: saponin, oil adjuvant, ground *Dermatobia hominis,* or 5% mucin. Each preparation was inoculated into 6 cattle. Twelve more cattle, 6 of which received dexamethasone, were inoculated with bacterial suspension alone. Abscesses but no lechiguana was produced in all 36 cattle. After abscess regression, 12 cattle were reinoculated with a suspension of mouse-passed *P. granulomatis*. Only abscesses were produced. The intralymphatic inoculation of *P. granulomatis* in 6 cattle did not produce the disease. Eleven cattle infected naturally with *D. hominis* had lesions containing dead larvae. These lesions were inoculated with *P. granulomatis*. Nine cattle were experimentally infected with larvae of *D. hominis* that had been contaminated with the bacteria. No lechiguana lesions were produced in these 20 cattle. Six cattle with severe natural *D. hominis* infection were inoculated in the larval lesions with *P. granulomatis*. One developed lesions indistinguishable from those of natural lechiguana. The lesions regressed after treatment with chloramphenicol. *D. hominis* larvae and exudate from lesions caused by the fly were collected from 7 cattle on 3 farms and examined bacteriologically. *P. granulomatis* was isolated from the larvae and the exudate of a healthy calf from a farm where lechiguana had never been observed. These results suggest that *P.*

granulomatis has a causal role in lechiguana, and that *D. hominis* may be a carrier of the bacterium. These observations suggest that lechiguana occurs when severe *D. hominis* lesions are infected with *P. granulomatis.* The apparent long incubation period, the negative results obtained in the other experiments, and also the infrequent occurrence of the natural disease suggest that lechiguana is a disease for which Koch's postulates are not easily fulfilled.

REFERENCES

1. RIET-CORREA, F., M. C. MÉNDEZ, A. L. SCHILD, G. A. RIBEIRO & S. M. ALMEIDA. 1992. Bovine focal proliferative fibrogranulomatous panniculitis (lechiguana) associated with *P. granulomatis.* Vet. Pathol. **29:** 93-103.
2. RIBEIRO, G. A., G. R. CARTER, W. FREDERIKSEN & F. RIET-CORREA. 1989. *Pasteurella haemolitica*-like bacterium from a progressive granuloma of cattle in Brazil. J. Clin. Microbiol. **27:** 1401-1402.
3. LADEIRA, S. L. 1993. Participação de *Pasteurella granulomatis* e *Dermatobia hominis* na etiologia da lechiguana (paniculite fibrogranulomatosa proliferativa) em bovinos. Ms Thesis. University of Pelotas.
4. ALMEIDA, S. M. 1986. Estudo sobre a epidemiologia, patologia e etiologia de um granuloma em bovinos conhecido como lechiguana. Ms Thesis. University of Pelotas.

Serological Survey of Leptospirosis in Livestock Animals in the Lesser Antilles[a]

P. N. LEVETT,[b,c] C. U. WHITTINGTON,[c] AND
E. CAMUS [d]

[b]*University of the West Indies*
Faculty of Medical Sciences
Barbados

[c]*Leptospira Laboratory*
Enmore no. 2, Lower Collymore Rock
St. Michael
Barbados

[d]*IEMVT-CIRAD*
Post Office Box 1232
97 184 Pointe-à-Pitre Cedex
Guadeloupe, French West Indies·

INTRODUCTION

Leptospirosis is an important disease of animals and humans. It has a worldwide distribution, but tropical climates are particularly favorable for the transmission of this zoonosis.[1] The most important mode of transmission to animals and man is contact with the urine of infected animals. The disease was diagnosed in the Caribbean shortly after the first description of its etiology, but was certainly present many years before.[2] The prevalence of leptospirosis in wild and feral animals in the Caribbean has been studied extensively.[3] Somewhat less emphasis has been placed on livestock animals, but several studies have shown widespread exposure to the infection throughout the Caribbean basin.

In the islands of the Eastern Caribbean, the seroprevalence in cattle ranged from 25-55%.[3] In Guadeloupe, a large survey of cattle revealed exposure to several serogroups.[4] However, economically important serovars apparently are not distributed widely through the Caribbean islands, and long-term shedding of leptospires in urine of animals appears to be uncommon.[3]

Studies on sheep and goats in the Lesser Antilles were reviewed recently.[3] Autumnalis was the most common serogroup detected, with Ballum, Pyrogenes, and Icterohaemorrhagiae also detected. The seroprevalence ranged from 17-65% in sheep and from 19-63% in goats. Several epizootics were detected with evidence of infection

[a] Supported in part by a grant from the Campus Research Fund of the University of the West Indies Cave Hill Campus.

TABLE 1. Antigens Used in the Microscopic Agglutination Test

Serogroup	Serovar
Ballum	*ballum*
Ballum	*arborea*
Canicola	*canicola*
Pyrogenes	*pyrogenes*
Icterohaemorrhagiae	*copenhageni*
Bataviae	*bataviae*
Bataviae	*brasiliensis*
Grippotyphosa	*grippotyphosa*
Autumnalis	*fortbragg*
Autumnalis	*bim*
Pomona	*pomona*
Sejroe	*hardjo*
Sejroe	*sejroe*
Mini	*georgia*
Tarassovi	*tarassovi*
Panama	*panama*
Panama	*mangus*
Australis	*bratislava*
Australis	*bajan*
Australis	*barbadensis*
Grippotyphosa	*canalzonae*
L. biflexa Samaranga	*patoc*

in almost all animals within the affected flock or herd. More recently, in St. Croix, 26% of goats and 32% of sheep were reported to be seropositive.[5] Antibodies to serogroup Autumnalis were found in approximately 30% of seropositive animals.

This study was performed in order to determine the current seroprevalence of leptospirosis in ruminant livestock animals in the islands of the Lesser Antilles.

METHODS

Animals were studied from the islands of St. Maarten, St. Kitts, Nevis, Antigua, Montserrat, Guadeloupe, Dominica, Martinique, St. Lucia, St. Vincent, Grenada, Carriacou, and Barbados. One percent of the herds in each municipality or parish were selected randomly, and all animals within selected herds were sampled. Sera were separated and stored at −20°C until tested.

Sera were tested for leptospiral agglutinins by the standard microscopic agglutination test (MAT),[6] using the panel of 22 antigens shown in TABLE 1. Multiple serovars of some serogroups were included to represent locally prevalent antigens. Each serum sample was initially screened for agglutinins at dilutions of 1:50 and 1:100, and any sera reacting at either dilution were then titrated by doubling dilutions from 1:50 to 1:25,600. Samples were recorded as positive if they reacted at a titer of ≥100. The antigen against which the highest titer was detected was recorded as the predominant serogroup. When the highest titer was recorded against more than one serogroup, this was recorded as a mixed equal titer.

RESULTS

A total of 1788 animals were studied, including 767 cattle, 442 sheep, and 579 goats. Previous exposure to leptospirosis (at a titer of ≥ 100) was evident in 101 (5.6%) animals (TABLE 2). The seroprevalence was highest in cattle and goats (7.2%) and lowest in sheep (1.7%). Thirteen animals (0.7%) exhibited titers of ≥ 800, consistent with acute infection. These included 11 cattle on Martinique and single sheep on St. Vincent and Dominica.

The distribution of antibodies was uneven; there was no evidence of leptospirosis on the islands of St. Maarten, Nevis, Montserrat, and Carriacou, and only a few animals with a titer of 50 were detected from Antigua. The highest seroprevalence among cattle was found on Martinique, where 20% of animals had titers of ≥ 100, and 6% showed evidence of recent infection. The highest seroprevalence in both goats and sheep was detected on St. Vincent, where 23% of goats and 22% of sheep had titers of ≥ 100.

The predominant serogroups in cattle were Sejroe, Autumnalis, Icterohaemorrhagiae, and Cynopteri (TABLE 3). In goats Autumnalis, Grippotyphosa, Cynopteri, and Icterohaemorrhagiae were the most common serogroups (TABLE 4), while in sheep, Grippotyphosa and Cynopteri were equally common, followed by Icterohaemorrhagiae and Autumnalis (TABLE 5).

DISCUSSION

Evidence of previous exposure to leptospirosis was found in 5.6% of cattle, goats, and sheep sampled from 13 islands of the Lesser Antilles. Earlier studies recorded seroprevalences ranging from 25-55% in cattle, and from 20-65% in goats and sheep.[3,7] The sampling procedure used in the present study may have underestimated the seroprevalence in some islands. In addition, the animals studied were mostly young (less than one year old) and this also may have contributed to the lower seroprevalence, since older animals are likely to have been exposed more to leptospirosis and therefore to have more evidence of exposure. However, similarly low seroprevalences, in both goats (8.9%) and sheep (5.6%), have been reported previously from Guyana.[8]

Against the background of a relatively low seroprevalence, the results from animals on two islands stand out. On Martinique, 20% of cattle were seropositive. Almost half of these cattle showed evidence of exposure to serogroup Sejroe, and at least 6% of cattle showed evidence of recent infection with Sejroe. This serogroup is uncommon in other animals in the Caribbean,[3] and it is probable that serogroup Sejroe was introduced to Martinique with the cattle stock, which is mostly Brahman, imported from the US. In the absence of leptospiral isolates from these cattle, it is impossible to determine the infecting serovar, but cattle are recognized as maintenance

TABLE 2. Seroprevalence and Range of Titers in Livestock Animals in the Lesser Antilles

		Number of Animals	MAT Antibody Titers								Number of Animals Seropositive at Titer	
			Negative	50	100	200	400	800	1600	3200	>50	>100
St. Maarten	Bovine	22	22	0	0	0	0	0	0	0	0	0
	Ovine	8	8	0	0	0	0	0	0	0	0	0
	Caprine	8	8	0	0	0	0	0	0	0	0	0
Antigua	Ovine	117	114	3	0	0	0	0	0	0	3	0
St. Kitts	Bovine	21	20	0	1	0	0	0	0	0	1	1
	Ovine	62	61	0	1	0	0	0	0	0	1	1
	Caprine	21	16	1	3	1	0	0	0	0	5	4
Nevis	Bovine	17	17	0	0	0	0	0	0	0	0	0
	Ovine	35	35	0	0	0	0	0	0	0	0	0
Montserrat	Bovine	61	61	0	0	0	0	0	0	0	0	0
	Ovine	27	27	0	0	0	0	0	0	0	0	0
	Caprine	47	47	0	0	0	0	0	0	0	0	0
Guadeloupe	Caprine	203	190	8	5	0	0	0	0	0	13	5
Dominica	Bovine	98	92	3	2	1	0	0	0	0	6	3
	Ovine	68	65	0	1	1	0	0	0	1	3	3
	Caprine	52	52	0	0	0	0	0	0	0	0	0
Martinique	Bovine	175	130	10	10	8	6	6	5	0	45	35
St. Lucia	Bovine	181	174	1	5	0	1	0	0	0	7	6
	Caprine	115	112	0	3	0	0	0	0	0	3	3

St. Vincent	Bovine	155	147	0	6	2	0	0	0	0	8	8
	Ovine	31	22	2	4	1	1	1	0	0	9	7
	Caprine	48	35	2	3	6	2	0	0	0	13	11
Grenada	Bovine	37	32	3	1	0	1	0	0	0	5	2
	Ovine	71	65	3	3	0	0	0	0	0	6	3
	Caprine	36	29	1	2	2	2	0	0	0	7	6
Carriacou	Caprine	6	6	0	0	0	0	0	0	0	0	0
Barbados	Ovine	23	22	1	0	0	0	0	0	0	1	0
	Caprine	43	39	1	2	1	0	0	0	0	4	3
Total		1788	1648	39	52	23	13	7	5	1	140	101
Percent		100.0	92.2	2.2	2.9	1.3	0.7	0.4	0.3	0.1	7.8	5.6

TABLE 3. Serogroup Predominance in Cattle in the Lesser Antilles[a]

| | | Number of Animals | Serogroup | | | | | | | | | | | | | Number of Mixed Reactors |
			Icterohaemorrhagiae	Autumnalis	Pyrogenes	Bataviae	Grippotyphosa	Canicola	Pomona	Sejroe	Cynopteri	Tarassovi	Ballum	Panama	
St. Kitts	Bovine	20	0	0	0	0	0	0	0	0	1	0	0	0	0
Dominica	Bovine	98	0	5	0	0	0	0	0	1	1	0	0	0	1
Martinique	Bovine	175	11	8	0	3	0	0	0	20	1	1	0	1	0
St. Lucia	Bovine	181	2	1	0	0	0	0	0	2	2	0	0	0	0
St. Vincent	Bovine	155	1	1	0	0	0	0	0	0	7	0	0	0	1
Grenada	Bovine	37	0	0	2	0	0	0	1	2	0	0	0	0	0
Total		666	14	15	2	3	0	0	1	25	12	1	0	1	2
Percent Prevalence		100.0	2.1	2.3	0.3	0.5	0.0	0.0	0.2	3.8	1.8	0.2	0.0	0.2	0.3

[a] Data shown only for islands with evidence of leptospirosis.

TABLE 4. Serogroup Predominance in Goats in the Lesser Antilles[a]

| | | Number of Animals | Serogroup | | | | | | | | | | | | Number of Mixed Reactors |
			Icterohaemorrhagiae	Autumnalis	Pyrogenes	Bataviae	Grippotyphosa	Canicola	Pomona	Sejroe	Cynopteri	Tarassovi	Ballum	Panama	
St. Kitts	Caprine	21	0	2	0	0	3	0	0	0	0	0	1	0	0
Guadeloupe	Caprine	203	0	3	0	0	0	0	0	1	2	0	0	0	1
St. Lucia	Caprine	115	0	3	0	0	0	0	0	0	0	0	0	0	0
St. Vincent	Caprine	48	5	2	0	1	8	1	1	0	0	0	0	0	4
Grenada	Caprine	36	0	5	0	0	0	0	0	0	2	0	0	0	0
Barbados	Caprine	43	0	0	0	0	0	0	0	0	3	0	0	0	0
Total		466	5	15	0	1	11	1	1	1	7	0	1	0	5
Percent Prevalence		100.0	1.1	3.2	0.0	0.2	2.4	0.2	0.2	0.2	1.5	0.0	0.2	0.0	1.1

[a] Data shown only for islands with evidence of leptospirosis.

TABLE 5. Serogroup Predominance in Sheep in the Lesser Antilles[a]

		Number of Animals	Icterohaemorrhagiae	Autumnalis	Pyrogenes	Bataviae	Grippotyphosa	Canicola	Pomona	Sejroe	Cynopteri	Tarassovi	Ballum	Panama	Number of Mixed Reactors
Antigua	Ovine	117	0	1	0	0	0	0	0	0	2	0	0	0	0
St. Kitts	Ovine	62	0	0	0	0	0	0	0	0	1	0	0	0	0
Dominica	Ovine	68	1	1	0	0	0	0	0	1	0	0	0	0	0
St. Vincent	Ovine	31	3	0	0	0	6	0	0	0	0	0	0	0	0
Grenada	Ovine	71	0	1	1	2	0	0	2	0	2	0	0	0	2
Barbados	Ovine	23	0	0	0	0	0	0	0	0	1	0	0	0	0
Total		372	4	3	1	2	6	0	2	1	6	0	0	0	2
Percent Prevalence		100.0	1.1	0.8	0.3	0.5	1.6	0.0	0.5	0.3	1.6	0.0	0.0	0.0	0.5

[a] Data shown only for islands with evidence of leptospirosis.

hosts of serovar *hardjo*. Other serogroups to which cattle in Martinique was exposed included Autumnalis and Icterohaemorrhagiae, both of which are commonly isolated from rats and mongooses throughout the Caribbean.

Antibodies to the serogroups Autumnalis, Icterohaemorrhagiae, and Cynopteri were distributed widely in sheep and goats on the islands studied. In St. Vincent, seroprevalence reached 23% in goats and 22% in sheep, the predominant serogroup in both animals being Grippotyphosa. Grippotyphosa is a common infecting serogroup in sheep and goats,[1] but in this study evidence of infection was almost confined to St. Vincent.

This study has demonstrated the continuing endemicity of leptospirosis in livestock animals on these islands, albeit at a lower prevalence than previously reported. The serogroups detected are those expected, based on previous studies. However, the high seroprevalence of serogroup Sejroe among cattle on Martinique requires further investigation, because of the economic and veterinary significance of this serogroup elsewhere.

SUMMARY

A serological survey was performed of 1788 cattle, goats and sheep on 13 islands in the Lesser Antilles. Sera were tested by microscopic agglutination (MAT) using a panel of 22 live antigens. Evidence of past exposure, at a titer of ≥ 100, was found in 101 animals (5.6%). Antibodies were more common in cattle and goats (7.2% in each) than in sheep (1.7%). Seroprevalence was highest in cattle in Martinique (20%) and in goats in St. Vincent (23%). The predominant serogroups were Sejroe (largely confined to cattle in Martinique), Autumnalis, Icterohaemorrhagiae, Grippotyphosa, and Cynopteri. Eleven cattle from Martinique and 2 sheep with titers of ≥ 800 showed evidence of recent infection.

REFERENCES

1. FAINE, S. 1994. *Leptospira* and leptospirosis. CRC Press. Boca Raton, FL.
2. BAYLEY, H. H. 1939. An investigation of the infectious jaundice of Barbados. Carib. Med. J. **1:** 135–142.
3. EVERARD, J. D. & C. O. R. EVERARD. 1993. Leptospirosis in the Caribbean. Rev. Med. Microbiol. **4:** 114–122.
4. TISSOT, D., M. MAILLOUX & Y. L. CORROLLER. 1975. Enquête sérologique sur les leptospiroses bovines en Guadeloupe. Bull. Soc. Path. Exot. **4:** 420–425.
5. AHL, A. S., D. A. MILLER & P. C. BARTLETT. 1992. *Leptospira* serology in small ruminants on St. Croix, U.S. Virgin Islands. Ann. N.Y. Acad. Sci. **653:** 168–171.
6. FAINE, S. 1982. Guidelines for the control of leptospirosis. Offset publication No. 67. World Health Organization. Geneva, Switzerland.
7. EVERARD, C. O. R., J. C. ROACH, Y. SHI MING & D. G. CARRINGTON. 1990. Leptospirosis in sheep and goats in the Eastern Caribbean. Unpublished Report.
8. MOTIE, A. & D. M. MYERS. 1986. Leptospirosis in sheep and goats in Guyana. Trop. Anim. Hlth. Prod. **18:** 113–114.

Pore-Forming Activity of *Coxiella burnetii* Outer Membrane Protein Oligomer Comprised of 29.5- and 31-kDa Polypeptides

Inhibition of Porin Activity by Monoclonal Antibodies 4E8 and 4D6

NIRUPAMA BANERJEE-BHATNAGAR,[a,b]
CHRISTOPHER R. BOLT,[a] AND JIM C. WILLIAMS [a,c,d]

[a]*Bacteriology Division*
Pathogenesis and Immunology Branch
United States Army Medical Research Institute of Infectious
Diseases
Fort Detrick
Frederick, Maryland 21702

[b]*Division of Bacterial Products*
[c]*Division of Vaccines and Related Products Application*
Office of Vaccine Research and Review
Food and Drug Administration
Rockville, Maryland 20582-1448

Coxiella burnetii, the agent of Q fever, is a gram-variable, spore-forming bacterium that survives and multiplies within phagolysosomes of various nucleated cells.[29] Like many other bacteria, cell-surface components of *C. burnetii* are believed to play an active role in pathogenesis. A major function of the bacterial outer membrane is to control the passage of small, hydrophilic molecules between the environment and the periplasmic space.[2] This nonspecific permeability is due to the presence of pore-forming proteins (porin) that produce channels across the outer membrane. The ability of *C. burnetii* to transport small molecules under acidic conditions has been proposed as a pathogenic mechanism for the regulation of microbial metabolism by H^+-dependent stimulation of cell function.[6] The ability of *C. burnetii* to carry out transport functions *in vitro* under conditions that approximate the pH of the phagolysosome provided impetus for testing the pore-forming activity of outer membrane proteins (OMP).

The results of several studies[28,30,31] indicate that the dominant OMP of *C. burnetii* have apparent molecular masses ranging from 29 to 31 kDa and that these proteins

[d] Author to whom correspondence should be addressed at: Connaught Laboratories, Inc., Regulatory Affairs Department, Route 611, P.O. Box 187, Swiftwater, PA 18370.

are major immunogens. Studies of the exposure of *C. burnetii* proteins on the bacterial surface by using radioiodination and immunoprecipitation with murine polyclonal antibodies indicate that the primary immunogens range from 29 to 31 kDa.[28] An immunoblot analysis of murine polyclonal antibodies reactive with antigenic determinants of *C. burnetii* cells also revealed immunodominant protein species form 29 to 31 kDa.[30] Serum antibodies from humans infected with *C. burnetii* also react in immunoprecipitation and immunoblot assays with surface protein antigens ranging from 28 to 31 kDa.[31] The induction of humoral immune responses to the 29- to 31-kDa OMP suggests that these proteins may be important targets for vaccine and diagnostic reagent development.

In an attempt to understand the role of *C. burnetii* immunogenic cell-surface proteins in pathogenesis, Snyder and Williams[19,20] purified a 29.5-kDa OMP by solubilizing envelope proteins in the presence of dithiothreitol (DTT) and/or 2-mercaptoethanol (2-ME), followed by ion-exchange chromatography and preparative sodium dodecyl sulfate-polyacrylamide gel electrophoresis (SDS-PAGE). Murine mAbs 4E8 and 4D6 were produced by immunizing mice with the excised region of SDS-polyacrylamide gels containing the 29.5-kDa protein.[21] In recent studies, *C. burnetii* large-cell variants (LCV) were used to prepare cell envelopes containing the 29.5- to 31-kDa OMP species recognized by mAbs 4E8 and 4D6.[15] The small cell variants lacked these two proteins when examined by immunoelectron microscopy, and assayed by SDS-PAGE, and immunoblot.[15] From such studies, McCaul *et al.*[15] concluded that the 29.5- and 31-kDa proteins are components of the outer membrane of only vegetative cells that initiate outgrowth and elongation. These observations led us to ask whether the 29.5- and 31-kDa OMPs function as porin at pH 7 and 4.5, and, if so, do polyclonal antibodies and mAb to these proteins inhibit porin function? In this study, we purified and characterized the pore-forming activity of a *C. burnetii* OMP oligomer.

MATERIALS AND METHODS

Microorganism and Growth Conditions

C. burnetii (Nine Mile clone 7, phase I) were grown in yolk sacs of specific-pathogen-free fertile hen's eggs (Truslow Farms, Inc., Chestertown, MD). The infected yolk sacs were harvested, and the microorganisms separated from host components through a series of differential centrifugation, filtration, and finally by Renografin-gradient centrifugation steps.[33] Baby hamster kidney (BHK) cells were infected with *C. burnetii* and cultured as previously described.[34] Microorganisms from infected BHK cells were released either by the infectious process or mechanically by scraping the monolayer of infected cells from the culture flask. *C. burnetii* were harvested from the cell culture fluid and separated from host cell components by Renografin-gradient centrifugation.

Preparation of Large-Cell Variant (LCV) Envelopes

Separating *C. burnetii* from host cell components yields a mixture of morphological cell types that vary in size and ultrastructure. The pleomorphic cells also vary in their sensitivity to lysis after suspension in water and from the shear forces generated by a French pressure cell. The LCV were used to prepare cell envelopes as previously described.[15] This preparation contained the outer and the inner (cytoplasmic) membranes, the PG layer, and the 29.5- to 31-kDa OMP species recognized by mAbs 4E8 and 4D6.

Preparation of PG-Associated Porin

C. burnetii aggregate OMP oligomer with pore-forming activity was purified as described by Nakae *et al.*[16] Briefly the 29.5- to 31-kDa OMP monomers were specifically detected in an immunoblot assay with mAbs 4E8 and 4D6.[15] LCV envelopes of *C. burnetii* containing 35 mg of protein, determined by the method of Lowry *et al.*,[13] were suspended by vortex mixing for 30 s and sonication (Bransonic 12 Ultrasonic Cleaner, Branson Cleaning Equipment Co., Shelton, CN.) for 15 s in 5 ml of buffer A [100 mM Tris HCl, pH 7.4; 10 mM ethylenediaminetetraacetic acid (EDTA); 1% SDS; 1% deoxycholate (DOC); 5 mM phenylmethylsulfonyl fluoride (PMSF); and 0.02% NaN_3]. The preparation was shaken in a 37°C water bath on a reciprocating (200 rpm) platform. After incubating for 1 h, particulates were separated from soluble components by centrifugation at 4°C for 1 h at $100,000 \times g$. Protein concentration in the supernatant was assayed. The protein composition was analyzed by SDS-PAGE and Coomassie brilliant blue R250 staining (see below). The pellet containing the PG-protein complex (PG-PC) with covalently linked and noncovalently associated proteins was washed twice with sterile water. The PG was digested with 600 μg of lysozyme (Sigma Chemical Co., St. Louis, MO) in 2 ml of 5 mM Tris-HCl buffer (pH 8) at 37°C for 2 h with periodic 15 s sonication. The suspension was centrifuged at 4°C for 1 h at $100,000 \times g$, and the supernatant was discarded. Pellets were treated two more times with lysozyme and washed twice.

Extraction and Purification of OMP Oligomer

Proteins associated with the above pellet were extracted in buffer B (20 mM Tris-HCl buffer pH 7.4, containing 0.1% SDS, 0.1% DOC, 0.1% Triton X-100, NaN_3 0.02%, and 5 mM PMSF) at room temperature for 30 min. The suspension was centrifuged at 4°C and $100,000 \times g$ for 60 min, and protein in the supernatant was assayed and its composition analyzed by SDS-PAGE and Coomassie brilliant blue R250 staining (see below). The pellet containing 16 mg of protein was suspended in 2 ml of buffer A and held overnight at 4°C.

A portion of the above solution containing 5 mg of protein was loaded onto a Sepharose 4B column (2.5 cm × 55 cm bed volume) which was equilibrated with buffer A. The elution profile, monitored at 280 nm, indicated that components eluted near the void volume in two separate peaks (FIG. 1). Peaks 1 (P1) and 2 (P2) were collected as 2 ml/tube fractions. P1 contained fractions 46 to 72 (54 ml), and P2 contained fractions 75 to 100 (52 ml). Proteins eluting from the column in P1 and P2 were precipitated by adding three volumes of ethanol at 4°C to collect the proteins and to remove associated detergents. Alternatively, detergents were removed from proteins by dialysis for 5 days at 4°C against water containing 3-mM NaN_3 with

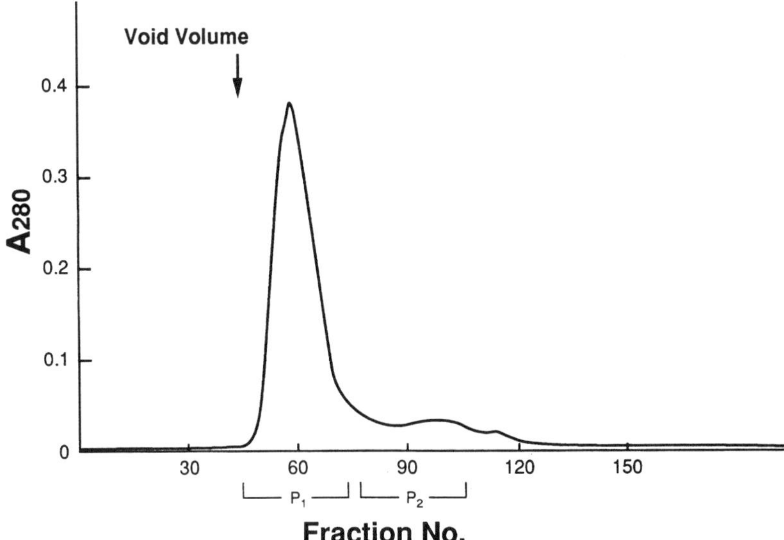

FIGURE 1. Elution profile of *C. burnetii* OMP aggregate from a Sepharose 4B column. Cell envelopes after detergent extraction and lysozyme treatment were solubilized in 100 mM Tris-HCl buffer, pH 7.4, containing 1% SDS and 1% DOC, and applied to a Sepharose 4B column. Peak 1 (P1) was the major peak near the void volume. P2 was a minor peak which trailed P1. P1 (54 ml) and P2 (52 ml) were analyzed for pore-forming activity and composition by SDS-PAGE and immunoblotting. Elution of proteins in or near the void volume of the Sepharose 4B column represents a molecular mass of approximately 2×10^4 kDa.

several changes of water. Detergent removal by either method rendered the protein insoluble, which appeared as a white precipitate. The protein suspensions from P1 and P2 were frozen in a dry ice–ethanol bath, and the protein was concentrated under vacuum. After thawing at room temperature, P1 and P2 were stored as a suspension in water at 4°C until used.

SDS-PAGE and Immunoblotting

SDS-PAGE was performed essentially as described by Laemmli[12] with the following modifications. Two-millimolar EDTA was added to the separating gel and 100-mM DTT was included in the sample preparation buffer. After electrophoresis, the proteins in the polyacrylamide gel were stained with Coomassie brilliant blue R250 (Sigma Chemical Co.). The molecular mass markers (Bio-Rad Laboratories, Inc., Melville, NY) were phosphorylase b, 97.4 kDa; bovine albumin, 66.2 kDa; ovalbumin, 42.7 kDa; carbonic anhydrase, 31 kDa; trypsin inhibitor, 21.5 kDa; and lysozyme, 14.4 kDa.

Immunoblotting was carried out according to the method of Swezyl and Kozloff[22] with modifications. After SDS-PAGE, the proteins were electrophoretically transferred at 500 mA overnight onto a nitrocellulose membrane in 25-mM Tris and 192-mM Tricine buffer, adjusted to pH 10.4 with NaOH. The nitrocellulose paper was washed with Tris-buffered saline (TBS) as described by Towbin et al.[26] and blocked with 3% (w/v) gelatin (60 bloom) in TBS for 30 min. mAbs were diluted and suspended in 1% gelatin (60 bloom) solution in TBS. The nitrocellulose blot was incubated in the diluted mAbs for 2 h at room temperature with gentle shaking, washed three times with TBS containing 0.05% (v/v) Tween 20, and incubated with ^{125}I-protein A (New England Nuclear, Albany, NY) for 2 h. The blots were washed three times with TBS containing 0.05% Tween 20, air dried, and exposed to XR film. The ^{14}C-labeled molecular mass markers (Bethesda Research Laboratories, Gaithersburg, MD) were phosphorylase b, 97.4 kDa; bovine serum albumin, 68 kDa; ovalbumin, 43 kDa; carbonic anhydrase, 29 kDa; beta-lactoglobulin, 18.4 kDa; and lysozyme, 14.3 kDa.

Liposome-Swelling Assay for Porin Activity

The permeation rate for various solutes was determined from the swelling rate of proteoliposomes minus the swelling rate of control liposomes as described by Luckey and Nikaido[14] and modified by Takada et al.[23] Egg yolk phosphatidylcholine (2.6 μmol; Sigma Chemical Co.) and dicetylphosphate (0.1 μmol; Sigma Chemical Co.) were mixed, sonicated for 60 s, and dried in vacuo to form a film at the bottom of a conical test tube.

Control liposomes were prepared like test mixtures, but without protein. An amount for each experiment of the OMP aggregate was added to prepare proteoliposomes. The suspension was first mixed by shaking and then by brief sonication. This material was dried under a stream of nitrogen at 45°C until no liquid was visible and held in vacuo for 1 h in a desiccator. Three-tenths milliliter of 15% (w/v) dextran T-40 (Sigma Chemical Co.) in 5 mM Tris-HCl buffer (pH 7.4 or pH 4.5) was added and mixed by vortex at room temperature. Thirty microliters of control liposomes or proteoliposomes and 0.7 ml of a 30 mM solution of test saccharides (Sigma Chemical Co.)—L-arabinose (150.1 M_r), D-glucose (180.2 M_r), sucrose (342.3 M_r), and stachyose (666.6 M_r)—in 5 mM Tris-HCl buffer (pH 7.4 or pH 4.5) were added to a separate semimicrocuvette. The penetration of solute into the vesicles produced liposome-swelling, which was measured as a decrease in the absorbance at A_{400} nm. The permeation rate of the solute was determined from the difference between the initial value at $t = 10$ s and each value thereafter periodically for 120 s. The results for control liposomes and proteoliposomes are reported graphically as liposome swelling, which reflects the change in absorbance from the initial value during 2 min. In a control experiment, an aliquot of the porin fraction was heated in the presence of SDS at 100°C for 10 min to inactivate the porin. After removing the SDS by ethanol precipitation, we used the proteins in the liposome-swelling assay.

Equilibrium Method

Porin activity was also measured by the equilibrium method, as described by Nikaido.[17] One micromole of egg phosphatidyl choline (Sigma Chemical Co.) in chloroform was added to a 15-ml glass, graduated, conical centrifuge tube; the solvent was removed under a stream of N_2, and the tube was placed in a vacuum desiccator for 20 min. The dried film of phospholipid was suspended in 0.2 ml of aqueous solution containing 0.15 μmol *Escherichia coli* lipopolysaccharide (LPS) (Serotype N00127 : B8, Sigma Chemical Co.) and variable amounts of OMP aggregate protein. The mixture was sonicated for 10 to 20 s, dried at 45°C under N_2, and kept in a desiccator at room temperature for approximately 3 h. The contents of the tube were suspended in 0.2 ml of aqueous solution containing 100 mM NaCl, 10 mM $MgCl_2$, 10 mM *N*-[2-hydroxyethyl]piperazine-*N'*-[2-ethanesulfonic acid] (HEPES, pH 7.4; Sigma Chemical Co.), 100 μg/ml chloramphenicol (Sigma Chemical Co.), 1.2×10^6 dpm dextran-carboxyl ([carboxyl-^{14}C]-, $5 - 7 \times 10^4$ mol. wt., DuPont Co., Wilmington, Del.), and 3.4×10^5 dpm ^{3}H-sucrose ([6,6' (n) − ^{3}H]sucrose, Amersham, Arlington Heights, IL.). The mixture was kept at 37°C for 30 min and cooled overnight at room temperature. The samples were applied to a Sepharose 4B column (1.27 cm × 16 cm) and eluted with the above buffer. One-milliliter (1-ml) fractions were collected through the void volume, which contained the radioactive liposomes. One hundred microliters from each fraction of liposomes was added to 5 ml of Hydrofluor™ (National Diagnostics, Manville, NJ), and the dpm counted in an LS 5801 liquid scintillation counter (Beckman Instruments, Inc., Palo Alto, CA). A normalized ratio was calculated by dividing the ^{14}C-dextran dpm/^{3}H-sucrose dpm ratio in the void volume by the ^{14}C-dextran dpm/^{3}H-sucrose dpm ratio in the liposome mixture applied to the Sepharose 4B column.

Inhibition of ^{14}C-Glucose Uptake by C. burnetii *with Monoclonal and Polyclonal Antibodies*

C. burnetii released from infected BHK cells was harvested by centrifuging the fluids at 4°C and 9,820 × *g* for 20 min. The residual BHK cells in the pellet were lysed by suspending them in buffer C [50-mM potassium phosphate buffer (pH 7) containing 152.5-mM KCl, 15-mM NaCl, and 100-mM glycine] and blending them in an Omni-Mixer™ 17150 (Sorvall, Inc., Newtown, CN) at top speed for 30 s. The suspension was centrifuged at 4°C and 9,820 × *g* for 20 min. The cell pellet was suspended in P-25 buffer (50-mM potassium phosphate, 152.5-mM KCl, 15-mM NaCl, 100-mM glycine) at the desired pH and adjusted to an optical density, in Klett units, equivalent to 1 mg/ml dry weight of cells (1 mg = 3.8×10^{10} microorganisms).[33] The transport medium (buffer D) contained 40-mM potassium phosphate, 122-mM KCl, 12-mM NaCl, 80-mM glycine, and 25-μM uniformly labeled ^{14}C-glucose (160 mCi/mmol, Amersham Corp.). Uptake of ^{14}C-glucose by *C. burnetii* cells was measured as described by Hackstadt and Williams.[6]

The following antibodies were heat treated at 56°C for 30 min and diluted in p-25 buffer containing 1% bovine serum albumin (BSA): a 1:100 dilution of ascites fluid containing murine mAbs to a 29.5-kDa OMP purified from *C. burnetii*;[20] or a 1:500 dilution of mouse hyperimmune (HI) serum against *C. burnetii* (HICB);[30] or a 1:100 dilution of mouse HI serum against *R. typhi* (HIRT)[5] without detectable antibodies to *C. burnetii,* which was used as a control serum. The diluted antibodies were added to the cell suspension, gently mixed, and were incubated at room temperature for 2 h. Unbound serum components were removed by centrifuging at 4°C and 12,400 × g (Microfugle™ 11, Beckman Instruments, Inc., Palo Alto, CA) for 10 min. The pellet was washed twice with P-25 buffer at pH 7 and finally suspended in 0.2 ml of P-25 buffer (pH 4.5). The uptake of ^{14}C-glucose by 1 mg (dry weight)[6] of *C. burnetii* was carried out in 1 ml of buffer D. The suspension of microorganisms was thermoequilibrated at 37°C for 5 min. Transport was initiated by rapidly adding the ^{14}C-glucose to the test tube during vortex mixing of the suspension. One-hundred-microliter samples were removed at 30, 60, 120, 300, and 600 s, and filtered by vacuum through prewetted Millipore filter (0.22 μm) pads, which were subsequently washed with 15 ml of buffer D without glucose at pH 7 and room temperature. The filter discs were placed in 7 ml of Hydrofluor™. The background radioactive cpm were measured by adding substrate to buffer D without cells, filtering 100 μl, washing, and counting the cpm per filter.

Radioiodination of Antibodies

Monoclonal antibodies 4E8 and 4D6, HICB serum, and HIRT serum were diluted 1:10 in P-25 buffer at pH 7 and radioiodinated with ^{125}I catalyzed by Iodobeads™ (Pierce Chemical Co., Rockford, IL) as recommended in the Pierce protocol. Radioactive ^{125}I-labeled proteins were determined by precipitating an aliquot with 10% TCA, washing, and counting the cpm.

Binding of ^{125}I-Labeled Antibody to **C. burnetii**

C. burnetii purified from infected BHK cells was prepared as described above. To reduce nonspecific binding of immunoglobulin to the cells, buffers were prepared in 1% BSA and adjusted to the desired pH for the antibody binding studies. Fifty microliters of a 1-mg-per-ml suspension of *C. burnetii* cells was mixed with a volume of ^{125}I-labeled antibody containing 2×10^6 cpm. Final dilutions of the ascites fluid or the sera were adjusted to 1:100 with P-25 buffer. The tubes were incubated on a reciprocating platform at room temperature for 2 h. Unbound serum components were removed by centrifuging at 4°C and 12,400 × g for 10 min. The pellet was washed three times with P-25 buffer at pH 7 or 4.5, as required, and finally suspended in 100 μl of P-25 buffer of the same pH. The cpm of 10-μl portions were determined in triplicate.

Statistical Analysis

Data were analyzed using descriptive statistics for a parametric t-test with a general statistics program (STAT QUICK, Version 4.0) for IBM personal computer (LUNDON Software Inc., Chagrin Falls, OH).

RESULTS

Isolation and Purification of the OMP Oligomer

The *C. burnetii* LCV envelope was the starting material for the purification of the OMP aggregate. An SDS-PAGE protein profile of the LCV envelope revealed, among other proteins, a broad, heavily stained protein band ranging in molecular mass from 29.5 kDa to about 33 kDa where neither the 29.5- nor 31-kDa polypeptides were discernable (FIG. 2, lane 1). After treating the LCV envelope preparation with 1% SDS, 1% DOC, and EDTA at 37°C, we recovered soluble proteins with molecular masses in the 29.5- and 31-kDa range in the high-speed supernatant (FIG. 2, lane 2). Treating the PG-PC pellet with lysozyme facilitated detergent solubilization (0.1% SDS and 0.1% deoxycholate at 22°C) of the associated proteins of high molecular mass and some of the 29.5- and 31-kDa proteins (FIG. 2, lane 3). The particulate material remaining after lysozyme digestion and detergent solubilization was collected by centrifugation, divided into two samples, and treated with either 1% SDS, 1% DOC, EDTA (buffer A) at 4°C overnight, or Laemmli's sample buffer (LSB), containing dithiothreitol (DTT) and 2-mercaptoethanol (2-ME) without dye. The proteins solubilized under these conditions were fractionated on a Sepharose 4B column as described in Materials and Methods (FIG. 1). P1 and P2 eluted near the void volume (apparent molecular mass of approximately 2×10^4 kDa) and were collected and analyzed for pore-forming activity and composition by SDS-PAGE and immunoblotting. For solubilization without reducing agent, P1 (FIG. 1) contained the OMP aggregate comprised of the 29.5- and 31-kDa polypeptides (FIG. 2, lane 4, P1); and P2 (FIG. 1) contained small amounts of the 31-kDa polypeptide and high and low molecular mass proteins (FIG. 2, lane 5). The 29.5-kDa polypeptide was more discernable in P1 than in P2. When the PG-associated proteins were solubilized in the presence of reducing agent and fractionated on the Sepharose 4B column, the 29.5- and 31-kDa proteins were not found in the peaks corresponding to P1 or P2, but were eluted from the Sepharose 4B column as monomers in later fractions (data not shown).

Monoclonal antibodies 4E8 and 4D6 reactive to the 29.5- and 31-kDa polypeptides were used in immunoblots to probe the LCV envelopes and the various fractions during purification of the OMP aggregate. To verify the presence of the OMP polypeptides of the LCV envelopes, various fractions were probed with 4E8 and 4D6 separately [FIG. 3, lanes 1 and 9, LCV-envelopes (30 µg); 2 and 6, detergent extract after lysozyme treatment of the PG-PC (15 µg); 3 and 7, P1 (10 µg); 4 and 8, P2 (10 µg)]. Although the protein concentration was too high to resolve the 29.5- and 31-kDa polypeptides, both mAbs reacted with each fraction in a similar profile. In another immunoblot with reduced protein loading of the LCV-envelope (10 µg), and 5 µg each of P1 and P2, the polypeptides reacted with mAbs 4E8 and 4D6 [FIG. 4, lanes 1 (P1), 2 (P2), and 3 (LCV-envelope)]. The dominant reactions in both immunoblots occurred over a broad band, suggesting the presence of intact and breakdown products of the 29.5- and 31-kDa proteins. For the proteins in the LCV envelope (FIG. 4, lane 3), the immunoblot revealed the region containing the 29.5- and 31-kDa proteins and molecular mass forms below and above the 43-kDa marker. This protein banding profile was not seen in the detergent extract after lysozyme treatment of the PG-PC

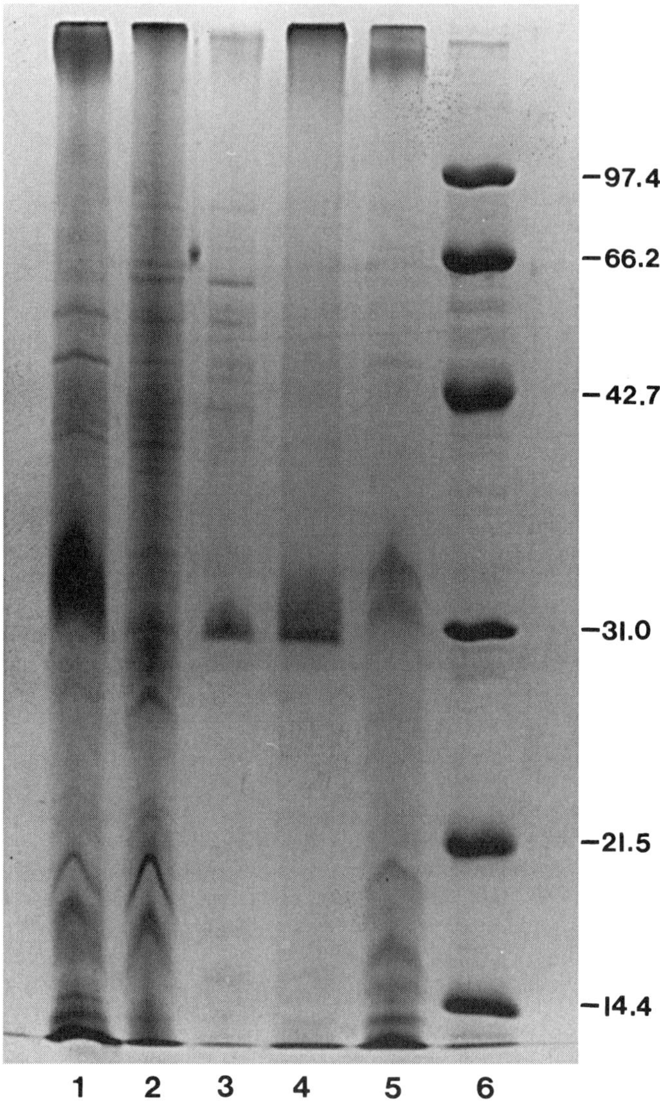

FIGURE 2. SDS-PAGE of various fractions obtained during the purification of the OMP aggregate from the LCV envelopes of *C. burnetii*. Lanes: 1, LCV envelope; 2, proteins solubilized from the LCV envelope suspended in EDTA, 1% SDS, and 1% DOC at 37°C; 3, proteins solubilized from the lysozyme-treated peptidoglycan-protein complex (PG-PC) suspended in 0.1% SDS, 0.1% DOC, 0.1% Triton X-100 at 22 °C for 30 min; 4, P1 proteins (FIG. 1) solubilized from the lysozyme-treated PG-PC suspended in EDTA, 1% SDs, and 1% DOC at 4°C overnight; 5, P2 proteins (FIG. 1) solubilized from the lysozyme-treated PG-PC suspended in EDTA, 1% SDS and 1% DOC at 4°C overnight; 6, protein molecular mass markers were phosphorylase b, 97.4 kDa; bovine albumin, 66.2 kDa; ovalbumin, 42.7 kDa; carbonic anhydrase, 31 kDa; trypsin inhibitor, 21.5 kDa; and lysozyme, 14.4 kDa.

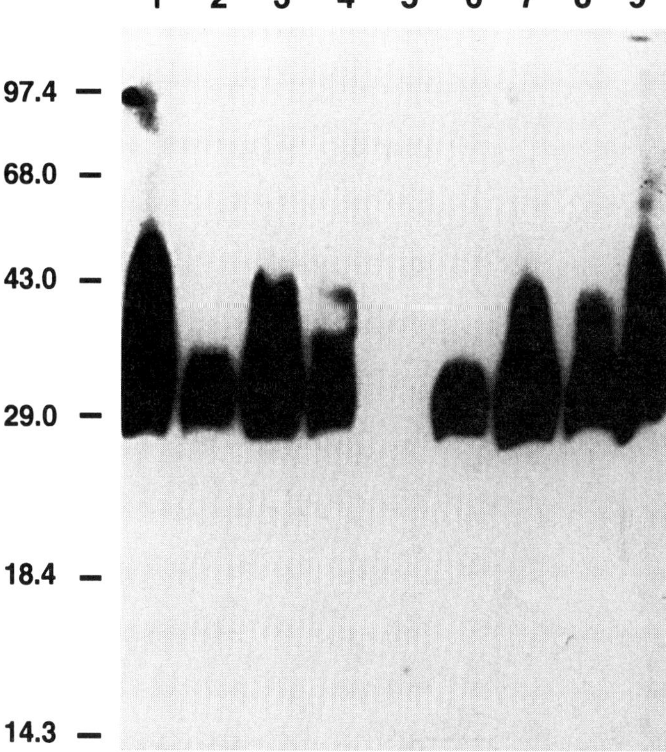

FIGURE 3. Immunoblot of various OMP-containing fractions during the purification of the OMP aggregate from the LCV envelopes of *C. burnetii*. Lanes: 1 and 9, LCV-envelopes (30 μg protein); 2 and 6, detergent extract after lysozyme treatment of the PG-PC (15 μg protein); 3 and 7, P1 (10 μg protein); 4 and 8, P2 (10 μg protein). Lanes 1 through 4 were blotted with mAb 4E8. Lane 5 was a blank lane. Lanes 6 through 9 were blotted with mAb 4D6. Proteins were stained by Coomassie brilliant blue R250. Protein standard molecular mass markers were ^{14}C-labeled (Bethesda Research Laboratories, Gaithersburg, MD) phosphorylase b, 97.4 kDa; bovine serum albumin, 68 kDa; ovalbumin, 43 kDa; carbonic anhydrase, 29 kDa; beta-lactoglobulin, 18.4 kDa; and lysozyme, 14.3 kDa.

(FIG. 3, lanes 2 and 6), P1 or P2. The amount of protein loaded in each lane did not allow the resolution of the 29.5-kDa and 31-kDa proteins (compare FIG. 2, lane 4).

Resistance of OMP Oligomer to SDS Denaturation at 100°C and Trypsin Treatment

To test whether SDS alone would dissociate the OMP aggregate into monomers, we extracted proteins from either cells or various subfractions with SDS at 100°C. On fractionating the 100°C SDS extract on a Sepharose 4B column, the OMP aggregate

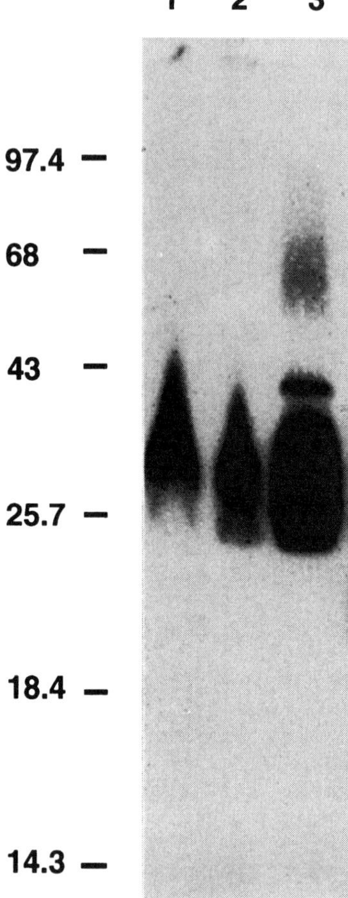

FIGURE 4. Immunoblot of various fractions obtained during the purification of the OMP aggregate from the LCV-envelopes of *C. burnetii*. Lanes: 1, P1 (5 μg protein); 2, P2 (5 μg protein), and 3, LCV envelope (10 μg protein). Monoclonal antibodies 4E8 and 4D6 were mixed and used to identify the antigenic polypeptides. Protein standard molecular mass markers were ^{14}C-labeled (Bethesda Research Laboratories, Gaithersburg, MD) phosphorylase b, 97.4 kDa; bovine serum albumin, 68 kDa; ovalbumin, 43 kDa; carbonic anhydrase, 29 kDa; beta-lactoglobulin, 18.4 kDa; and lysozyme, 14.3 kDa.

eluted near the void volume and with an elution profile shown in FIGURE 1. This suggested that the aggregate was not dissociated by SDS.

Because porin proteins are characteristically resistant to trypsin digestion, we tested the *C. burnetii* proteins by treating LCV envelope and purifying OMP aggregate with trypsin (1:1, wt/wt) for 5 h at 37°C. Evaluation of the protein profile by reducing SDS-PAGE indicated that the electrophoretic migration of the polypeptide monomers did not differ from fractions that were not treated with trypsin (data not shown).

TABLE 1. Porin Activity of *C. burnetii* OMP Oligomer by the Equilibrium Distribution Assay

Vesicles[a]	Protein (μg)	Normalized Value[b]
Without protein	0	0.8
LCV envelopes	10	4.8
	50	7.0
Proteins solubilized at 37°C in 1% SDS	50	0.5
Proteins solubilized at 100°C in 1% SDS	200	0.0
PG-PC fraction	50	59.0
P1 from Sepharose 4B	200	50.0
P2 from Sepharose 4B	200	25.0

[a] Described in the text.

[b] The activity is expressed as the ratio of dextran [carboxyl-^{14}C] (50–70 kDa) to sucrose [6,6′(n) – ^{3}H]. The normalized ratio was calculated by dividing the dextran:sucrose ratio in the void volume of the Sepharose 4B column by the dextran:sucrose ratio in the liposomes before application to the column.

However, the OMP aggregate was completely digested by proteinase K after boiling 5 min in LSB (data not shown).

Porin Activity of LCV Envelope and OMP Oligomer by the Equilibrium Distribution Assay (TABLE 1)

Liposomes prepared without protein had a normalized ratio (^{14}C/^{3}H) of 0.8. The LCV envelope suspended at 10 μg and 50 μg of protein per ml had concentration-dependent porin activity. Normalized ratios similar to those of protein-free liposomes were obtained with proteoliposomes prepared from proteins solubilized from LCV envelope at either 37°C or 100°C in 1% SDS. The SDS was removed by either ethanol precipitation or dialysis before the assay of porin activity. This indicated that 1% SDS at 37°C or 100°C did not solubilize active porin. Proteoliposomes prepared from 50 μg of the PG-PC remaining after the first detergent extraction at 37°C had a normalized ratio of 59. The proteins associated with the lysozyme-treated PG-PC were solubilized in EDTA, 0.1% SDS, and 0.1% DOC at 4°C and separated into P1 and P2 by Sepharose 4B column chromatography. When 200 μg of protein from P1 and/or P2 were each used in the porin assay, the normalized ratios were 50 and 25, respectively. Although P2 contained less of the 29.5-kDa polypeptide than the 31-kDa polypeptide as well as other polypeptides below and above the 31-kDa marker (FIG. 2, lane 5), this fraction exhibited about 50% of the porin activity observed for the P1 fraction (FIG. 2, lane 4).

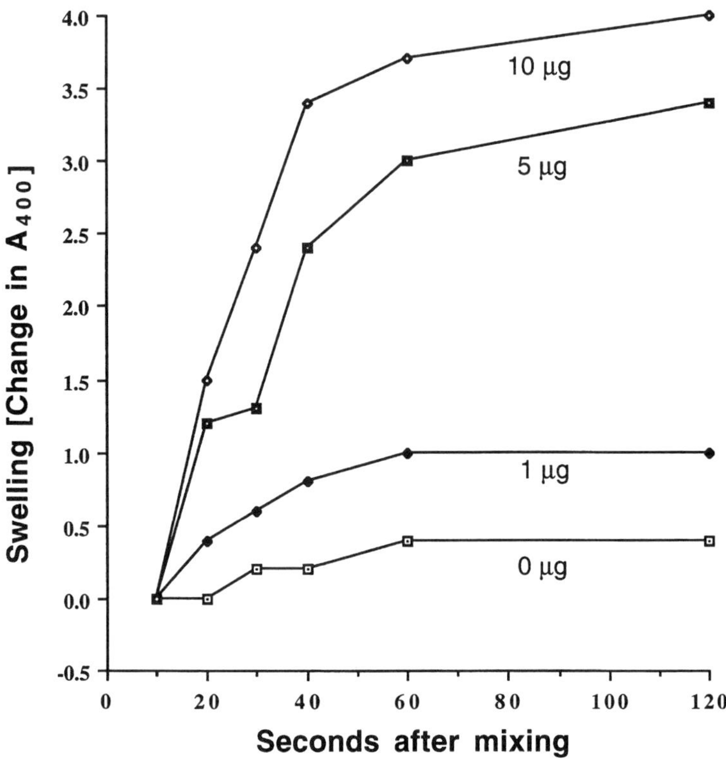

FIGURE 5. Swelling of proteoliposomes prepared from LCV envelopes of *C. burnetii*. Liposomes were prepared as described in Materials and Methods. A 30-µl portion of the liposome suspension was added to 0.7 ml of 30-mM arabinose solution in Tris-HCl buffer at pH 7.4. The change in absorbance (A_{400}) was recorded as indicated in seconds after mixing. Each successive decrease in absorbance was subtracted from the initial absorbance and used to calculate the fold change. The fold change in absorbance was plotted as change in swelling at A_{400} against time in seconds after mixing. Liposomes with 0, 1, 5, and 10 µg of protein were tested for porin activity. Each point is the mean of two samples.

Liposome Swelling by Using the LCV Envelope, P1, and P2

The LCV envelope was tested for porin activity as defined by the liposome-swelling assay. The swelling rates of liposomes prepared from LCV-envelopes were greater than swelling rates of control liposomes (FIG. 5). In addition, the proteoliposomes had protein-concentration-dependent swelling. A fivefold increase in protein resulted in a threefold increase in proteoliposome swelling.

The pore-forming activity of the OMP oligomer in P1 and P2 was also tested in the liposome-swelling assay (FIG. 6). The pore-forming activity of proteoliposomes prepared from the OMP oligomer in P1, which contained both the 29.5- and 31-kDa polypeptides in apparently similar amounts (FIG. 2, lane 4), was slightly greater than

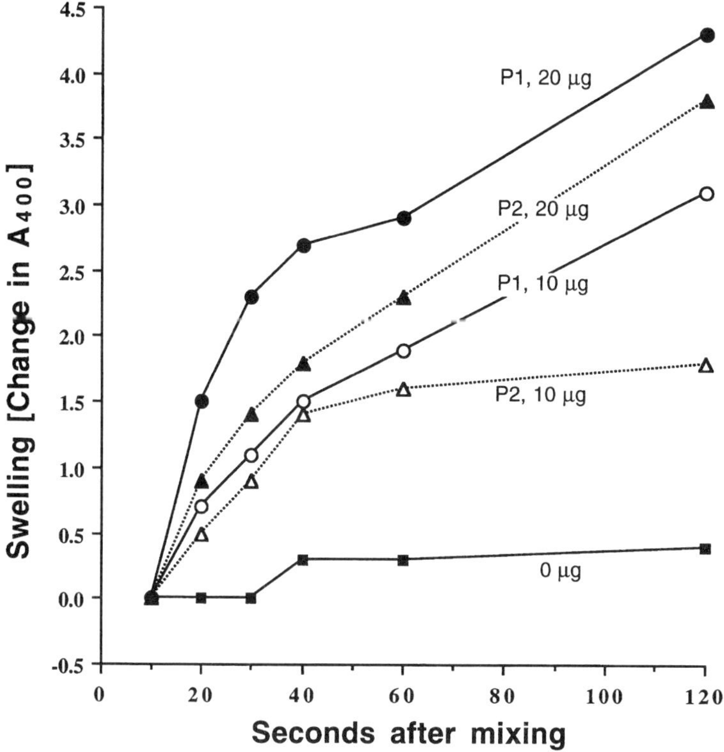

FIGURE 6. Swelling of proteoliposomes prepared with purified OMP oligomer obtained from a Sepharose 4B column in fractions of P1 and P2. A 30-µl portion of the proteoliposome suspension was mixed with 0.7 ml of arabinose solution in a cuvette, and the A_{400} was recorded. Each value was determined and plotted as described in the legend to FIGURE 5. Liposomes prepared with, 0 µg protein; P1 with 10 µg; P1 with 20 µg; P2 with 10 µg; and P2 with 20 µg. Each point is the mean of two samples.

the activity of those prepared from P2, which contained the 31-kDa polypeptide as well as other polypeptides below and above the 31-kDa marker (FIG. 2, lane 5). In addition, the pore-forming activity of proteoliposomes from either P1 or P2 had protein-concentration-dependent swelling.

Sizing the Porin Channel

We estimated the size of the porin channel by evaluating the dependence of proteoliposome swelling on the size of the hydrophilic solutes (FIG. 7). The diffusion of arabinose (150.1 M_r), into proteoliposomes was 1.5-fold greater than glucose (180.2 M_r), which was 2.6-fold greater than sucrose (342.3 M_r). The diffusion of stachyose (666.6 M_r), a tetrasaccharide, into proteoliposomes was insignificant by comparison to control liposomes and 15.5-fold less than in arabinose.

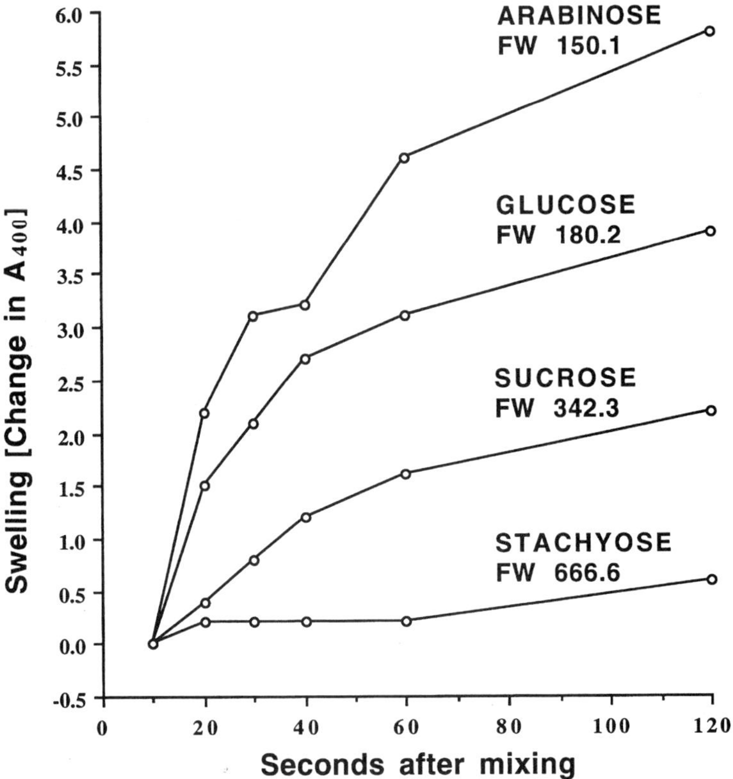

FIGURE 7. Swelling of proteoliposomes in the presence of various hydrophilic solutes. Proteoliposomes were prepared with 5 μg of protein from P1 of a Sepharose 4B column. A 30-μl portion of the proteoliposomes was mixed with different saccharide solutions which had been adjusted to the same osmotic pressure as dextran inside the proteoliposomes. Each of the values was determined as described in the legend to FIGURE 5 and plotted against time in seconds after mixing.

Effect of pH on the Pore-Forming Activity of the OMP Oligomer

Because *C burnetii* is an acidophile that grows in the phagolysosome of eukaryotic cells and transits from cell to cell at physiological pH, we tested the porin activity of the OMP oligomer at pH 7 and 4.5 in the liposome-swelling assay. To approximate the pH conditions of the physiological environment of the phagolysosome, proteoliposomes and control liposomes were prepared and tested at both pH 7 and 4.5 (FIG. 8). The leakage of solute by control liposomes prepared at pH 7 or 4.5 was tested at both pH values. The swelling rates of control liposomes prepared at pH 7 and tested at pH 7 and 4.5 were identical (FIG. 8, panels A1 and A2). The swelling rates of control liposomes prepared at pH 4.5 and tested at pH 7 and 4.5 (FIG. 8, panels B1 and B2) were similar, but they were greater than control liposomes prepared at

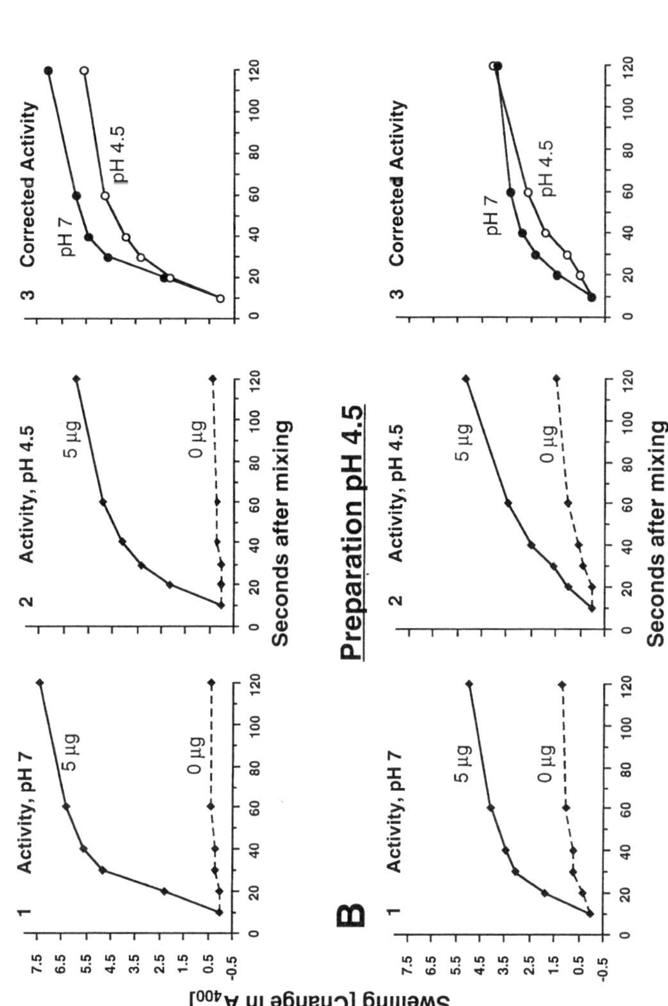

FIGURE 8. The pH dependence of swelling of proteoliposomes prepared and assayed at different pH values. Proteoliposomes were prepared with 5 µg of protein from P1 in Tris-HCl buffer at pH 7 and 4.5. Control liposomes were prepared in the same buffer without protein. An aliquot of the control liposomes and proteoliposomes were mixed with 30 mM arabinose in 5 mM Tris-HCl buffer, pH 7 and 4.5. Changes in swelling of the control liposome and proteoliposome were calculated and plotted as described in Figure 5. Panel A: swelling of liposomes and proteoliposomes prepared at pH 7 and tested at pH 7 (panel A1) and 4.5 (panel A2). Panel B: swelling of liposomes and proteoliposomes prepared at pH 7 (panel B1) and 4.5 (panel B2). Panels A3 and B3 show the absorbance values after subtracting the control liposome values form the proteoliposome values to correct for the leakage of solute by control liposomes.

pH 7 and tested at pH 7 and 4.5. Comparing the swelling rates of control liposomes prepared at pH 7 and tested at pH 7 and 4.5 with those prepared at pH 4.5 and tested at pH 7 and 4.5 revealed significant ($p = 0.048$ and $p = 0.024$, by Student's t-test) leakage of solute by the latter.

Proteoliposomes prepared at pH 7 and tested at pH 7 and 4.5 had 4.6- and 3.3-fold increases in swelling rates at 30 s, respectively, compared to those of control liposomes (FIG. 8, panels A1 and A2). There was an insignificant ($p = 0.290$) 28% reduction in channel function for proteoliposomes prepared at pH 7 and tested at pH 4.5 (FIG. 8, panel A3). Proteoliposomes prepared at pH 7 and assayed at pH 7 had slightly greater swelling rates than those assayed at pH 4.5.

Proteoliposomes prepared at pH 4.5 and tested at pH 7 and 4.5 had 2.3- and 1.2-fold increases in swelling rates at 30 s, respectively, compared to control liposomes (FIG. 8, panels B1 and B2). There was an insignificant ($p = 0.310$) 48% increase in channel function for proteoliposomes prepared at pH 4.5 and tested at pH 7 (FIG. 8, panel B3). Proteoliposomes prepared at pH 4.5 and assayed at pH 4.5 had slightly less swelling activity than the preparation assayed at pH 7.

The preparation of proteoliposomes at pH 7 or 4.5 significantly influenced the swelling activity with the greater activities residing in the proteoliposomes prepared at pH 7 and tested for swelling at pH 7 or 4.5 ($p = 0.012$ and 0.031, respectively) (FIG. 8, panels A3 and B3).

Binding of ^{125}I-Labeled mAbs and Polyclonal Antibodies to C. burnetii *Cells*

The influence of pH on the binding of ^{125}I-labeled antibodies to *C. burnetii* cells is shown in TABLE 2. The binding of ^{125}I-labeled HIRT serum components to *C. burnetii* at pH 7 or 4.5 followed by washing at pH 4.5 served as the controls for nonspecific binding of ascites or serum components. Radioactive counts due to the binding of ^{125}I-labeled serum components in the control HIRT serum at pH 7 and 4.5 were similar. Only 29% of the polyclonal antibodies to *C. burnetii* were bound at pH 7 and remained bound after washing at pH 4.5. Of the ^{125}I-labeled 4E8 mAbs bound at pH 7 or 4.5, 96% and 98%, respectively, remained bound after washing the cells at pH 4.5. However, of the ^{125}I-labeled 4D6 mAbs bound at pH 7 or 4.5, only 58% and 54%, respectively, remained bound after washing at pH 4.5.

Inhibition of Glucose Uptake

Binding monoclonal and/or polyclonal antibodies to the surface-exposed epitopes of the 29.5- and 31-kDa polypeptides of *C. burnetii* cells inhibited the uptake of ^{14}C-glucose (TABLE 2). ^{14}C-glucose uptake by *C. burnetii* cells was measured at pH 4.5 after the monoclonal and polyclonal antibodies were bound at pH 7 or 4.5. To determine the optimum concentration of antibodies required to inhibit ^{14}C-glucose transport, mAbs 4E8 and 4D6 were mixed and diluted serially before binding to the surface of *C. burnetii* at pH 7. For the serial dilutions 1:50, 1:100, 1:200, and 1:800, the inhibitions in uptake of ^{14}C-glucose were 71%, 85%, 34%, and 0.5%, respectively, as compared to control.

TABLE 2. Antibody Binding to *C. burnetii* and Inhibition of Glucose Uptake

Antibody[a] in Serum or Ascites	Binding pH	Wash pH	Percent Antibody Bound[b]	Percent Inhibition of Uptake[c]
HIRT	7.0	4.5	0	0[d]
None	7.0	4.5	ND[e]	+16
HICB	7.0	4.5	29	43
4E8	7.0	4.5	96	88
4D6	7.0	4.5	58	91
4E8 + 4D6	7.0	4.5	ND	94
4E8 + 4D6	4.5	4.5	ND	0
4E8	4.5	4.5	98	ND
4D6	4.5	4.5	54	ND

[a] The dilution of specific murine antibodies either ^{125}I labeled or unlabeled was described in the text. HIRT = hyperimmune *R. typhi* serum; None = buffer and 1% BSA; HICB = hyperimmune *C. burnetii* serum; 4E8 and/or 4D6 mAb ascites fluids.

[b] ^{125}I-labeled antibodies were bound to cells at pH 7 or 4.5, and the cells were washed at pH 7 or 4.5. After adjustments were made for nonspecific binding, the percent ^{125}I-labeled antibody remaining bound to the cells after washing at pH 4.5 was determined by dividing the final CPMs at pH 4.5 by the initial CPMs at pH 7 or 4.5 and multiplying by 100.

[c] The uptake of glucose by cells in the absence of specific antibodies, but in the presence of HIRT serum, was used as the control for calculating the percent inhibition of uptake activity by antibodies.

[d] Compared to the uptake of glucose in buffer containing only BSA, the HIRT serum inhibited glucose uptake by 16% (data not shown).

[e] ND = not determined.

C. burnetii suspended with only 1% BSA in buffer transported ^{14}C-glucose 16% better than when control murine serum was added to the buffer (TABLE 2). Because the non-cross-reactive HIRT serum inhibited ^{14}C-glucose uptake by 16%, this was used as the control value for the calculation of inhibition of ^{14}C-glucose uptake by HICB serum and the 4E8 and 4D6 mAbs. The HICB serum antibodies inhibited ^{14}C-glucose uptake by 43%. When either 4E8 or 4D6 were bound to cells separately, they inhibited ^{14}C-glucose uptake by 88% and 91%, respectively (TABLE 2). By combining the two antibodies before mixing with the cells at pH 7, the inhibition of ^{14}C-glucose uptake was 94%.

No inhibition of glucose uptake was observed when *C. burnetii* cells were exposed to 4E8 and 4D6 at pH 4.5 and the ^{14}C-glucose transport activity measured at pH 4.5. The lack of inhibition of glucose uptake was not due to an inability of the antibodies to bind to the cells at pH 4.5. In fact, the binding of either 4E8 or 4D6 separately to the cells was 98% and 54%, respectively, under these pH conditions (TABLE 2).

mAb binding to the surface of *C. burnetii* at pH 7 inhibited porin activity at pH 4.5. In contrast, mAb binding to the surface of *C. burnetii* at pH 4.5 did not inhibit porin activity at pH 4.5.

DISCUSSION

An OMP aggregate containing oligomeric polypeptides with pore-forming activity was isolated from the LCV envelope of *C. burnetii*. Extraction of the OMP aggregate from the PG-PC was facilitated by lysozyme digestion of the PG and SDS solubilization of the OMP aggregate at low temperature under nonreducing conditions. Under nonreducing conditions, the SDS-solubilized OMP aggregate was eluted in two peaks (P1 and P2) near the void volume of a Sepharose 4B column. The exclusion of the aggregated proteins by Sepharose 4B (exclusion limit 2×10^4 kDa) indicates that the OMP aggregate had a molecular mass slightly less than this value. The disruption of the OMP aggregate by reducing agents before loading the Sepharose 4B column resulted in the elution of the oligomeric polypeptides in fractions other than P1 and P2. The polypeptides of the OMP oligomer migrated as a doublet of 29.5 and 31 kDa under the reducing conditions of SDS-PAGE (FIG. 2, lane 4). Copurification of the 29.5- and 31-kDa polypeptides as an OMP oligomer supports a strong association between them.

The OMP oligomer of *C. burnetii* was not dissociated by SDS at room temperature or 100°C, and it was resistant to the proteolytic activity of trypsin, a common property of envelope proteins.[10] In another study, Amano *et al.*[1] showed that trypsin-resistant proteins associated with the PG of *C. burnetii* were not extracted from intact PG at 100°C in the presence of SDS or reducing agents. Lysozyme digestion of the PG was necessary for the effective dissociation of the PG-PC. In the current study, the proteins released from the PG-PC after lysozyme treatment were resistant to trypsin. On the other hand, proteinase K completely digested the porin after it had been denatured by reducing agents and SDS. The trypsin resistance of the OMP oligomer probably enables survival of *C. burnetii* in the hostile hydrolytic environment of the phagolysosome.

Purification of the 29.5- and 31-kDa polypeptides was monitored by immunoblotting with mAbs 4E8 and 4D6 (FIGS. 3 and 4), which were previously selected on the basis of their reactivity with a highly purified and homogenous 29.5-kDa polypeptide.[21] A similar disulfide-linked doublet has been shown to occur in the *Legionella pneumophila* outer membrane.[4] The *Legionella* oligomer is composed of 28- and 31-kDa subunits with the latter subunit being covalently bound to the PG. Another study revealed that the 31-kDa subunit of *Legionella* oligomer was actually the 28-kDa subunit with covalently bound PG.[9] Thus, the protein moiety had a common origin, but the 28- and 31-kDa molecular forms were the result of covalent modification after synthesis. Whether there is a similar association between the 29.5- and 31-kDa polypeptides of the *C. burnetii* oligomer remains to be determined.

Reducing SDS-PAGE consistently gave discrete bands interspersed in a background of smeared protein above and between the dominant 29.5-kDa and 31-kDa bands (FIG. 2). The protein bands and the smeared protein were recognized by mAbs 4E8 and 4D6 (FIGS. 3 and 4). Rosenbusch[18] has suggested for *E. coli* that such protein banding represents apparent heterogeneity, which is primarily due to incomplete hydrolysis of strongly bound PG resulting in fragments of different lengths remaining attached to the protein. The proteins may be aggregates of monomers that are not dissociated by the reducing conditions of SDS-PAGE. Also, unusual binding of the detergent to the protein monomers or oligomer may also cause this effect. Artifactual carbamoylation has been suggested to contribute, in part, to aberrant banding profiles.[3] The contributing factor to the development of multiple banding of the *C. burnetii* proteins is likely due to the covalent association of PG fragments with protein aggregates.

The OMP oligomer of *C. burnetii* produces diffusion channels at pH 7 in proteoliposomes with an exclusion limit of approximately 700 M_r (FIG. 7), which is similar to other bacterial porins.[14] This suggested that the porin of *C. burnetii* should function similarly to other bacterial porins in controlling the passage of small hydrophilic molecules between the environment and the periplasmic space. Because *C. burnetii* actively transports exogenously supplied substrates optimally at pH values near 4.5,[24] we were interested in determining how an acid pH would affect porin channel function in the proteoliposomes. Therefore, the channel function of proteoliposomes prepared at either pH 7 or pH 4.5 were tested at pH 7 and 4.5 or pH 4.5 and 7, respectively. Although we did not determine the amount of protein incorporated by proteoliposomes prepared at pH 7 or 4.5, the channel function of porin in proteoliposomes prepared at a given pH was compared to the preparation pH and the porin assay pH. Proteoliposomes prepared at pH 7 and tested at pH 4.5 or pH 7 had greater porin activity than those prepared at pH 4.5 and tested at pH 4.5 or pH 7 (FIG. 8, panels A3 and B3). Maximal channel function was observed for proteoliposomes prepared at pH 7 and tested at pH 7 (FIG. 8, panel A3). Clearly, the pH values at which proteoliposomes were prepared affected the porin activity, and the porin channel functioned best at pH 7 regardless of the preparation pH. Inferentially, the physiological significance of porin channel function at pH 7 and 4.5 may be to allow hydrophilic molecules into the periplasmic space of *C. burnetii* where the molecules would be actively transported across the cytoplasmic membrane only at acid pH.[24]

Recent studies have shown that the diffusion of hydrophilic molecules of ≤700 M_r through the outer membrane of gram-negative bacteria occurs by pore-forming proteins that have two or more pH-dependent substates for open channels with variable sizes that correspond to unique structural conformations.[3] The changes in *E. coli* porin channel function at acid pH indicate the involvement of pH-dependent conformational changes or protonation of amino groups in the channel.[25] The functional changes observed for the porin of *C. burnetii* upon a pH change in the environment needs further study to determine the mechanism of the pH dependence.

The physiological significance of pH-dependent changes in porin channel function is relevant for the acidophilic, phagolysosomal growth of *C. burnetii*. The acid pH dependence for the triggering of active transport of various metabolites and their subsequent metabolism is well documented for *C. burnetii*.[8,24] An increase in [H⁺] has been shown to be the key environmental stimulus for the activation of substrate transport by *C. burnetii*. However, in previous studies,[6,8,24] the role of porin as a permeability barrier during pH-dependent activation of active transport was unknown. We studied the pH dependence of porin function for *C. burnetii* by determining the conditions for mAbs to inhibit uptake of glucose. Because the mAbs recognized porin epitopes on the surface of *C. burnetii* at pH 4.5 and 7, we were able to determine that porin channel permeability at pH 4.5 was inhibited by mAbs bound to the surface of cells at pH 7. Importantly, porin channel permeability at pH 4.5 was not inhibited by mAbs bound to the surface of *C. burnetii* at pH 4.5. This indicates that mAb binding at pH 7 prevented either pH-dependent conformational changes or protonation of amino acids in the porin channel at pH 4.5. The fact that mAb binding at pH 4.5 did not inhibit glucose transport at pH 4.5 suggests that either pH-dependent conformational change and/or protonation of amino acids in the porin channel had occurred before the binding of mAb. The inhibition of porin channel function at pH 7 by mAbs confirms the role of porin as a permeability barrier for the subsequent active transport of glucose by *C. burnetii* at pH 4.5.

One may argue that inhibiting porin channel function with mAbs explains,[31] in part, the protective effects of passively administered anti-*C. burnetii* antibodies.[27] Whether porin-specific antibodies bound to the surface of *C. burnetii* survive the proteolytic activities of the phagolysosome long enough to allow the microbicidal milieu to kill the microorganism is unknown. We know that acid activation of *C. burnetii* in the absence of substrate leads to depletion of ATP and cell death.[7] We propose that inhibition of porin permeability function by porin-specific serum antibodies may facilitate the phagolysosomal killing of *C. burnetii*. In another study, we showed that the 29.5-kDa OMP antigen induced active immunity against virulent challenge.[32] This information, combined with the recent confirmation that porins are important antigens in the induction of specific protective immune responses against infection by gram-negative bacteria,[11] suggests that humoral immunity directed against *C. burnetii* porins might play an important role in immunity against Q fever.

SUMMARY

Envelopes of large-cell variant *Coxiella burnetii,* the agent of Q fever, were the starting material for purification of an outer membrane protein (OMP) oligomer with aggregate molecular mass of approximately 2×10^4 kDa. The oligomer was resistant to trypsin and dissociation by SDS at 100°C. Reducing agents dissociated the oligomer into monomers of 29.5 and 31 kDa, which migrated as a doublet during SDS-polyacrylamide gel electrophoresis. Both monomers were reactive in an immunoblot assay with monoclonal antibodies (mAbs) 4E8 and 4D6, which were previously selected for their reactivity with purified and SDS-denatured 29.5 kDa protein. Proteoliposomes were functional in an equilibrium assay at pH 7 and a swelling assay at pH 7 and 4.5. The pores in proteoliposomes allowed the passage of arabinose, glucose, and sucrose, but restricted stachyose. Polyclonal antibodies to *C. burnetii* cells and the mAbs were able to bind *C. burnetii* at pH 7 and 4.5. The uptake of ^{14}C-glucose at pH 4.5 was inhibited by polyclonal antibodies and mAbs after binding to cells at pH 7. The mAbs did not inhibit ^{14}C-glucose uptake at pH 4.5 after binding to cells at pH 4.5. Although the mAbs bind *C. burnetii* porin epitopes before and after acid activation, the mAbs bound under acidic conditions were unable to inhibit porin function. The inhibition of porin channel function by mAbs confirms the role of porin as a permeability barrier for the subsequent active transport of glucose by *C. burnetii*. In another study, we showed that the 29.5 kDa OMP antigen induced active immunity against virulent challenge. This information, combined with the recent confirmation that porins are important antigens in the induction of specific protective immune responses against infection by gram-negative bacteria, suggests that humoral immunity directed against *C. burnetii* porins might play an important role in immunity against Q fever (human infection) and coxiellosis (animal infection), global enzootic diseases.

ACKNOWLEDGMENTS

This work was performed during the tenure of Nirupama Banerjee-Bhatnagar as a National Research Fellow at USAMRIID. The assistance provided by the Medical Illustrations Department (USAMRIID) is gratefully appreciated. We thank Drs. Byrne, Anderson, Waag, Ezzell, Novak, and Chandler for reviewing this manuscript.

DISCLAIMER

The views, opinions, and/or findings contained in this manuscript are those of the authors and should not be construed as an official Department of the Army position, policy, or decision unless so designated by other documentation.

REFERENCES

1. AMANO, K., J. C. WILLIAMS, T. F. McCAUL & M. G. PEACOCK. 1984. Biochemical and immunological properties of *Coxiella burnetii* cell wall and peptidoglycan-protein complex fractions. J. Bacteriol. **160:** 982-988.
2. BENZ, R. & K. BAUER. 1988. Permeation of hydrophilic molecules through the outer membrane of gram-negative bacteria. Eur. J. Biochem. **176:** 1-19.
3. BRAGG, P. D. & C. HOU. 1972. Organization of proteins in the native and reformed outer membrane of *Escherichia coli*. Biochim. Biophys. Acta **274:** 478-488.
4. BUTLER, C. A. & P. S. HOFFMAN. 1990. Characterization of a major 31-kilodalton peptidog-lycan-bound protein of *Legionella pneumophila*. J. Bacteriol. **172:** 2401-2407.
5. DASCH, G. A., J. R. SAMMS & J. C. WILLIAMS. 1981. Partial purification and characterization of the major species-specific protein antigens of *Rickettsia typhi* and *Rickettsia prowazekii* identified by rocket immunoelectrophoresis. Infect. Immun. **31:** 276-288.
6. HACKSTADT, T. & J. C. WILLIAMS. 1981. Biochemical stratagem for obligate parasitism of eukaryotic cells by *Coxiella burnetii*. Proc. Natl. Acad. Sci. USA **78:** 3240-4344.
7. HACKSTADT, T. & J. C. WILLIAMS. 1981. Stability of the adenosine 5'-triphosphate pool in *Coxiella burnetii;* influence of pH and substrate. J. Bacteriol. **148:** 419-425.
8. HACKSTADT, T. & J. C. WILLIAMS. 1983. pH dependence of the *Coxiella burnetti* glutamate transport system. J. Bacteriol. **154:** 598-603.
9. HOFFMAN, P. S., J. H. SEYER & C. A. BUTLER. 1992. Molecular characterization of the 28- and 31-kilodalton subunits of the *Legionella pneumophila* major outer membrane protein. J. Bacteriol. **174:** 908-913.
10. INOUYE, M. & M.-L. YEE. 1972. Specific removal of proteins from the envelope of *Escherichia coli* by protease treatments. J. Bacteriol. **112:** 585-592.
11. ISIBASI, A., V. ORTIZ-NAVARRETE, J. PANIAGUA, R. PELAYO, C. R. GONZALEZ, J. A. GARCIA & J. KUMATE. 1992. Active protection of mice against *Salmonella typhi* by immunization with strain-specific porins. Vaccine **10:** 811-813.
12. LAEMMLI, U. D. 1970. Cleavage of structural proteins during the assembly of the head of bacteriophage T4. Nature **227:** 680-685.
13. LOWRY, O. H., N. J. ROSEBROUGH, A. L. FARR & R. J. RANDALL. 1951. Protein measurement with the Folin phenol reagent. J. Biol. Chem. **193:** 265-275.
14. LUCKEY, M. & H. NIKAIDO. 1980. Specificity of diffusion channels produced by lambda phage receptor protein of *Escherichia coli*. Proc. Natl. Acad. Sci. USA **77:** 167-171.
15. McCAUL, T. F., N. BANERJEE-BHATNAGAR & J. C. WILLIAMS. 1991. Antigenic differences between *Coxiella burnetii* cells revealed by postembedding immunoelectron microscopy and immunoblotting. Infect. Immun. **59:** 3243-3253.

16. NAKAE, T., J. ISHII & M. TOKUNAGA. 1979. Subunit structure of functional porin oligomer that form permeability channels in the outer membrane of *Escherichia coli*. J. Biol. Chem. **254:** 1457–1461.

17. NIKAIDO, H. 1983. Proteins forming large channels from bacterial and mitochondrial outer membrane porins and phage lambda receptor protein. Methods Enzymol. **97:** 85–100.

18. ROSENBUSCH, J. P. 1974. Characterization of the major envelop proteins from *Escherichia coli*. J. Biol. Chem. **249:** 8019–8029.

19. SNYDER, C. E., JR. & J. C. WILLIAMS. 1985. Isolation of envelope proteins from *Coxiella burnetii:* use of detergent solubilization and ion-exchange chromatography. Abstr. Annu. Meet. Am. Soc. Microbiol. K202.

20. SNYDER, C. E., JR. & J. C. WILLIAMS. 1986. Purification and chemical characterization of a major membrane protein from *Coxiella burnetii*. Abstr. Annu. Meet. Am. Soc. Microbiol. K6.

21. SNYDER, C. E., JR. & J. C. WILLIAMS. 1986. Production of monoclonal antibodies reactive with envelope components of *Coxiella burnetii*. Abstr. Annu. Meet. Am. Soc. Microbiol. K303.

22. SWEZYL, B. & L. M. KOZLOFF. 1985. A method for the efficient blotting of strongly basic proteins from sodium dodecyl sulfate–polyacrylamide gels to nitrocellulose. Anal. Biochem. **150:** 403–407.

23. TAKADA, H., T. OGAWA, F. YOSHIMURA, K. OTSUKA, S. KOKEGUCHI, K. KATO, T. UMEMOTO & S. KOTANI. 1988. Immunobiological activities of a porin fraction isolated from *Fusobacterium nucleatum* ATCC 10953. Infect. Immun. **56:** 855–863.

24. THOMPSON, H. A. & J. C. WILLIAMS. 1991. pH and membrane physiology in *Coxiella burnetii, In* Q Fever: the Biology of *Coxiella burnetii.* J. C. Williams & H. A. Thompson, Eds.: 117–129. CRC Press. Boca Raton, FL.

25. TODT, J. C., W. J. ROCQUE & E. J. MCCROARTY. 1992. Effects of pH on bacterial porin function. Biochemistry **31:** 10471–10478.

26. TOWBIN, H., T. STAEHELIN & J. GORDON. 1979. Electrophoretic transfer of proteins from polyacrylamide gels to nitrocellulose sheets: procedure and some applications. Proc. Natl. Acad. Sci. USA **76:** 4350–4354.

27. WAAG, D. M., C. R. BOLT, T. J. MARRIE & J. C. WILLIAMS. 1991. Immune responses of humans to *Coxiella burnetii:* specific antibody and cell-mediated responses after vaccination or infection. *In* Q Fever: the Biology of *Coxiella burnetii.* J. C. Williams & H. A. Thompson, Eds.: 157–173. CRC Press. Boca Raton, FL.

28. WILLIAMS, J. C. 1981. Iodination of *Coxiella burnetii* cells: optimal conditions for the iodination of exposed surface proteins. *In* Rickettsiae and Rickettsial diseases. W. Burgdorfer & R. L. Anacker, Eds.: 473–482. Academic Press. New York, NY.

29. WILLIAMS, J. C. 1991. Infectivity, virulence, and pathogenicity of *Coxiella burnetii* for various hosts, *In* Q Fever: the Biology of *Coxiella burnetii.* J. C. Williams & H. A. Thompson, Eds.: 21–71. CRC Press. Boca Raton, FL.

30. WILLIAMS, J. C. & S. STEWART. 1984. Identification of immunogenic proteins of *Coxiella burnetii* phase variants. Microbiology: 257–262.

31. WILLIAMS, J. C. & D. M. WAAG. 1991. Antigens, virulence factors and biological response modifiers of *Coxiella burnetii:* strategies for vaccine development, *In* Q Fever: the Biology of *Coxiella burnetii.* J. C. Williams & H. A. Thompson, Eds.: 175–224. CRC Press. Boca Raton, FL.

32. WILLIAMS, J. C., T. A. HOOVER, D. M. WAAG, N. BANERJEE-BHATNAGAR, C. R. BOLT & G. H. SCOTT. 1990. Antigenic structure of *Coxiella burnetii:* a comparison of lipopoly-

saccharide and protein antigens as vaccines against Q fever. Ann. N.Y. Acad. Sci. **590:** 370-380.

33. WILLIAMS, J. C., M. G. PEACOCK & T. F. McCAUL. 1981. Immunological and biological characterization of *Coxiella burnetii* phase I and II, separated from host components. Infect. Immun. **32:** 840-851.

34. ZUERNER, R. L. & H. A. THOMPSON. 1983. Protein synthesis by intact *Coxiella burnetii* cells. J. Bacteriol. **156:** 186-191.

An Apparent Role of Dogs in the Transmission of *Toxoplasma gondii*

The Probable Importance of Xenosmophilia

J. K. FRENKEL[a,c] AND B. B. PARKER[b]

[a]*Department of Biology*
University of New Mexico

[b]*Arrighetti Animal Hospital*
Santa Fe, New Mexico

Unlike in cats[1,2] *Toxoplasma* does not replicate in the gut of dogs and no oocysts are shed. Because of these findings, dogs were considered to be uninvolved in the transmission of *Toxoplasma*.[1,2] However, during a prospective epidemiologic study of the transmission of *Toxoplasma* in Panama, in which 72 of 571 children seroconverted, dog contact was more frequent than cat and soil contact in the children that had developed antibody.[3] These results were difficult to reconcile with facts known to one of us (JKF) who had never owned a dog. Subsequently it was pointed out to him that dogs were coprophagous and might have ingested cat feces. During the investigation of this lead, he was informed by the second author (BBP) that dogs also rolled in feces. This paper presents the evidence obtained so far that dogs may play a role in the transmission of *Toxoplasma*.

Children between 1 and 6 years of age were studied in Panama, after their passively transferred maternal antibody had decayed. The children were examined quarterly by pediatricians, who noted intercurrent illnesses, diet, and environmental conditions on questionnaires. Children were bled quarterly, and their serologic status was determined in the direct *Toxoplasma* agglutination test. During the five years, 72 of 571 children developed antibody, for a cumulative total of 12.6%.[3] Thirty-eight environmental factors were examined for their role as risk factors for seroconversion. The variables that were highly associated with a risk for seroconversion are listed in TABLE 1, together with comparable variables.

Whereas contact with cats was associated with a relative risk for seroconversion between 2.01 and 3.04, contact with nursing and young dogs showed a relative risk between 3.43 and 5.75. Cats and dogs over 1 year and itinerate dogs did not show a correlation with seroconversion of children.

After becoming aware of that some dogs ate and rolled in cat feces, we inquired whether these traits had also been observed in Panama. One of the physicians associated with the study and a veterinarian associated with Gorgas Memorial Laboratory confirmed of having observed that dogs ate and rolled in cat feces.

[c] Address correspondence to: 1252 Vallecita Dr., Santa Fe, NM 87501-8803.

TABLE 1. Numbers and Percentages of Panamanian Children With and Without Selected Variables, Arranged by Relative Risks of Seroconversion[3]

Categorical Variables	With Variable		Without Variable		Relative Risk of Seroconversion	95% Confidence Interval
	Total Number	Percent Seropositive	Total Number	Percent Seropositive		
Dogs 1+ years	546	11.9	25	28.0	0.43	0.22–0.83
Dogs visiting	436	12.2	135	14.1	0.86	0.53–1.41
Cats 1+ years	512	12.9	59	10.2	1.26	0.57–2.80
Wood floor	66	22.7	505	11.2	2.01	1.21–3.35
Soil cont. unspecif.	28	32.1	543	11.6	2.77	1.54–4.98
Many roaches	145	21.4	426	9.6	2.22	1.45–3.40
Much garbage	128	22.7	443	9.7	2.44	1.52–3.58
Many flies	100	31	471	8.7	3.56	2.35–5.39
Cats visiting	332	16	239	7.9	2.01	1.22–3.30
Cats nursing	96	25	475	10.1	2.47	1.60–3.83
Cats 6–12 mo.	109	25.7	462	9.5	2.69	0.57–2.80
Cats weaned	94	28.7	477	9.4	3.04	1.99–4.65
Dogs 6–12 mo.	65	33.8	506	9.9	3.43	2.23–5.26
Dogs weaned	46	45.7	525	9.7	4.70	3.17–7.08
Dogs nursing	27	59.3	544	10.3	5.75	3.86–8.58

We conducted a survey of the frequency with which dogs consumed and rolled in feces in Santa Fe, New Mexico (TABLES 2a, 2b, and 2c). Fifty-two owners that visited a veterinarian (BBP) for a variety of reasons were questioned about their dogs' behavior. Of 52 dogs, 44.2% ate cat feces and 55.7% ate any feces. In the latter group, ingestion of feces from cattle and elk were most frequent. Dogs will also eat their own feces, but apparently will not roll in them.

Twenty-three percent of dogs rolled in cat feces, and 48% rolled in a number of foul substances. When comparing the habits of individual dogs, eating cat feces was statistically more frequent than rolling in cat feces, but not eating or rolling in other foul substances. Dogs exhibit either one or both behaviors with similar frequency. Twenty-three to 56% of dogs engage in either eating or rolling in feces, or show both behaviors.

What might be the relative importance of the two behaviors for the transmission of *Toxoplasma?* When mice ingest oocysts, the majority excyst and enter the gut wall, and only about 10% are reshed in the feces.[4] Although this has not yet been studied in dogs, a qualitatively similar behavior of oocysts in the gut of dogs could be expected, with low-grade contamination of the perianal area. Whereas the survival of oocysts on the fur of dogs has not yet been studied, we found oocysts to persist in dry bedding for at least 4 days and in moist bedding for a least 7 days.[4] Qualitatively similar survival time could be expected on the fur of dogs, the contamination of which could be expected to decline slowly.

These considerations suggest, that contamination of fur, after rolling in cat feces containing oocysts, might make these accessible to children who pet dogs. Contamina-

TABLE 2a. Coprophagia and Rolling in Feces by 52 Dogs in Santa Fe, NM

	Never	Occasionally	Often	Percent Yes
Dog eats feces of cats:	29	14	9	44.2
Dog eats other feces:	23	20	9	55.7
Dog rolls in cat feces:	40	8	4	23.1
Dog rolls in other foul substances:	27	18	7	48.1

TABLE 2b. Coprophagia vs. Rolling in Cat Feces by 52 Dogs in Santa Fe, NM

		Eats Cat Feces		
		Never	Occasionally	Often
Rolls in cat feces	Never	28	9	3
	Occasionally	1	5	2
	Often	—	—	4

Summary:

Neither:	28	Both:	11
Either:	13	Total:	52

Statistically significant by McNemar's test. $\chi^2 = 11.2$ with p adjusted from 0.005 to 0.01 because of use of data in two tests.

TABLE 2c. Coprophagy vs. Rolling in Other Odoriferous Substances by 52 Dogs in Santa Fe, NM

		Eats other Feces		
		Never	Occasionally	Often
Rolls in foul substance	Never	17	9	1
	Occasionally	6	11	1
	Often	—	—	7

Summary:

Neither:	17	Both:	19	
Either:	16	Total:	52	McNemar's test: NS

tion of the fur of head and body would tend to be greater, covering a larger area, and present for a longer time, than the low-grade perianal contamination after ingestion of such material.

Cats defecate usually outdoors, and not necessarily close to houses. Although children could pick up oocysts by playing in such soil, the degree of soil contamination was low in Panama, although higher rates have been observed in Costa Rica.[3,5] The fur of cats is generally free of oocyst contamination.[6] Dogs that have rolled in such feces could bring the infectivity closer to children than cats. Likewise the greater tolerance of dogs than cats to be petted by children could favor contamination of the hands of children. The smell of such fecal contamination might not be a deterrent to small children, who although conscious of odors, do not usually have a strong aversion to fecal smells, as do adults.

Are eating and rolling in cat feces abnormal behaviors of dogs that might be amenable to veterinary medical treatment? In the veterinary literature coprophagia is often considered an abnormal behavior,[7] suggestive of protein, amylase, lipase, or vitamin deficiency, or a psychoneurosis. Because the diet of dogs has changed from high protein to high carbohydrate during their association with man, an amylase deficiency has been postulated, with coprophagy representing an attempt to conserve or acquire this enzyme.[8] However, even well-fed and well-adjusted dogs are said to eat feces. We know of no studies comparing the frequency of coprophagy in dogs, either on a high-protein or a high-carbohydrate diet. A textbook of animal behavior states that dogs roll in strong-smelling substances, but that the function of this behavior was unknown.[9]

We distinguish here maternal coprophagia from coprophagia by adult dogs. Eating the feces of pups is widespread among mammals, and could be thought of as conferring a biological advantage, by removing smells that might attract a predator. However, eating the feces of other animals is uncommon among mammals and unlikely to have been a factor of natural selection. If it were a "scavenging behavior," one would expect that good nutrition would diminish it; but this does not appear to be the case.

The literature on olfactory behavior of feral canids, including wolves, coyotes, and foxes, yielded references to rolling in feces and "scent marking"; however no reference to eating feces was found. Fox[10] mentions a tendency of the more sociable canids (wolves and coyotes) to roll in novel odors, such as carrion or deer musk, whereas the less sociable canids (foxes) tended to urinate or defecate on such odors. However, it is not clear to us that the latter, or "scent marking," is biologically or functionally comparable to eating and rolling in feces. Scent marking appears largely to be territorial, whereas rolling in feces would seem to serve an ambulatory dog, possibly by confusing a potential predator. No experimental studies have been encountered, and the literature found was limited to observational studies.

Fecal smells associated with secretion of the perianal glands appear to be utilized also to recognize other wolves and probably their sex and if female, stage in the estrus cycle. Females "scent mark" objects in their environment to draw the attention of male wolves.[11] The latter often add their scent to that of the females. Because wolf packs are highly territorial, it is reasonable that "scent marking" is also used to delineate a territory. Dogs also appear to scent mark objects with their urine. However, "dogs do not scent-mark consistently until they are eight or nine months old, and under the influence of the male hormone."[12]

After a description of "scent marking" by wolves, the following observation was made. "Wolves will also rub in almost any strong odor, from that of rotten fish to that of expensive perfume, and this behavior occurs even in animals only three months old,"[12] hence probably before they are sexually mature.

Lopez[13] mentioned that both dogs and wolves roll in putrid substances, such as animal feces and carcasses of fish and mammals, and he refers to this behavior as "scent rolling." Because we have not seen this term used for indexing, we would like to suggests "xenosmophilia" as a more general and challenging term, to refer to fondness for foreign smells.

Coyotes have been observed to lower their head, rubbing the side of their muzzles across smelly objects, and to roll in meat of another animal or in rabbit fur.[14] The significance of this behavior is unknown. In a study of foxes and coyotes, Robert L. Harrison, Department of Biology, University of New Mexico (personal communication) found plenty of coyote scats, many with fox scats on top, making it unlikely that foxes ate or rolled in coyote feces, and supporting Fox's generalization.[10]

The observation that both wild canids and domestic dogs roll in strong odors suggests that this habit is instinctual, and not usually a pathologic behavior of domestic dogs. It has been suggested that dogs only eat fresh feces deposited in their territory, and roll only in fresh feces, which may be a clue as to its significance. This needs to be investigated experimentally.

Our studies suggest that these behaviors of domestic dogs, to roll in or eat cat feces, allow the transmission of *Toxoplasma* from the fur of dogs, in environments contaminated with cat feces and *Toxoplasma* oocysts, such as Panama.[3] Although this behavior is fairly common in the US, it remains to be proven whether it is important for the transmission of *Toxoplamsa* in the US.

As a precaution, hand washing after contact with dogs should be added to the recommendations on the prevention of *Toxoplasma* transmission[15] until the importance of dog behavior in the transmission of *Toxoplasma* can be evaluated in the US.

ACKNOWLEDGMENTS

Research in Panama was supported by grant AI-23730 of the National Institute of Allergy and Infectious Diseases, United States Public Health Service to JKF. We thank Prof. Evangelos A. Coutsias, University of New Mexico, for determining the correct formulation of xenosmophilia.

REFERENCES

1. WEILAND, G., D. KÜHN & C. SAAR. 1971. Untersuchungen von Sekreten und Exkreten sowie von Lymphknoten oral und parenteral mit *Toxoplasma*-Zysten infizierter Hunde. Tierärzfl. Wochenschrift **84:** 181-184.

2. MILLER, N. L., J. K. FRENKEL & J. P. DUBEY. 1972. Oral infections with *Toxoplasma* cysts and oocysts in felines, other mammals, and in birds. J. Parasitol. **58:** 928-937.

3. FRENKEL, J. K., K. M. HASSANEIN, R. S. HASSANEIN, E. BROWN, P. THULLIEZ & R. QUINTEREO-NUÑEZ. 1995. Transmission of *Toxoplasma* in Panama City, Panama: a five-year prospective cohort study of children, cats, rodents, birds, and soil. Am. J. Trop. Med. Hyg. **53:** 458-468.

4. DUBEY, J. P. & J. K. FRENKEL. 1973. Experimental *Toxoplasma* infection in mice with strains producing oocysts. J. Parasitol. **59:** 505–512.

5. FRENKEL, J. K. & A. RUIZ. 1981. Endemicity of toxoplasmosis in Costa Rica. Am. J. Epidem. **113:** 254–269.

6. DUBEY, J. P. 1995. Duration of immunity to shedding of *Toxoplasma gondii* oocysts by cats. J. Parasitol. **81:** 410–415.

7. HOUPT, K. A. 1991. Feeding and drinking behavior problems. *In* The Veterinary Clinics of North America. Small Animal Practice: 281–298 Saunders. Philadelphia, PA.

8. McCUISTION, W. R. 1971. Prevention of coprophagy. In Current Veterinary Therapy. R. W. Kirk, Ed.: 62–63. W. B. Saunders. Philadelphia, PA.

9. GRIER, J. W. 1984. Biology of Animal Behavior. Times Mirror/Mosby. St. Louis, MO.

10. FOX, M. W. 1975. Evolution of social behavior in canids. *In* The Wild Canids: Their Systematics, Behavioral Ecology and Evolution. M. W. Fox, Ed.: 429 460. Van Nostrand Reinhold Co. New York, NY.

11. ZIMEN, E. 1982. A wolfpack sociogram. *In* Wolves of the World. Perspectives of Behavior, Ecology and Conservation. F. H. Harrington & P. C. Paquet, Ed.: 282–322. Noyes Publications. Park Ridge, NJ.

12. MECH, L. D. 1970. The wolf: the ecology and behavior of an endangered species. University of Minnesota, Minneapolis, MN.

13. LOPEZ, B. H. 1978. Of Wolves and Men. Charles Scribner's Sons. New York, NY.

14. LEHNER, P. N. 1978. Coyote communication. *In* Coyotes. Biology, Behavior and Management. M. Bekoff, Ed.: 127–162. Academic Press. New York, NY.

15. FRENKEL, J. K. 1995. Prevention of *Toxoplasma* infection in pregnant women and their fetuses. Clin. Infect. Dis. **20:** 728–729.

Ovarian Follicular Dynamics during the Estrus Cycle in Buffalo (*Bubalus bubalis*)

Preliminary Research

P. S. BARUSELLI,[a] R. G. MUCCIOLO,[b]
J. A. VISINTIN,[b] W. G. VIANA,[b] R. P. ARRUDA,[b]
E. H. MADUREIRA,[b] AND J. R. MOLERO-FILHO[c]

[a] *Experimental Farm*
Institute of Zoothechny
[b] *Department of Animal Reproduction*
Faculty of Veterinary Medicine
University of São Paulo
11.900, Registro, São Paulo, Brazil

INTRODUCTION

In Brazil, the buffalo population is estimated at 3 million head. The main concentration is located in the northern part of the country, the Amazonic region, with approximately 1.5 million buffalo. Next, we have the southeast region with 400,000 head of animals, and it was in this area, more specifically in Sao Paulo State, that this work was performed, during the breeding season, influenced positively by the shortening of the daylight period in autumn from March to July.

The main buffalo breeds raised in Brazil are Murrah, Mediterranean, Jafarabadi (river buffalo), and Carabao (swamp buffalo).

In Brazil, as well as in other countries of South America, there is a widespread habit among farmers, who begin buffalo raising, to handle the animals in the same way as cattle. This procedure will certainly bring difficulties to the management of the animals and will make impossible the use of reproductive biotechnologies. Comprehension of the phenomena related to buffalo ovarian physiology and its follicular dynamics during the estrous cycle is fundamental for the orientation of the technical procedures aimed at the improvement of a herd's fertility, and of reproductive management, as well as the introduction of biotechnologies such as artificial insemination (AI) and embryo transfer (ET). These phenomena have been studied in buffalo by several researchers.[1–4]

[c] Private veterinarian.

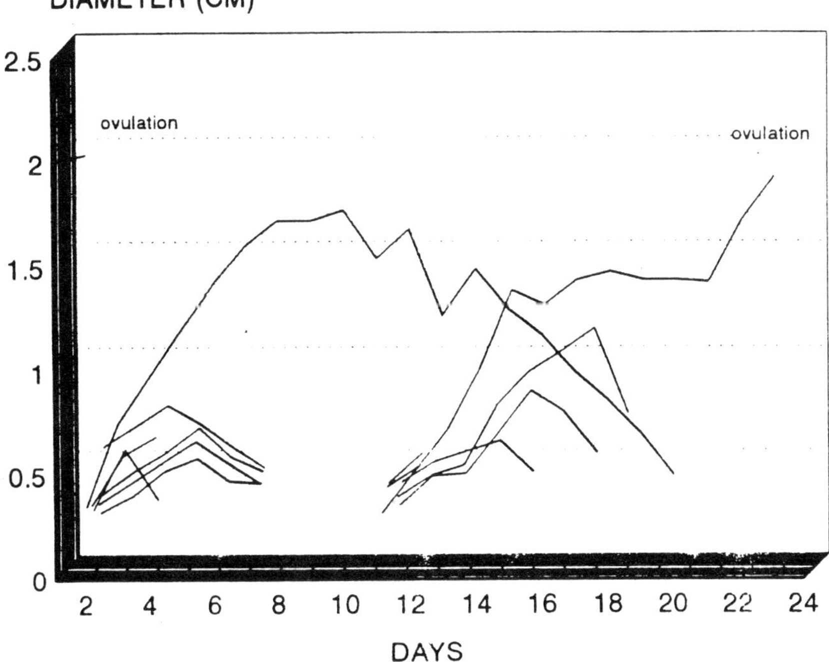

FIGURE 1. Buffalo presenting two waves of follicular development. São Paulo, Brazil, 1994.

MATERIAL AND METHODS

In this research, 30 Murrah buffalo were used belonging to a herd from a private farm situated in São Paulo State, in the southeast of Brazil, during the autumn season, in the period from March to July.

As an aid for estrous detection, two vasectomized bulls were used and two androgenized female buffalo, fitted with chin ball markers. The reproductive behaviors were observed twice a day at 6:00 AM and 6:00 PM.

Female follicular development was examined and followed using a scanner 480, Pie Medical 5.0/7.5 MHZ frequency transducer designed for intrarectal placement, with a printer. This examination was performed every morning, after the heat detection procedure, using the 7.5 MHZ frequency transducer monitoring the growth, ovulation, and regression of ovarian follicles during the estrous cycle.

Skill of operation and a clear contact point in which to use the probe were necessary in ultrasonographic examination. Fecal materials from the rectum were removed before inserting the transducer. For orientation, the transducer was moved along the dorsal surface of the reproductive tract, then moved laterally to examine the ovaries. In the active involving position, the image was retained on the screen and the size of the follicle was determined and recorded in the printing outlet of scanner. Diagrams of the relative positions of the follicle as well as ovarian structure

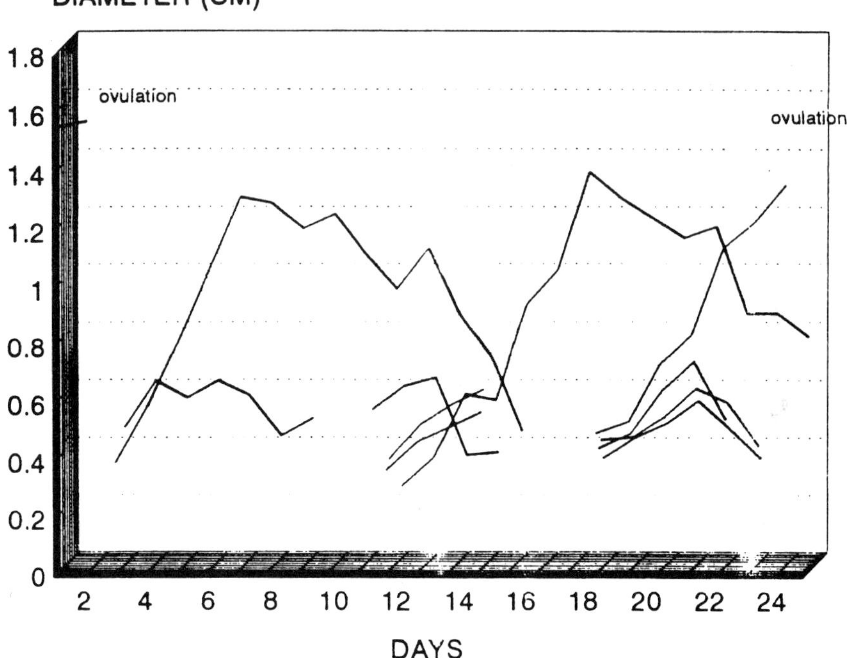

FIGURE 2. Buffalo presenting three waves of follicular development. São Paulo, Brazil, 1994.

were drawn. In order to reduce error, all examinations were carried out by the same operator. The ovarian activities of all animals were examined, once a day, during the period of an estrus cycle, beginning after the end of estrus until next heat.

RESULTS

The results indicated the presence of three types of waves of follicular growth per cycle. Two waves of follicular growth were observed in 19 buffalo (corresponding to 63.33%), while in 10 females, three waves were detected (33.33% of the total). In only one animal was observed the presence of one single wave (3.33%).

Each wave was characterized by the development of one dominant (large) follicle and a variable number of smaller follicles.

For those females with two follicular waves, the first and second waves started on day 1.16 ± 0.50 and 10.83 ± 1.09, respectively (FIG. 1). While in those cows with three waves, that started on day 1.10 ± 0.32, 9.30 ± 1.25 and 16.80 ± 1.22, respectively (FIG. 2). In our paper, day zero corresponds to ovulation day.

The estrus cycle with two dominant follicles had an average (± SD) duration of 21.84 ± 1.01 days, while those with three dominant follicles had an average (± SD) duration of 24.00 ± 2.21 days and 13 days for one wave cycle.

TABLE 1. Frequency, Length of Estrus Cycle, and Day of Start of the Follicular Wave, According to Number of Waves, São Paulo, Brazil, 1994.

	1 Wave	2 Waves	3 Waves
Frequency (%)	3.33	63.33	33.33
First wave (day)	1	1.16 ± 0.50	1.10 ± 0.32
Second wave (day)	—	10.83 ± 1.09	9.30 ± 1.25
Third wave (day)	—	—	16.80 ± 1.22
Estrus cycle (days)	13	21.84 ± 1.01	24.00 ± 2.21

The estrus cycle with one dominant follicle had a short duration of 13 days. This cycle probably was the first cycle observed in the postpartum period in buffalo.[5] Results are showed in TABLE 1.

CONCLUSIONS

Results supported the two-wave hypothesis, as the Hajakowsk theory for cattle,[6] but also indicated that buffalo may have three waves of follicular growth. More seldom, there are also those with only one dominant follicle and respective wave, when the cycle is short.

REFERENCES

1. DANELL, B. 1987. Oestrous behavior, ovarian morphology and cyclical variation in follicular system and endocrine pattern in water buffalo heifers. Thesis Faculty of Veterinary Medicine, Swedish University of Agricultural Sciences. Uppsala, Sweden.
2. LE VAN TY, D. CHUPIN & D. A. DRIANCOURT. 1989. Ovarian follicular population in buffaloes and cows. Anim. Reprod. Sci. **19:** 171-178.
3. SINGH, G., G. B. SINGH, S. S. SHARMA & R. D. SHARMA. 1984. Studies on oestrous symptoms of buffalo heifers. Theriogenology **21:** 849-858.
4. TANEJA, M., M. S. TOTEY, C. H. PAWSHE, G. SHING, A. ALI & G. P. TALWAR. 1991. Ultrasonographic monitoring of ovarian follicular dynanics in superovulated buffaloes. *In* World Buffalo Congress, **3:** 581-85. Sofia, Bulgaria.
5. BARUSELLI, P. S., W. G. VIANA, R. G. MUCCIOLO, C. A. OLIVEIRA & E. H. MADUREIRA. 1994. Ovarian activity and sexual behaviour during postpartum period in buffaloes. *In* IV Word Buffalo Congress: 440-442. São Paulo, Brasil.
6. RAJAKOSKI, E. 1990. Ovarian follicular system in sexually mature heifers with special reference to seasonal cyclical and left-right variations. Acta Endocrinol. **34:** (Suppl. 52): 379-392.

Herbage Density of Third-Stage Larvae of Goat Strongyles during the Dry Season in Guadeloupe

R. SIMON,[a] G. AUMONT,[a] R. APRELON,[b] AND
N. BARRÉ[b]

[a]Institut National de la Recherche Agronomique
Unité de Recherches Zootechniques
BP 515
97165 Pointe à Pitre cédex
Guadeloupe, French West Indies

[b]Centre International de Recherche Agronomique pour le
Développement
Elevage et Médecine Vétérinaire des pays Tropicaux
BP 515
97165 Pointe à Pitre cédex
Guadeloupe, French West Indies

INTRODUCTION

Gastrointestinal strongyloses are usually regarded as a major problem for the development of small ruminant breeding in the tropics. The widespread prevalence of *Haemonchus* and to a lesser extent of *Trichostrongylus* induces losses due to death, retarded growth, and unthriftiness.[1-3] It has been claimed that there is no doubt that gastrointestinal strongyloses are the most common and serious diseases in small ruminants of the Caribbean.[4] The importance of hemonchosis in goat breeding in terms of prevalence and pathology was shown in an experimental farm in Guadeloupe (French West Indies, FWI).[5,6] However, epidemiological data on these helminthiases of small ruminants in farms of Guadeloupe were not available.

Due to the increase of resistance of *Haemonchus* and other strongyles to anthelmintics in the West Indies[7-9] pasture management is now regarded as a potential and alternative approach to controlling strongylosis in small ruminants. In Guadeloupe, infestation risk was simulated by Aumont *et al.*[10] with models fitted on marginal kinetics of third-stage larvae (L3) on herbage after experimental dung deposition in pasture, according to the irrigation practice or herbage mass availability.[11,12] These models showed that the pasture management system, particularly the rotating system, may be effective to control helminths in small ruminant breeding. However, no studies in farms were performed to support these simulations.

Finally, no clear evidence was given on how third-stage larvae can survive to the drastic conditions of the dry season in the Caribbean. No studies have been conducted to support evidence that *Haemonchus* and *Trichostrongylus* can achieve their complete life cycles in such conditions. Therefore, the objective of this study was to determine the main sources of variation in herbage densities of infective third-stage larvae of goat strongyles in farms during the dry season in Guadeloupe. These observations were collected during a complete epidemiological survey of goat strongylosis in different climatic and geographic parts of Guadeloupe in 1993 and 1994. It allowed estimation of the L3 population size according to pasture management and amount of deposed eggs, in different locations characterized by contrasted different microclimates that are commonly encountered in the West Indies: (a) calcareous areas on coral limestones with marked dry season and different typographies of small hills, (b) windward or leeward regions of acid soil areas, (c) high-altitude volcanic region with large annual rainfalls. The design of the survey was fitted in order to discriminate environmental effects (locations and climates), pasture management system effects (continuous grazing, rotating grazing, tethered grazing animals) and eggs deposition effects on L3 densities, egg hatching, and development during the marked dry season of 1994.

MATERIAL AND METHODS

Experimental Design

The epidemiological survey was carried out in Guadeloupe (lat. 16°0′N, long. 61°3′ W) between January and May 1994. Out of 397 breeders surveyed in 1992, 21 farms were chosen among 397 breeders spread on the main islands and regions of the archipel of Guadeloupe, for their goodwill of collaboration with research, in order to be allocated in an unequilibrated 5 × 3 design as follows:

- 5 locations: Northern Grande Terre (GT) (coral limestone area with marked dry season and rainfall of 1250 mm/year); Hills region of GT (small hills of 30–70 m of height on coral limestones area, rainfall of 1500 mm/y, contrast between humid valley and dry top of hills); Windward Basse Terre (BT) (acid soils, altitude ranging between sea level to 200 m; rainfall ranging between 3000–4000 mm/y), Leeward BT (acid soils annual, altitude ranging between sea level to 200 m; rainfall ranging between 2500–3000 mm/y with a marked dry season), "Soufrière" region (windward part of volcano region, altitude ranging between 300 and 1500 m, annual rainfall ranging between 4000 to 6000 mm/y).

- 3 pasture management systems: continuous grazing, rotating grazing, tethered grazing animals. This factor was crossed with the use of specific night stocking paddocks and the irrigation practice.

- 4 sampling periods (SP).

Animals ($n = 365$) were sampled 4 times each 4 weeks for fecal egg counts (FEC). Animals were drenched after sampling during the second period with ivermectin (IVOMEC, ND). The age of animals was estimated by examination of the teeth. Three category of animals were retained: lower than 3 months, between 3 and 18 months, adults. Paddocks ($n = 58$) were sampled at the same time for dry matter content of soil (SDM) and for L3 density in herbage (LD) determinations according to Aumont and Gruner.[12] Stocking rate and pasture management events were recorded at each period of sampling. No drenching was carried out by breeders during a 2-month period before the beginning of the survey. Climatic data (rainfall and global radiation) were collected daily in meteorological stations no farther than 2 km from each farm.

Parasitological Methods

FECs were carried out by suspension in water, centrifugation, and examination in NaCl dense solution in a MacMaster slide according to Aumont *et al.*[13] Culture of fecal material and Baerman method for the determination of genus of L3[14] were carried out according to sex, class of age, and sampling period. L3 density (LD) in herbage was assessed by a sedimentation and sucrose/water interface method[15] and expressed per kg of dry matter (DM). LD were estimated for the three genera encountered i.e., *Haemonchus, Trichostrongylus, Oesophagostomum.*

Calculations and Statistical Methods

An index of egg hatching and development in larvae (IEDL) was calculated as the ratio of LD to an estimation of the number of deposed eggs per area unit during the 4th, 3rd, and 2nd weeks before sampling as follows:

$$i = 4 \ a = 3$$

$$\text{IEDL} = \text{LD}/\Sigma \ (\Sigma \ (\text{FEC}_{ia}*\text{BW}^{0.75}/\text{S})$$

$$i = 2 \ a = 1$$

where LD was the L3 density, i was the number of weeks before herbage sampling, a was the class of age from suckling young ($a = 1$) to adult ($a = 3$), FEC_{ia} was the mean FEC for each week and class of age, $\text{BW}^{0.75}$ was the metabolic bodyweight as an indicator of total fecal output, and S was the surface of paddock.

LD and IEDL variables were analyzed by general linear models[16] after logarithmic transformation to normalize the distributions:

$$Y_{ijk} = \text{REG}_i + \text{PM}_j + \text{SP}_k + \text{SP*REG}_{ik} + e_{ijk}$$

where Y were the LD or IWDL variables, REG was the region factor, PM was the pasture management factor, SP was the sampling period factor, and e_{ijk} was the residual.

Two groups of variables were constituted: (a) the breed and environmental variables (type of breed, pasture management, irrigation, stocking rate, SDM, climatic data) and (b) the parasitological variables (frequency of genera in L3 population, LDs, IEDLs). Continuous variables were transformed in ordinal categorical variables of 5 levels defined by the quartiles plus 1 level for nil values. Multiple component factorial analysis[17,18] was performed on breed and environmental variables. Then,

TABLE 1. Ranges of Monthly Rainfall (mm/Month) Recorded in Surveyed Farms during the Dry Season in the Five Sampled Locations in Guadeloupe[a]

Month	Jan 1994	Feb 1994	Mar 1994	Apr 1994
Northern GT	51−63	15−52	8−23	7−24
Hills of GT	58−125	51−63	13−37	16−53
Windward BT	63−113	58−97	28−97	34−87
Leeward BT	49−148	54−153	18−36	17−41
"Soufrière" region	318−643	213−725	17−310	93−167
Means Calculated over a 10 Year Period				
Northern GT	48−65	46−55	29−58	56−101
Hills of GT	55−78	48−57	53−85	77−95
Windward BT	61−120	48−113	85−109	94−123
Leeward BT	52−104	24−48	16−55	38−75
"Soufrière" region	312−653	218−460	314−578	285−572

[a] GT: Grande Terre island; BT: Basse Terre island.

correlations between parasitological variables and principal components were calculated.

RESULTS AND DISCUSSION

Observed monthly rainfall compared with the mean over a 10-year period (TABLE 1) clearly supported that 1994 was a year of marked dry season in the different regions of Guadeloupe. The drought increased with the number of sampling periods. Therefore, it can be assumed that this epidemiological survey was conducted during a dry period that became more and more pronounced.

Worm burden in tracer animals may be used as an indicator of infectivity of pastures. But it has been shown that pasture sampling for L3 determination is more precise than tracer animal methods particularly for detection of high larval density detection.[19] Furthermore, the direct sampling technique is more adapted to field studies in small farms, for which the use of tracer animals may be expensive and may disturb pasture management system and stocking rates. *Haemonchus* and *Trichostrongylus* were the main observed genera of parasites. *Oesophagostomum* sp. were observed but with a frequency lower than 12%. These observations confirmed that *Haemonchus contortus* and *Trichostrongylus* sp. remained the most important gastrointestinal parasites in small ruminants in the West Indies, as was suggested by other limited epidemiological studies in the lesser Antilles.[5,20,21] It must be emphasized that *Trichostrongylus axei* was also present with *Trichostrongylus colubriformis* (results not shown), although the other studies in Guadeloupe and Martinique reported only *Trichostrongylus colubriformis* in sheep and goats.

TABLE 2. Distribution of Third-Stage Larvae Density in Herbage (LD), Egg Hatching and Development in Larvae (IEDL) Recorded in Guadeloupe (FWI) during the Dry Season, 1994

Variables:	LD	LD	LD	IDHL	IDEL	IDEL
Genus:	AG[a]	*Hae*	*Tric*	AG	*Hae*	*Tric*
Unit:	L3/ kgDM	L3/ kgDM	L3/ kgDM	[b]	[b]	[b]
Nil value (%)	7.7	00.5	11.2	10.4	14.4	17.3
n^c	143	128	127	103	89	86
Mean[c]	8,111	4,699	4,153	450	660	975
Minimum[c]	42	21	21	0.1	0.1	0.3
Median[c]	3,397	1,853	1,410	41	33	48
Maximum[c]	114,748	64,259	50,489	17,335	21,200	29,239

[a] AG: All genera of parasites including *Haemonchus* (Hae), *Trichostrongylus* (Tric), and *Oesophagosptomum.*

[b] $L3/kgDM/eggs.kg^{0.75}/g/m^2$.

[c] Results based on no nil values.

Distribution of LD, IEDL, and genus frequency is reported in TABLE 2. The method for LDS determination has a high accuracy (good recovery of added L3)[15] and a low detection limit (190 L3/kg DM). This characteristic might explain the very great size of L3 population that was recorded in some paddocks (higher than 50,000 L3/kg DM). It must be underlined that despite the marked dry season, the frequency of nil values remained low for the two main genera, *Haemonchus* and *Trichostrongylus* (TABLE 2). These data suggest that in the West Indies, even in a dry area, the complete life cycle of strongyles might be achieved even during a marked dry season.

The sampling period, the region, and the irrigation practice were the main sources of variation of LD variables. Total explained variance by the models of LD variables was 41.5%, 46.7%, 17.3%, respectively for all genera, *Haemonchus* and *Trichostrongylus,* supporting evidence that the sampling design was adapted to analyze the variations in LD variables. The LD recorded in irrigated paddocks of the Northern GT and in the windward region were higher than in the other regions (TABLE 3). As the dryness lasted according to the sampling periods, the LD significantly decreased (TABLE 3; p < 0.01), but it remained important (496 L3/kg DM) despite the drought. The size of L3 population on herbage was sometimes very important (higher than 10,000 L3/kg DM). Surprisingly, these high values were not related either to pasture management systems or to the night stocking paddock. The IEDL for the two main genera of parasites were higher in the irrigated paddock of the Northern GT region ($p < 0.01$).

The frequency of *Trichostrongylus* in L3 population increased from 32% to 64% from the first sampling period to the last one. This result can be explained in part by a better development of *Trichostrongylus* which is not strongly inhibited during the dry sampling period compared with other genera (TABLE 3). As a matter of fact, IEDL of *Trichostrongylus* was higher than that of *Haemonchus* in the last two sampling periods ($p < 0.05$).

TABLE 3. Estimated Means According to the Regions and the Sampling Period (from Jan 1994 to April 1994) of Third-Stage Larvae Density in Herbage (LD), Egg Hatching, and Development in Larvae (IEDL) Recorded in Guadeloupe (FWI) during the Dry Season, 1994

Variables:	LD	LD	LD	IDHL	IDEL	IDEL
Genus:	AG	*Hae*	*Tric*	AG	*Hae*	*Tric*
Unit:	L3/kgDM	L3/kgDM	L3/kgDM	[a]	[a]	[a]
Regions						
Northern GT no irri	624^b	154^b	895^b	49^b	27^b	39^b
Northern GT irri	4537^b	$1095^{b,c}$	2057^c	307^c	141^c	424^c
Hills of GT	566^b	263^b	32^b	12^b	22^b	17^b
Windward BT	6195^c	3217^c	377^b	23^b	20^b	28^b
Leeward BT	2040^b	$959^{b,c}$	427^b	18^b	6^b	84^b
"Soufrière" region	1694^b	$1177^{b,c}$	453^b	28^b	15^b	40^b
Sampling periods						
January	3286^b	2000^b	823^b	—	—	—
February	3676^b	3422^b	1004^b	34^b	55^b	44^b
March	$1760^{b,c}$	$560^{b,c}$	435^b	40^b	20^b	97^c
April	496^c	77^c	351^b	32^b	12^b	39^b

[a] $L3/kgDM/eggs.kg^{0.75}$ g/m^2.

[b,c] Means within factor with different subscript letter are significantly different ($p < 0.01$).

No clear relationship was found by correlation procedure between parasitological variables and rainfall data. In contrast, there were negative coefficients of correlation between LD and global radiation variables ranging between −0.21 and −0.40 according to species of parasite and the sampling period. These results confirm the influence of global radiation on L3 dynamics, as was shown for cattle L3 population around pats in Guadeloupe.[22] The factorial analysis of breed and environmental variables showed that the different samplings in the different locations could be sorted from the dryness situations (Northern GT, no irrigated paddock, last sampling period) to humid situations (irrigated paddock, first sampling period, windward GT). In fact, the factorial axis can be described as an axis of dryness. The correlation of the first factorial axis, which can be described as an axis of dryness, with the frequency of the different genera of parasite was −0.39, 0.45 and −0.24, for *Haemonchus, Trichostrongylus,* and *Oesophagostomum,* respectively. In contrast, the correlation of the first component with the LD variables was −0.33, 0.04 and −0.30, for *Haemonchus, Trichostrongylus,* and *Oesophagostomum,* respectively.

These different facts supported evidence that dry conditions and high temperature have less constraint for *Trichostrongylus* eggs hatching and larval development than for *Haemonchus.* Eggs of *Trichostrongylus* seem to better resist dessication than those of the other strongyles. Such results have already been shown in the field[23–25] and in the laboratory.[26,27]

CONCLUSIONS

The use of a very accurate and precise method to extract and count L3 in herbage permitted the measurement of very high level of LD, even in very dry conditions. Our results suggested that the complete life cycle of gastrointestinal strongyles can be achieved during marked dry season in the West Indies. Irrigation is one of the major factor that improves egg hatching and development in the field, confirming previous experimental studies[10] and the importance of this practice for the infestation risk of pasture management in the West Indies.[11] The windward region of Basse Terre appeared as a location with a greater infestation risk than the other regions and islands of Guadeloupe, in agreement with its high annual rainfall and slight dry season. However, some excess of rainfall as it might happen in very humid regions like the "Soufrière" region, appeared unfavorable to egg hatching and larval survival in herbage. *Haemonchus* and *Trichostrongylus* were the main parasites encountered. The respective importance of these genera depended on the dryness conditions. However, studies on animals with worm burden examination at necropsy, larval examination after fecal culture, and animal biological indicators are required to confirm the significance and the importance of these sources of variation in the incidence and prevalence of gastrointestinal strongyles in goats of Guadeloupe.

SUMMARY

The objective of this study was to determine the main sources of variation in herbage densities of infective third stage larvae of goat strongyles during the marked dry season of 1994 in Guadeloupe (FWI). Herbage samples were collected for L3 density (LD) determination by an accurate method, 4 times at 4-week intervals in 58 paddocks of 21 farms spread out in five regions of the archipel of Guadeloupe. At the same time, FEC of each grazing animal and fecal culture for parasite genus determination according to sex and age were carried out. Stocking rate, dry matter content of soil, and daily climatic data were also recorded. An index of egg development in larvae (IEDL) was calculated as the ratio of LD to the eggs deposed during the 4th, 3rd, and 2nd weeks before sampling. Medians of LD in herbage were 3397, 1853, 1410, and 324 L3/kg DM for all parasites, *Haemonchus, Trichostrongylus,* and *Oesophagostomum,* respectively. Date of sampling, region, and irrigation practice in the northern windward region were the main sources of variation in LD and in frequency of each parasite. LD decreased as the dryness lasted, but it remained important (500 L3/kg DM) despite the drought. LD in windward regions were higher than in other regions. The region, the farm, and the paddock were the main sources of variation of IEDL. LD of each parasites were inversely correlated to global radiation recorded 1 to 3 weeks before herbage sampling, but no relation was found with rainfall data. *Trichostrongylus* frequency in L3 population increased as the dryness lasted. A dryness axis was extracted from environmental variables (climatic data, dry matter of soil, duration of dryness) by a multiple factorial procedure. LD and *Haemonchus* frequency in L3 population were inversely correlated to dryness axis ($p < 0.01$). In contrast, *Trichostrongylus* frequency was positively correlated to the dryness component.

REFERENCES

1. EYSKER, M. & R. A. OGUNSUSI. 1980. Observations on epidemiological and clinical aspects of gastro-intestinal helminthiasis of sheep in North Nigeria during the rainy season. Res. Vet. Sci. **28:** 58–62.

2. FABIYI, J. P. 1987. Production losses and control of helminths in ruminants of tropical regions. Int. J. Parasitol. **17:** 435–442.

3. TILLARD, E. 1991. Prophylaxies chez les petits ruminants au sénégal: évaluation technico économique de leurs effets en milieu villageois, mémoire de fin d'étude, DESS Maison Alfort. Paris, France.

4. WILLIAMS, H. 1990. Diseases of ruminants in the Caribbean with special focus on ticks and tick-associated diseases. *In* Cowdriosis and Dermatophilosis of Livestock on the Caribbean region, STA Seminar Proceedings, 12–14 Nov. 1990. St. John (Antigua), CARDI and CTA Ed.: 31–62.

5. GRUNER, L., F. PEROUX & P. CHEMINEAU. 1984. Distribution et rôle de l'haemonchose dans un élevage semi-intensif de chevreaux de race créole en Guadeloupe. In Les Colloques de l'INRA. **28:** 705–715. INRA Versailles, France.

6. PEROUX, F. 1982. Epidémiologie des parasitoses gastro-intestinales des caprins à la Guadeloupe. Maisons-Alfort. ENVA. Thesis.

7. GRUNER, L., D. KERBOEUF, C. BEAUMONT & J. HUBERT. 1986. Resistance to benzimidazole of *Haemonchus contortus utkalensis* in sheep on Martinique. Vet. Rec. **118:** 276.

8. BASTIEN, O., D. KERBOEUF, F. LEIMBACHER, J. GEVREY, J. A. NICOLAS, J. HUBERT & O. HEINRICH. 1991. Recherches des causes d'échecs thérapeuthiques de la lutte contre les strongyloses gastro-intestinales des ovins en martinique. Rev. Elev. Méd. Vét. pays Trop. **141:** 117–122.

9. HUBERT, J., J. KERBOEUF & G. AUMONT. Resistance of *Haemonchus contortus* field strain to netobimin in Guadeloupe. Vet. Rec. (In press.)

10. AUMONT, G., L. GRUNER & P. BERBIGIER. 1991. Dynamique des populations des stades infestants de strongles gastrointestinaux des petits ruminants en milieu tropical humide. Conséquences sur la gestion des pâturages. Rev. Elev. Méd. Vét. pays Trop. **141:** 123–131.

11. GRUNER, L., P. BERBIGIER, J. CORTET & C. SAUVE. 1989. Effects of irrigation on appearance and survival of infective larvae of goat gastro-intestinal nematodes in Guadeloupe (French West Indies). Int. J. Parasitol. **19:** 409–415.

12. AUMONT, G. & L. GRUNER. 1989. Population evolution of the free living stage of goat gastrointestinal nematodes on herbage under tropical conditions in Guadeloupe. Int. J. Parasitol. **19:** 539–546.

13. AUMONT, G., N. BARRE, S. DIAW, L. OUATTARA, R. POUILLOT, E. TILLARD & G. VASSILIADES. 1995. Preservation of fecal samples and laboratory sources of variation in fecal strongyles egg counts of small ruminants from tropical regions. International Conference on Novel Approaches to the Control of Helminth Parasites of Livestock, Armidale, Australia, 18–21 April.

14. GRABER, M. & C. PERROTIN. 1983. Helminthes et helminthoses des ruminants domestiques d'Afrique tropicale. Maisons-Alfort, IEMVT: 378.

15. AUMONT, G., D. FRAULI, R. SIMON, R. POUILLOT, S. DIAW & N. MANDONNET. A method for counting infective third stage larvae of gastrointestinal nematodes of small ruminants in pastures under tropical conditions. Vet Parasitol. (In press.)

16. SAS INSTITUTE INC. 1987. SAS/STAT Guide for Personal Computers. Version 6 Edit. NC: SAS Institute Inc.

17. BENZECRI, J. P. 1972. L'analyse des données, Tome 2: L'analyse des correspondances. 2nd Edit. Dunod. Paris, France.

18. GIFI, A. 1981. Non Linear Multivariate Analysis. Department of Data Theory. University of Leiden. Leiden, The Netherlands.

19. WALLER, P. J., R. J. DOBSON, A. D. DONALD & R. J. THOMAS. 1981. Populations on strongyloid nematode infective stages in sheep pastures: comparisons between direct

pasture sampling and tracer lambs as estimators of larval abundance. Int. J. Parasitol. **11:** 359-367.

20. ESTERRE, P. & M. J. MAITRE. 1985. Les affections parasitaires des ruminants en Guadeloupe. Rev. Elev. Méd. Vét pays Trop. **1:** 49-53.

21. POUILLOT, R. 1993. Epidémiologie des strongyloses gastro-intestinale des caprins créoles en Guadeloupe, saison humide. DESS Maison Alfort. Paris, France.

22. AUMONT, G., D. GAUTHIER, L. GRUNER & G. MATHERON. 1992. Dynamics of the free-living populations of gastro-intestinal trichostrongyles of cattle in a natural grazing system in Guadeloupe (French West Indies). Prev. Med. Vet. **12:** 245-258.

23. LEVINE, N. D., J. R. TODD & P. A. BOTTMAN. 1974. Development and survival of *Haemonchus contortus* on pasture. Am. J. Vet Res. **35:** 1413-1422.

24. BERBIGIER, P., L. GRUNER, M. MAMBRINI & S. A. SOPHIE. 1990. Faecal water content and egg survival of goat gastro-intestinal strongyles under dry tropical conditions in Guadeloupe. Parasitol. Res. **76:** 379-385.

25. BANKS, D. J. D., R. SINGHJ, I. A. BARGER, B. PRATAP & L. F. LE JAMBRE. 1990. Development and survival of infective larvae of *Haemonchus contortus* and *Trichostrongylus colubriformis* on pasture in a tropical environment. Int. J. Parasitol. **20:** 155-160.

26. WALLER, P. J. & A. D. DONALD. 1970. The response of dessication of eggs of *Trichostrongylus colubriformis* and *Haemonchus contortus* Nematoda. (Trichostrongylidae). Parasitology **61:** 195-204.

27. LEVINE, N. D. & J. R. TODD. 1975. Micrometeorological factors involved in development and survival of free-living stages of the sheep nematodes *Haemonchus contortus* and *Trichostrongylus colubriformis.* Int. J. Biometeor. **19:** 174-183.

Genetic Variability in Resistance of Creole Goats to Natural Infection with Trichostrongylids in Guadeloupe[a]

N. MANDONNET,[b] G. AUMONT,[c] J. FLEURY,[d]
L. GRUNER,[b] J. BOUIX,[e] AND J. VU TIEN KHANG[e]

[b]INRA Station de Pathologie Aviaire et de Parasitologie
37380 Nouzilly, France

[c]INRA Unité de Recherches Zootechniques
BP 515
97 165 Pointe-à-Pitre Cedex
Guadeloupe, French West Indies

[d]INRA Domaine Expérimental de Gardel
97160 Le Moule
Guadeloupe, French West Indies

[e]INRA Station d'Amélioration Génétique des Animaux
BP 27
31326 Castanet-Tolosan Cedex, France

INTRODUCTION

Helminthosis constitutes a widespread pathology of grazing small ruminants in tropical climates with a prevalence of near 100%. The zootechnical losses are obvious but hard to quantify. Until now, anthelmintic treatments were the most common methods of control but the increasing frequency of drug resistance makes them less and less efficient.[1] Therefore, alternative approaches to control internal parasites have to be taken into account such as pasture management and host genetic variations in resistance. Evidence for genetic variation in goats for resistance to endoparasites is little documented although goats are numerous, specially in tropical areas. In Kenya, Rohrer *et al.*[2] studied resistance in Kenyan Dual Purpose goats aged between 10 and 12 months, infected mainly with *Haemonchus contortus*. They published heritability estimates very similar to those for sheep. In contrast, Woolaston *et al.*[3] reported very low heritabilities for Fidjian goats naturally infected with *Haemonchus contortus* and *Trichostrongylus colubriformis*.

In Guadeloupe, *Haemonchus contortus* limited the growth of kids in a semiintensive breeding system,[4] and development of anthelmintic resistance obliged owners

[a]This project was supported by the INRA AIP "Gènes de résistance aux maladies" and by the CIRAD ATP "Strongyloses gastro-intestinales des petits ruminants en milieu tropical."

TABLE 1. Number of Male and Female Kids from the 4 Cohorts Measured at 6 and 10 Months of Age

		Number of	
Cohort	Age	Male Kids	Female Kids
1	6 months	—	—
	10 months	25	23
2	6 months	27	21
	10 months	26	21
3	6 months	26	22
	10 months	26	26
4	6 months	27	23
	10 months	21	19

to take care of pasture management.[5,6] The objective of our experiment was to estimate the genetic variation of Creole goats for resistance to helminth infections in an experimental flock in Guadeloupe (French West Indies). Parasitological, hematological, and production parameters were recorded on those animals during fattening.

MATERIALS AND METHODS

Location and Experimental Design

The experimental flock was located in the dry part of Guadeloupe (16 0′N, 61.3 W). The climate of this area is characterized by an annual rainfall of 1280 mm and a marked dry season (December–June, 67 mm per month). The genetic variation of the resistance of 13 bucks to internal parasites was assessed by the mean performance of their offspring, fattened between 3 and 11 months of age. The genetic origins of the 13 sires were chosen as different as possible in the flock, in which resistance was never selected and treatment frequency was high (10 per year). The 13 bucks were mated to 149 goats, and 203 offsprings were used for this experiment (TABLE 1). Each buck was represented by 12 six-month-old kids on average.

Kids were bred on an intensive pasture (0.9 ha) composed of Pangola grass (*Digitaria decumbens*) irrigated (when rainfall was nil during 10 successive days) and receiving 300 kg N/ha/year. The rotational system consisted of 7 days of grazing and 21 days of herbage regrowth. A high stocking rate was achieved: 1400 kg/ha. The flock was organized in 3 lambing times every 2 years, with two groups of goats. So two cohorts followed one another with a 4-month interval. Four cohorts were included in this experiment. They were suckled on pasture and weaned at 3 months of age on average. At weaning, kids were separated according to sex. After weaning, our principle was to leave the animals in contact with parasites during 5 to 6 weeks before treatment. Then they were reinfected by grazing during 5 to 8 weeks up to the individual monitoring at 6 months of age. A drenching was carried out to end the first experimental period. The animals were then 7 months old. A second experimental period followed with the same principle. Individual tests were done when animals were 10 months old. The experiment was conducted from June 1993 to September 1994 and involved 4 periods of measures (A = August-September 1993, B = December 1993-January 1994, C = April-May 1994, and D = July-August 1994). A and B occurred during the rainy season, C and D during the dry season.

Weight Gains, Parasitological Records, and Blood Analyses

Animal and pasture infection levels were performed in order to avoid death of kids and to compare parasite pressures between groups and between seasons. The infection level, estimated as fecal egg count, was weekly estimated on 12 randomly choosen kids in each group. The larval stage three (L3) population density on herbage was sampled every week on the paddock grazed by the animals. L3 extractions from herbage were performed as previously described,[7] and counts were realized according to the method of Eysker and Kooyman.[8] Individual fecal egg counts (FEC) were recorded at 6 and 10 months of age, with a repetition a week later during the third period. The FEC was estimated by a modified McMaster's technique[9] using flotation in saturated magnesium sulfate solution (density $d = 1.27$). At each period, the proportions of the different genera and species of Trichostrongylids were estimated on L3 from fecal cultures (one culture per sire and per sex) and on adult worms collected on 2 necropsied kids per sex period. The larvae were extracted from feces using the modified Baermann method.[10] Blood samples were collected from each kid at 6 and 10 months of age for hematologic analyses. The packed cell volume (PCV) was measured by the capillary microhematocrit method.

Kids were weighed at birth, twice a month between birth and weaning, at weaning, at the beginning of experiment (two weeks after weaning), then monthly between 3 and 11 months of age. These records enabled us to calculate average daily gains between different dates: weaning and the first test at 6 months, weaning and second test at 10 months of age.

Statistical Methods

Packed cell volume was studied as rough data; a square root transformation was applied to FEC in an attempt to normalize the data distribution. Data definition and distributions are shown in TABLE 2. Data were analyzed using the computer package SAS.[11] Data were analyzed including the following fixed variation factors when significant in the model: a combining litter size/rearing status (5 levels), a combining group/cohort effect (GROUP: 8 levels), an age effect (AGE: 2 levels; 6 and 10 months). Two random effects were included in the repeatability models: a worm burden effect nested in animal describing the variability of the infections at 6 and 10 months of age for each animal, and an animal effect nested in GROUP describing the variability of the infection between 6 and 10 months of age. Repeatability within a one-week interval was calculated as the following ratio:

$$(\sigma^2_A + \sigma^2_{WB})/(\sigma^2_A + \sigma^2_{WB} + \sigma^2_e)$$

where: σ^2_A was the animal effect variance, σ^2_{WB} was the worm burden variance, and σ^2e was the residual variance.

TABLE 2. Definition and Distribution of the Variables[a]

Variable	Number of Observations	Units	Mean	Standard Deviation
Age at "6 month" record	149	days	192	8.9
Age at "10 month" record	197	days	306	11.8
FEC at 6 months of age	98	epg	32.47	14.41
FEC at 10 months of age	140	epg	26.66	13.47
PCV at 6 months of age	145	percent	24.91	3.14
PCV at 10 months of age	190	percent	29.47	3.36
Live weight at 6 months of age	149	kg	11.71	1.93
Live weight at 10 months of age	190	kg	16.82	2.74
Weight gain between weaning and 6 months of age	126	g/day	28.97	11.12
Weight gain between weaning and 10 months of age	190	g/day	35.68	9.09

[a] FEC: fecal egg count.

Repeatability between 6 and 10 months was calculated as the ratio

$$(\sigma^2_A)/(\sigma^2_A + \sigma^2_{WB} + \sigma^2_e)$$

RESULTS

Contamination of the pastures and infection of animals and infection levels reached on pastures and for male and female kids are represented in FIGURE 1. The paddocks were old mowing fields turned into pasture for the experiment, so the infection of kids remained moderate during the first period of observations. The infection rhythm seemed to be quick and explosive. During the second period, a self cure occurred: 80% of the 10-month-old kids had nil EPG and 75% of the 6-month-old kids had less than 200 EPG. At this time no worm was found in the abomasum of 3 necropsied male kids,and only L4 were observed in 2 necropsied female kids. This phenomenon seemed to be related to a heavy ingestion of larvae. FEC data recorded at this time have been suppressed in the statistical analysis. The frequency of *Haemonchus contortus* in L3 populations issued from fecal cultures and in worm burdens was high (FIG. 2) during the rainy season: 77% to 99%. *Trichostrongylus colubriformis* settled down gradually during the dry season and became the main species in the last period: 67% to 93%.

Effect of Age on FEC, PCV, and Body Weight Gains

The 6-month-old kids were in general more infected than older kids (TABLE 3). They excreted more eggs, and they were more anemic. The weight of 10-month-old kids was lower in August 1994, corresponding to the end of an exceptionally long

Male kids

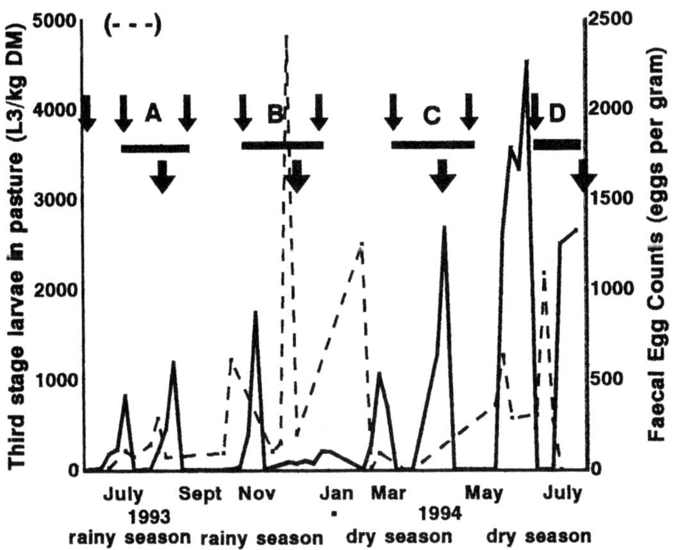

Female kids

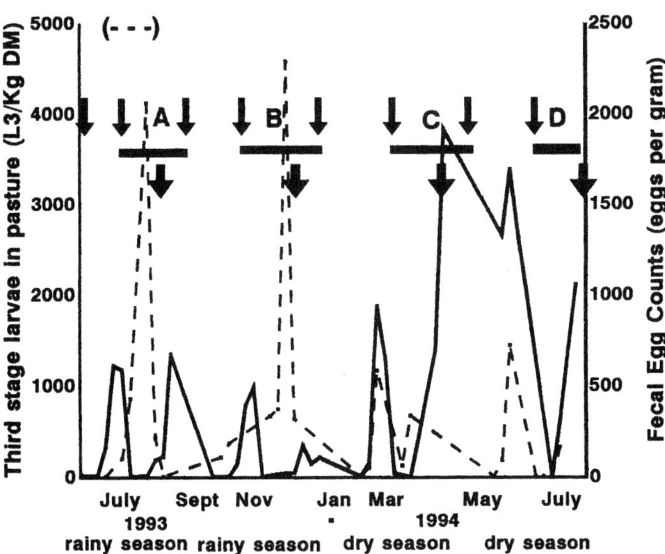

FIGURE 1. Evolution of male and female kid infection (EPG) during the 4 experimental periods and evolution of parasite pressure on pasture measured as infective larvae per kg of dry matter (DM) of grass. A, B, C, D are related to the 4 experimental periods, initiated and ended by a drenching (thin arrow). The large arrow indicates the time of individual measures.

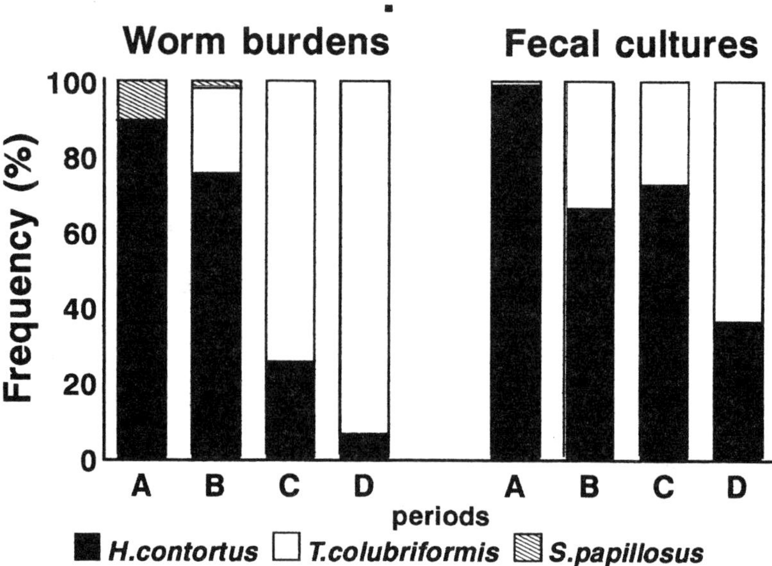

FIGURE 2. Proportions of *Haemonchus contortus, Trichostrongylus colubriformis,* and *Strongyloides papillosus* recovered in worm burdens and fecal cultures during the 4 experimental periods.

dry season. A complementary cohort of half-sib offspring was bred on an intensive system and regularly drenched to have an estimate of the growth without parasite infection. Those animals weighed 18.8 kg on average at the end of fattening. The infected kids were 1 to 3 kg lighter than the control ones.

Phenotypic Correlations

At 6 months of age, the correlation between FEC and PCV measures was −0.46 ($p < 0.001$). At 10 months of age, it was −0.29 ($p < 0.01$). Egg output was not significantly correlated with growth parameters: at 6 months of age, the correlation coefficients were −0.09 between FEC and weight and −0.07 with weight gain between weaning and 6 months. At 10 months of age, these coefficients were respectively −0.06 and −0.01. In contrast, PCV measured at 6 months of age was related to weight and weight gain between weaning and 6 months: 0.36 ($p < 0.001$) and 0.39 ($p < 0.001$) respectively. At 10 months of age, the relations between PCV measures and growth parameters were not significant any more: 0.12 and 0.10 respectively.

Variability of FEC and PCV

Repeatabilities at one-week intervals and at 4-month intervals were computed on FEC and PCV measures (TABLE 4). High values were obtained on 2 measures within

TABLE 3. Means of Fecal Egg Counts (FEC), Hematocrit (PCV), and Weight Adjusted for Sex at the 4 Experimental Periods

		Mean for	
Period	Variable	6-Month-Old Kids	10-Month-Old Kids
August–Sept.	FEC[a]	1156	264
1993	PCV	24	30
	Weight	12.4	17.9
Dec. 1993	FEC[a]	288	48
Jan. 1994	PCV	25	30
	Weight	11.7	16.3
Apr.–May	FEC[a]	582	688
1994	PCV	26	29
	Weight	10.9	17.0
Jul.–Aug.	FEC[a]	–	1018
1994	PCV	–	30
	Weight	–	15.8

[a] Geometric means of variables normalized by logarithmic transformation

an age, but values between 6 and 10 months of age were less important. Individual effect was significant on the 2 parameters. Repeatability estimates were very high for PCV, but we did not demonstrate any genetical variability for this parameter. On the contrary, FEC measures were moderately repeatable at a short-term interval and nearly nil at the 4-month interval. The sire effect tested on FEC measured at 6 months of age was significant ($p < 0.03$); sire variability reached 17.29 EPG vs. 179.38 EPG for residual variability. The sire effect tested on FEC measured at 10 months of age was nearly significant ($p < 0.15$): sire variability reached then 4.15 EPG vs. 112.88 EPG for residual variability. The ranking of the 13 bucks on 6-month-old kid performances was slightly correlated to their ranking on 10-month-old kid performances (FIG. 3): Spearman correlation coefficient was 0.38 ($p < 0.10$).

DISCUSSION

The experiment ran over 14 months and successively covered a rainy season and a particularly long dry one (February–September 1994). The seasons affected the

TABLE 4. Repeatability Estimates on FEC and PCV Measures at One-Week Interval and at 4-Month Interval

			Repeatability for a	
Variable	Individual Effect Significance	Number of Observations	1-Week Interval	4-Month Interval
FEC	$p < 0.04$	201	0.47	0.06
PCV	$p < 0.001$	200	0.99	0.41

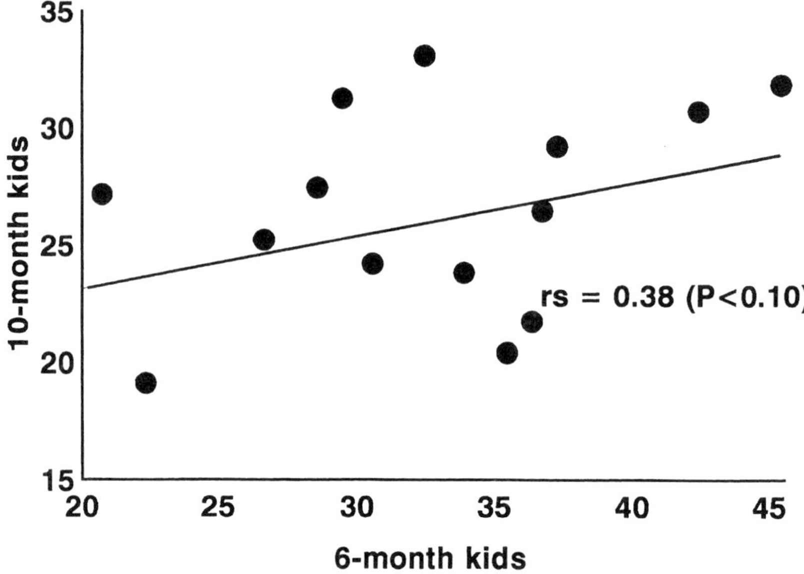

FIGURE 3. Comparison of sire ranking when performances were estimated on FEC at 6 or 10 months of age.

species composition of nematode parasites, populations of *H. contortus* decreasing and those of *T. colubriformis* increasing with time to be predominant mainly during the last observation period D. The irrigation had not totally compensated for the lack of rainfall, and the growth of the kids was lower during the last period D. The rotation system of one week in and three weeks off was chosen from previous ecological studies and management modelization,[6] animals coming back when maximal number of infective larvae occurred. It would permit a rapid infection of the kids. It was confirmed by the high level of L3 populations on herbage at the entrance of the animals on each paddock. As a consequence, the FEC of the groups of kids quickly increased after a drenching, this all along the experiment. Very low values of heritability were found[3] in goats of the Fidji islands. In this study, the evaluation of FEC repeatability was performed by using animals from different flocks. To limit the environmental variability, we have chosen to work in one experimental flock, in order to have all the genealogy information, and to estimate the genetic variability of resistance in animals bred in an environment as constantly as possible. The irrigated pangola pasture allowed significant sire effects when the offspring were evaluated at 6 months of age. Actually, only 13 sires were evaluated, a number too low to propose a good estimate of the heritability of FEC. The sire effect was less significant on performances at 10 months of age. A moderate correlation between sire performances at 6 and 10 months of age was estimated. Our results are not consistent with the other studies.[14] Baker *et al.*[14] showed a genetic correlation between 2 FEC measures at a 2-month interval not significantly different from unity. However, our repeatability estimate of FEC was nearly nil between 6 and 10 months of age. In fact, the 6-month-old kids were more susceptible than older ones. As all animals were tested at the same time, the infection level of 10-month-old kids was probably not high enough to express their resistance level. The important parasitological pressure created in our experimental system permitted during the rainy season (during period B) a huge larval ingestion, and self-cure occurred as verified by FEC and in necropsied animals. At that time, FEC was not usable to test animal resistance to Trichostrongylids. In the same period, hematocrit remained as low as during the other periods. A correlation was shown between FEC and PCV measures. PCV also seemed to be a sensitive parameter to growth variation in infected animals. At 6 months of age, the phenotypic correlation between PCV and growth parameters was similar to those reported by Blackburn *et al.*[12] for goats artificially infected with *H. contortus,* and to those reported by Baker *et al.*[13] for sheep naturally infected with *H. contortus.* For example, they respectively reported 0.46 and 0.27 between PCV and live weight. PCV was a highly repeatable measure, but no genetic control of this parameter was evidenced under our experimental conditions. Rohrer *et al.*[2] reported a heritability estimate for PCV values of dual-purpose goats naturally infected with *H. contortus* as high as the 0.22 found for sheep. Further investigations are needed in Creole goats to understand this lack of genetic variability of hematocrit.

In conclusion, a genetic variability of fecal egg count is shown in Creole goats at 6 months of age naturally infected with *H. contortus* and *T. colubriformis.* This was less evident at 10 months of age. Further studies are running to confirm this genetic determinism of resistance to natural infection with strongyles in Creole goats, and to investigate the variability of hematocrit values.

SUMMARY

The objective of this study was to show the existence of genetic variability in resistance of Creole goats to natural infection with intestinal nematodes. Four successive cohorts of male and female kids were reared from weaning during 9 months in intensive pangola pastures (stocking rate 1.4 t/ha, regrowth of 21 days, 300 kg N/ha/year of fertilization). All cohorts considered, 203 offsprings from 13 bucks and 149 goats were used in the experiment. Animals were treated with an oral dose of ivermectin every seven weeks, from 3 to 11 months of age. They were sampled for fecal egg counts (FEC) and packed cell volume (PCV), 6 weeks after drenching, when 6 and 10 months old. *Haemonchus contortus* and *Trichostrongylus colubriformis* were the predominant species. Data of one cohort when 6 months of age and another one when 10 months old were not included in calculations because a self-cure occurred during a rainy season. Estimated means of FEC after square-root transformation were 1054 and 711 respectively for 6-month-old and 10-month-old kids. Phenotypic correlations between FEC and PCV were -0.46 ($p < 0.001$) and -0.29 ($p < 0.01$) when 6 and 10 months old, respectively. FEC repeatabilities estimated as variance ratio were 0.47 and 0.06 for within-age measures (data recorded at one week interval) and between age measures (6 vs. 10 months) respectively. No sire effect was significant for PCV. Sire effect on FEC was more significant when 6 months old ($p < 0.03$) than when 10 months old ($p < 0.15$). Further studies are now being carried out to confirm this genetic determinism of resistance to natural infection with strongyles in Creole goats.

ACKNOWLEDGMENTS

Thanks are due to the breeding team of Gardel, and especially to J. C. Mainaud, for their contribution in the organization of this experiment, to H. Varo, the personnel of the laboratory, and the team of the slaughterhouse for their technical assistance, to Dr. N. Barre for his hospitality in his laboratory at CIRAD-EMVT, Centre de Recherches des Antilles-Guyane, Guadeloupe.

REFERENCES

1. PRITCHARD, R. 1994. Anthelmintic resistance. Vet. Parasitol. **54:** 259–268.
2. ROHER, G. A., J. F. TAYLOR, S. K. DAVIS, R. M. WARUIRU, F. RUVUNA, B. A. J. MWANDOTTO, T. MCGUIRE & F. RURANGIAWA. 1991. The use of randomly amplified polymorphic DNA markers in analysis of susceptibility to *Haemonchus* and *Coccidia*. Proceedings of the 9th Scientific Workshop of the Small Ruminant Collaborative Research Support Program: 71–85.
3. WOOLASTON, R. R., R. SINGH, N. TABUNAKAWAI, L. F. LE JAMBRE, D. J. D. BANKS & I. A. BARGER. 1992. Genetic and environmental influences on worm egg counts of goats in the humid tropics. Proceedings of the Australian Association for Animal Breeding and Genetic. **10:** 147–150.
4. GRUNER, L., F. PEROUX & P. CHEMINEAU. 1984. Distribution et rôle de l'haemonchose dans un elevage semi-intensif de chevreaux de race creole en Guadeloupe. *In* Les Maladies de la Chevre. Ed. INRA Publ. 705–715.
5. GRUNER, L. 1985. Contrôle des strongyloses digestives des petits ruminants aux Antilles Francaises: développement de résistances aux benzimidazoles et intérêt d'une gestion raisonnée des paturages. Rev. Elev. Med. Vet. pays Trop. **38:** 386–393.
6. AUMONT, G., L. GRUNER & P. BERBIGIER. 1991. Dynamique des populations de larves infestantes des strongles gastro-intestinaux des petits ruminants en milieu tropical humide. Conséquences sur la gestion des pâturages. Rev. Elev. Med. Vet. pays Trop. **141:** 123–131.
7. GRUNER, L. & J. P. RAYNAUD. 1980. Technique allégée de prélévement d'herbe et de numération pour juger de l'infestation des pâturages de bovins par les larves de nématodes parasites. Rev. Med. Vet. **131:** 521–529.
8. EYSKER, M. & F. N. J. KOOYMAN. 1993. Notes on necropsy and herbage processing techniques for gastrointestinal nematodes of ruminants. Vet. Parasitol. **46:** 205–213.
9. RAYNAUD, J. P. 1970. Etude de l'efficacité d'une technique de coproscopie quantitative pour le diagnostic de routine et le contrôle des infestations parasitaires des bovins, ovins, équins et porcins. Ann. Parasitol. Hum. Comp. Paris **45:** 321–342.
10. GRUNER, L. 1986. Strongyle larval recovery from ovine faeces sampled on pasture: efficiency of the Baermannization and epidemiological interest of the technique. IV International Symposium of Veterinary Laboratory Diagnosticians, 2–6 June, Amsterdam, the Netherlands: 186–189.
11. SAS INSTITUTE INC. 1989. SAS/STAT Guide. Version 6 Edit. SAS Institute Inc. NC.
12. BLACKBURN, H. D., J. L. ROCHA, E. P. FIGUEIREDO, M. E. BERNE, L. S. VIEIRA, A. R. CAVALCANTE & J. S. ROSA. 1992. Interaction of parasitism and nutrition in goats: effects on haematological parameters correlations, and other statistical associations. Vet. Parasitol. **44:** 183–197.

13. BAKER, R. L., L. REYNOLDS, D. M. MWAMACHI, J. OUMA, M. MAGADI & J. E. MLLET. 1993. Genetic resistance to gastro-intestinal parasites in Dorper and red Massai × Dorper lambs in coastal Kenya. Proceedings of the 11th SR-CRSP Scientific Workshop, 3-4 March, Nairobi, Kenya.

14. BAKER, R.L., T. G. WATSON, S. A. BISSET, A. VLASSOF & P. G. C. DOUCH. 1991. Breeding sheep in New Zealand for resistance to internal parasites: research results and commercial application. *In* Breeding for Disease Resistance in Sheep. G. D. Gray & R. R. Woolaston, Eds.: 19-32. Australian Wool Corporation. Melbourne, Australia.

Incidence of Bovine Myiasis (*Cochliomyia hominivorax*) in the Central Area of Argentina

O. S. ANZIANI[a] AND M. M. VOLPOGNI[b]

[a]*Estación Experimental Agropecuaria Rafaela*
Instituto Nacional de Tecnología Agropecuaria
CC 22, 2300 Rafaela, Santa Fe, Argentina

[b]*Facultad de Agronomía y Veterinaria*
Universidad Nacional del Litoral
Kreder 2805, 3080 Esperanza, Santa Fe, Argentina

The fly *Cochliomyia hominivorax* is a common pest in central and northern Argentina, but very little is known about the incidence of myiasis in the livestock production of this country. The present study was designed to survey the seasonal incidence of bovine cutaneous myiasis at the dairy farm of the Rafaela Experimental Station, National Institute of Agriculture Technology, located in central Argentina (31° 11′ S; 61° 30′ W) from August 1992 to October 1993. The dairy herd has approximately 230 milking purebred Holstein cows together with 90 heifers, 70 steers, and 140 calves. Parturitions occur in two periods, one from July to September and the other from February to April. During the 15 months of the census, animals were inspected daily and the body region with active myiasis, the age of the animal, and observation date were recorded. A minimum of 10 larvae were collected for laboratory identification from approximately 40% of the animals showing active myiasis.

A total of 111 cases of myiasis were observed in the warmest months of the year, with 9.9% occurring in the spring, 50.5% during the summer, and 39.6% in the fall. All collected larvae were classified as *C. hominivorax*. The body regions most frequently infested were the ears injured by ear tagging procedures (27%), the navel of the newborn calves (24%), and the vulva of postparturient cows (21.7%). The highest incidence was reported in March with 33 cases of myiasis observed. The March peak in screwworm infestations could be explained not only by natural increase of the fly population, but also by the large number of calvings and newborn calves that occurred in this month providing many attractive sites to ovipositing female flies. This is also supported by the fact that navel of newborn calves and vulva of postparturient cows were responsible for more than 45% and 72% of the overall and March strikes respectively. During the period of study, temperature was in the normal pattern for the region according to the weather data from 30 years.

Low winter temperatures seem to be the most important factor affecting screwworm populations in the central area of Argentina. No cases of myiasis were observed from June to October, although parturitions and others predisposing husbandry procedures also occurred at this time. These data suggest that in this area the survival of *C. hominivorax* flies could be uncertain in winter. The new cases produced in the spring might be largely the result of repopulation of this insect from overwintering areas to the north of the country where screwworm is a problem throughout the year.[1] Migration could be carried out by flight of the adults and/or by passage of larvae in infested livestock. At present, there are no regulations to control the transport of infested livestock.

REFERENCES

1. CARDONA LOPEZ, G. A., O. BALBUENA & C. A. LUCIANI. 1994. Miasis del ganado en la región noreste de la Provincia del Chaco. Vet. Arg. **11:** 304–313.

The Role of Botfly Myiasis Due to *Dermatobia hominis* L.Jr. (Diptera: Cuterebridae) as a Predisposing Factor to New World Screwworm Myiasis (*Cochliomyia hominivorax* Coquerel) (Diptera: Calliphoridae)[a]

I. RUÍZ-MARTÍNEZ,[b,e] F. GÓMEZ,[c] J. M. PÉREZ,[c] AND
F. A. POUDEVIGNE [d]

[b]*United States Department of Agriculture (USDA)*
Agricultural Research Services (ARS)
Screwworm Research Laboratory
San José, Costa Rica

[c]*Department of Animal Biology and Ecology*
Area of Parasitology
Faculty of Experimental Sciences
Box 62
University of Jaén
E-23071-Jaén, Spain

[d]*United States Department of Agriculture (USDA)*
Agricultural Research Services (ARS)
Screwworm Research Program
Panama City, Panama

INTRODUCTION

Known as "Tórsalo," myiasis due to the botfly, *Dermatobia hominis*, (BF) or dermatobiasis[1] is easily identified by its characteristic foruncular cutaneous lesions. Botflies are responsible for most myiasis in Central and South America (from Mexico to Argentina).[2] The effect on the wild fauna is unknown; however, BF maggots are a serious pest to livestock and humans. The literature on epidemiology is extensive[3]

[a]This project has been supported by a postdoctoral grant from the MEC/Fulbright Commission (Ref. Fu92-26737749-Res.Jul92).

[e]Address correspondence to Department of Animal Biology and Ecology, Area of Parasitology, Parasitic Diptera Research Unit, Box 62, Faculty of Experimental Sciences, University of Jaén, E-23071-Jaén, Spain.

but too reiterative and often not clear. In Costa Rica, some comprehensive studies on dermatobiasis have been done,[4,5] but altogether, little information about BF biology is available.[6]

According to Catts,[6] adult BF do not visit hosts. Instead, they oviposit on a carrier, usually a zoophilic fly or mosquito.[f] The carrier behavior allows BF eggs to hatch on a reduced surface on the host's skin provoking a great concentration of botfly lesions (BFL). These extended lesions could be an important source of proteins for screwworm flies and possibly oviposition sites.

Ranchers frequently report by that BFL are a predisposing factor to myiasis due to the primary parasite New World Screwworm, *Cochliomyia hominivorax* (NWS). Field observations in Costa Rica and Mexico did not agree with the cattle owners' observations: of the 1500 BFL we studied in "Hacienda Montezuma" (Cañas, Guanacaste, Costa Rica) and the 1000 BFL observed in Yucatan (Mexico),[7] none served to promote primary NWS myiasis or secondary myiasis *Cochliomyia macellaria.* Nevertheless, both Thomas' observations in Mexico and ours in Costa Rica needed to be confirmed.

The main objectives of this study are (1) to evaluate the precise role of BFL in cattle as a predisposing factor to NWS myiasis, (2) to evaluate the importance of BFL as a food supply to NWS gravid females and (3) to determine experimentally whether NWS larvae could actually survive and grow in BFL.

MATERIALS AND METHODS

This study was conducted in Costa Rica from May to October 1993 at the Animal Husbandry Central America (Escuela Centroamericana de Ganadería) in Atenas, Costa Rica.

NWS Flies

NWS flies proceeded from our strain "Cañas & Ochomogo 92" (Costa Rica) and were collected nine days after emergence.[8] NWS wild flies were individually captured with insect nets while they fed on rotten beef liver placed on the ground.[9] They were transferred to our laboratory at "El Alto de Ochomogo" (San José, Costa Rica) and held for 2 days in screened cages containing separate dishes of water and protein medium to insure a high rate of gravidness. On the third day, the NWS females (wild and laboratory strains) were individually transferred into a 50-ml test tube tapped with cotton until the oviposition field assays on *Dermatobia* lesions (BFL). After oviposition tests (with positive or negative results), females were dissected in the laboratory under stereoscopic microscope to determine the number of eggs or ovarian developmental stage.

---

[f]However, several times we observed adult BF on sheep, cattle, or horses. The reason for these visits remains unknown.

BFL

During a period of 5 weeks, 15 cows, each with 20 to 50 BFL were observed. An anatomical chart was prepared for each animal, and the number of BFL and their location were recorded as well as the animal's identification number, color, sex, and breed.

pH

In order to determine the possible interference of pH with NWS oviposition on BFL,[10,11] the pH evolution was recorded from 400 BFL before each test by means of a portable pH-meter and 0.6 mm diameter electrode (Porter™) and cross-checked using pH paper.

Test

In field conditions, only late third instar BFL can be considered as attractive to NWS. The first and second instars are barely noticeable. We sorted the BFL into three groups:

Group 1: BFL with active larvae of third instar (experiment no. 1). At this stage, BFL present persistent serous discharge and an opening of 3 to 5 mm diameter; 160 tests were performed in this oviposition group.

Group 2: 12 to 24 hours after the larvae of third instar crawled off naturally (experiment no. 2). The BFL have serous or purulent exudate, and the opening is frequently closed by a scab. In this group, 135 tests were performed.

Group 3: the third instar larvae has been manually forced out of the lesions, in the way most owners do (experiment no. 3). One or 2 hours after that manipulation the exudate is still abundant, seropurulent, and mixed with blood. The opening is 3 to 7 mm diameter. In this group, 165 tests were carried out.

Following this classification, we designed three oviposition tests respectively on group 1, 2, and 3 of BFL (showed as "exp. 1" etc. in TABLES 1 and 2). The variable was the kind of BFL. In all three experiments, the BF myiasis was not disturbed in its natural development and was allowed to mature until all BF larvae reached the third instar and left the furuncle (naturally or forced out).

The tubes containing the gravid NWS flies (containing one fly per tube) were placed to cup each kind of BFL, and obscurity was obtained by covering the tube with the holding hand. Previously we recorded the pH and extruded some liquid exudate from the lesion.[12,13] Thomas and Mangan[12] observed that some NWS flies would divide their clutch into two or more masses. The fly would either leave the wound after the first clutch and come back a few minutes later to oviposit again, or lay another egg mass in a different wound. These authors reported that the mean time for oviposition of NWS was 15 minutes. We used the same lapse of time to measure the oviposition in our experiments. However, if a female began to oviposit before 15 minutes, the test was extended until oviposition was completed. In this study, we referred to a fly moving an ovipositor excitedly as "oviposition behavior," and we referred to a fly that fed on BFL as "feeding behavior." Both behaviors were recorded during the test, and the flies were identified and marked with a number. After the test, all flies were dissected under a stereoscopic microscope to determine the number of eggs that were left in the ovaries, or the stage of oogenesis. If any oviposition occurred, the eggs were counted *in situ* using a magnifying glass, and the development of the clutch was observed through 12-h periods.

RESULTS AND DISCUSSION

The NWS oviposition rates varied from 5.21% for wild flies to 7.41% for laboratory flies (TABLE 1). Moreover, for the 460 NWS females tested, this lack of attraction for oviposition was noticeable too in the ratio of eggs laid/eggs available (only 7.05%) (TABLE 1).

For clutch size, a maximum of 240 and a minimum of 18 eggs were recorded. The average number of eggs per fly varied between 156.2 and 176.6 (mean = 165.2). There were significant differences between laboratory flies (159.7) and wild flies (187.5) (TABLE 1). These data meet with the normal rates obtained in other studies such as (246 eggs per female),[14] (214 to 290 eggs per female),[15] or (mean: 200 eggs per clutch, range 28-492).[12] Only seven complete eggs masses were observed in our study (22% from the total positive cases).

Oviposition is preceded by grooming of the fly abdomen, exertion of the ovipositor, and a search for a suitable site.[12] Accordingly to this behavior, several hypothesis could explain the reluctance of NWS fly to oviposition on BFL.

Hypothesis A

pH in the foruncle is not suitable for NWS oviposition. Recently Guerrini *et al.*[16] showed the role of pH in gravid female selection and larval development in the Australian sheep blowfly *Lucilia cuprina* Wied. The pH rate associated to NWS myiasis in artificially infested sheep, rabbits, and guinea pigs (Poudevigne, Welch and Ruíz-Martinez, unpublished data) was substantially alkaline, while furuncle fluids (whether aqueous or purulent) from BFL were physiologically acidic. No significant differences in pH values were recorded in BFL within the period of the study. pH varied between 6.57 ± 0.05 and 6.91 ± 0.046 (6.5-6.9). The NWS does not oviposit on pus.[7] However, we observed that the active BF myiasis is rarely purulent. It produces an aqueous exudate until 48 hours after the third instar larva has left the lesion (naturally or forced out). Only 2 to 5% of lesions had purulent exudate (always without larvae), and they generally had a scab and presented a dry surface. In our opinion, the pH in BF myiasis is the main reason for the NWS repulsion and for the poor development of the larvae whenever an oviposition does occur.

Hypothesis B

The microflora developed in BFL might be unsuitable for attraction and stimulation of NWS gravid females. Recently we studied the microflora associated with BF

TABLE 1. Results of Oviposition for NWS *C. hominivorax* on BFL by *D. hominis*[a]

	Exp. 1	Exp. 2	Exp. 3	Total
Total no. tested	160	135	165	460
No. ovipositions	11	10	11	32
Percent oviposition	6.87	7.40	6.67	6.95
Maximum no. eggs layed	138	230	240	240
Minimum no. eggs layed	18	38	18	18
Average eggs layed	85.8 ± 34.5	114.6 ± 67.2	104.9 ± 72.6	−
Total no. eggs layed	942	1146	1154	3,242
Percent resp. total available	29.06	35.35	35.59	7.05
No. flies without ovarian St = 10^b	34	22	26	82
Maximum no. eggs in ovaries	325	325	387	387
Minimum no. eggs in ovaries	12	45	35	12
Average eggs in ovaries	156.2 ± 62.5	176.6 ± 62.3	160.5 ± 68.8	165.2
No. lab. flies tested	124	105	135	364
No. ovipositions	10	8	9	27
Percent ovipositions	8.06	7.61	6.67	7.41
Maximum no. eggs layed	135	170	240	240
Minimum no. eggs layed	18	38	18	18
Average eggs layed	87.0 ± 36.2	92.6 ± 53.2	89.6 ± 71.0	−
No. flies without ovarian St = 10^b	30	15	21	66
Maximum no. eggs in ovaries	302	325	324	325
Minimum no. eggs in ovaries	12	45	35	12
Average eggs in ovaries	143.9 ± 57.1	175.9 ± 64.6	153.4 ± 62.2	−
No. wild flies tested	36	30	30	96
No. ovipositions	1	2	2	5
Percent ovipositions	7.83	6.52	6.52	5.21
Maximum no. eggs layed	74	230	194	230
Minimum no. eggs layed	74	175	153	74
Average eggs layed	74	202.5 ± 38.8	173.5 ± 29.0	−
No. flies without ovarian St = 10^b	4	7	5	16
Maximum no. eggs in ovaries	325	267	387	387
Minimum no. eggs in ovaries	71	74	49	49
Average eggs in ovaries	192.6 ± 64.3	179.3 ± 53.7	192.7 ± 87.6	−
No. larvae at L-I developed	0	0	82	82
Percent resp. total eggs layed	−	−	7.11	2.53
No. larvae at L-II developed	0	0	0	0
Percent resp. total eggs layed	−	−	−	−

[a] NWS: new world screwworm; BFL: botfly lesions.
[b] Females without mature eggs.

(Sancho, Caballero and Ruíz-Martínez, unpublished data) and the prevalent species isolated were *Staphylococcus* spp. while NWS associated microflora is mainly of the *Enterobacteriaceae* family.[17,18,13] The *Proteus* spp., *Providencia* spp. group also plays an important role in the attraction of NWS gravid females.[18] In relation to hypothesis A, it is noticeable that *Staphylococcus* spp. produce acidic fluids where *Enterobacteriaceae* produce alkaline fluids.

TABLE 2. Behavior of Female *C. hominivorax* on Contact with *D. hominis* Lesions

	Exp. 1	Exp. 2	Exp. 3	Total
Total no. tested	160	136	165	460
No. lab. flies tested	124	105	135	364
Percent resp. total	77.5	77.78	81.82	79.13
No. wild flies tested	36	30	30	96
Percent resp. total	22.5	22.22	18.18	20.87
Total no. ovipositions	11	10	11	32
No. flies moved ovipositor[a]	27	35	29	91
Percent resp. total flies tested	16.86	25.93	17.57	19.78
No. lab. flies moved ovipositor	16	24	16	56
Percent resp. total lab. flies	12.90	22.86	11.85	15.38
No. wild flies moved ovipositor	11	11	13	35
Percent resp. total wild flies	30.55	36.67	43.33	36.46
No. flies feeding on BFL[b]	129	108	137	374
Percent resp. total flies tested	80.62	80.00	83.03	81.30
No. lab. flies feeding on BFL	102	85	112	299
Percent resp. total lab. flies	82.78	80.95	82.96	82.14
No. wild flies feeding on BFL	27	23	25	75
Percent resp. total wild flies	75.00	76.67	83.33	78.12
No. simultaneous feed & moved ovip.	26	35	23	83
No. nonsimult. feed & moved ovip.	2	0	6	8

[a] Oviposition stimuli.
[b] Nutritional stimuli.

Hypothesis C

NWS association with BFL might only be for food; 81.30% of the NWS flies tested fed on the BFL—we refer to this as a "feeding behavior." There was no significant difference between laboratory flies (82.14%) and wild flies (78.12%) (TABLE 2). Both wild and laboratory flies were maintained and fed for a few days in cages to allow a proper maturation of eggs.[8,9] This insures that the flies were not hungry and that feeding behaviors were similar. Guillot *et al.*[19] concluded that visits to wounds fulfilled two functions: oviposition and nutrition. They reported that females fed on wound fluids, but the visits to hosts were primarily for oviposition. Thomas and Mangar's[12] findings contradicted this conclusion, observing that feeding without oviposition occurred in 62.35% of the cases. They concluded that feeding was the primary purpose of many visits to wounds. The nutritive value of the wound exudates is unknown, and the contribution to reproduction is unclear. Dermatobiasis is a prevalent problem in Central and South America,[4,5,20] and BFL might be a natural and abundant protein source for the NWS fly (in Costa Rica we observed many times NWS wild flies feeding on BFL in natural conditions). The results obtained in our study meet with these observations (TABLE 2).

A fly moving an ovipositor excitedly is referred to as displaying an "oviposition behavior," 19.78% of the tested flies showed oviposition behavior. There was a noticeable difference between laboratory flies (15.38%) and wild flies (36.46%) as far as oviposition behavior was concerned: 92.21% of the flies presenting an oviposition behavior simultaneously presented a feeding behavior. Only 22% of feeding behavior was combined with oviposition behavior. These observations confirmed little interest of NWS gravid flies in BFL as an oviposition site (TABLE 2).

Hypothesis D

The BFL might be too small and hard to be a suitable oviposition site for NWS. The importance of wound size is unknown, but the structure of the wound surface seems to be important. According to Laake,[21] 31% of NWS infestations occur in man-made wounds, and Thomas[7] reports that the NWS fly prefers bleeding wounds. Our recent study in Costa Rica (Ruíz-Martínez *et al.*—unpublished data) suggests that the role of wound size in the NWS larval development could be determinant. BFL offer a small and hard surface to the larval development of NWS. Nevertheless, it is common to see a high concentration of BFL on a small area of cattle skin.[22] When a cattle owner manipulates and pulls out the larvae, the erosion produced is very important (abundant serous exudates and blood) and could be attractive to NWS females (only one suspicion of primary NWS myiasis developed in such conditions was reported in Montezuma Tropical Rain Forest, Costa Rica). Of the 3242 eggs deposited on various BFL, only 82 (2.53%) reached the first instar larva and never developed a second instar even when the size of the clutch was important (200 to 230 eggs) and would have helped for a good development (TABLE 1).

CONCLUSION

This conclusion is supported by Thomas' field data.[7] BFL cannot be considered a predisposing cause for NWS myiasis in the tropics (either under dry or rainy season), and the phenomenon is rarely observed.

SUMMARY

In the tropics, the botfly *Dermatobia hominis* and the NWS *Cochliomyia hominivorax* are the most important myiasis agents in cattle. It is frequently reported that furuncular lesions due to *D. hominis* are a predisposing cause for screwworm myiasis. Our results pointed out that only 5.2 to 7.4% of *C. hominivorax* gravid females oviposited in the offered furuncular lesions. Of 3242 eggs layed on botfly lesions (BFL), only 82 (2.5%) developed to second instar and died. In the flies tested, the furuncular lesions due to *Dermatobia* were used as food supply in 81.3% of the cases. In our opinion, the role of pH, the microflora associated with BFL, and the foruncular structure were the reasons for this lack of attraction. BFL do not serve as a predisposing factor for screwworm myiasis in the tropics.

ACKNOWLEDGMENTS

The authors wish to express their gratitude to Escuela Centroamericana de Ganadería (ECAG) of Costa Rica and Mr. Angel Richmond for technical support and for research and installations facilities. Specially valuable was the helpful and technical support of Mr. G. Chinchilla in the ECAG and Hacienda Montezuma (Gte, Costa Rica). Our gratitude to Dr. Frank D. Parker (Research Leader, Costa Rica), Dr. John D. Welch (Research Leader, Panama), and Dr. Robert B. Matlock (USDA-ARS-Screwworm Program, Costa Rica). Thanks to Mr. M. Murillo (USDA-ARS-Screwworm Program, Costa Rica) for computer assistance. We thank our technicians José J. Richmond, Juan Bosco Granados, and Fernando Corrales for their field assistance, capture of the flies, and especially for their good work.

REFERENCES

1. KASAI, T., M. CORDERO DEL CAMPILLO, J. EUZEBY, S. GAAFAR, T. HIEPE & C. A. HIMONA. 1988. Standarized nomenclature of animal parasitic diseases (SNOAPAD). Vet Parasitol. **29:** 229-326.
2. GUIMARAES, J. H., N. PAPAVERO & A. P. DoPRADO. 1983. As miiases na regiao neotropical (identificaçao, biologia, bibliografia). Rev. Bras. Zool. **1:** 239-416.
3. GUIMARAES, J. H. & N. PAPAVERO. 1966. A tentative annotated bibliography of *Dermatobia hominis* (Linnaeus Jr., 1781) (Diptera: Cuterebridae). Arq. Zool. **14:** 223-294.
4. SANCHO, E., C. BOSCHINI & J. BOLAÑOS. 1983. Estudio del tórsalo en ganado vacuno. III. Distribución de la parasitosis en Costa Rica por zonas geográficas y en el cuerpo del hospedero (ganado vacuno). Cienc. Vet. **5:** 69-78.
5. SANCHO, E. 1987. *Dermatobia,* the neotropical warble fly. Parasitol. Today **4**(9): 242-246.
6. CATTS, E. P. 1982. Biology of New World bot flies: Cuterebridae. Annu. Rev. Entomol. **27:** 313-338.
7. THOMAS, D. B., JR. 1987. Incidence of screwworm (Diptera: Calliphoridae) and torsalo (Diptera: Cuterebridae) myiasis on the Yucatan Peninsula of Mexico. J. Med. Entomol. **24**(4): 498-502.
8. Crystal, M. M. 1967. Reproductive behaviour of laboratory-reared screwworm flies (Diptera: Calliphoridae). J. Med. Entomol. **4:** 443-450.
9. PARKER, F. D. & J. B. WELCH. 1991. Alternative to sentinel animals for collecting egg masses from wild females of the screwworm (Diptera: Calliphoridae). J. Econ. Entomol. **84**(5): 1476-1479.
10. LAAKE, E. W. 1939. On the hydrogen-ion concentration of myiotic wounds and its relation to the oviposition stimulus in *Cochliomyia americana* C. and P. Am. J. Trop. Med. **19:** 193-197.
11. LAAKE, E. W. & C. L. SMITH. 1938. The hydrogen-ion concentration of myiotic wounds in sheep and goats. J. Econ. Entomol. **31:** 441-443.
12. THOMAS, D. B. & R. L. MANGAN. 1989. Oviposition and wound-visiting behavior of the screwworm fly *Cochliomyia hominivorax* (Diptera: Calliphoridae). Ann. Entomol. Soc. Am. **82** (4): 526-534.
13. HAMMACK, L. 1990. Protein feeding and oviposition effects on attraction of screwworm flies (Diptera: Calliphoridae) to host fluids. Ann. Entomol. Soc. Am. **83:** 97-102.
14. CRYSTAL, M. M. & H. H. MEYNERS. 1965. Influence of mating on oviposition by screwworm flies. J. Med. Entomol. **2:** 214-216.

15. DE VANEY, J. A. & J. J. GARCIA. 1975. Longevity, oviposition and fertility of several strains of the screwworm *Cochliomyia hominivorax* (Diptera: Calliphoridae). J. Med. Entomol. **12:** 511–513.
16. GUERRINI, V. H., G. M. MURPHY & M. BROADMEADOW. 1988. The role of pH in the infestation of sheep by *Lucilia cuprina* larvae. Int. J. Parasitol. **18**(3): 407–409.
17. EDDY, G. M., J. A. DE VANEY & B. D. EKSTEE. 1975. Response of the adult screwworm (Diptera: Calliphoridae) to bacteria-inoculated and incubated bovine blood in olfactometer and oviposition tests. J. Med. Entomol. **12**(3): 379–381.
18. BROMEL, M., F. M. DUH, G. R. ERDMANN, L. HAMMACK & G. GASSNER. 1983. Bacteria associated with the screwworm fly *Cochliomyia hominivorax* (Coquerel) and their metabolites. *In* Endocytobiology **2:** 791–800. Walter de Gruyter & Co. Berlin, Germany.
19. GUILLOT, F. S., J. R. COPPEDGE, J. L. GOODENOUGH, E. AHRENS & T. S. ADAMS. 1977. Behaviour and reproductive status of native female screwworms attracted to a host. Ann. Entomol. Soc. Am. **70:** 588–590.
20. THOMAS, D. B. JR. 1988. The pattern of *Dermatobia* (Diptera: Cuterebridae) myiasis in cattle in tropical Mexico. J. Med. Entomol. **25**(2): 131–135.
21. LAAKE, E. W. 1936. Economic studies of screw worm flies, *Cochliomyia* species (Diptera: Calliphorinae), with special reference to the prevention of myiasis of domestic animals. Iowa State Coll. J. Sci. **10:** 345–360.
22. RIBEIRO, P. B. & C. M. OLIVEIRA. 1983. Fase parasitaria da *Dermatobia hominis* (L.Jr., 1781) (Diptera: Cuterebridae) sobre bovinos. Arq. Bras. Med. Vet. Zool. **35:** 691–698.

Larvicidal Activity of Abamectin against Natural *Cochliomyia hominivorax* Larva Infestation

O. S. ANZIANI,[a] A. A. GUGLIELMONE,[a] AND
D. H. AGUIRRE[b]

[a]*Estación Experimental Agropecuaria Rafaela*
Instituto Nacional de Tecnologíe Agropecuaria
CC 22, 2300 Rafaela, Santa Fe, Argentina

[b]*Estación Experimental Agropecuaria Rafaela*
Instituto Nacional de Tecnología Agropecuaria
CC 228, 4400 Salta, Argentina

The prophylactic activity of some members of the avermectin family against cutaneous myiasis caused by *Cochliomyia hominivirax* larvae has been recently reported.[1-3] This paper describes the results of a field trial carried out during December 1994 in a beef farm of the central area of Argentina (31° 27′ South 60° 56′ West) to evaluate the efficacy of abamectin in the prevention of myiasis caused by *Cochliomyia hominivorax* larvae. Fifty-six bull calves (*Bos taurus* × *Bos indicus*), 4 to 6 months old, and with a minimum body weight of 100 kg were allocated by ranked pairs to a control (CG) or to a treated group (TG) of the same number of animals on the basis of bodyweight. On day 0, all experimental animals were surgically castrated, and calves of TG were treated subcutaneously with abamectin (''Duotin'' Merck Sharp & Dohme AGVET) at a minimum dose of 200 µg/kg. Calves of CG remained as control animals. After castration, the calves of both groups were run together on the same pasture, and inspections for screwworm larva infestation were carried out on days 3, 5, 7, 10, 12, and 14. A minimum of 10 larvae were collected for laboratory identification from each infested animal. During the observation period, *C. hominivorax* egg masses were found on the scrotal wounds of most of the calves of both groups, but 27 animals of the CG (96.4%) and 1 of the TG (3.5%) developed active myiasis ($p <$ 0.01, chi-square). All collected larvae were classified as *C. hominivorax*. The results seem to indicate that abamectin did not interfere with the oviposition of gravid *C. hominivorax* females, and this drug exerted a larvicidal action after egg hatching. All scrotal strikes in the CG occurred during the first 5 days postcastration, while in the TG, calf myiasis was observed on day 10 (TABLE 1).

TABLE 1. Postcastration Myiasis in Calves Treated with Abamectin and Controls. Number of Calves with Active Myiasis

Treatment/No. of Calves	Days of Inspection Posttreatment						
	3	5	7	10	12	14	Total
Abamectin/28[a]	0	0	0	1	0	0	1[b]
Controls/28	9	18	0	0	0	0	27

[a] Minimum dose = 200 μg per kg of body weight.
[b] $p < 0.001$.

REFERENCES

1. ANZIANI, O. S. & C. LOREFICCE. 1993. Prevention of cutaneous myiasis caused by screw worm larvae (*Cochliomyia hominivorax*) using ivermectin. J. Vet. Med. B. **40:** 287-290.
2. BENITEZ USHER, C., J. CRUZ, L. CARVALLO, A. BRIDI, J. EAGLESON, D. FARRINGTON & R. A. BARRICK. 1993. Efficacy of ivermectin against cattle myiasis caused by *Cochliomyia hominivorax*. *In* 14th International Conference of the World Association for the Advancement of Veterinary Parasitology: 198. Cambridge, United Kingdom.
3. MUNIZ, R. A., O. S. ANZIANI, J. ORDONEZ, J. O. ERRECALDE, J. MORENO & R. S. REW. Efficacy of doramectin in the protection of neonatal calves and post-parturient cows against field strikes of *Cochliomyia hominivorax*. Vet. Parasitol. (In press.)

Factors of Variation of *Amblyomma variegatum* Infestation on Creole Cattle in Guadeloupe

M. NAVES,[a] N. BARRÉ,[b] M. FARGETTON,[a]
R. APRELON,[b] AND C. SHEIKBOUDOU[b]

[a]INRA
Unité de Recherches Zootechniques
BP 515
97165 Pointe à Pitre Cedex, Guadeloupe

[b]CIRAD-EMVT
BP 515
97165 Pointe à Pitre Cedex, Guadeloupe

INTRODUCTION

The direct incidence of ticks on animal production as well as their role in the dissemination of severe diseases raises interest in methods able to limit animal infestation.[1] In Guadeloupe, a Caribbean island where *Amblyomma variegatum* is widespread, little impact on the local Creole cattle has been noticed. But the danger this tick represents[2] led us to study the variations of the level of infestation on cattle in different conditions.

Five herds of cows and four lots of steers bred in an experimental farm were submitted to a long-term survey of tick infestation. The main purpose of the survey was to evaluate the efficacy of different acaricides.[2] In this study, we analyzed the effect of environmental and individual factors that could have an impact on tick control policy.

MATERIAL AND METHODS

Animal Management

The survey reached a total of 83 cows (C) weighing 368 kg and 7.5 years old, and 61 steers (S) between 8 and 18 months old, from 150 to 300 kg (TABLE 1). All the animals were grazing, either on native savannah (SAV) or on irrigated pastures of *Digitaria decumbens* (PAN). The stocking rate ranged between 2 and 4.7 heads/ ha according to the herd for the cows, and was 5.5 heads/ha on average for the steers.

Data Collected

On each animal, ticks were counted every 28 days, on the left-hand side, separately for adult male and female ticks. Cattle weights were registered with the same interval. Climatic data were also collected during the experimental period.

TABLE 1. Experimental Conditions[a]

Herd	n	Sex	Weight (kg)	Age (year)	Pasture	Head/ha	Period
1	19	C	326	8.3	SAV	3.0	February
2	16	C	338	8.0	SAV	3.4	to
3	18	C	412	7.0	PAN	4.7	May 90
4	15	C	387	7.2	SAV	2.0	February
							to
5	15	C	382	7.1	SAV	2.0	April 92
6	17	S	268	1.4	PAN	5.7	March 90
7	15	S	214	1.1	PAN	5.6	to
8	16	S	230	1.0	PAN	5.4	May 91
9	13	S	168	0.8	PAN	5.2	

[a] C: cows; S: steers; SAV: native savannah; PAN: *Digitaria decumbens* pastures.

In order to minimize the effect of the treatments applied, only tick counts over an average of 5 adult ticks per animal were taken into account. For the cows, the data concerned a period of 4 months without treatment, prior to the test of different acaricides. For the steers, the experimental period lasted 16 months, when a slight control of the tick infestation was only performed by spraying the animals every month. Four counts were recorded on cows, and four to eleven counts on steers.

Statistical Analysis

The tick count variable was normalized using a cube-root transformation. Statistical analyses were conducted using SAS software,[3] separately on male and female ticks counts, on cows and steers. Correlations and general linear model analysis were performed. Factors included in variance analysis were:

For cows: herd (H), rank of control (C), and their interaction (H*C), physiological stage (PS), as fixed factors; liveweight (LW) as covariable.

For steers: herd (H), season and rank of control intraseason [C(S)].

In genetic analysis, a sire effect (S) was introduced as random effect, for both sexes.

The repeatability was estimated by analysis of variance, with the only fixed factor (H) for cows or (H) and (S) for steers, and the random effect animal intraherd. It was also estimated on steers, by the correlations between counts made at three-month intervals.

RESULTS

Correlations between Variables

The correlations between tick counts and LW were only significant for the cows ($R^2 = 0.123$, prob. $= 5\%$), not for the steers. No correlation was found between tick counts and live weight gain. Male counts and female counts were highly correlated ($R^2 = 0.570$ on cows; 0.740 on steers; Prob. $= 0.1\%$).

Nongenetic Factors of Variation

All the factors were significant, unless PS and LW on female counts on cows (TABLES 2a and b). The fixed effects explained 25 to 40% of the variability of tick counts.

The main differences are due to herd management. For the cows, herds 4 and 5 were the most infested ones, while herds 2 and 3 had the lowest counts (FIG 1). Pregnant cows had more male ticks (2.32) than lactating or dry cows (2.00 and 1.98 respectively).

On the steers, climatic conditions (FIG. 2) had a strong influence on tick counts (FIG. 3), which followed approximately the variations of temperature.

Genetic Effects

The sire effect was significant on cows, but a tendency also exists on steers (TABLE 3).

On cows, repeatability was good for male counts, probably due to the stability of tick population within a season in absence of treatment, but poor for females ticks (TABLE 4).

Repeatability was low on steers, for measures repeated at different seasons (TABLE 4). But, estimated by the correlation between counts made at a three-month interval, repeatability was high when counts were made respectively in spring (March to May) and summer (June to August): $R^2 = 0.25$, (Prob. $= 8\%$) for male ticks and $R^2 = 0.33$ (Prob. $= 2\%$) for female ticks, or in summer and fall (September to November): $R^2 = 0.32$ (Prob. $= 0.3\%$) for male ticks and $R^2 = 0.22$ (Prob. $= 4\%$) for female ticks. It was very low and not significant between controls made in winter (December to February) and other seasons.

DISCUSSION AND CONCLUSION

Difficulties occurred in the analysis due to the presence of nil counts. The tick population was very low and obliged us to transform the data available. Conclusions have to be made with precautions.

Known effects such as seasonal fluctuations or effects of liveweight or physiological stage are verified in this study.[2] As previously observed in Creole cattle,[4] we failed in finding effect of tick counts on liveweight gain, as on other breeds.[5]

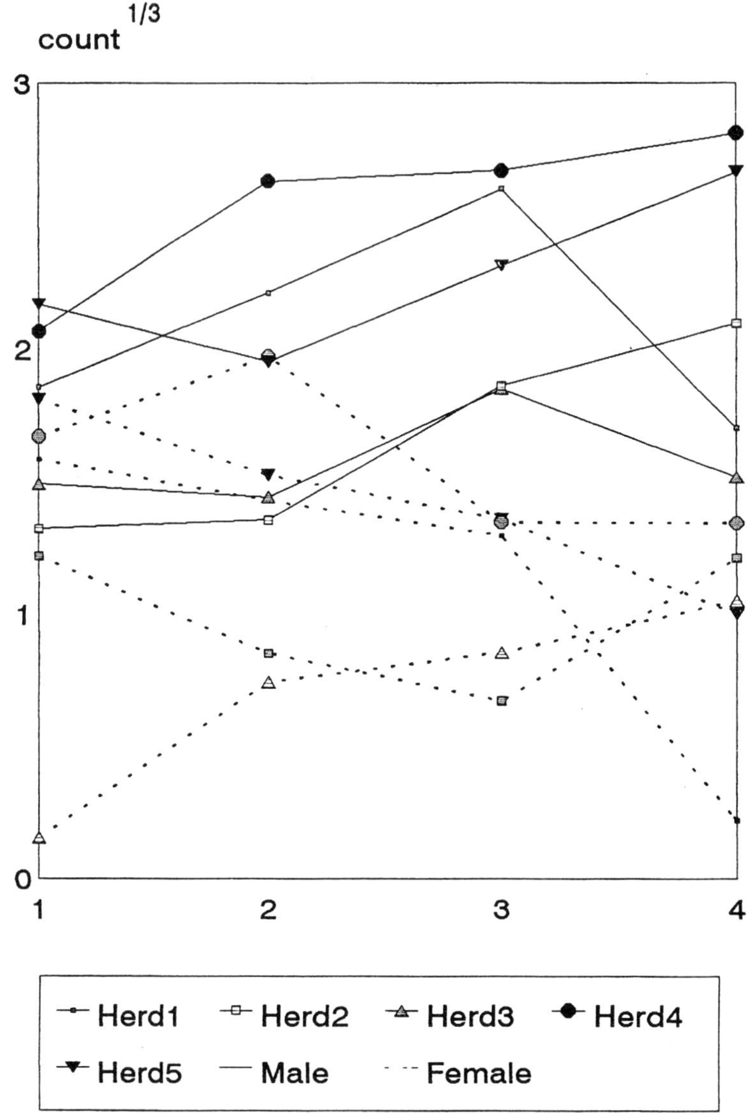

FIGURE 1. Evolution of tick counts on cows at different controls.

A genetic resistance to *Amblyomma variegatum* ticks seems to exist, as has been observed previously mainly with *Boophilus microplus*.[6] But this study points out the importance of control conditions, and especially the season, when assessment of individual differences is required. The procedure used on steers (monthly treatments and counts) seems to be efficient for their evaluation, in favorable seasons.

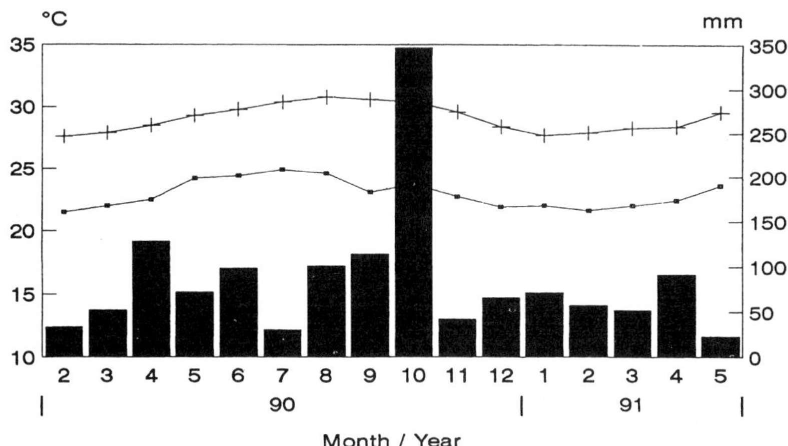

FIGURE 2. Climatic conditions during the survey of tick counts on steers.

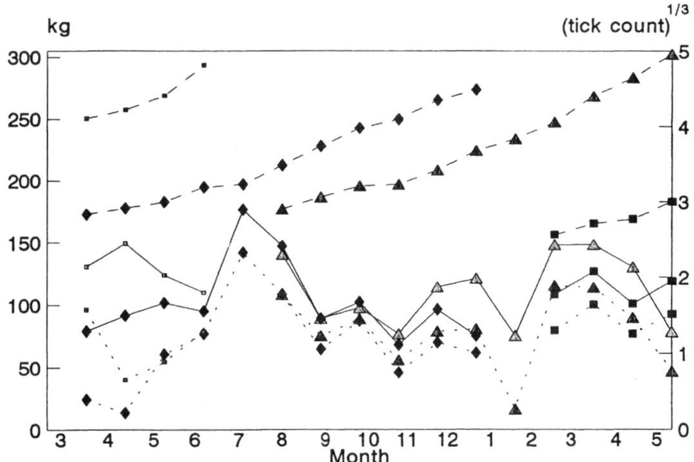

FIGURE 3. Evolution of liveweight and tick counts on steers.

TABLE 2. Results of Analyses of Variance for Fixed Effects. a. Tick Counts on Cows[a]

		Male Counts$^{1/3}$		Female Counts$^{1/3}$	
Effect	DF	Mean Square	Prob. (%)	Mean Square	Prob. (%)
H	4	4.59	0.1	2.93	0.1
C	3	3.75	0.1	2.04	0.2
H*C	12	1.27	0.1	2.97	0.1
PS	2	1.15	6.0	0.27	n.s.
LW	1	4.12	0.1	0.76	n.s.
Error	308	0.42		0.38	
			$R^2 = 0.29$		$R^2 = 0.39$

[a] H: herd effect; C: effect of the rank of control; H*C: interaction between H and C; PS: physiological stage effect; LW: liveweight covariable. DF: degrees of freedom.

TABLE 2b. Tick Counts on Steers[a]

		Male Counts$^{1/3}$		Female Counts$^{1/3}$	
Effect	DF	Mean Square	Prob. (%)	Mean Square	Prob. (%)
H	3	4.49	0.1	1.80	0.8
Season	4	6.63	0.1	10.9	0.1
C(S)	11	3.96	0.1	4.59	0.1
Error	417	0.49		0.45	
			$R^2 = 0.25$		$R^2 = 0.31$

[a] H: herd effect; C(S): effect of the rank of control intraseason. DF: degrees of freedom.

TABLE 3. Sire Effect on Tick Counts

		Residual Mean Square	Sire Variance	Prob. (%)
Cows	Male count$^{1/3}$	0.36	0.07	0.1
(13 sires)	Female count$^{1/3}$	0.36	0.05	0.1
Steers	Male count$^{1/3}$	0.48	0.01	11
(3 sires)	Female count$^{1/3}$	0.44	0.01	8

TABLE 4. Individual Effect on Tick Counts

		Residual Mean Square	Individual Variance	Repeatability (%)
Cows	Male count$^{1/3}$	0.27	0.23	46
	Female count$^{1/3}$	0.45	0.06	11
Steers	Male count$^{1/3}$	0.51	0.07	12
	Female count$^{1/3}$	0.50	0.05	10

SUMMARY

The level of infestation of "Creole" beef cattle of Guadeloupe by the tick *Amblyomma variegatum* was recorded during a long-term survey in an experimental farm: 61 steers and 83 cows were distributed in different lots according to sex and management. They grazed continuously either on irrigated *Digitaria decumbens* pastures or on dry native savannahs. Tick numbers as well as animal weights were registered monthly. Climatic data were also recorded. Different acaricide treatments were tested during the survey. But in order to minimize their effect in the data analysis, only tick counts over an average of 5 adult ticks per cattle were taken into account.

The level of infestation is analyzed with respect to environmental factors (season, management) and individual factors (sex, weight, physiological stage, genetic effect). The effects of these factors are discussed with regard to alternative tick-control methods, such as the selection of resistant hosts.

REFERENCES

1. PEGRAM, R. G. & H. G. B. CHIZYUKA. 1991. Prospects for tick control in Zambia. Zimbabwe Vet. J. **22**(1): 27–36.
2. BARRE, N. & G. GARRIS. 1990. Biology, ecology and management of the tick *Amblyomma variegatum* in the Caribbean. *In* Proc. "Cowdriosis and Dermatophilosis of Livestock in the Caribbean Region," Antigua. Nov. 12–14, 1990: 63–84.
3. SAS INSTITUTE INC. 1988. SAS/STAT User's Guide. Release 6.03 Edition. Cary, NC.
4. STACHURSKY, F., N. BARRE & E. CAMUS. 1988. Incidence d'une infestation naturelle par la tique *Amblyomma variegatum* sur la croissance de bovins et caprins Créoles. Rev. Elev. Med. vet. Pays trop. **41**(4): 395–405.
5. STACHURSKY, F., E. N. MUSONGE, M. D. ACHU-KWI & J. T. SALIKI. 1993. Impact of natural infestation of *Amblyomma variegatum* on the liveweight gain of male Gudali cattle in Adamawa (Cameroon). Vet. Parasitol. **49**: 299–311.
6. WILLADSEN, P. 1974. Mechanism and genetics of hosts resistance to ticks. 2nd World Congress on Animal Genetics Applied to Livestock Production. Madrid. Spain: 571–580.

Development of a Computer Model of the Population Dynamics of *Amblyomma variegatum* and Simulations of Eradication Strategies for Use in the Caribbean

THOMAS W. POPHAM,[a] GLEN I. GARRIS,[b] AND
NICOLAS BARRÉ [c]

*[a]United States Department of Agriculture
Agricultural Research Service
Southern Plains Area Biometrics
1301 North Western
Stillwater, Oklahoma 74075*

*[b]United States Department of Agriculture
Animal and Plant Health Inspection Service
Veterinary Services
National Animal Health Programs
4700 River Road, Unit 43
Riverdale, Maryland 20737*

*[c]Centre de Coopération Internationale en Recherche Agronomique
pour le Développement
Institut d'Elevage et de Médicine Vétérinaire des Pays Tropicaux
BP 515
97165 Pointe-à-Pitre
Guadeloupe, French West Indies*

INTRODUCTION

The tropical bont tick, *Amblyomma variegatum,* has been in the Caribbean region for over 150 years. It is a three-host tick originally imported about 1830 on cattle shipped from Senegal to Guadeloupe.[1] It is a vector of heartwater, a rickettsial disease of ruminants caused by *Cowdria ruminantium,* and it is implicated in the epidemiology of dermatophilosis, an acute skin disease caused by *Dermatophilosis congolensis.*

In early 1992, the tick was found on 14 islands. In 1991, one parish in Barbados was found infested with *A. variegatum;* by 1992 there were four infested parishes. Also, on 18 May 1992, 29 of 33 animals on one farm in the northwest section of Puerto Rico were found infested with *A. variegatum.* These two examples emphasize the danger that, if unchecked, *A. variegatum* and its associated diseases have the potential to spread to noninfested islands. Of particular concern are the islands of Trinidad/Tobago, Grenada, and the islands that make up the Grenadines. These islands are the last link to the mainland of South America. If the tick is allowed to reach the mainland countries of South and North America, eradication may be impossible.

Recent evidence indicates that migratory birds, especially the cattle egret (*Bubulcus ibis*), may be important in the spread of the tick in the Caribbean. Cattle egrets appeared in the Caribbean in the late 1950s and became well established as breeding colonies on a number of islands during the 1960s and 70s. The tick began to spread to a number of islands during the 1970s, which circumstantially suggests that cattle egrets may play an important role in the spread of the tick. Recent data indicate that not only do cattle egrets migrate frequently, but may do so for long distances.[2]

LIFE CYCLE AND ENVIRONMENTAL FACTORS

Free-Living Stages

After the blood meal, engorged female ticks detach from their host and fall to the ground where eggs are oviposited. After hatching or molting, the flat stages wait (larvae and nymphs) or seek a new host (adults). Survival of the free-living life stages is influenced by environmental conditions, particularly temperature and relative humidity.

Under experimental conditions, Barré[3] showed that surviving larvae per female is very low outside of 20 to 30°C temperature (T) and relative humidity (RH) below 75%. At 25°C, and 85 to 95% RH, more than 3000 larvae were produced per gram of female. All stages succumb at high T and low RH.

Rate of survival is greater at higher humidities and at lower temperatures. Seventy-five percent of nymphs maintained at 97% RH and 15°C were still alive after 7 months. This loss of 25% occurs in less than 3 months at 97% RH and 35°C.[2] In the field, adults survive longer than nymphs and nymphs longer than larvae. The longest survival periods were recorded for ticks maintained in sheltered environments, under bushes or high grasses. In suitable conditions, over 75% of adults remain alive after one year,[4] and the maximum length of life observed in 10 releases at different seasons was 9.5 months for larvae, 15 months for nymphs, and 20 months for adults.[5]

Ticks have few natural enemies in Guadeloupe. Mongoose and mice occasionally feed on engorged ticks. Birds, such as cattle egrets (*Bubulcus ibis*), Carib grackles (*Quiscalus lugubris*), and chickens eat ticks. Of 5324 organisms identified in crop contents of birds, only 103 (1.9%) were ticks. Of the ticks, only 30% were *A. variegatum.* Fire ants, *Solenopsis geminata,* may be important as predators. An average of 9% of released engorged ticks are attacked by this common and widespread predator.[6]

Host Finding

It seems that CO_2, alone, is sufficient to stimulate attraction and attachment if it is produced in large amounts, i.e., by large host animals such as cattle. The probability of ticks finding a host depends on the abundance and mobility of suitable hosts. Experiments were conducted in Guadeloupe with increasing stocking rates of cattle grazing on artificially infested pasture.[3] The number of ticks attached on cattle after one day was used to evaluate the kinetics of extraction of adult ticks from pastures by normally managed cattle. Without reinfestation, i.e., if cattle are regularly treated with acaricides, a stocking rate of four head of cattle per ha removes all the adults in 3 weeks of grazing. This was obtained in 5 weeks with two head of cattle per ha and 9 weeks with one cow per ha. The removal rate in 1 week was 85% with 4 cows/ha, 70% with 2 cows/ha, and 45% with 1 cow/ha.[3]

Parasitic Stages

Although the parasitic stages of the life cycle are very short in duration relative to the free-living stages, they are critical to maintaining and disseminating populations of ticks. The quality of the blood meal determines fertility and molting success.[3] The parasitic stages also offer the best opportunities for control and eradication of tick populations.

Of 432 animals examined in Guadeloupe belonging to 12 species of birds and mammals, no rodents were found with ticks while 100% of the cattle were infested. Heaviest infestations were in goats (580 ticks/animal), cattle (350 ticks/animal), and dogs (110 ticks/animal), while small mammals and birds were inconsistently and weakly infested (average 1-4 ticks per animal).[7]

Because of their suitability and abundance, ungulates are hosts of 95.5% of all stages of ticks in Guadeloupe, the remaining ticks feeding on other mammals (4%) and birds (0.5%). From a survey conducted in 1984-85,[3] it appeared that adults were found only on ungulate hosts, consistent with the fact that adults need large amounts of CO_2 for activation.

However, a dog was recently found infested with 53 males and 10 females (from flat to semiengorged) and a cattle egret with one male attached on the sole of the foot. The cattle egret is potentially an efficient disseminator of *A. variegatum:* out of 80 birds examined in Guadeloupe, 26% (nestlings) to 30% (adults) were infested with 1-101 larvae and 1-3 nymphs at different states of engorgement, including 9% fully engorged ticks.[7] Barré[1] demonstrated experimentally that immature *A. variegatum* fed on cattle egrets are as viable as immatures fed on goats: larval weight 2.6 mg, 81% of them moult (2.8 mg, 94% on goats); nymphal (female) weight 73 mg, 95% moult (73 mg, 96% on goats).

Attachment of females (and to a lesser extent males) depends on pheromones produced by previously attached males. Females in cloth bags glued on goats without males, together with males, or one day after males, do not attach. About 25% of females will attach in 6 hours when exposed to attached males that have fed for 3 days, 50% with males that have fed for 4-5 days, and 100% with males that have been fed for 6 days, respectively. Pheromones remain attractive on the skin or in the hair after manual removal of males: the site is still attractive for females 7 days (but not 10-14 days) after detachment of males. These pheromones can probably be used to lure females and to induce their attachment in the absence of males or attract them to a site containing slow-release acaricides.[3]

Females do not grow during the first 3 days after attachment, then regular and slow growth is observed for 7-10 days, and, finally, a sudden and important growth [the idiosoma (body) doubles its size] occurs during the day preceding the detachment.

An orange-colored female is ready to detach in the next few hours; a black-brown semiengorged female will detach in 12 hours to 5 days.[3]

On Guadeloupe, the blood meal of larvae fed on goats takes 5-11 days, 4-10 days for nymphs, and 7-15 days for females.[3] On Puerto Rico, the life cycle in the laboratory was similar.[8] Under field conditions in the Caribbean, only one generation may be possible, and it would probably take one and one-half years to complete one generation.

If the minimum time of feeding, 7 days for females, is added to the time necessary for males to become attractive for females (3 days), the minimum parasitic period for adults is 10 days. This has implications for the choice of the treatment cycle of the hosts with acaricides.[5] Detachment from the host is essentially diurnal: 98% of larvae, 95% of nymphs, and 90% of females drop off the host between 6 AM and 6 PM.[4] This biological feature may be of use for decreasing pasture infestation by keeping animals in a stable during the day and letting them graze at night.

Production of viable eggs by females, which must hatch and survive the larval stage, is a key factor for the maintenance of a flourishing population. Production of viable offspring by *A. variegatum* is influenced by host size (females do not attach and feed properly on small hosts),[3] duration of attachment of males, external temperatures, and the range of infestation. Fertility is altered slightly or not at all by the sex ratio, the range of mating of males (at first to fifth mating, hatchability remains over 80%), the age of females (no influence of age until at least 400 days), and the degree of infestation (five females per goat are as fertile as 40).[3]

It must be considered that this three-host tick, despite the great fecundity of females, is strongly disadvantaged compared with *Boophilus*. At an average temperature of 25°C, assuming a host finding period of 30 days for each stage, Barré[3] calculated that *B. microplus* produces eight generations in 2 years and 10^{25} descendants, whereas *A. variegatum* could produce a maximum of three generations in 2 years with 10^{12} descendants. *A. variegatum* has a much lower capability of multiplying than *Boophilus,* which can be confirmed by the fact that *Boophilus* entirely reinvaded Puerto Rico in 2 years in 1978-79[9] whereas *A. variegatum,* introduced into Martinique in 1948,[10] is still geographically restricted 50 years later. A similar situation occurred in Puerto Rico, where *A. variegatum* did not spread far from its original foci even though it was on the island for more than 6 years before eradication efforts were started.

This has considerable practical importance: after some ticks have been detected in a farm or group of farms, there is time to organize the eradication of this new focus before it spreads elsewhere. Information and surveillance are of great value in threatened islands where domesticated animals are predominant. It will not be sufficient in areas where wild hosts of adult ticks, deer for example, may hide, support tick reproduction, and spread them unchecked for a long period.

In Guadeloupe, under field conditions, three different pour-on pyrethroids have been tested: flumethrin (1%), deltamethrin (1%), and permethrin (6%), each at 1 ml per 10 kg on three different herds of Creole cattle treated every 14 days. Adult ticks present were counted every month, and the tick burden compared with that of a herd sprayed monthly with coumaphos. The decrease in infestation was rapid in the flumethrin-treated group with mean infestation reduced to zero after the fourth treatment (2 months). The same result was obtained after the eighth treatment with deltamethrin and permethrin. In the flumethrin group, no engorged females were observed on any of the treated animals after the beginning of the experiment. Considering the residual activity of the compound, a cycle of 14 days should prevent any attached female from engorging and laying eggs. A similar result was obtained, but only after the seventh treatment, with the two other pyrethroids.

A. variegatum is well adapted to climatic conditions in the Caribbean where for most of the year it finds suitable humidity (>75%) and temperature (20-30°C) in open, exposed pastures, and all year in sheltered habitats. In such environments, adults may survive for almost 2 years still able to infest animals. This species is also well adapted to the hosts present, especially cattle, which represent 60% of the total livestock in the region and on which about 96% of adult ticks attach and feed.

COMPUTER SIMULATION PROGRAM

The purpose of the computer simulation of *A. variegatum* populations was to obtain a better understanding of how tick numbers increase, the effects of environment and host animals on the growth or decline of populations of different life stages, to determine how sensitive population numbers are to changes in assumptions, and to evaluate the effects of a variety of eradication strategies on simulated populations. It must be remembered when using a computer simulation of biological processes that the results are produced by a set of instructions developed from the best knowledge biologists have of the process. Because this knowledge is necessarily incomplete, we cannot say that nature will emulate the results of the simulation every time. The simulation results are predictive of what will usually happen in nature if most of the biological knowledge that went into the program is correct. Validation for this model consisted of verifying that simulation results are consistent with the observations of tick populations by Garris[7] in Puerto Rico and Barré[3] and Barré *et al.*[7] in Guadeloupe.

The computer program to accomplish simulations was written to minimize programmed assumptions concerning parameters of all mathematical or probability distribution functions. The forms of the functions and their effect are included in the programming, but are identified in the source code so they may be easily altered. A parameter data input file contains information that might be changed infrequently such as temperature, relative humidity, and probability distribution parameters. Input supplied by the user through the keyboard would likely change often and includes information such as herd size, pasture area, size of introduced infestation of ticks, and treatment parameters.

In order to incorporate weather information, the calendar day when simulation begins is determined and incremented each day of simulation until day 365 is reached and then reset to 1. The calendar day is matched with the input weather data, and the temperature and relative humidity values for the month are supplied to the life-stage subprocedures. In Guadeloupe, if the daily maximum temperature at the microenvironmental level reached 32°C or if the relative humidity was ≤66%, engorged female ticks and larvae released in this environment died.[3] Threshold temperature and relative humidity values can be set by the user so all of the populations of off-host life stages die when the threshold is exceeded.

For all simulations, it was assumed that the only way ticks can occur in a group of animals and environment is by introduction of infested animals. Simulations beginning with an existing established population both on host animals and in the field cannot be done with this computer model. The number of animals in the herd is the sum of the original herd plus the introduced animals. Initial populations of attached larvae, nymphs, and adults are obtained as the product of the number of introduced animals and the initial numbers per animal. To simplify calculations, the introduced ticks are assumed to be one day old. For each day of the simulation, the numbers of ticks that survive or change to the next life stage for each age cohort within a life stage are computed by calling subprocedures.

The program computes the probability of an age cohort member dying during the time interval (PD) and the probability (PC) of an age cohort member changing to the next life stage during the time interval using a three-parameter Weibull probability density function with user-supplied parameters. The Weibull function is a flexible frequency distribution which can be used as a model of change of state.[11] The cumulative distribution function (CDF) for the Weibull is $F(x) = 1 - \exp [-(x - a)/b]^c$, where $0 \le a \le x,\ b > 0$, and $c > 0$. It can be used to summarize survival information[12] and in modeling distributions of insect development times.[13,14] The parameter a is the estimate of the earliest time when $F(x) > 0$. The b and c parameters of the Weibull distribution function define the scale and shape, respectively, of the curve.[15] The probability density function (PDF), the first derivative of the CDF is $f(x) = \{(c/b) * [(x - a)/b]^{c-1}\} * \exp \{-[(x - a)/b]^c\}$. It may be used to obtain the probability of change of state during the time interval x.

The program then calculates the number of each age cohort of a life stage surviving the time interval as initial number (IN) less the products of PD * IN and PC * IN. Survivors are moved to the next age cohort. The number changing to the next life stage is the sum of the products of PC * IN over all age cohorts. Lastly, the number of each life stage at the end of each time interval is obtained by summing all of the age groups in each life stage.

The exceptions to the above general procedure have to do with the sex ratio of adults, and pickup rate of free-living life stages by host animals. The program obtains the proportion of females as a normally distributed random variate with mean and standard deviation supplied by the user.

Pickup rate is computed using information about host density (number of host animals/pasture area) supplied by the user and an S-shaped function with an estimable upper asymptote, $P(d) = a * \{1 - \exp [-r * d]\}^c$, where a is an upper asymptote, r is a parameter that determines how rapidly the asymptote is approached, and d is host density. As r becomes smaller, the asymptote is reached with small numbers of hosts and remains uniform over the range of host values. This results in an essentially uniform pickup rate such as would be expected with tethering.

Output includes five graphs for run time output to the screen: (1) egg and free larvae, (2) attached larvae and free nymphs, (3) attached nymphs, free adult males, and free adult females, (4) attached adult males, attached adult females, and engorged females, and (5) maximum temperature and minimum relative humidity. The user may select four of these by making the appropriate entry in the parameter input file. Two output files are produced; one has original parameter information and the total numbers of each life stage for each day in a format for printed output. The second output file has only the total numbers of each life stage for each day in a format for use with plotting software.

TABLE 1. Input Parameters for All Simulations

Number of animals in the herd	10
Pasture size (hectares)	0.4
Number of introduced animals	3
Attached larvae/introduced animal	250
Maximum attached larvae/animal	5,000
Nymphs/introduced animal	36
Maximum attached nymphs/animal	1,000
Adult males/introduced animal	23
Adult females/introduced animal	20
Maximum adults/animal	400
Maximum engorged females/animal	60
Mean proportion females	0.5215
Std dev proportion females	0.03
Mean size of egg mass	14,600
Std dev size of egg mass	2,000

SIMULATION AND RESULTS

Four different scenarios were simulated. The input parameters for these are shown in TABLES 1, 2, 3, and 4. The results are in FIGURES 1, 2, 3, and 4. Briefly, these may be described as follows.

Simulation 1, Description

An *A. variegatum* population introduced into a herd of 10 animals on 0.4 hectares by 3 new animals. No treatment is applied in order to establish the pattern for introducing a population, its growth, and its stabilization.

TABLE 2. Different Input Parameters for Four Simulations. Differences are in the Number of Days Run, and the Treatment Intervals

Simulation	1	2	3	4
Beginning date	March 1	March 1	March 1	March 1
Number of days	1000	1000	1500	1500
Treatment				
Beginning date	None	May 1	May 1	May 1
Years since introduction of ticks		0	1	0
Interval between applications (days)		14	14	21
Effectiveness (percent)		98	98	98
Residual effect (days)		10	10	10
Number of applications		52	52	35

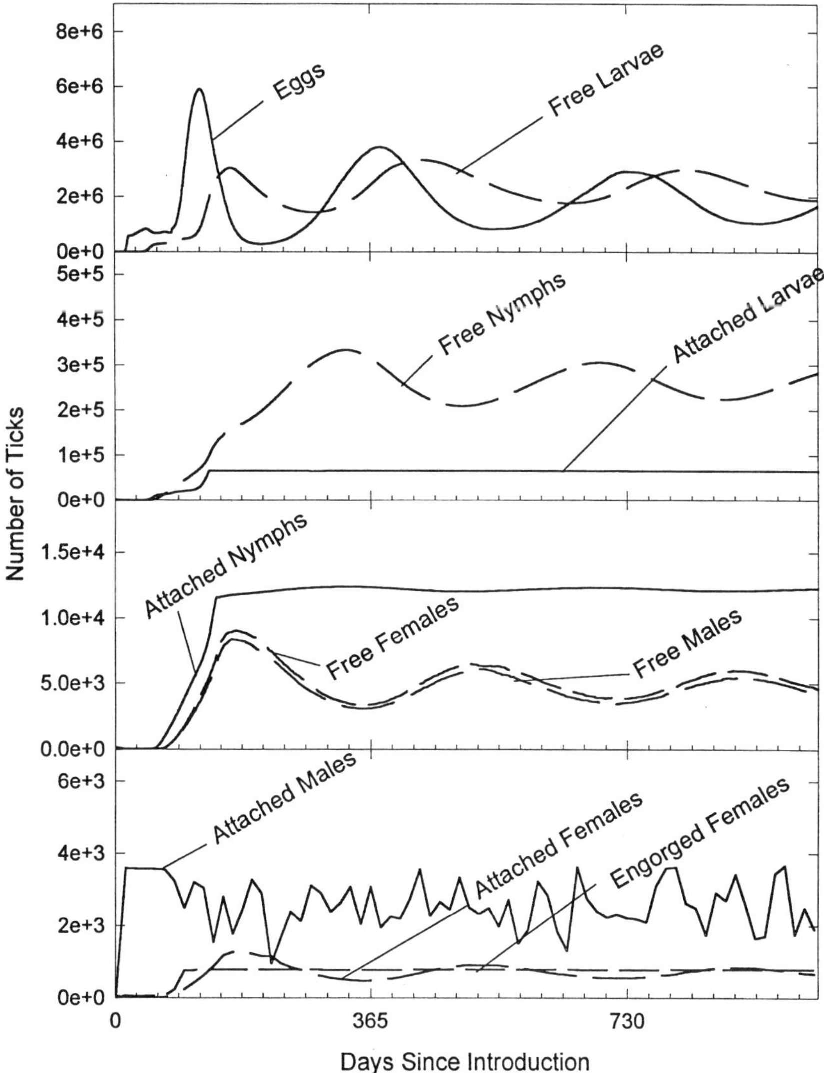

FIGURE 1. Introduction on March 1 with no treatment.

Simulation 1, Results

About one year after introduction of the tick, the niche is apparently fully occupied. All the life stages are cycling over about 14 months. Attached stages with a flattened curve indicate the maximum number that could be attached on the number of animals in the herd. If host animals are available, there is a reservoir of flat free-living stages that can be picked up. If large feral animals are available, the ticks will be able to

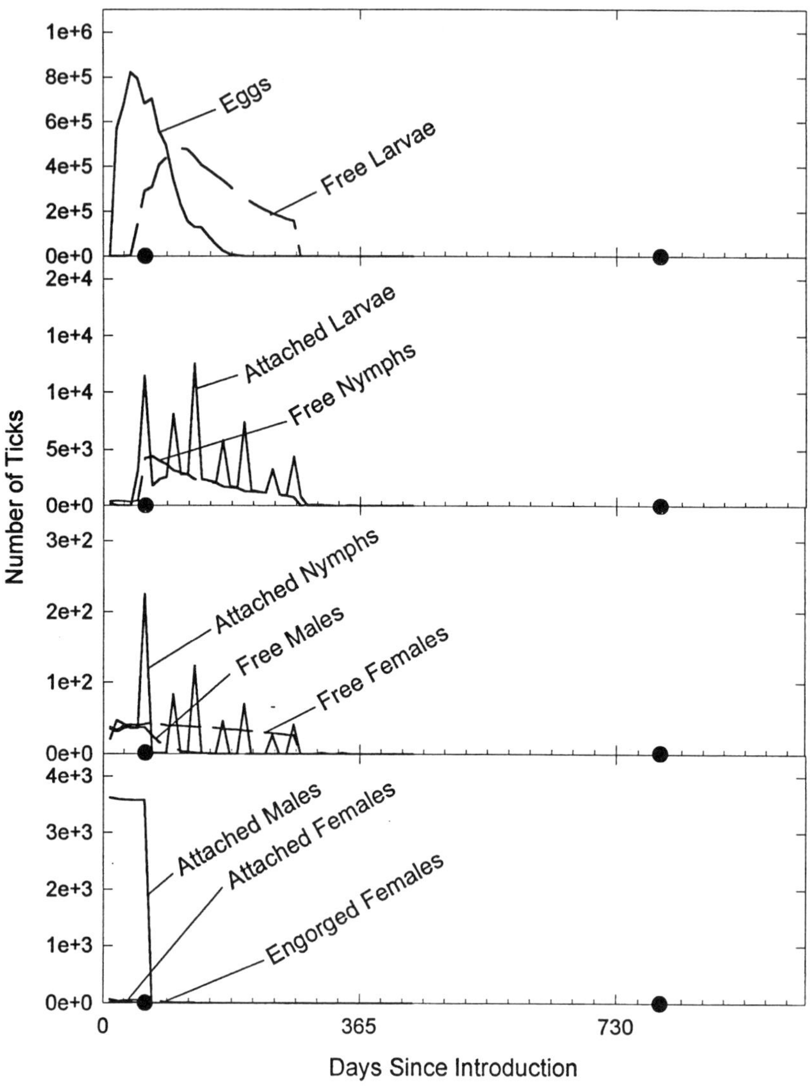

FIGURE 2. Introduction on March 1 with treatment for 730 days at 14-day intervals beginning 60 days after introduction. (Dots on horizontal axes indicate start and end of treatments.)

sustain higher population levels. Host availability is the limiting factor here, but is realistic for the region being simulated.

Simulation 2, Description

An *A. variegatum* population introduced into a herd of 10 animals on 0.4 hectares by 3 new animals. A treatment is applied every 14 days for 2 years. The treatment

TABLE 3. Climatic Parameters for All Simulations. Maximum Temperature Threshold for All Life Stages Was 32°C and Minimum Relative Humidity Threshold for All Life Stages was 66%

Parameter: Month	Maximum Temperature	Minimum Relative Humidity
January	27.7	75
February	27.8	78
March	28.2	77
April	28.5	78
May	29.2	72
June	30.0	73
July	30.1	75
August	30.2	77
September	30.1	78
October	30.0	71
November	29.2	70
December	28.5	68

begins 60 days after introduction. This is to determine the effect of quick response to an infestation using the currently recommended eradication strategy.

Simulation 2, Results

Prompt response with the recommended acaricide results in the elimination of all life stages in less than a year. Continued treatment and inspection should eliminate any remaining free-living stages. If there is a possibility of reintroduction by small animals, continued treatment and inspection are required.

Simulation 3, Description

An *A. variegatum* population introduced into a herd of 10 animals on 0.4 hectares by 3 new animals. The treatment begins 430 days after introduction. Treatment is applied every 14 days for 2 years to determine the effect of treating a well-established population using the currently recommended eradication strategy.

Simulation 3, Results

By waiting a year and 60 days before beginning treatment, the population has been allowed to occupy the niche and stabilize. It is still possible to eradicate all life stages from the herd animals and pasture area in less than a year using the recommended eradication method.

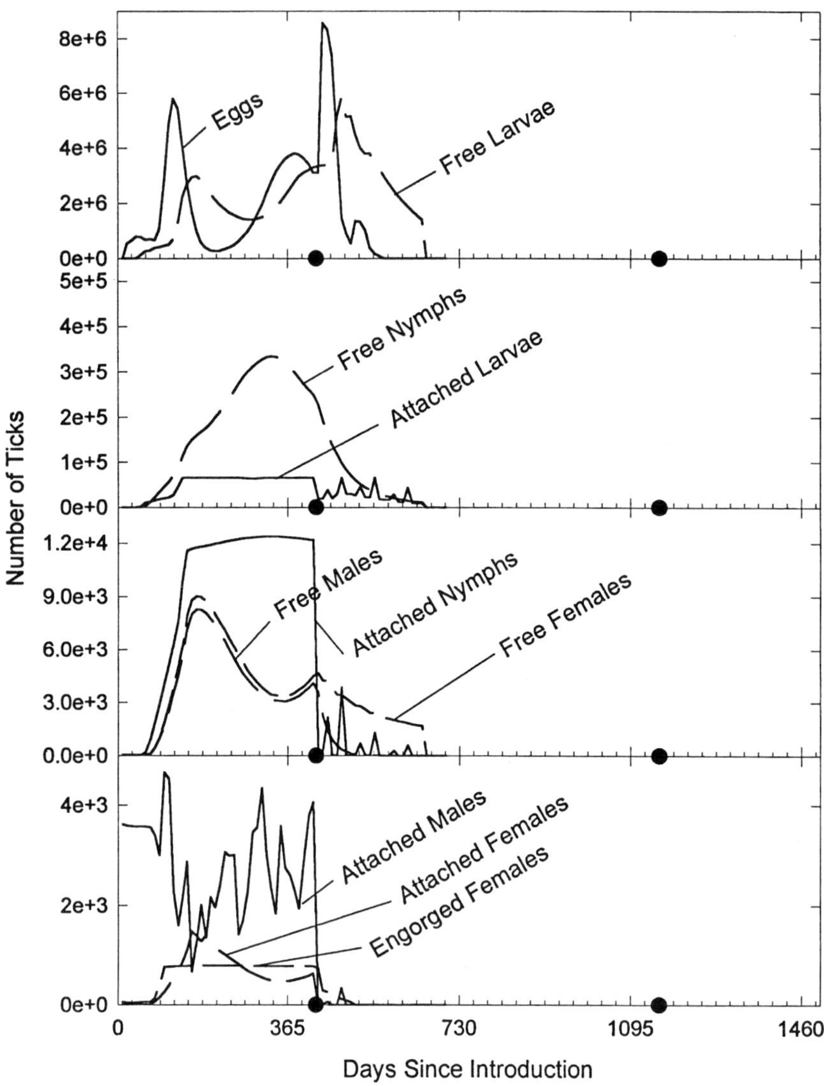

FIGURE 3. Introduction on March 1 with treatment for 730 days at 14-day intervals beginning 430 days after introduction. (Dots on horizontal axes indicate start and end of treatments.)

Simulation 4, Description

An *A. variegatum* population introduced into a herd of 10 animals on 0.4 hectares by 3 new animals. The treatment begins 60 days after introduction. Treatment is applied every 21 days for 2 years to determine the effect of lengthening the treatment

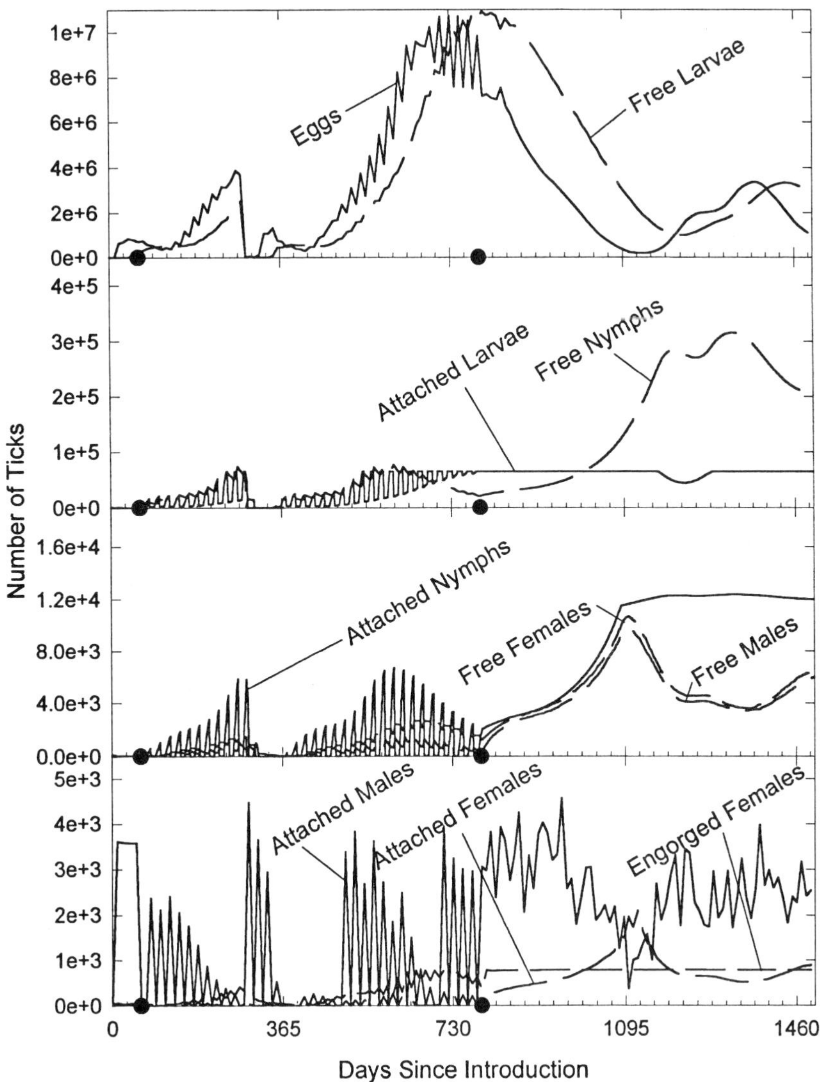

FIGURE 4. Introduction on March 1 with treatment for 730 days at 21-day intervals beginning 60 days after introduction. (Dots on horizontal axes indicate start and end of treatments.)

interval to reduce treatment costs, assuming a quick response, using the currently recommended eradication strategy.

Simulation 4, Results

Even with prompt response, none of the life stages are eradicated in the herd with the extended treatment interval of 21 days. The residual population of free-

TABLE 4. Parameters for the Pickup Functions ($P(d) = a * \{1 - \exp [-r * d]\}^c$) for Life Stages that Are Picked Up by Hosts and Parameters for Survival and Change of State Functions (Weibull, $F(x) = 1 - \exp [- (x\text{-}a)/b]^c$) Used in All Simulations

Function:	Pickup			Survival (Weibull)				Change to Next Stage (Weibull)		
Parameters	a	r	c	a	b	c	Max age	a	b	c
Free larvae	1	1E-14	0.18	0	50	2.5	111	5	16	1.2
Attached larvae				0	5	2.5	12	4	1.5	2.5
Free nymphs	1	1E-12	0.18	0	8	2.2	17			
Attached nymphs				0	8	2.2	17	7	6	2.2
Free adult males	1	1E-09	0.18	0	7	2	23			
Free adult females	1	1E-09	0.18	0	7	2	23	6	5	2.8
Engorged females				0	10	3	23	2	2.5	2.2
Eggs				0	25	1.5	146	19	20.5	0.6

living stages are able to find and attach to host animals, then drop off in the period between the end of effective killing of on-host stages and the next acaricide application. An acaricide with a longer effective residual period would be required to extend the treatment cycle.

SUMMARY

Control or eradication efforts should cover wide areas because of long-range migration potential of immature ticks attached to birds, such as cattle egrets. Careful follow-up inspections are also imperative. Simulation 1 shows how rapidly the population numbers may expand from small numbers, either introduced or missed. Simulation 4 indicates extending the treatment cycle beyond the residual effectiveness of the acaricide to save labor or funds would be false economy. The result could be to maintain the tick population indefinitely, with an accompanying increase in long-term control costs. Simulations 2 and 3 indicate that eradication is possible if response is prompt and the treatment cycle is less than the period of residual effectiveness of the acaricide used.

REFERENCES

1. CURRASSON, G. 1943. Traite de protozoologie veterinaire et comparee. Tome 1, Chap. XII, *Trypanosoma vivax* et varietes. Vigot Freres Ed. **1:** 267–279. Paris, France.
2. CORN, J. L., N. BARRÉ, B. THIEBOT, T. E. CREEKMORE, G. I. GARRIS & V. F. NETTLES. 1993. Potential role of cattle egrets, *Bubulcus ibis* (Ciconiformes : Ardeidae), in the dissemination of *Amblyomma variegatum* (Acari : Ixodidae) in the eastern Caribbean. J. Med. Entomol. **30:** 1029–1037.
3. BARRÉ, N. 1989. Biology et ecologie de la tique *Amblyomma variegatum* (Acarina : Ixodina) en Guadeloupe (Antilles Francaises). These Doc. es Sciences, Paris-sud.

4. BARRÉ, N. 1988. Mesures agronomiques permettant une diminution des populations de la tique *Amblyomma variegatum.* Rev. Elev. Med. vet. Pays Trop. **41**(4): 387–393.

5. BARRÉ, N. & G. I. GARRIS. 1990. Biology and ecology of *Amblyomma variegatum* (Acari : Ixodidae) in the Caribbean. Implications for a regional eradication program. J. Agric. Entomol. **7**(1): 1–9.

6. BARRÉ, N., H. MAULEON, G. I. GARRIS & A. KERMARREC. 1991. Predators of the tick *Amblyomma variegatum* (Acari : Ixodidae) in Guadeloupe (FWI). Exp. Appl. Acarol. **12**(2): 163–170.

7. BARRÉ, N., G. I. GARRIS, G. BOREL & E. CAMUS. 1988. Hosts and population dynamics of *Amblyomma variegatum* (Acari : Ixodidae) on Guadeloupe, French West Indies. J. Med. Entomol. **25**(2a): 111–115.

8. GARRIS, G. I. 1984. Colonization and life cycle of *Amblyomma variegatum* (Acari : Ixodidae) in the laboratory in Puerto Rico. J. Med. Entomol. **21**: 86–90.

9. COMBS, G. P. 1987. Economics of tick eradication in Puerto Rico. Expert consultation on the eradication of ticks, with special reference to Latin America. Mexico, June 1987.

10. MOREL, P. C. 1966. Etude sur les tiques du betail en Guadeloupe et Martinique. I. Les tiques et leur distribution. (Acariens, Ixodoidea). Rev. Elev. Med. Vet. Pays Trop. **3**: 307–321.

11. DELL, T. R., J. L. ROBERTSON & M. I. HAVERTY. 1983. Estimation of cumulative change of state with the Weibull function. Bull. Entomol. Soc. Am. **29**: 38–40.

12. PINDER, J. E., III, J. G. WIENER & M. H. SMITH. 1978. The Weibull distribution: a new method of summarizing survivorship data. Ecology **59**: 175–179.

13. WILLIAMS, L. H. & T. W. POPHAM. 1983. Synchronizing mass emergence of *Xyletinus peltatus* (Harris) (Coleoptera : Anobiidae) adults. Environ. Entomol. **12**: 1664–1668.

14. GARRIS, G. I., T. W. POPHAM & R. H. ZIMMERMAN. 1990. *Boophilus microplus* (Acari : Ixodidae): oviposition, egg viability, and larval longevity in grass and wooded environments of Puerto Rico. Environ. Entomol. **19**: 66–75.

15. JOHNSON, N. L. & S. KOTZ. 1970. Distributions in Statistics: Continuous Univariate Distributions. John Wiley and Sons. New York, NY.

Use of Polymerase Chain Reaction in Bovine Babesiosis Research[a]

G. M. BUENING[b] AND J. V. FIGUEROA [c]

[b]*Department of Veterinary Pathobiology*
UMC
Columbia, Missouri 65211

[c]*CENID-PAVET*
INIFAP-SARH
Jiutepec, Mor., 62500 Mexico

INTRODUCTION

The introduction of the polymerase chain reaction (PCR) technique,[1] and its reportedly high sensitivity when used for the detection of infectious organisms[2-5] in diagnostic procedures, prompted us to test a PCR approach for the simultaneous detection of the hemoparasites *Babesia bigemina* and *B. bovis* in carrier cattle.

The technical aspects of the procedures for the *B. bigemina* and *B. bovis* PCR DNA probe assays have been described and summarized in the publication describing a multiplex PCR system for the simultaneous detection of *B. bigemina*, *B. bovis*, and *Anaplasma marginale* in experimentally infected cattle and samples collected in the field.[6,7] Use of the primer groups in the assay is described in detail[6] (see TABLE 1).

DISCUSSION AND CONCLUSIONS

The duplex PCR/nonradioactive DNA probe assay has been applied to several epidemiological field studies in Mexico and has the following applications: (1) differentiation of *Babesia* species of infection in cattle; (2) differentiation of *Babesia* species of infection in ticks; (3) less time-consuming than extensive examination of slides; (4) evaluation of infection of cattle with live attenuated strains of *B. bigemina* and *B. bovis;* and (5) molecular epidemiology studies to define when susceptible cattle become infected when introduced into an endemic area.

The advantages of the procedure are (1) highly sensitive and specific; (2) rapid *Babesia* species differentiation; and (3) a large number of samples can be batch processed.

The disadvantages of the procedure are (1) investigator must have a research laboratory equipped to do molecular biology; (2) personnel must follow detailed protocols; and (3) the cost of the test procedure is moderately expensive.

[a]This work was supported in part by USAID Prime Agreement No. DAN 4178-A-00-7056-00. Julio V. Figueroa was partially supported by INIFAP-SARH, Mexico.

TABLE 1. Nucleotide Sequence of Primers Used in the *Babesia* Duplex PCR Assay

Code	Sequence	Hemoparasite	Group
BiIA	5′ CATCTAATTTCTCTCCATACCCCTCC 3′	*B. bigemina*	I
BiIB	5′ CCTCGGCTTCAACTCTGATGCCAAAG 3′		I
BiIAN	5′ CGCAAGCCCAGCACGCCCCGGTGC 3′		II
BiIBN	5′ CCGACCTGGATAGGCTGTGTGATG 3′		II
BoF	5′ CACGAGGAAGGAACTACCGATGTTGA 3′	*B. bovis*	I
BoR	5′ CCAAGGAGCTTCAACGTACGAGGTCA 3′		I
BoFN	5′ TCAACAAGGTACTCTATATGGCTACC 3′		II
BoRN	5′ CTACCGAGCAGAACCTTCTTCACCAT 3′		II

SUMMARY

The polymerase chain reaction (PCR) in babesiosis research was originally developed to detect *Babesia bovis* or *Babesia bigemina* in blood samples containing infected erythrocytes. These preliminary studies led to development of a sensitive PCR/DNA probe assay to detect the following hemoparasites: *B. bigemina* and *B. bovis* in a single sample. This modified procedure, referred to as a duplex PCR/nonradioactive probe assay, has an analytic sensitivity of 0.00001% for *B. bigemina,* and 0.00001% infected erythrocytes for *B. bovis*. This procedure has been modified to detect *Babesia* DNA in tick tissue and hemolymph.

The above procedures can be performed in central laboratories that have access to a thermocycler and quality reagents. Precautions must be observed to prevent cross-contamination of samples. At the present time the procedure has application in epidemiology studies to detect carriers and the species of *Babesia* in the bovine population. Preliminary studies are in progress by various research groups to utilize this technique in studying the biology of the *Babesia* protozoans in tick vectors. The applications, advantages, and disadvantages of the technique are presented.

ACKNOWLEDGMENTS

The authors acknowledge the technical assistance provided by Karen K. McLaughlin and the secretarial assistance of Ellen Swanson.

REFERENCES

1. SAIKI, R. K., D. H. GELFAND, S. STOFFEL, S. J. SCHARF, R. HISUCHI, G. T. HORN, K. B. MULLIS & H. A. EHRLICH. 1988. Primer-directed enzymatic amplification of DNA with a thermostable DNA polymerase. Science **239**: 487–491.
2. FAHRIMAL, Y., W. L. GOFF & D. P. JASMER. 1992. Detection of *Babesia bovis* carrier cattle by using polymerase chain reaction amplification of parasite DNA. J. Clin. Microbiol. **30**: 1374–1379.
3. FIGUEROA, J. V. 1991. *Babesia bigemina:* detection by DNA-based assays and expression in *Escherichia coli* of a parasite surface antigen. M.S. Thesis. University of Missouri. Columbia, Missouri.

4. FIGUEROA, J. V., L. P. CHIEVES, G. S. JOHNSON & G. M. BUENING. 1992. Detection of *Babesia bigemina*-infected carriers by polymerase chain reaction amplification. J. Clin. Microbiol. **30:** 2576–2582.

5. TIRASOPHON, W., M. PONGLIKITMOGKOL, P. WILAIRAT, V. BOONSAEN & S. PANYIM. 1991. A novel detection of a single *Plasmodium falciparum* in infected blood. Biochem. Biophys. Res. Commun. **175**(1): 179–184.

6. FIGUEROA, J. V., L. P. CHIEVES, G. S. JOHNSON & G. M. BUENING. 1993. Multiplex polymerase chain reaction-based assay for the detection of *Babesia bigemina, Babesia bovis* and *Anaplasma marginale* DNA in bovine blood. Vet. Parasitol. **50:** 69–81.

7. FIGUEROA, J. V., J. A. ALVARES, J. A. RAMOS, C. A. VEGA & G. M. BUENING. 1993. Use of a multiplex polymerase chain reaction-based assay to conduct epidemiological studies on bovine hemoparasites in Mexico. Rev. Med. Vet. Pays Trop. **46**(1–2): 71–75.

Light Microscopy Diagnosis of *Babesia bovis* and *B. bigemina* Kinetes in the Hemolymph of Artificially Infected *Boophilus microplus* Engorged Female Ticks[a]

A. A. GUGLIELMONE, A. B. GAIDO,[b] AND
A. J. MANGOLD

Estación Experimental Agropecuaria Rafaela
Instituto Nacional de Tecnología Agropecuaria
CC 22, CP 2300 Rafaela, Santa Fe, Argentina

[b]*Consejo Nacional de Investigaciones Científicas y Tecnológicas*
CC 228, CP 4400 Salta, Argentina

Boophilus microplus is the main vector of cattle babesiosis (*Babesia bovis* and *Babesia bigemina*). The cycle of both *Babesia* species includes sporokinetes (kinetes) that infect tick tissues via tick hemolymph. Babesial infection in engorged female ticks can be diagnosed by light microscopy of hemolymph samples after Giemsa staining. However, the specific diagnosis is difficult using this technique.[1] In this report, we compare the morphology of Argentinean *B. bovis* and *B. bigemina* kinetes from monospecific infections in *B. microplus* engorged female ticks.

B. microplus larvae from a colony free of *Babesia* were fed on splenectomized calves infected with either *B. bovis* or *B. bigemina* pathogenic strains. Engorged female ticks detached on day 23 after the larval infestation were collected and maintained in darkness at 27°C, 83-86% relative humidity. Six ticks showing an infection of at least 10 mature kinetes of *B. bovis* or *B. bigemina* in a sample obtained on day 5 after detachment were also monitored on day 7, 9, and 10 after collection. Length, width, nucleus position, and kinete tail shape were recorded.

An increasing tick mortality from day 7 to day 10 postdetachment made impossible to analyze the same number of kinetes in the sampling dates. Most nuclei were in a central position in both species of *Babesia*. The mean length $\pm$ standard deviation and range for *B. bovis* kinetes were 14.3 $\pm$ 0.92 μm (11.9-16.3 μm) and 11.3 $\pm$ 0.90 μm (9.0-13.1 μm) for *B. bigemina* ($p < 0.001$, "*t*" test). The width was 3.3 $\pm$ 0.31 μm (2.6-4.0 μm) for *B. bovis* and 2.2 $\pm$ 0.29 (1.5-2.8 μg) for *B. bigemina* kinetes ($p < 0.001$). A total of 58% of *B. bovis* kinetes showed the typical curved

---

[a] Presented with kind permission from Elsevier Science, B.V., Amsterdam. Source of information: Guglielmone *et al.*, 1995. Light microscopy diagnosis of *Babesia bovis* and *Babesia bigemina* kinetes in the haemolymph of artificially infected *Boophilus microplus* engorged female ticks. Veterinary Parasitology. In press.

TABLE 1. Mean Length (μm), Mean Width, and Their Ranges (in parentheses) of *Babesia bigemina* Kinetes in the Hemolymph of *Boophilus microplus* Engorged Female Ticks Obtained in Previous Studies

Length	Width	Strain Origin and Reference Number
11.1 (9.0–13.0)	2.6 (2.0–2.9)	Australia[2]
11.8 (11.5–12.1)	2.6 (2.5–2.7)	Thailand[3]
12.6 No data	No data No data	Kenya[4]

tail. No effect of time postcollection and individual host tick on the kinetes of *B. bigemina* was detected, while an unexpected influence of individual host tick in the width of *B. bovis* kinetes was found ($p < 0.01$, analysis of variance).

The length and width of kinetes of both species of *Babesia* were statistically different. However, sizes overlapped and a proportion of *B. bovis* kinetes did not present the typical curved tail. Moreover intraspecific size variation of specimens from different geographical origin has been found (see TABLE 1 for *B. bigemina* kinetes). It will be difficult to apply the laboratory results to study babesial infection in *B. microplus* engorged female ticks from the field. There is a need of more accurate techniques to identify *B. bovis* and *B. bigemina* kinetes in the vector tick.

REFERENCES

1. GUGLIELMONE, A. A. 1995. Epidemiology of babesiosis and anaplasmosis in South and Central Am. Vet. Parasitol. **57:** 109–119.
2. RIEK, R. F. 1964. The life cycle of *Babesia bigemina* (Smith & Kilborne, 1893) in the tick vector *Boophilus microplus* (Canestrini). Aust. J. Agr. Res. **15:** 802–821.
3. MORZARIA, S. P. & D. W. BROCKLESBY. 1977. A differential diagnostic criterion for *Babesia major* and *Babesia bigemina* vermicules from tick haemolymph. Z. Parasitenkd. **52:** 241–243.
4. BÜSCHER, G. 1988. The infection of various tick species with *Babesia bigemina*, its transmission and identification. Parasitol. Res. **74:** 324–330.

Enrofloxacin to Control *Anaplasma marginale* Infections

A. A. GUGLIELMONE,[a] O. S. ANZIANI,[a]
A. J. MANGOLD,[a] M. M. VOLPOGNI,[b] AND
A. VOGEL [c]

[a]*Estación Experimental Agropecuaria Rafaela
Instituto Nacional de Tecnología Agropecuaria
CC 22
CP 2300 Rafaela, Santa Fe, Argentina*

[b]*Facultad de Agronomía y Veterinaria
Universidad Nacional del Litoral
Kreder 2805
CP 3080 Esperanza, Santa Fe, Argentina*

[c]*Bayer Argentina S.A.
División Veterinaria
General Rivas 2466
CP 1417, Buenos Aires, Argentina*

Anaplasmosis (*Anaplasma marginale*) is a common disease in South American cattle raised north to 33°S. Oxytetracycline or imidocarb is usually applied to treat clinical cases of anaplasmosis. It has been recently reported that the fluorquinolon enrofloxacin was effective in controlling infections with *A. marginale* in South Africa.[1] This report describes an experiment to control anaplasmosis in the increasing parasitemia phase of the disease.

Fifteen Holstein steers (15 months old), free of antibodies against *A. marginale* (card agglutination test)[2] were inoculated with 200 million erythrocytes infected with the *A. marginale* strain S1P (no appendages)[3] of Argentinean origin. Eight steers were randomly allocated to the treated group (TG) and medicated with 10 mg of enrofloxacin per kg of body weight for two consecutive days when the parasitemia reached a minimum of 3% of the red cells. The other 7 steers formed the control group (CG). They were treated with 20 mg per kg of body weight of a long acting oxytetracycline when the hematocrit index reached a value of 15% to avoid deaths.

The steers of the TG developed a parasitemia of at least 3% of the red cells in a mean period of 22.9 ± 2.03 days, range: 21-27 days. The average parasitemia ± SD and range on the day of the first treatment with enrofloxacin was 4.5 ± 2.27% (3.0-10.0%). Seventy-four hours after the second treatment, the parasitemia values were 0.9 ± 0.76% (0.1-2.0%) ($p < 0.01$, "*t*" test with angular transformation; the figures presented are nontransform values). The parasitemia found 72 h after the second enrofloxacin treatment was also significantly lower ($p < 0.01$) than that found in the CG (10.0 ± 6.14%, 1.0-20.0%) on the equivalent day. Six of the steers of the CG were treated with oxytetracycline due to severe anemia on days 25 to 30 after *A. marginale* inoculation. The hematocrit index had an average value of 27% on day 27 postinfection in the TG, while this value for the CG was 17% on days 29 and 30 postinfection.

Enrofloxacin was able to inhibit the parasitemia increase of *A. marginale* S1P, hence avoiding a drastic drop in the hematocrit level in the TG. This confirms that enrofloxacin may have a potential role to control *A. marginale* cattle infections. Additional studies are needed to know if enrofloxacin is useful to control anaplasmosis in cattle under peremptory risk of death (hematocrit 15% or lower).

REFERENCES

1. SCHRÖDER, J., K. KOWOLLIK & A. F. VAN AMELSFOORT. 1991. The effect of enrofloxacin on *Anaplasma marginale* in cattle. *In* Abstracts of Twenty Fourth World Veterinary Congress: 60, Rio de Janeiro, Brazil.
2. AMERAULT, T. E. & T. O. ROBY. 1968. A rapid card agglutination for bovine anaplasmosis J. Am. Vet. Med. Assoc. **153:** 1828–1834.
3. DE RIOS, L. G., E. PIPANO, A. J. MANGOLD, D. H. AGUIRRE, A. B. GAIDO & A. A. GUGLIELMONE. 1988. *Anaplasma marginale* con apéndices aislados en el noroeste argentino. Rev. Med. Vet. (Buenos Aires) **69:** 248–250.

Sus scrofa domestica Endoparasitic Resistance in the Amazonas[a]

DEBORAH LEAL RODRIGUES[b] AND
MARIO HIRAOKA[c]

[b]*Apdo 285-7210 Guapiles, Costa Rica*

[c]*Museu Emilio Goeldi
CP 399-66000, Belem, Brasil*

INTRODUCTION

The first European pig was brought to Brazil in 1532 by Martim Afonso de Sousa, a Portuguese colonist.[1] Currently, the species survives in the Amazonic Estuary floodplain island (AEFI) (FIG. 1) under primitive management conditions, living on the residual vegetable production of the floodplain forest (*Mata de Varzea*). Now, it seems that the pig went back to a feral condition,[2] and brings economic and subsistence food resources to the local peasants.[3] These people partially survive by subsistence fishing and commercialization of palm fruits [i.e., *miriti* (*Mauritia flexuosa*) and *acai* (*Euterpe* spp.)]. The waters that innundate these islands have a daily regimen of flooding, and the temperature ranges from 24 to 35°C, with a rainy (Dec.-May) and a dry (June-Nov.) season. This paper reports on the prevalence and pig resistance in different age classes to coccidia and nematoda.[4–6]

METHODS

Between May and June 1993, we assessed the parasitic level and tolerance in 75 pigs, proceeding from 35 families separated by tooth eruption into three age classes: 0-3, 4-9, and 10-18 months.[7] Fecal samples were collected directly from the rectum and transported to our field lab at the Maracapucu-Miri River. We identified eggs and oocysts by Willis-Mollay and Hoffman methods,[8] and counted eggs and oocysts per gram of feces (EPG and OPG) by a modified McMaster method.[9] By culturing the infective larvae by Roberts O'Sulivan's method,[8] we distinguished Strongyloidea family eggs.[10] Carcasses destined for commercial use were submitted to *postmortem* inspections for collection of adult endoparasites.[11]

RESULTS

We assessed mixed infection by *Eimeria* spp., Strongyloidea and *Strongyloides ransomi* in 43% of the samples. *Eimeria* was present in 89% of the cases, with the

[a]Financial support was received from the Ford Foundation, New York Botanical Garden, and Bayer from Centroamerica.

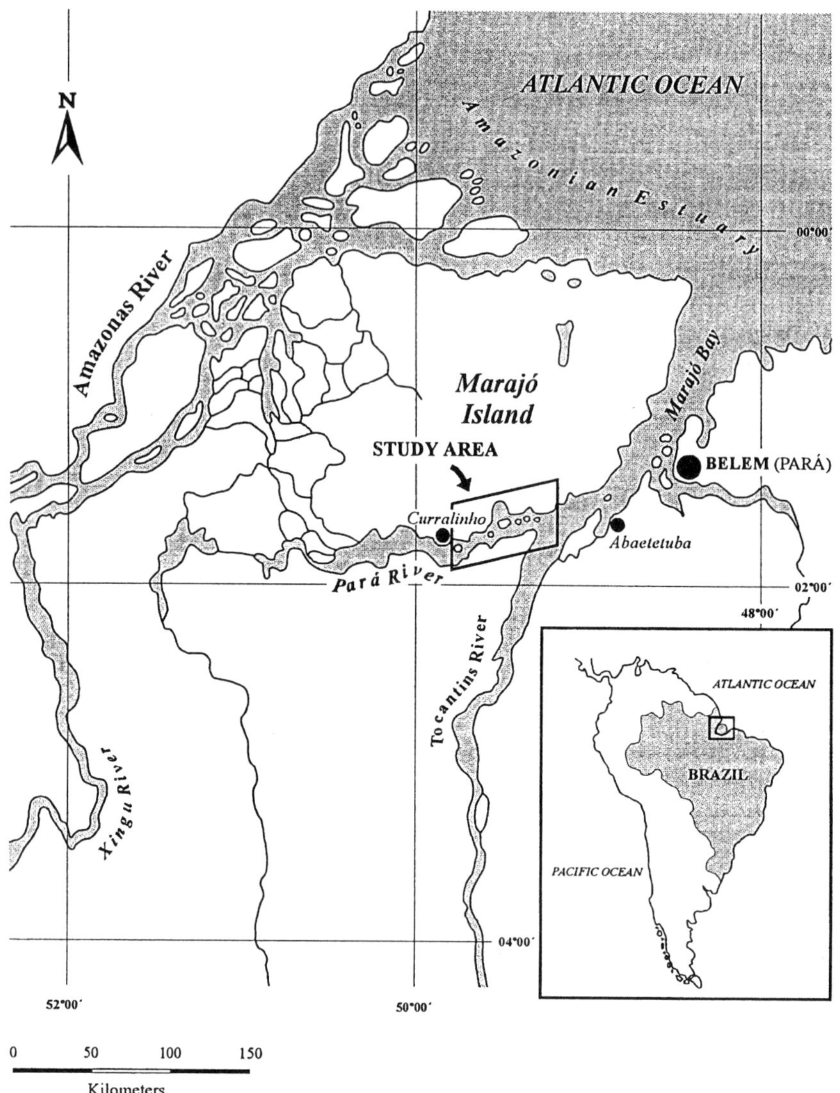

FIGURE 1. Location of the study area, between Curralinho and Abaetetuba towns, State of Para, Brazil. Floodplains Islands.

highest incidence in 4-9 mo animals. Eggs of *S. ransomi* and Strongyloidea were obtained in 60% of the cases, with the highest incidence in 0-3 and 4-9 mo animals, respectively (FIG. 2). We obtained the following means: 6204 OPG for *Eimeria*, 336 EPG for *S. ransomi*, and 520 EPG for Strongyloidea. Young animals (0-3 mo) always gave highest the means (OPG, EPG). Around 40% of cases positive for *S. ransomi*

TABLE 1. *Eimeria* spp. Oocysts per Gram (OPG), *Strongyloides ransomi,* and Strongyloidea Eggs per Gram of Feces (EPG) Found in 1993 in the Amazonic Estuary Islands

OPG and EPG Values	*Eimeria* spp.				*Strongyloides ransomi*				Strongyloidea			
	0-3	4-9	10-18	Total population	0-3	4-9	10-18	Total population	0-3	4-9	10-18	Total population
>300					13	04	05	22	07	12	04	23
>500					06	03	03	12	03	10	02	15
>1000	10	15	08	33	03	02	03	08	02	04	0	06
>5000	07	06	04	17								
>10000	06	02	03	11								
Mean	14.472	2.394	3.500	6.204	764	23	100	338	782	489	261	520
Maximum	97.600	13.000	22.500	97.600	4.500	2.500	22.500	22.500	10.000	1.900	800	10.000
Minimum	200	100	100	100	100	100	100	100	100	100	100	100

and Strongyloidea were up to 300 EPG. From 50 spp., adults of *Stephanurus dentatus* and *Macracanthorhynchus hirudinaceus* were isolated from 6 *postmortem* inspections.

DISCUSSION

The incidence of *Eimeria* in pigs from AEFI is high compared with Lowestein's[12] data from domestic pigs (62%), but similar to that obtained by him in wild boars (89%). In Papua New Guinea, Varghese's[13] results were: nondiarrheic free-ranging, and scavenging village piglets, 30% *Eimeria* incidence and piglets raised on concrete, 62%.

In this study, a subclinical coccidiosis was found with higher incidence and OPG values than those observed by Chhabra and Mafukidze[14] in pigs from Zimbabwe. Those pigs displayed diarrhea with an incidence of 25% and 150-11,000 OPG (the mean was <10,000 OPG). Varghese[13] observed coccidiosis incidence of 36% in diarrheic village piglets, and of 53% in piglets on concrete. Symptoms like dysentery, anorexia, lethargy, weight loss and fever were pointed out by Jones[15] in 7-16 week old pigs from Australia with 1500-51,600 OPG. However, we did not observe any symptoms in spite of the fact that 16% of our pigs presented values up to 10,000 OPG, with the highest in a 2-mo-old pig.

SUMMARY

The domestic pig (*Sus scrofa domestica* L.) has been wild breeding in the Varzea Amazonian Estuary area. Without dietary supplement or medical prevention, the pigs

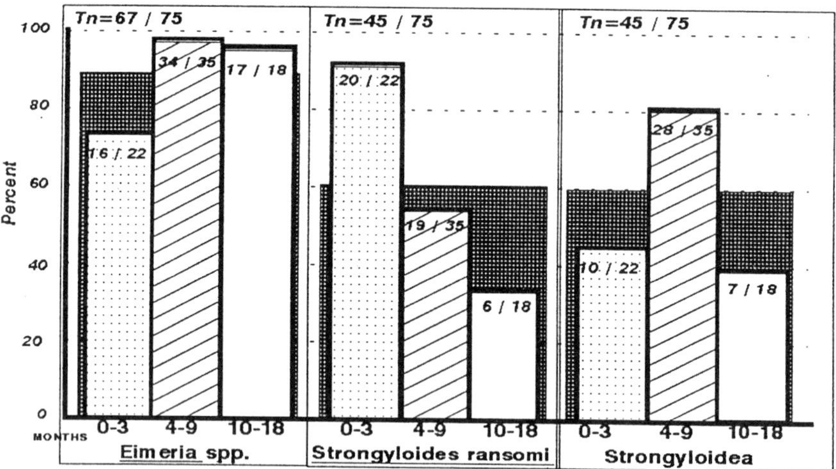

FIGURE 2. *Eimeria* spp., *Strongyloides ransomi,* and Strongyloidea incidences in 1993 in the pigs (gray bars) from the Amazonic Estuary floodplains islands. Tn = number of positive samples/total for each endoparasite. Incidences for each age class by months (0-3, dotted bars; 4-9, diagonal lines; and 10-18, white).

survive, bringing economic benefits to local peasants. Endoparasitic diseases generate economic losses because of neonatal death and clinical disease. We assessed the nematoda (O. Nematoda) and coccidia (O. Eucoccidia) parasitic level and tolerance in 75 pigs (0-18 months) from 35 families using the following methods: (i) larvae culture, (ii) Hoffman, (iii) Willis-Mollay (Hoffman 1987) and (iv) postmortem examinations. We identified Eimeria species Strongyloidea and *S. ransomi* present in 60% of the cases, giving a mean of 336 and 520 EPG (eggs per gram of feces), respectively. Culturing the larvae showed *S. ransomi* in 47% and *Oseophagostomum* sp. in 17% of the cases. We found *Stephanurus dentatus* and *Macracanthorhynchus hirudinaceus* adults in 77% and 23% of the postmortem tests, respectively. Pig resistance was demonstrated by lack of mortality and no clinical disease symptoms, in the presence of that high endoparasitic load.

ACKNOWLEDGMENTS

Isa Torrealba reviewed the manuscript and made useful suggestions. We thank Dr. Ralph Lainson from Evandro Chagas Institute and Dr. William G. Vale from Federal University of Para, and both institutions.

REFERENCES

1. 1980. Enciclopedia Delta Universal. Vol. **12.** Delta. Rio de Janeiro, Brazil.
2. RODRIGUES, D. L. & M. HIRAOKA. 1993. Dieta do proco domestico no Estuario Amazonico. Unpublished report. Apdo 285-7210, Guapiles, Costa Rica.
3. BODMER, R. E. 1992. Personal Communication. University of Florida, Center for Latin American Studies, 304 Grinter Hall, P.O. Box 115531, Gainesville, FL 32611-5531.
4. LEVINE, N. D., J. O. CORLISS, F. E. G. COX, *et al.* 1980. A newly revised classification of the Protozoa. J. Protozool. **27:** 37-58.
5. PELLERDY, L. P. 1974. Coccidia and coccidiosis, 2dn, Akad, Kiado. Paul Parey. Berlin, Hamburg & Budapest.
6. SOULSBY, E. J. L. 1968. Helminths, Arthropods and Protozoa of Domesticated Animals. Waverly Press. Philadelphia, PA.
7. GETTY, R. 1981. Anatomia dos Animais Domesticos. 5 ed. Ed. Interamericana. Rio de Janeiro, Brazil.
8. HOFFMAN, R. P. 1987. Diagnostico de parasitismo Veterinario Ed. Sulina. Porto Alegre, Brazil.
9. URQUHART, G. M., J. ARMOUR, J. L. DUNCAN, A. M. DUNN & F. W. JENNINGS. 1990. Parasitologia Veterinaria. Ed. Guanabara Koogan S.A. Rio de Janeiro, Brazil.
10. OGASSAWARA, S. 1983. Parasitologia Clinica. *In* Patologia Clinica Veterinaria. Sociedade Paulista de Medicina Veterinaria. Sao Paulo, Brazil.
11. VASCONCELOS, A. C. 1988. Necropsia e remessa de material para laboratorio. MEC/ ABEAS. Brasilia, Brazil.
12. LOWENSTEIN, M. 1987. A contribution to the knowledge of pig coccidia. Inaugural Dissertation. Vet. Med. Universitat Wien. Vienna, Austria.
13. VARGUESE, T. 1986. Porcine coccidia in Papua New Guinea. Vet. Parasitol. **21:** 11-20.
14. CHHABRA, R. C. & R. T. MAFUKIDZE. 1992. Prevalence of coccidia in pigs in Zimbabwe. Vet. Parasitol. **41:** 1-5.
15. JONES, G. W., R. J. PARKER & C. R. PARKER. 1985. Coccidia associated with enteritis in growing pigs. Aust. Vet. J. **62:** 319.

Subject Index